Carbon Meta-Nanotubes

Carbon Meta-Nanotubes

Synthesis, Properties and Applications

Marc Monthioux

A John Wiley & Sons, Ltd., Publication

This edition first published 2012
© 2012 John Wiley & Sons Ltd

Registered Office

John Wiley & Sons Ltd, The Atrium, Southern Gate, Chichester, West Sussex, PO19 8SQ, United Kingdom

For details of our global editorial offices, for customer services and for information about how to apply for permission to reuse the copyright material in this book please see our website at www.wiley.com.

Library of Congress Cataloging-in-Publication Data

Carbon meta-nanotubes : synthesis, properties, and applications / [edited by] Marc Monthioux.
     p.   cm.
  Includes bibliographical references and index.
  ISBN 978-0-470-51282-1
1. Nanostructured materials.   2. Nanotubes.   3. Organic compounds–Synthesis.   I. Monthioux, Marc.
  TA418.9.N35C336 2012
  620.1'17–dc23

                                           2011033599

A catalogue record for this book is available from the British Library.

Print ISBN: 978-0-470-51282-1 (HB)
ePDF ISBN: 978-1-119-95473-6
oBook ISBN: 978-1-119-95474-3
ePub ISBN: 978-1-119-96094-2
Mobi: 978-1-119-96095-9

Set in 10/12pt Times by SPi Publisher Services, Pondicherry, India
Printed and bound in Singapore by Markono Print Media Pte Ltd

Dedication

While this book was being prepared for publication, the news has hit us that our friend and colleague Jack Fischer had passed away on Tuesday, June 28th, 2011, at the age of 72. Of course, we know that we are all only briefly on this earth. Of course, we know that this will happen to every one of us eventually. Of course, we knew that this probably had to happen sooner for Jack as he had been struggling with complications arising from polycystic kidney disease for a decade. But the death of a dear person, relative or friend, is one of the hardest pieces of news to accept in life, as we subconsciously think of them as immortal. We just refuse to believe they can die.

We are not going to talk extensively about Jack's scientific skills and his dedication to science in general and carbon more specifically. The importance of Jack's inputs in carbon science, from graphite intercalation compounds to nanocarbons, has resulted in a highly successful career which accounts for more than 400 papers, more than 20 000 citations, a h-index of 65, 68 supervised PhD students and postdocs … the kind of achievements that all scientists would be proud of, and that most will not get. Detailed summaries about Jack's scientific path and career, from his first appointment to the University of Pennsylvania in Philadelphia, one of the eight prestigious universities of the so-called US 'Ivy League' only six years after graduating with a PhD degree in Nuclear Science and Engineering in 1966, can be found in several obituaries that have been published since then (for instance by P.K. Davies and Y. Gogotsi in *Carbon* 49 (2011) 4075).

We are proud and happy that this book contains Jack's last scientific contribution. And therefore, we wish to dedicate this book to him. We are friends grieving, and aside from the scientific merits that will survive him for many years to come, we would also like to state that Jack was not only a brilliant scientist, but that he was also a multicultural person, an oyster lover, a fan of classical music, a faithful friend, and a nice fellow. We will miss him.

Marc Monthioux
On behalf of the co-authors

About the Editor

Marc Monthioux has dedicated his scientific life to carbon materials for more than 30 years, with transmission electron microscopy as his main investigation tool. He graduated from the University of Orléans (France), where he also prepared his first two-year Doctorate thesis work (Specialty Thesis) on investigating the carbonization and graphitization mechanisms of heavy petroleum products, with Dr. Agnès Oberlin as his supervisor. He was then recruited by the *French National Center for Research* (CNRS) in 1982 and worked for five years at the *French Institute of Petroleum* (IFP, Rueil-Malmaison) while preparing his second five-year thesis (State Thesis) dedicated to experimentally mimicking the natural coalification and petroleum generation processes. After receiving this ultimate academic degree in 1986, he moved to the University of Pau, France along with Agnès Oberlin's team to work on advanced carbon-containing ceramic fibres and composites and pyrolytic carbons. After Dr. Oberlin retired, he led the laboratory for two years then merged it with a large CNRS laboratory in 1995, the Center of the Preparation of Materials and Structural Studies (CEMES) located at the University of Toulouse, France, renowned worldwide for its pioneering role in the development of transmission electron microscopy. While affiliated to CEMES, he has spent about two years working in the USA at the Dupont Experimental Station, Wilmington, Delaware as a consultant for Dupont de Nemours (carbon fibres) then the Conoco Company (pitch-based cokes), then at the University of Pennsylvania, Philadelphia under a NATO fellowship where he started getting involved in carbon nanoforms with David Luzzi and Jack Fischer. Once back in France, he combined working mainly on carbon nanoforms and pyrolytic carbons at CEMES with working as a consultant for three years for IMRA-Europe (Sophia-Antipolis, France) installing a laboratory and leading a team dedicated to hydrogen storage in carbon materials.

Dr. Monthioux has authored more than 190 papers in international journals and conference proceedings and has contributed to more than 15 books and topical journal special issues. Among his scientific achievements one can note: the demonstration of the existence of a *graphitizability continuum* between the formerly known 'hard' (non-graphitizable) carbons and 'soft' (fully graphitizable) carbons; the most complete and successful experimental duplication of the natural coalification processes ever, which has revealed the leading role of the pressurized confinement of the effluents; the discovery of the ability of single-wall carbon nanotubes (SWCNTs) in being filled with molecules, starting with the example of nanopeapods; the fabrication of the world smallest and most sensitive superconducting quantum interference device (the *nanoSQUID*) for magnetic measurements, based on a single SWCNT.

He is currently a CNRS Research Director at CEMES (Toulouse, France), Editor of *Carbon* journal (Elsevier), Chairman of the French Carbon Group (GFEC), and Chairman of the European Carbon Association (ECA).

Contents

List of Contributors

Revathi R. Bacsa
Laboratoire de Chimie de Coordination – ENSIACET, UPR-8241 CNRS, Université de Toulouse, BP 44362, F-31030 Toulouse, France
revathi.bacsa@ensiacet.fr

Stéphane Campidelli
Laboratoire d'Electronique Moléculaire, DSM/IRAMIS/SPEC, CEA Saclay, F-91191 Gif-sur-Yvette, France
stephane.campidelli@cea.fr

John E. Fischer
Dept. of Materials Science and Engineering, University of Pennsylvania, Philadelphia, PA 19104-6272, USA

Dmitri Golberg
Nanoscale Materials Center, National Institute for Materials Science, University of Tsukuba, Ibaraki 305-0044, Japan
golberg.dmitri@nims.go.jp

Marc Monthioux
Centre d'Elaboration des Matériaux et d'Etudes Structurales (CEMES), UPR-8011 CNRS, Université de Toulouse, BP 94347, F-31055 Toulouse, France
marc.monthioux@cemes.fr

Alain Pénicaud
Centre de Recherche Paul Pascal (CRPP), UPR-8641 CNRS, Université Bordeaux I, F-33600 Bordeaux, France
penicaud@crpp-bordeaux.cnrs.fr

Pierre Petit
Institut Charles Sadron, UPR-22 CNRS, Université de Strasbourg, BP 84047, F-67034 Strasbourg, France
petit@ics.u-strasbg.fr

Maurizio Prato
Dipartimento di Scienze Farmaceutiche, Universitá di Trieste, I-34127 Trieste, Italy
prato@units.it

Philippe Serp
Laboratoire de Chimie de Coordination – ENSIACET, UPR-8241 CNRS, Université de Toulouse, BP 44362, F-31030 Toulouse, France
philippe.serp@ensiacet.fr

Jeremy Sloan
Dept. of Physics, University of Warwick, Coventry, CV4 7AL, UK
j.sloan@warwick.ac.uk

Ferenc Simon
Dept. of Physics, Institute of Physics, Budapest University of Technology and Economics, PO Box 91, H 1521 Budapest, Hungary
simonf5@univie.ac.at

Mauricio Terrones
Dept. of Physics, PMB 136, The Pennsylvania State University, University Park, PA 16802, USA
Research Center for Exotic Nanocarbons (JST), Shinshu University, Nagano 380-8553, Japan
mut11@psu.edu and *mtterrones@shinshu-u.ac.jp*

Stanislaus S. Wong
Dept. of Chemistry, State University of New York at Stony Brook, Stony Brook, NY 11794-3400, USA
Condensed Matter Physics and Materials Science Dept., Brookhaven National Laboratory, Upton, NY 11973, USA
sswong@notes.cc.sunysb.edu

Foreword

Carbon is a fascinating element and one of the most abundant on Earth. It is still surprising us with all the variety of structures that it produces at the nanoscale. The birth of C_{60} (or Buckminsterfullerene) in the mid-1980s spawned a revolution in nanoscience and nanotechnology. Soon after the bulk synthesis of C_{60}, a new type of chemistry emerged and these studies resulted in the quest of novel nanocarbon forms such as carbon nanotubes (single-, double-, and multi-walled). Although single- and multi-walled carbon nanotubes were observed in the 1970s by Morinobu Endo, it was not until 1991, when Sumio Iijima reported the concentric tubular structure (nested concentric elongated fullerenes) using electron diffraction of the carbon residues produced in an arc discharge generator used to synthesize C_{60} soot, that carbon nanotubes became a major focus for nanoscience and nanotechnology.

These novel sp^2-like hybridized carbon structures (*fullerenes* and *nanotubes*) are unique due to their cage-like morphology. Therefore, it is not only possible to use their outer surface reactivity, but also their inner surface, in order to functionalize or encapsulate molecules or materials that could lead to novel applications. This book tries to emphasize the fact that the available surfaces of nanotubes could be used to obtain modified nanotubes by the adsorption or chemisorption of different materials including molecules, metals, clusters, and so on. These modified tubular carbon materials could be called *meta-nanotubes* and the objective of this book, the first of its kind, is to review the different approaches able to modify carbon nanotubes (single-, double- and multi-walled), in order to produce carbon-based materials that could be used in more efficient devices during the twenty-first century.

The book is divided into six chapters contributed to by leading scientists around the world. They deal with different ways of modifying the surfaces of carbon nanotubes. The first chapter provides an overview of pristine (pure) carbon nanotubes and how this tubular material (considered as a building block) can then be modified to give rise to meta-nanotubes. The following chapters of this book deal with the different ways of modifying nanotubes by doping, chemical molecule functionalization, cluster decoration, filling of nanotubes, and the synthesis of heteroatomic nanotubes. These chapters review the latest work in the field, pose some unsolved issues and propose new directions along the production of novel meta-nanotubes and their applications. Therefore, this book should be used as a guide to perform novel and innovative research in the area of carbon nanotubes, and some of this work could even be extrapolated to explore other graphene-like systems.

This book is also dedicated to the memory of John E. 'Jack' Fischer, an outstanding carbon scientist who passed away on June 28, 2011. He contributed enormously to the fullerene and nanotube fields publishing more than 400 papers in high impact journals.

His contributions triggered innovative work and applications, especially in the areas of batteries and energy storage devices. He was a great human being and had a spectacular sense of humour. This is another great loss for the world carbon community, after Richard Smalley (2005) who boosted the research on nanotubes, Adolphe Pacault (2008) who organised the research on carbon in France from the 60', Peter Eklund (2009) who was a major figure in the world of condensed matter physics and world famous for his experiments on nanotubes, P.L. Walker (2009) who pioneered the research on carbon at Penn State and Sugio Otani (2010) who first introduced and developed pitch as a precursor for carbon fibre. We are fortunate to inherit their scientific legacy.

Harold, W. Kroto (Nobel Laureate, FRS)
Florida State University
Tallahassee, Florida, USA

List of Abbreviations

AAC	azide-alkyne cycloaddition
AC	alternating current
AEPA	2-aminoethylphosphonic acid
AES/OES	atomic emission/optical emission spectroscopy
AFM	atomic force microscopy
ALD	atomic layer deposition
APTEOS	aminopropyltriethoxysilane
bcc	body centered cubic
BCS	Bardeen-Cooper-Schrieffer
bct	body-centred tetragonal
BEDT-TTF	bis-ethylenedithiotetrathiafulvalene
BNNT	boron nitride nanotube
BNNW	boron nitride nanowire
BSA	bovine serum albumin
BWF	Breit-Wigner-Fano
CAT	coaxial tube
CB	conduction band
CCVD	catalysed chemical vapour deposition
CESR	conduction electron spin resonance
CNF	carbon nanofibre
h-CNF	herringbone carbon nanofibre
p-CNF	platelet carbon nanofibre
CNT	carbon nanotube
CoPc	cobalt phtalocyanine
CVD	chemical vapour deposition
DC	direct current
DFT	density functional theory
DMF	dimethylformamid
DMSO	dimethylsulfoxide
DMTA	dynamic mechanical thermal analysis
DNA	deoxyribonucleic acid
DOS	density of states
DPP	diphenylporphyrin
DWCNT	double-wall carbon nanotube
EDAC	ethyl dimethylaminopropylcarbodiimide
EELS	electron energy loss spectroscopy

EPR	electron paramagnetic resonance
ESR	electron spin resonance
EXAFS	extended X-ray absorption fine structure
FAD	flavine adenine dinucleotide
fcc	face-centred cubic
FE	field emission
FEB	ferrocene-ethanol-benzylamine
FET	field effect transistor
FITC	fluorescein isothiocyanate
FL	Fermi liquid
FTIR	Fourier transform infrared spectroscopy
GC-MS	gas chromatography mass spectrometry coupling
GIC	graphite intercalation compound
GOx	glucose oxidase
HADF	high-angular dark field
HDP	homogeneous deposition precipitation
HOMO	highest occupied molecular orbital
HOPG	highly oriented pyrolytic graphite
HPLC	high pressure liquid chromatography
HRTEM	high resolution transmission electron microscopy
HV	hexyl-viologen
ICP	inductively coupled plasma
IgG	immunoglobulin G
IR-vis	infrared-visible
ITO	indium tin oxide
IUPAC	International Union of Pure and Applied Chemistry
LDA	local density approximation
LEEPS	low energy electron (point) source microscopy
LPC	lyso-phosphatidylcholine
LUMO	lowest unoccupied molecular orbital
LPG	liquefied petroleum gas
MAO	methylaluminoxane
MD	molecular dynamics
MOCVD	metal organic chemical vapour deposition
MRI	magnetic resonance imaging
MUA	mercaptoundecanoic acid
MS	mass spectrometry
MWCNT	multi-wall carbon nanotube
b-MWCNT	bamboo multi-wall carbon nanotube
c-MWCNT	concentric multi-wall carbon nanotube
h-MWCNT	herringbone multi-wall carbon nanotube
NDT	nonane dithiol
NMR	nuclear magnetic resonance
NMRP	nitroxide mediated radical polymerisation
ONIOM	our own n-layered integrated molecular orbital and molecular mechanics
PAA	polyacrylic acid

PABS	poly(m-aminobenzenesulfonic acid)
PAMAM	polyamidoamine
PDADMAC	poly(diallyldimethylammonium chloride)
PDDA	poly(diallyldimethylammonium)
PE	polyethylene
PEG	polyethyleneglycol
PEI	polyethyleneimine
PEM	proton exchange membrane
PES	photo-electron/photo emission spectroscopy (also named XPS)
PMMA	polymethylmetacrylate
PmPV	poly{(m-phenylenevinylene)-co-[(2,5-dioctyloxy-p-phenylene)vinylene]}
P3OT	poly-3-octylthiophene
PPM	pentagonal pinch mode
PS	polystyrene
PSA	prostate specific antigen
PSS	poly(sodium styrene-4-sulfonate)
PVA	polyvinylalcohol
PVP	polyvinylpyrollidone
QC	quantum computing
QCM	quartz crystal microbalance
RBM	radial breathing mode
RDF	radial distribution function
RPA	random phase approximation
SA	streptavidin
SDS	sodium dodecyl sulfate
SEM	scanning electron microscopy
SGO	similar graphene orientation
SHE	standard hydrogen electrode
SpA	staphylococcal protein A
SSA	specific surface area
STEM	scanning transmission electron microscopy
STM	scanning tunnelling microscopy
STS	scanning tunnelling spectroscopy
SWBNNT	single-wall boron nitride nanotube
SWCNT	single-wall carbon nanotube
TCE	tetracyanoethylene
TCNQ	tetracyanoquinodimethane
TCNQF$_4$	tetrafluorotetracyanoquinodimethane
TDAE	tetrakis(dimethylamino)ethylene
TEM	transmission electron microscopy
TEP	thermoelectric power
TG	thermal gravimetry
TGA	thermogravimetry analysis
TLL	Tomonaga-Lüttinger liquid
TM	tangential mode

TMTSF	tetramethyl-tetraselenafulvalene
TMTTF	tetramethyltetrathiafulvalene
TTF	tetrathiafulvene
TPD	thermally programmed desorption
TPO	triphenyl phosphine oxide
UV-vis-NIR	ultraviolet-visible-near infrared
VGCF	vapour-grown carbon fibre
vis-NIR	visible-near infrared
XANES	X-ray absorption near-edge structure
XPS	X-photoelectron spectroscopy (also named PES)
XRD	X-ray diffraction
ZnNc	zinc-naphtalocyanine
ZnP	zinc-porphyrin

Acknowledgements

Published with the kind assistance of the Groupe Français pour l'Etude du Carbone http:// www.gfec.net a member of the European Carbon Association *http://www.hpc.susx.ac.uk/ ECA*

Introduction to the Meta-Nanotube Book

Marc Monthioux
CEMES, CNRS, University of Toulouse, France

1 Time for a Third-Generation of Carbon Nanotubes

So-called multi-walled carbon nanotubes (MWCNTs) have been known since the 1950s (see [1] and references within, the very first evidence of their occurrence being dated 1952 [2]). Micron-size filamentous carbons (which are now understood as being catalytically-grown MWCNTs subsequently thickened by a catalyst-free chemical vapour deposition or CVD process) have been known from the early twentieth century [3], and even the late nineteenth century [4] where a patent was filed [5] to protect the intellectual property of a CVD-based process aiming to prepare carbon filaments to be used as wires in the just-born electric bulbs invented by Joseph Swan in 1878 (and subsequently improved, or more or less duplicated, by Thomas Edison a few months later). After carbon filaments were replaced in electric bulbs by tungsten wires, they were mostly seen as undesirable by-products to get rid of, for example, in the coke industry [3] where gas exhausts could be obstructed by the abundance of grown carbon filaments (resulting from the thermal crack-ing of the released gaseous hydrocarbons in presence of iron minerals, that are omnipresent in coal), or in the nuclear industry, where the formation and deposition of carbon filaments could adversely affect the efficiency of metallic-tubed heat exchangers in helium-cooled reactors [6]. Carbon nanotubes with diameters in the ~3 nm range were even incidentally synthesized as early as 1976 [7] but were merely considered as a material curiosity whose growth mechanisms were worth being elucidated.

Hence, it took decades and the landmark paper by Sumio Iijima in 1991 [8], followed after two years with the discovery of the so-called single-walled carbon nanotubes (SWCNTs) and, more importantly, a way to synthesize them in a reliable and reproducible manner [9,10], for the unique properties of CNTs to be finally figured out and their countless applicative potentialities to be acknowledged and explored. A second generation of carbon nanotubes was thus born, as the flagship of the worldwide activity on nanoscience and nanotechnology which started meanwhile.

Twenty years later, the relevant synthesis processes (i.e. those which are suitable for commercial scale production) are now all defined, most of CNT properties are predicted and demonstrated. Most of the current research efforts are focused on the actual use of CNTs, that is, their integration in devices and incorporation in advanced materials and crafts of practical interest. This is where the natural beauty of CNTs, that is, their structural perfection, which had made them ideal objects to study and had attracted the attention of physicists, caused their limitations. In many instances, CNTs appeared to be difficult to handle, purify, sort, disperse, mix, and so on. On the other hand, in many of the applications envisaged, CNTs were needed to be suitably combined with other phases. In other words, it was time to develop a third-generation of carbon nanotubes modified from the pristine ones by the various means that chemists could think of. That is to say, it was time to consider 'meta-nanotubes'.

2   Introducing Meta-Nanotubes

Carbon nanotubes can be modified in many ways, generally involving chemical treatments that make use of the polyaromatic nature of their skeleton. Because electronics was among the very first applications that CNTs were considered for, doping their electronic structure was rapidly thought to allow promising developments. The doping was achieved by following various routes, for example, grafting functions or metal nanoparticles, inserting dopants in or between SWCNTs, and/or substituting lattice carbon atoms by heteroatoms. The resulting materials were however often indifferently referred to as 'doped nanotubes', although they could be very different from each other from the point of view of materials. Furthermore, modifying CNTs and combining them with foreign components appeared to open many more promising developments than in merely electronics. CNTs had to no longer be considered as such but as starting materials to be used as host, substrate, mould, structural skeleton, and so on. A whole specific field of research was then opened, and it was necessary to find a way to differentiate modified CNTs from pristine ones and to accurately discriminate the various forms of modified CNTs. This is where we created the word 'meta-nanotubes' [11,12] (from ancient Greek *metá*, meaning 'beyond, after'), defined as modified nanotubes resulting from the transformation of pristine nanotubes by various ways, which leaves the nanotubes being associated with a foreign component X, where X can be atoms or molecules, chemical functions, or phases. Five different kinds of association of X with CNTs were identified and are listed below, tentatively roughly ranked according to the increasing structuration of the component X (from isolated atoms to phases), as well as the increasing intimacy and strength of the X-versus-CNT association.

2.1 Doped Nanotubes (X:CNTs)

These are carbon nanotubes which are associated with electron donor or acceptor elements such as Br_2, K, Rb, and so on. These substances (or compounds) more generally can be used to modify the overall electronic structure and behaviour. The association is intimate, that is, it is not merely a matter of coating or grafting the outer surface but does not involve a strong chemical bonding. Both the association mechanisms and the doping processes are related to that of intercalation compounds in graphite (GICs) and the inserted element (e.g. Li, K, etc.) is therefore found in between graphenes in MWCNTs, or in between SWCNTs in SWCNT bundles. The doping material does not make an individual phase, as opposed to the coating or filling phases mentioned next. The preferred techniques to identify the dopant relate to spectroscopy (e.g. EELS, XPS) or chemical analysis methods. The doping process is usually reversible, sometimes spontaneously in some conditions, as for Li-doped CNTs exposed to air for instance.

2.2 Functionalized Nanotubes (X-CNTs)

These are nanotubes at the surface of which various individual chemical functions are grafted, that is, where the foreign component X does not form a phase by itself. X-CNT bonds can be covalent or weaker, for example, through so-called π-stacking. Reversibility, that is, defunctionalization, is easy via either chemical or physical (e.g. thermal) procedures. The preferred ways to characterize the grafted functions are chemical analysis techniques such as FTIR spectroscopy or chemical titrations. It is a quite important topic mostly developed since to address some of the major issues of pristine CNTs such as reactivity, dispersability, solubility, and so on which are compulsory properties for many processes involving CNTs.

2.3 Decorated (Coated) Nanotubes (X/CNTs)

These are nanotubes at the surface of which the foreign component X is a genuine phase, from the point of view of chemistry and structure. It is deliberately bonded to the CNT surface, either as a continuous coating or a discontinuous decoration. They are most often nanoparticles but the category may also include CNTs onto which complex compounds are adsorbed or bonded such as polymers, structured biomolecules (e.g. DNA), and so on. Reversibility is generally easy and made possible via chemical or physical procedures, as for functionalized CNTs. In the related applications, the intrinsic properties of the CNT underneath may be of minor importance since the CNTs may basically be used as a substrate, while the needed properties are brought by the coating material. As the foreign component of those meta-CNTs is a phase, it may be characterized by diffraction techniques (e.g. XRD, electron diffraction), among others. It is worth noting that functionalizing the CNTs so that the nanotube surface attractiveness towards the nanoparticles is enhanced may be a previous step in the preparation of decorated (coated) CNTs.

2.4 Filled Nanotubes (X@CNTs)

These are carbon nanotubes which the inner cavity is fully or partially filled with foreign atoms, molecules, or compounds. They are probably the most versatile of the meta-nanotubes described here, and thereby are likely the ones which open the wider research field, both regarding scientific and technological aspects. As opposed to doping, the filling components generally gather and associate into a phase, enabling their identification by diffraction techniques, in addition to spectroscopic methods. The foreign component X is not chemically bonded to the CNT host, but possibly by van der Waals forces. However, because the encapsulation situation often corresponds to a situation of low free energy, the filling event exhibits a limited reversibility upon solvating methods, yet variable with the diameter of the CNT host cavity (the narrower the cavity, the more stable the contained phase). Removing the filling phases hence can require severe conditions such as decomposition upon high thermal treatments.

2.5 Heterogeneous Nanotubes (X*CNTs)

These are carbon nanotubes whose carbon atoms from the hexagonal graphene lattice are partially or even totally substituted with hetero-atoms, typically nitrogen and/or boron. Obviously, one of the major consequences of such a substitution is the modification of the electronic structure, and thereby the electronic behaviour, of the resulting nanotubes. It also allows the development of a specific side-wall chemistry using the reactivity of the substituting heteroatoms. Yet the amount of foreign component X can be low (e.g. in the range of few percent for the substitution of C by N or B). Hetero-CNTs are the only meta-CNTs for which X is no longer more or less independent from the host. It is integrated to the meta-CNT structure and because of this, the reversibility is not possible.

3 Introducing the Meta-Nanotube Book

Chapter 1 of this book is dedicated to an introduction to pristine carbon nanotubes. Because it is doubtful that any reader of this book would ignore what carbon nanotubes are in the first place, it was wondered whether such a chapter was necessary. We finally decided it was, because the world of carbon nanotubes, and nanocarbons generally speaking, is related to the complex and labyrinthic world of carbon materials, which has kept scientists and engineers busy for more than a century. The high potential and attractiveness of carbon nanotubes in the field of nanosciences and nanotechnology means that many researchers, specifically younger ones, get involved in the field without really knowing what a genuine graphitic material is, for example. Another reason is that official definitions, as stated by IUPAC for instance, of terms related to nanotube science are still lacking and it was necessary to define those terms at least in the context of this book.

Chapters 2 to 6 describe the five kinds of meta-nanotubes in the same order as presented in Section 2 above. Only Chapter 5 is split into two parts, due to the specific place that a peculiar type of filled carbon nanotube, namely the *peapod*, is taking in the field. Each of the chapters is authored by the pioneering and/or most active scientists in the related field. Hence, although authors have made an overview of their field as far as possible, the content of each chapter reflects in a various extent the field of interest and activity of the related authors. This has meant that some aspects, yet of significance, appear not to be covered. For instance, CNT functionalization aiming to optimize the interaction with the matrix in composites designed for structural (mechanical load transfer), electrical (charge carrier transfer), or thermal (phonon transfer) application was not specifically developed. Also, in spite of the clear discrimination we tentatively made between each meta-CNT categories, we could not prevent some overlapping between chapters.

Indeed, at some point, discriminating between a heavily doped CNT for which the dopant is located inside the CNT cavity and a CNT slightly filled with atoms whose low proportion barely allows a phase to build up could not be obvious. Likewise, functionalized CNTs and coated CNTs were found to meet when the foreign component X to consider was in large quantity and bonded to the CNT surface via van der Waals forces only. For instance, I would personally have categorized DNA-wrapped SWCNTs as coated-CNTs but the authors preferred to consider them functionalized-CNTs and I respected that. Another difficulty was that meta-CNTs may belong to several categories and this actually is an eventuality whose occurrence is going to increase as the field develops. For instance, decorating CNTs with nanoparticles often comes with a preliminary functionalization step to attract and retain the nanoparticles as well as to prevent their further coalescence. Even worse, hetero-CNTs can subsequently be functionalized then decorated, making the resulting materials belong to three out of five categories…

Hence, we beg the reader to be indulgent. This book is the first attempt of its kind. We hope to be given the opportunity to make further, revised editions of it, in which we will try to gradually fix every flaw that will be brought to our attention, thanks to the feedback we will receive from our readership.

References

[1] M. Monthioux and V.L. Kuznetsov, Who should be given the credit for the discovery of carbon nanotubes?, *Carbon*, **44**, 1621–1623 (2006).

[2] L.V. Radushkevich and V.M. Lukyanovich, O strukture ugleroda, obrazujucegosja pri termiceskom razlozenii okisi ugleroda na zeleznom kontakte, *Zurn. Fisic. Chim.*, **26**, 88–95 (1952).

[3] C. Pélabon and H. Pélabon, Sur une variété de carbone filamenteux, *C. R. Acad. Sci. Paris,* **137**, 706–708 (1903).

[4] P. Schützenberger and L. Schützenberger, Sur quelques faits relatifs à l'histoire du carbone, *C. R. Acad. Sci. Paris*, **111**, 774–778 (1890).

[5] T.V. Hughes and C.R. Chambers, Manufacture of carbon filaments, *US Patent Office*, Letters Patent 405480, June 18 (1889).

[6] M.R. Everett, D.V. Kinsey and E. Römberg, Carbon transport studies for helium-cooled high-temperature nuclear reactors, in *Chemistry and Physics of Carbon*, **Vol. 3**, (ed. P.L. Walker Jr.), 289–436, Marcel Dekker, New York, USA (1968).

[7] A. Oberlin, M. Endo and T. Koyama, Filamentous growth of carbon through benzene decomposition, *J. Cryst. Growth*, **32**, 335–349 (1976).

[8] S. Iijima, Helical microtubules of graphite carbon, *Nature,* **354**, 56–58 (1991).

[9] S. Iijima and T. Ichihashi, Single-shell carbon nanotubes of 1-nm diameter, *Nature*, **363**, 603–605 (1993).

[10] D.S. Bethune, C.H. Kiang, M.S. de Vries, G. Gorman, R. Savoy, J. Vazquez and R. Beyers, Cobalt-catalysed growth of carbon nanotubes with single-atomic-layer walls. *Nature*, **363**, 605–607 (1993).

[11] M. Monthioux, E. Flahaut and J.-P. Cleuziou, Hybrid carbon nanotubes: Strategy, progress, and perspectives, *J. Mater. Res.*, **21**, 2774–2793 (2006).

[12] M. Monthioux and E. Flahaut, Meta-and hybrid-CNTs: a clue for the future development of carbon nanotubes, *Mater. Sci. Eng. C,* **27**, 1996–2101 (2007).

1

Introduction to Carbon Nanotubes

Marc Monthioux
CEMES, CNRS, University of Toulouse, France

1.1 Introduction

This chapter does not intend to tell us everything about carbon nanotubes. Its purpose is to merely provide the basics of what is needed to understand the other chapters, regarding the various kinds of carbon nanotubes that are mentioned, the various possibilities of treatments to make them 'meta-nanotubes', and the properties that are expected from them. Many books, book chapters, and review papers have been written already on carbon nanotubes within the past 15 years and the reader who wishes to learn more than what is told here is invited to refer to them. For example, most of what is given in this chapter can be found in References [1–10], although it is the author's belief that some aspects, for example, regarding nanotube typology and structure, are treated in a personal way. Hence, not all the related references will be cited and the reader is again invited to consult the review literature cited [1–10] for further bibliographic details.

1.2 One Word about Synthesizing Carbon Nanotubes

Synthesizing carbon nanotubes requires a carbon source, an energy source and, most usually, a catalyst. Three main methods share this principle, cited below in decreasing order of importance in terms of overall contemporary production capacity. Others do

Carbon Meta-Nanotubes: Synthesis, Properties and Applications, First Edition. Edited by Marc Monthioux.
© 2012 John Wiley & Sons, Ltd. Published 2012 by John Wiley & Sons, Ltd.

exist but they are miscellaneous and do not contribute significantly in this regard (the template-based method will, however, be cited afterwards):

1) The catalytic chemical vapour deposition method (CCVD) and the variety of related methods, for example, plasma enhanced, radio frequency enhanced, fluidized bed, spray pyrolysis, and so on;
2) The electric arc plasma method;
3) The pulsed laser vaporization method, which is now abandoned when preparing carbon nanotubes for commercial purpose.

For the former methods (CCVD and kin/electric arc), the carbon source is usually a liquid or gaseous hydrocarbon, more rarely carbon monoxide when specifically considering the principle of CO disproportionation (patented as the HipCO process). Additives may be added (e.g. H_2, H_2O...) or organic precursors others than hydrocarbons (e.g. ethanol) may be used, aiming to increase the yield and purity of the nanotube products formed by preventing both the formation of amorphous carbon and the deactivation of the catalyst particles upon their encapsulation in carbon shells. They are classified as 'low temperature' methods, as involving temperatures ranging from ~400°C to ~1200°C, depending on a variety of parameters including the chemical nature and particle size of the catalysts and the composition of the gas phase, among others. In CCVD-related methods, each catalyst particle most generally generates a single carbon nanotube, following either a so-called base-growth (if the interaction of the catalyst particle with the substrate underneath is strong enough to maintain the former on the latter) or tip-growth (if the particle/substrate interaction is weak enough for allowing the catalyst particle to lift off) mechanism. More rarely and mostly in the specific case of some herringbone-type multi-walled carbon nanotubes and nanofibres (*h*-MWCNTs and *h*-CNFs respectively, see Section 1.4), a single catalyst particle can generate two nanotubes at once (growing in opposed directions). Growing more than one (possibly two) nanotubes per catalyst particle is therefore exceptional, whereas it is rather common when growing platelet-type CNFs (*p*-CNFs), for energetic reasons [11]. Hence, the common CCVD growth mechanisms imply that the CNT diameters are more or less templated by the catalyst particle size, meaning that forming single-walled carbon nanotubes (SWCNTs) requires nanosized (i.e. in the range 1–3 nm) catalyst particles to be formed first. The catalyst particle geometry can also impose some features to the nanotubes, for example, round particles more frequently generate SWCNTs (if small enough) or concentric-type MWCNTs (*c*-MWCNTs, see Section 1.4), whereas facetted particles tend to systematically generate *h*-MWCNTs or *h*-CNFs whose graphenes making the tube wall or the fibre body are partly parallel to the particle faces. Products from CCVD methods can exhibit a large variety of features (regarding length, inner and outer diameters, texture, etc.) but are usually relatively homogeneous within the same batch, with a limited amount of impurities, typically catalyst remnants and amorphous or poorly organized polyaromatic carbon soot. Large-scale production of nanotubes with reasonably good quality (as far as 'quality' is measured through the degree of perfection of the graphenes making the nanotube walls, what is called 'nanotexture' in this chapter, see Section 1.4) uses catalysts selected among the principal transition metals (i.e. Fe, Ni, or Co) but a large variety of other metals (e.g. Cu, Cs, Pd) have demonstrated some efficiency in catalysing nanotube and nanofibre formation. An overall statement is that the quality and yield of the nanotubes synthesized by CCVD methods is variable and depends tightly on the set of parameters used.

For the two latter methods (arc and laser), the carbon source is a graphite anode (arc) or a graphite target (laser) previously loaded with the catalyst(s). The whole is then atomized by applying a strong electric field between the anode and a cathode (arc) or by irradiating with a laser beam. A plasma is thus created, whose temperature is in the range of several thousand degrees Celsius, hereby justifying why these methods are classified as 'high temperature' methods (yet such plasmas are considered as cold ones by plasma specialists). However, this does not mean that the nanotubes are all formed at such high temperatures. As opposed to CCVD methods, thermal gradients are huge in plasma methods and it is not ascertained yet at which temperature the nanotube growth specifically occurs in plasma reactors. Plasma methods both generate MWCNTs and SWCNTs all at once, in different areas of the reactor. MWCNTs, always of the concentric type when 'regular' conditions are used (typically, the conditions used to prepare fullerenes, as in [12] and [13], or to prepare SWCNTs, as in [14]), are formed following a catalyst-free process involving the contribution of small carbon entities (e.g. C_2) as building blocks [15], giving rise to high quality nanotubes from the nanotextural and purity points of view, yet with lengths barely better than micrometric as a consequence of the steep temperature gradients. Impurities are limited to facetted, graphene-based multi-walled carbon shells (often termed as 'carbon onions', which is inappropriate as they are hollow) with high nano-texture, possibly partially filled with metal from the catalyst (although not active for the MWCNT formation). In the electric arc reactor, the area for the MWCNT formation is located at the vicinity of the cathode, resulting in a solid deposit attached to it. On the other hand, SWCNTs are found away from the plasma zone and are probably formed out of this zone as well. The growth can be so successful that the whole SWCNT-containing product can form a kind of spider web (possibly attached to the cathode by a SWCNT-rich collaret material) hanging on the inner parts of the reactor. The growth mechanism is different from that in CCVD, and it is admitted that liquid droplets form first (with sizes in the range of several tens of nm), containing carbon species and metal catalyst atoms. As the droplets cools down, the solubility of carbon into the metal decreases, expelling the carbon atoms toward the former droplet surface from which SWCNTs grow radially, usually gathered into bundles [16]. Whether SWCNTs grow open or closed is still debated, yet the latter is more likely. SWCNT diameters distribute following a relatively narrow range usually centred on ~1.38 nm, as a probable consequence of energetic constrains. The impurity content is high, and the types of impurities are numerous (i.e. catalyst remnants, fullerenes and fullerenoids including the so-called nanohorns, amorphous carbon, carbon shells, and poorly organized, microporous polyaromatic carbon soot). Hence, the homogeneity of the SWCNT materials is attractive, but getting rid of the whole cortege of the various impurities makes a problem which is tentatively solved by applying a variety of chemical and physical purification procedures [17] whose severity is not without consequence on the SWCNT integrity [18]. In this regard, the most challenging issue is to remove the carbon impurities whose sp^2 character, and then the reactivity, is about the same as that of the nanotubes.

Table 1.1 summarizes some of the information given above.

Although the three synthesis methods briefly described above cover, say, 95% of the carbon nanotube production all combined, it is worth mentioning the template-based method as an example of catalyst-free synthesis processes. Based on the pioneering work by Kyotani et al., [19], The principle of it is to deposit a carbon film onto the walls of

Table 1.1 The basic guideline for the relation between various catalyst-assisted synthesis conditions and the nanotube material obtained. (c,h,b) in column 3 refers to concentric, herringbone, and bamboo texture for MWCNTs respectively. Modified from [9].

A QUICK GUIDELINE for the RELATIONSHIP between SYNTHESIS CONDITIONS and CARBON NANOFILAMENT TYPE and FEATURES	Increasing temperature… …and physical state of catalyst			Presence of a substrate (for solid or molten catalyst)		Thermal gradient	
	solid (crystallized) (1)	liquid from melting (2)	liquid from condensing clusters (3)	yes	no	low	high
< ~3 nm	SWCNT	SWCNT	?	base-growth			
Catalyst particle size > ~3 nm	MWCNT (c,h,b)	c-MWCNT	SWCNT	tip-growth	tip-growth	long length	short length
> ~80 nm	p-CNF	–		–	base-growth		
Nanotube diameter	heterogeneous (related to catalyst particle size)		homogeneous (independent from particle size)				
Nanotube/particle(4)	1-2 nanotube(s)/particle		several SWNT bundles/particle				

(1),(2): as in CCVD
(3): as in arc plasma and PLV
(4): for particle size < ~80 nm.
Otherwise, for large crystallised catalysts, as many p-CNFs may grow on as many crystal facets

channel-like pores of a suitable porous substrate (e.g. anodic alumina, or zeolite), starting from a gaseous (more rarely liquid) carbon precursor (hydrocarbon) to be thermally cracked within the substrate porosity. The process is thus merely CVD-based, not CCVD-based. The inorganic substrate is then dissolved, resulting in releasing a bunch of carbon nanotubes. As it is difficult to control the deposition of a single layer of carbon atoms onto a surface from cracking hydrocarbons, it is much more common to produce MWCNTs than SWCNTs this way. The interest of this synthesis route is that it is able to produce very uniform nanotubes, whose length and diameter are controlled by the features of the substrate, whose the number of graphenes in the tube wall can be controlled by the CVD parameters, which do not need any purification process as they are naturally catalyst-free and do not contain by-products (but possibly soot and amorphous carbon), and which can be closed at one end or be maintained open at both ends. The graphene stacking in the wall may exhibit a poor to good nanotexture (see Section 1.4) depending on the CVD conditions, and whether an additional annealing step is carried out. Large scale synthesis (e.g. by tons) is, however, prohibited by this method.

1.3   SWCNTs: The Perfect Structure

Regardless of the way they actually form, there are two ways to figure out what a single-wall carbon nanotube (SWCNT) looks like. One is to consider a fullerene, for example, C_{60}, and to add a ring of 10 carbon atoms to the lattice at its diameter thereby forming a C_{70} molecule, then to keep adding such a ring and subsequently form a C_{80}, and then a C_{90}, and so on, more and more elongated until an infinite length is reached with respect to the diameter value (0.7 nm in that case). The other way is to consider a single-atom-thick hexagonal lattice of sp^2 hybridized carbon atoms, the so-called graphene, then to roll it into a cylinder, then to close both cylinder ends with hemifullerenes with the appropriate diameter. The interest of this description is to illustrate the dual nature of SWCNTs, which may be regarded either as macromolecules or as nanoobjects, and which can actually exhibit behaviours and properties from both. For instance, as molecules, SWCNTs have atom positions perfectly defined, and they can be put in solution, be chiral, and exhibit a specific chemical reactivity. As nanoobjects, they can be put in suspension, be filtered, functionalized, and exhibit a specific surface area.

The overall *morphology* of isolated SWCNTs, whatever the growth process, is unique, that is, a hollow filament whose diameter can range from ~0.4 nm to 4–5 nm and whose length can range from few nanometres to several centimetres [20]. Of course, the boundary values of those dimension ranges are extreme and correspond to specific and rare cases. Constrains related to process conditions (e.g. temperature gradients) and energetics make that SWCNTs most often exhibit lengths in the range of few micrometres and diameters in the ~0.7 to 2 nm range.

With respect to the ideal *structure* inherited from graphene, SWCNTs can be defective only by the presence of vacancies or by the replacement of regular hexagons by heterocycles, usually pentagons and heptagons. However, as they significantly affect the planarity of the graphene (for instance, adding one pentagon deforms the plane into a Chinese hat, whereas adding one heptagon deforms the plane into a horse saddle) heterocycles have to

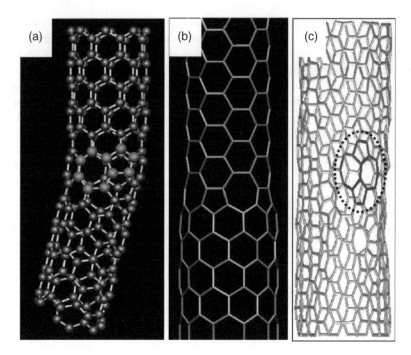

Figure 1.1 *Examples of SWCNTs whose structure includes various heterocycle combinations. (a) pentagon-heptagon ring pair with its symmetry axis oriented oblique to the nanotube axis induces a knee-like distortion of the tube. It is worth noting than this case induces a change in the tube structure, from a (8,0) zigzag tube (top) to a chiral (7,1) tube (bottom) thereby forming a heterojunction. See Figure 1.2 and related comments (reprinted with permission from [21] Copyright (1996) American Physical Society). (b) A ring of alternate pentagons and heptagons induces a bottleneck-like diameter change. It is worth noting than this case also induces a change in the tube structure, from armchair (top) to zigzag (bottom), thereby also forming a heterojunction. Image supplied by A. Rochefort (Polytechnique Monréal). (c) The circled area shows a combination of two heptagon-pentagon adjacent pairs symmetrically displayed, which does not affect the morphology or the structure further than by merely inducing a slight and local distortion (reprinted with permission from [22] Copyright (1998) American Physical Society). Such a combination of two 5–7 ring pairs is very frequently referred to as a 'Stone-Wales defect' in the literature, which is wrong. Stone and Wales actually discussed the case of a combination of two 5–6 ring pairs [23] (in fullerene structure). The combination involving two 5–7 ring pairs should better be referred to as the 'Dienes defect', from the author who first, as far as we can tell, discussed it [24] (in graphene from graphite).*

combine adequately in order to exist within the SWCNT lattice while maintaining the tube integrity. Typically, combining one pentagon and one heptagon makes graphene recover the planar morphology. Hence, this is also true for SWCNTs and depending on the number, nature, and location of the heterocycles present in the structure, SWCNTs may exhibit various morphological deformations, from slight to severe (Figure 1.1).

Yet close to perfection, SWCNTs have their Achilles heels with respect to graphene, namely heterocycles and curvature. Curvature makes that the bonds in which a given carbon atom, yet supposedly sp^2, is involved are no longer located in the same plane. A partial sp^3 character is therefore brought to the carbon atom hybridization, hence making them more reactive than in genuine, flat, graphene and as more reactive as the curvature radius is short [25]. Heterocycles (here, pentagons) are necessary to close the end caps (six pentagons at each end), and constrains on bond angles again make them more reactive than hexagons. In addition, the π electrons of the carbon atoms involved in the pentagons are no longer delocalized. All this makes that a specific SWCNT chemistry can be developed without which meta-nanotubes would nearly not exist, nor this book. On the other hand, the cylinder geometry in capped SWCNTs eliminates the own Achilles heel of unrolled graphene, that is, the edge carbon atoms which offer dangling bonds and a facile route to high chemical reactivity.

Of course, graphene can be rolled up in many ways into a cylinder, making as many different SWCNTs, from a structural, and eventually behavioural, point of view. Finding an easy and practical way to unambiguously describe the various possible SWCNTs was needed. This was achieved by using the indices proposed by Hamada et al. [26]. For making the tubular part of a SWCNT, it can be considered that a graphene has to be rolled up so that a carbon atom from a given hexagon (for instance, the atom labelled O from the left-hand shaded ring in Figure 1.2) is made exactly overlap an equivalent carbon atom from another ring anywhere in the graphene (for instance, the atom labelled A from the right-hand shaded hexagon in Figure 1.2).

The so-called helicity vector C_h is equal to OA in the example and become the circumference of the nanotube thus created. It may be decomposed into two vectors parallel to the graphene lattice vectors a_1 and a_2 respectively. The Hamada's indices n and m count the number of hexagons crossed by each vector, namely $n = 4$ for the dashed vector, and then $m = 2$ for the dotted vector in Figure 1.2. The tube thus formed is a (4,2) SWCNT. Simple geometry rules apply to relate n and m values to the helicity angle θ on the one hand and to the SWCNT diameter d (Figure 1.2), considering a C=C bond distance equal to ~0.142 nm, that is, a bit longer than in graphite (to account for the partial sp^3 character generated by the curvature). If the direction of the helicity vector C_h is selected such as it is parallel to one of the symmetry planes of the graphene, typically one of the six directions perpendicular or parallel to C=C bond directions, the helicity angle θ is nil or equal to 30° respectively. This generates two major families whose names are based on the aspect of the rolled graphene edge at the tube mouth, that is, 'zigzag' (for $\theta = 0$) and;'armchair' (for $\theta = 30°$). In the former case, the tube axis is parallel to the x direction in the figure, and it is parallel to the y direction in the latter case. Due to the geometrical rules with which Hamada's indices are defined, zigzag tubes are all tubes whose one of the n or m indices is nil, whereas armchair tubes are all tubes whose n and m indices are equal. Because both zigzag and armchair tubes are achiral (consecutively to the fact that the graphene is rolled up with respect to a symmetry plane) whereas other tubes (both n and $m \neq 0$) are all chiral (Figure 1.3), a frequent mistake in literature is to refer to θ and C_h as the 'chiral angle' and the 'chiral vector' respectively. As it would be contradictory having SWCNTs with an achiral character while exhibiting a chiral angle whose value is not equal to 0, 'helicity angle' and 'helicity vector' are preferred. It will be shown later on that this variety of helical character goes with a variety of electronic behaviour, typically metallic or large gap semiconducting, or small gap semiconducting (see Section 1.5).

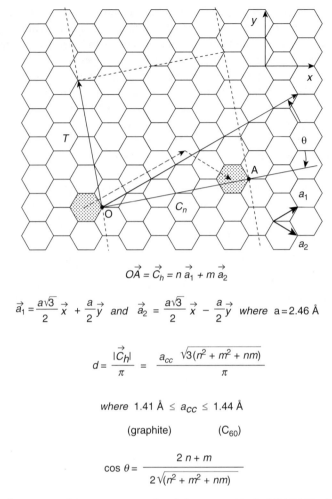

$$\overrightarrow{OA} = \overrightarrow{C_h} = n\,\overrightarrow{a_1} + m\,\overrightarrow{a_2}$$

$$\overrightarrow{a_1} = \frac{a\sqrt{3}}{2}\,\overrightarrow{x} + \frac{a}{2}\,\overrightarrow{y} \quad and \quad \overrightarrow{a_2} = \frac{a\sqrt{3}}{2}\,\overrightarrow{x} - \frac{a}{2}\,\overrightarrow{y} \quad where \;\; a = 2.46 \text{ Å}$$

$$d = \frac{|\overrightarrow{C_h}|}{\pi} = \frac{a_{cc}\;\sqrt{3(n^2 + m^2 + nm)}}{\pi}$$

$$where \;\; 1.41 \text{ Å} \le a_{cc} \le 1.44 \text{ Å}$$

$$(graphite) \qquad\qquad (C_{60})$$

$$\cos\theta = \frac{2n + m}{2\sqrt{(n^2 + m^2 + nm)}}$$

Figure 1.2 *Illustration of the geometric rules defining any kind of SWCNT with the Hamada's indices, starting from a flat graphene subsequently rolled into a cylinder. Adapted from [1].*

Except in some rarely claimed cases, as-prepared SWCNT materials are actually a mixture of several types. However, probably upon energetic reason, some types are sometimes prevalent. For instance, The (10,10) armchair SWCNT, with a diameter of 1.38 nm, is the most frequent SWCNT found in SWCNT materials synthesized by means of one of the so-called 'high temperature' methods (see Section 1.2).

Because SWCNTs are isolated graphenes by nature, they somewhat behave like flat graphenes in graphite (and many other polyaromatic solids), that is, they tend to associate face-to-face as much as possible in order to share their π electrons and thereby bond through weak van der Waals forces. However, due to the high curvature of the graphene surfaces, the most SWCNTs can do is to align parallel as far as possible, hence forming bundles (Figure 1.4a). If all the SWCNTs within bundle exhibit similar

(13, 0)   (13, 1)   (10, 4)   (8, 6)   (8, 8)

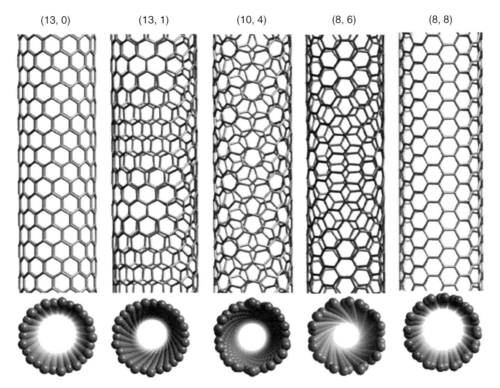

Figure 1.3 *Examples of SWCNTs for each of the three types. Full* **left-hand**: *a zigzag tube. Full* **right-hand**: *an armchair tube.* **In between**: *three examples of chiral tubes. Modified with permission from [27] Copyright (2006) American Physical Society.*

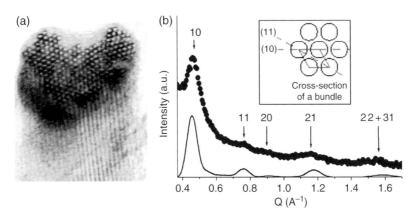

Figure 1.4 *(a) Transmission electron microscopy image of a bundle of SWCNTs obtained by the pulsed laser vaporization method. The scale is given by the SWCNT diameters which all are in the 1.3–1.4 nm range. (b) X-ray diffraction pattern of a SWCNT material in which the SWCNTs are assembled into bundles.*

diameters, as it frequently occurs for high temperature synthesis processes, the most favourable display is for them to pack according to a 2D hexagonal lattice (Figure 1.4b, inset). The diffraction pattern of SWCNT bundles (Figure 1.4b) exhibits specific peaks that differentiate it from other crystallized carbon forms, as an evidence for a new carbon allotrope that was named nanotubulite, which joined the formerly known graphite, rhombohedral graphite, diamond, lonsdaleite, and fullerite.

All the analytical methods which are useful in some extent to the study of carbon nanotubes, SWCNTs or MWCNTs, are not going to be described here. Electron Energy Loss Spectroscopy (EELS), Electron Spin Resonance (ESR), photo-electron emission spectroscopy (XPS), adsorption isotherms, vibrational spectroscopies (IR, UV, etc.) are examples of methods that are worth considering for gathering data about, for example, the presence of vacancies, that of grafted functions, agglomeration, and so on. However, Raman spectroscopy, which is based on the inelastic scattering of photons, has to be mentioned, as the other most important method for describing carbon nanotubes, beside TEM.

As in any other periodic solid, various vibrations related to the collective behaviour of atom pairs and the direction of vibrations with respect to remarkable axes (typically perpendicular to or aligned with the nanotube axis) generate energy absorptions for specific frequency ranges of an incoming laser light. Low frequency vibration modes, say below 500 cm^{-1}, so-called *radial breathing mode (RBM)*, reflect the radial component of the collective vibration of the C-C bonds in the direction perpendicular to the tube axis, making the tube diameter pulse around an average value. Hence, it is a mode which is highly sensitive to the SWCNT diameter (the smaller the diameter, the higher the frequency) and helicity through the excitation wavelength (Figure 1.5a), and which allows the type of SWCNT to be identified by applying the proper equation. Several versions of this equation can be found in the literature, which includes a constant whose actual numerical value is still debated. One recent and reliable one is believed to be found in [28]. To exploit better this interesting relation, a plot was made available that relates the (n,m) indices of a given SWCNT with its specific energy adsorption in the RBM area (often referred to as the 'Kataura plot' [29,30], from the name of the person who proposed it for the first time).

Vibration modes at higher frequencies ($\sim$1300–1600 cm^{-1}) are tangential ('in-plane'). One is around 1350 cm^{-1}, which is complex and relates to the presence of several kinds of defects (hence it was named the *'D' band*) which involve a backscattering of an electron, whereas another band around 1580 cm^{-1} relates to the collective vibration of the in-plane C-C bonds. The latter is as higher as the graphene lattice is more perfect, that is, this band is very intense for a perfect graphene structure as in genuine graphite (hence it was named the *'G' band*). As the pristine SWCNT structure is perfect as well, the G band of SWCNTs is also pretty intense while the D band is minimized, as in graphite. However, because of its anisotropic character with respect to graphite or graphene (except ribbon-shaped graphene), the G band of SWCNTs splits into a G$^-$ and a G$^+$ sub-bands which are representative of the related vibration in the direction perpendicular and parallel to the SWCNT elongation axis respectively for semiconducting tubes (Figure 1.5a). For metallic tubes, due to electron-phonon coupling, the opposite attribution has been proposed (see Section 1.5 for the metallic/semiconducting character discrimination for SWCNTs). The position of G$^+$ is nearly constant, whereas that of G$^-$ is affected by the nanotube diameter and helicity (through the metallic or semiconducting nature of the SWCNT).

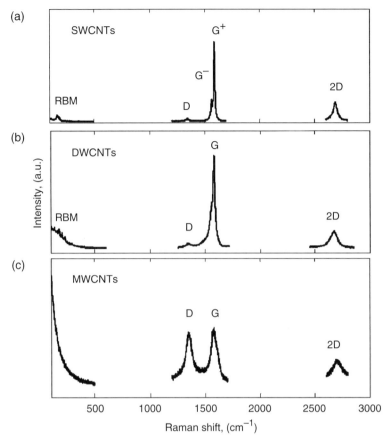

Figure 1.5 *Raman spectra obtained for a laser wavelength of 514 nm on various kinds of carbon nanotubes. (a) Only one RBM peak is seen for the SWCNTs, which does not mean that the bunch of SWCNTs being analysed exhibited a single helicity and diameter. Other SWCNT types could be present which could resonate with other wavelengths. In addition to the presence of RBM, typical features of SWCNTs are the very low intensity of the D band, the splitting of the G band, and the sharpness of the 2D band. The related SWCNT material was prepared by electric arc. (b) Raman spectra for DWCNTs are similar to that of SWCNTs, however with some discrepancies such as the merging of the G⁻ and G⁺ components into a single, larger G band and the widening of the 2D band as well. Both result from the combination and merging of the various contributions from each tube (two SWCNTs per DWCNT). RBMs are more numerous than in (a) because the chances for the presence of resonating SWCNTs are enhanced by the fact that each DWCNT stands for two SWCNTs with different diameters. It is worth noting that, if a single DWCNT is analysed, the respective contribution of the outer and the inner tubes to the 2D band can be discriminated. The related DWCNT material was prepared by CCVD, which is another reason for the presence of multiple RBM peaks as geometrical features of CCVD-prepared nanotubes are more dispersed than for arc-prepared SWCNTs. (c) RBM peaks for MWCNTs (concentric type, see Section 1.4) are no longer visible, mostly because diameters are too large to resonate. The 2D band is large, and the D/G intensity ratio is commonly used to quantify the nanotexture (see Figure 1.12 and related comments). Image supplied by V. Tishkova, CEMES-CNRS.*

Finally, around 2750 cm^{-1} is a band formerly named the "G' band", as it was present in genuine graphite but now more and more consensually named the '2D band' as it is explained as a second order of the D band, yet it does not behave by merely mimicking it.

What is important with respect to this general scheme is that all the Raman bands are able to be affected in their frequency and/or in their intensity upon various factors. This is specifically observable for carbon nanotubes with a limited number of graphenes, ideally SWCNTs or DWCNTs. For instance, RBM peaks can be upshifted under the effect of outer pressure [31], or downshifted upon partial substitution of lattice carbon atoms by nitrogen atoms [32]; the G bands can be up- or downshifted and their intensity can be decreased upon the effect of p-type or n-type doping respectively, possibly with the G$^-$ and G$^+$ being affected differently; the G band can discriminate isolated nanotubes from bundled ones [33]; the G band or its components can be downshifted upon heating [34] or mechanical stress [35,36], and so on. This makes Raman spectroscopy a very useful characterization tool for studying carbon nanotubes and an abundant literature can be found which is currently dealing with such kind of effects. On the other hand, considering the number of interactions the Raman bands can be sensitive too, and assuming that not all aspects are elucidated yet, a high expertise is required for exploiting Raman data at their highest possibilities.

1.4 MWCNTs: The Amazing (Nano)Textural Variety

Multi-walled carbon nanotubes (MWCNTs) are built from several graphenes and, because of this, describing them does not require considering *morphology* and *structure* only, as for SWCNT, but also *texture* and *nanotexture*. MWCNTs may be defined as nanosized, hollow, carbon filaments whose wall is made up with more than one graphene, with the intergraphene distance being equal to the regular ~0.34 nm van der Waals distance for turbostratic, polyaromatic carbons (see the following). However, this definition is not consensual. It is often considered in the literature that the term 'MWCNT' should be restricted to the concentric assembly of a series of n SWCNTs (with $n \geq 2$, and no limitation for the upper value) with increasing diameter (Figure 1.6a) which corresponds to the typical texture of arc-grown MWCNTs which were shown for the first time in the 1991 Iijima's landmark paper [13]. In such a case, any other carbon nanofilament, hollow or not, specifically those exhibiting the so-called 'herringbone' texture where the graphenes making the wall are oblique (Figures 1.6b and 1.6c) and sometimes transverse with respect to the tube axis, is preferably named carbon nanofibre (CNF).

We will not adopt such a definition line here. We prefer to follow the reasonable opinion that, provided it is made of carbon, hollow, with a nanosized diameter (i.e. < 100 nm), and with a wall made up with an assembly of several graphenes, a filament may be named 'MWCNT', regardless of the orientation of the graphenes making the tube wall with respect to the tube axis (in this regard, the 'multi-wall' denomination is not really appropriate and 'multi-graphene wall' would have been more suitable, but the use of the former is now too much widespread). We thus consider that carbon nanofilaments divide into carbon nanotubes and carbon nanofibres. The latter do not exhibit any empty core, whereas the former are hollow, whether the inner cavity is continuous or not, that is, whether it is incidentally or periodically obstructed with graphenes oriented transversally with respect to the tube axis.

(a)   (b)   (c)

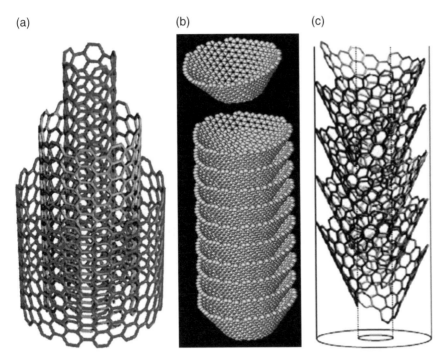

Figure 1.6   (a) Model of a so-called 'concentric' texture. (b) Model of a so-called 'herringbone' texture, for which truncated graphene cones are piled up as cups (cup-stake model, by courtesy of M. Endo). (c) Another model for the herringbone texture, for which a single graphene ribbon is helically wrapped (adapted from [11]). The existence of carbon nanotubes or nanofilaments corresponding to either models (b) or (c) was long debated, but both have been finally demonstrated (see Figure 1.7). On the other hand, the existence of carbon nanotubes or nanofilaments corresponding to model (a) is demonstrated for long; see for instance Figures 1.9a and 1.9b.

As soon as a carbon material is built up with stacked graphenes, accurately describing it requires a multi-scale analysis procedure so as to determine, and quantify when possible, the various description parameters that are morphology, texture, nanotexture and structure (Table 1.2). This makes the description of MWCNTs much more complex to analyse and describe than for SWCNTs.

The *morphology* describes features affecting the overall aspect of any kind of carbon nanofilaments, whether it is external (of which Figure 1.8 is showing some examples) or internal (i.e. the nanofibre versus nanotube discrimination, according to the definition given above) regardless of the actual presence of graphenes and their orientation. For instance, the same morphological terms as listed in Table 1.2 could apply, hypothetically, to amorphous carbon nanofilaments. Although they have been found for either CNFs or MWCNTs, all kinds of outer morphologies are not necessarily all possible for both CNFs and MWCNTs. For instance, the coiled morphology (Figure 1.8a) has never been found for

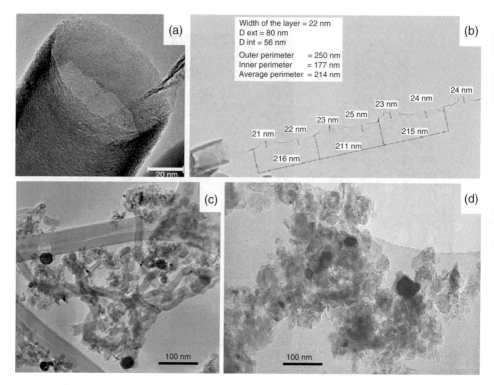

Figure 1.7 *(a) and (b) correspond to the unambiguous demonstration of the existence of the single, helically wrapped graphene model shown as Figure 1.6c (reprinted with permission from [38] Copyright (2006) Elsevier). (c) and (d) correspond to the demonstration [11], yet more indirect, of the existence of the cup-stacked model shown as Figure 1.6b. Both are TEM images of the same herringbone-MWCNT material prepared by CCVD (image supplied by L. Noé, CEMES-CNRS). The only difference between both images is that, for image (d), the material was gently grinded in ethanol then deposited as a suspension droplet onto the TEM grid whereas, for image (c), the material was merely sprinkled from the dry, raw powder over the TEM grid. This has resulted in having the filamentous morphology being maintained in (c), while it was destroyed in (d). Platelet-CNFs behave the same way. On the other hand, h-MWCNTs obeying the single, helically wrapped graphene model are able to withstand the grinding–based TEM preparation (see Figure 1.10c, for instance).*

MWCNTs with concentric texture (neither for SWCNTs, actually), which relates to their difference in growth mechanisms.

The *texture* is a feature which is common to all kinds of polyaromatic solids, as soon as they involve graphene stacking. It describes the inner display of the graphenes (or the graphene stacks) with respect to each other and/or to a peculiar direction, hence it includes any description related to isotropy/anisotropy. For CNFs or MWCNTs, one peculiar direction to consider is typically the elongated axis. The 'concentric' (*c*-) texture is described with the model of Figure 1.6a, and can apply to both MWCNTs (Figure 1.9a) and CNFs (Figure 1.9b). In the case of *c*-CNFs, the absence of inner cavity is declared as soon

Table 1.2   *Chart listing the five-step procedure for the full description of carbon nanofilaments, and the imaging methods to be used. See text for a detailed description of each step. This is valid for multi-graphene CNFs or CNTs only (i.e. SWCNTs are excluded).*

The description chart of multigraphene carbon nanofilaments		
1. Outer morphology (SEM)	Regular, coiled, branched, conical, …	
2. Inner morphology (TEM)	Nanofibre (*CNF*)	Nanotube (*MWCNT*)
3. Texture (TEM)	Platelet (p-) Concentric (c-) Herringbone (h-) Bamboo (b-)	
4. Nanotexture (HRTEM)	L_1, L_2, N, β parameters	
5. Structure (diffraction)	turbostratic $\rightarrow$ graphitic d_{002}, hkl with h and/or $k \neq 0$, and $l \neq 0$	

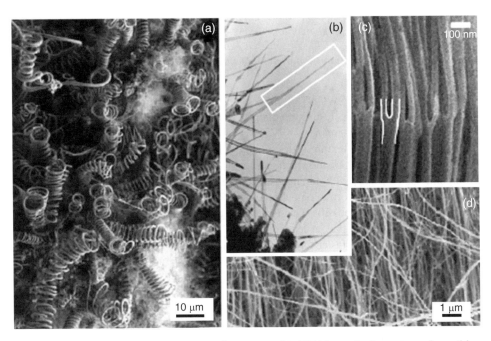

Figure 1.8   *Scanning electron micrographs (except (b): TEM image) of a variety of possible morphologies, all found affecting carbon nanofilaments in literature. (a) Coiled texture (modified with permission from [39] Copyright (2006) Elsevier); (b) Conical texture. The white-framed part is in the micrometre range (modified with permission from [40] Copyright (2007) Elsevier); (c) Branched texture (modified with permission from [41] Copyright (2005) National Academy of Sciences, USA). This particular example was produced by a catalyst-free templating method (see Section 1.2), but CCVD methods are also able to produce this peculiar morphology (e.g. [42]); (d) regular texture.*

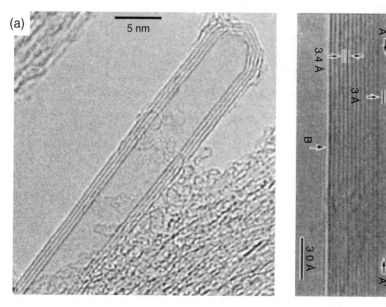

Figure 1.9 *TEM images of examples of the concentric texture. (a) c-MWCNT. (b) c-CNF. Between A and A', a 0.3 nm large nanotube is discernable, yet with a lesser contrast than the other graphenes making the CNF wall, which is explained by the very small radius of curvature. Reprinted with permission from [43] Copyright (2004) American Physical Society.*

as the ultimate, smallest SWCNT exhibits a diameter in the range of the intergraphene distance (~0.34 nm, Figure 1.9b), or a diameter of twice the intergraphene distance and is filled with a linear carbon chain.

The 'herringbone' (*h*-) texture is described by the models of Figures 1.6b and 1.6c, and corresponds to the case where the graphenes making the filament are not parallel but oblique to the tube or fibre axis, with an angle $\alpha/2$ (referring to the related Figure 1.10a) which varies according to the morphological features of the catalyst particles, which most often exhibit a conical part, yet possibly facetted, whose cone angle is α (Figure 1.10b). It is worth noting that, when the crystal (at least the conical part) is facetted, there is a chance that graphenes build slower or faster for a given crystal facet with respect to the others, which is responsible for the occurrence of the coiled morphology (Figure 1.8a). Again, the herringbone texture may affect both CNFs (Figures 1.10a) and MWCNTs (Figures 1.10c–d).

In addition, two textures, namely 'platelet' (*p*-) and 'bamboo' (*b*-), are specific to CNFs and MWCNTs respectively. *p*-CNFs are basically CNFs for which the angle $\alpha = 180°$ (Figure 1.11a). As opposed to *h*-CNFs or *h*-MWCNTs which grow from several catalyst crystal facets or a whole conical part of the catalyst particle (if not facetted), such *p*-CNFs grow each from a single, necessarily large catalyst crystal facet. On the other hand, the most frequent occurrence of the bamboo texture is found with *h*-MWCNTs, for which transverse walls obstruct the inner cavity, somewhat periodically (Figure 1.11b). It is worth noting that the bamboo texture may also affect *c*-MWCNTs (Figure 1.11c). Hence, two textures may combine, for example, bamboo with concentric (*cb*-), or bamboo with herringbone (*hb*-).

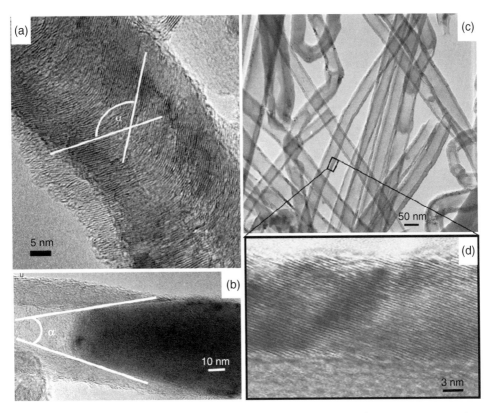

Figure 1.10 *TEM images of examples of the herringbone texture. (a) h-CNF. No inner cavity is seen. α is the angle made by the opposed directions of the graphenes making the CNF. As the graphenes are bowl-shaped, the herringbone angle α is not constant across the CNF, so only the graphene direction close to the surface is considered (reprinted with permission from [11] Copyright (2007) Royal Society of Chemistry). (b) Image of a h-MWCNT taken at the tube/catalyst junction. The conical shape of the catalyst is emphasized, with the cone angle α being equal to the herringbone angle (image credit: L. Noé, CEMES-CNRS). (c) A bunch of carbon nanofilaments whose a majority are h-MWCNTs, that is, showing no transverse wall obstructing the inner cavity (image credit: H. Allouche, CEMES-CNRS). (d) Detail of the tube wall portion framed in (c) to demonstrate the herringbone texture, that is, the oblique orientation of the graphenes making the wall with respect to the tube axis (image credit: H. Allouche, CEMES-CNRS).*

One now understands better why it is not appropriate calling *h*-MWCNTs 'carbon nanofibres' regardless of the actual presence of an inner cavity all along. There is no reason why most of the nanofilaments shown in Figure 1.10c should not be called 'nanotubes', if there is not a single transverse wall obstructing their inner cavity. On the other hand, how many transverse walls should a hollow, herringbone filament contain for being called '*h*-CNF' instead of '*hb*-MWCNT'? Strictly speaking, one should be enough, which shows how inappropriate such a principle of denomination is.

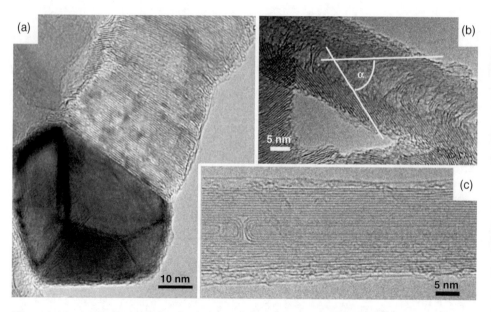

Figure 1.11 *(a) Example of the platelet texture. Note the facetted morphology of the catalyst crystal, on one facet of which the p-CNF has grown (reprinted with permission from [11] Copyright (2007) Royal Society of Chemistry). (b) Example of a combined herringbone/bamboo texture (hb-MWCNT) (modified with permission from [11] Copyright (2007) Royal Society of Chemistry). (c) Example of a combined concentric/bamboo texture (cb-MWCNT) in an arc-grown MWCNT. The surface of the tube seems to be defective merely because of the presence of physisorbed SWCNTs (image supplied by L. Noé, CEMES-CNRS).*

The *nanotexture* is also a feature which is common to all kinds of polyaromatic solids, as soon as they involve graphene stacking. It describes the quality of face-to-face stacking and side-to-side association of graphenes within areas of overall similar graphene orientation (SGO), whatever the texture. In most of polyaromatic solids, such SGO areas are spatially limited and separated from each other by grain boundaries in which structural defects concentrate upon thermal annealing, resulting in large, nonhealable disorientations between neighbouring SGO areas. However, in specific cases involving highly anisotropic carbons, there is no such grain boundary in one or several directions, for example, along the elongated axis in in c-MWCNTs. The nanotexture is quantified thanks to various parameters obtained from high resolution TEM images of the SGO areas (Figure 1.12a). Parameter values may vary depending on the tube synthesis process and conditions. For a given tube, they may also vary upon thermal annealing, ultimately possibly resulting in $L_2 = 0$ and $\beta = 0$ while L_1 and $[(N–1) \times d_{002}]$ are equal to the dimensions of the SGO area respectively (Figure 1.12b). It is worth noting that nanotextural parameters may also somehow quantify graphene alteration, for example, resulting from thermal oxidation or acidic attack, as the parameters can also retrogress.

Raman spectra for most of MWCNTs only exhibit the D, G, and 2D bands (Figure 1.5c). Other bands can be present in MWCNT spectra that are not present in SWCNT spectra, for example, the D' band at a few tens of cm^{-1} higher frequency than G, which relates to

(a)

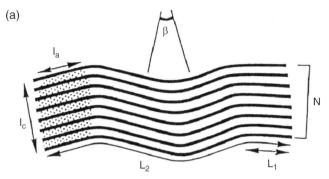

L_1 Average length of the stiff fringes
L_2 Average length of the continuous (yet distorted) fringes
N   Average number of fringes within the coherent graphene stacks
β   Average misorientation angle

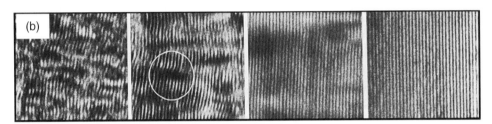

Figure 1.12 *(a) Sketch of the aspect of a SGO area as seen in HRTEM lattice fringe imaging, and definition of the various parameters used to describe the related nanotexture.*
Fringes are actually graphenes seen edge-on. Distortions are expected to relate to the local presence of in-plane defects (heterocycles, single or combined) or, for the least, to local bond angle strains. One example of coherent graphene stack is provided as the shaded area [$l_a \times l_c$]. It follows the definition of a coherent crystal domain for diffraction, that is, all the graphenes involved in the stack scatter the interacting coherent beam (electrons or X-ray photons) in the same directions exactly. As seen by HRTEM, $l_a = L_1$ and $l_c = [(N-1) \times d_{002}]$. (b) From left to right, sequence of lattice fringe TEM images of a portion of SGO area with increasing nanotexture quality, as it may happen upon thermal annealing in the range 1500–2800°C. This approach was developed for long, see for instance [44–46].

graphene stacking, and also very high energy modes such as at 3240 cm^{-1}, which correspond to overtones. But the exploitation of the latter in the characterisation of MWCNTs is still subordinate so far and therefore they will not be commented on further.

The variations in the coherent domain size are observable in Raman spectroscopy by the variations in the intensity of the D and G bands. This has been acknowledged for long, and the D/G band intensity ratio was proposed to characterize the structural order in polyaromatic carbons as early as in 1970 [47]. From what is said in this section, we can understand that the D/G band intensity ratio correlates with the nanotextural rather than structural variations. Hence, the nanotexture quantification from TEM as described above relates well with the D/G band intensity ratio obtained by Raman spectroscopy (See Figure 1.5c) at least for graphene extension larger than 2 nm [48]. All the other effects mentioned in Section 1.3 (SWCNTs) that are supposedly able to affect the Raman spectrum features are actually hindered by the large presence of defects and the high number of graphenes in the wall which are common features for MWCNTs. Only if the nanotexture is perfect (no D band) and the number N of graphenes stacked is < 10, the G band position can show a slight sensitivity to the number N, and an upshift from 1582 to 1584 cm^{-1} is observed when N decreases from ~10 to 2 [49]. Likewise, only c-MWCNTs with N = 2 or 3 (DWCNTs or triple-wall carbon nanotubes) are able to show RBM peaks, due the increasing diameter of the concentric shells that make them resonate at very low frequencies that are out of the range of Raman spectroscopy. This means that, in most cases, Raman spectra for MWCNTs are not differentiated from that of other multi-graphene-based carbon materials. For the latter as for most of MWCNTs, Raman is not sensitive to the variations in texture as it is defined above and is barely sensitive to the variations in structure as it is defined below.

Finally, the *structure* relates to the various ways graphenes can pile up within a coherent stack. There are several remarkable configurations, whose some are gathered in Figure 1.13 and below listed by increasing chance of occurrence:

- AA stacking sequence (Figure 1.13a): two graphenes are superimposed exactly. This is energetically not favoured and is never found but sometimes in isolated, few-graphene flakes (few = e.g. two or three) of limited lateral extension (in the micrometre range).
- ABC stacking sequence: three graphenes are exactly superimposed but each of them is shifted over a distance equal to one C-C bond length in the direction parallel to the bond with respect to the lattice underneath. Hence only a fourth graphene D stacked with the same sequence will be back to a position exactly similar to that of the bottom graphene A. This configuration is also unstable but less than AA. It corresponds to the rhombohedral structure. It has never been isolated as a single crystal but can be found as stacking faults in genuine graphite or, again, in small-size few-graphene flakes.
- AB stacking sequence (Figure 1.13b, also called Bernal stacking, from the name of the person who first described the structure in 1924): two graphenes are superimposed while their lattices are shifted the same way as they are for the ABC sequence above. The difference is that the third graphene is stacked back to the position of the bottom graphene A instead of being further shifted to reach position C. This configuration is the most stable and therefore corresponds to the hexagonal structure of genuine graphite. The conditions for the stacking to occur are not always encountered, depending on the thermal and steric conditions the graphene stacks have been subjected to. Yet the most stable, this structure remains rare with respect to the next structure.
- The turbostratic stacking sequence: graphenes are superimposed while being randomly rotated with respect to each other (Figure 1.13c). It is by far the most frequent configuration, yet energetically metastable, even after very high temperature treatments (3000°C in argon) in the case of nongraphitizable carbons. Among the infinite number of turbostratic stacking possibilities, some situations of commensurability are possible for specific rotation angle values, resulting in the occurrence of a supercell (Figure 1.13d).

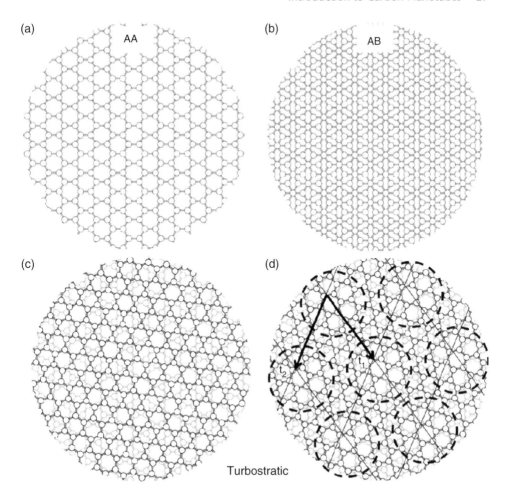

(a) AA

(b) AB

(c) Turbostratic

(d)

Figure 1.13 *Sketches of various possible stacking sequences for two graphenes. (a) AA stacking. (b) AB (Bernal) stacking. (**c**) Turbostratic stacking, for a rotation angle between the two graphene lattices of 23°. (d) Commensurate turbostratic stacking, for a rotation angle between the graphene lattices of 13°. The commensurability is revealed within the circled areas, in which the two graphenes are locally nearly under the AA stacking sequence within areas the size of a coronene molecule. This occurs periodically, giving rise to a superstructure, whose supercell vectors are indicated as bold black arrows. Images supplied by L. Ortolani, IMM-CNR.*

Changing the structure from turbostratic (2D structure) to graphitic (3D structure) is possible, and it is actually the regular path for graphitization upon thermal annealing (Figure 1.14a). The reversal is also possible, upon any structure alteration process such as partial oxidation. Likewise, locally changing from the hexagonal, graphitic structure (AB sequence) to the rhombohedral structure (ABC sequence) is a common consequence of applying severe shearing stresses. There is no demonstration yet whether other kinds of structure transformation (e.g. AB into AA, or turbostratic into commensurate

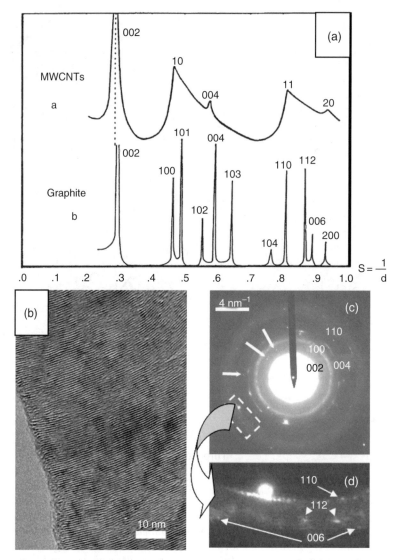

Figure 1.14 (a) Typical diffraction pattern of a turbostratic structure (top) as a bunch of regular MWCNTs can exhibit compared to the diffraction pattern of genuine graphite (bottom). As opposed to graphite whose structure is 3D, hence showing all the hkl diffraction spots, the turbostratic structure is 2D only, hence showing only hk and 00l diffraction spots. For graphitizable carbons, 2D may progressively change to 3D structure upon thermal annealing. The asymmetric shape of the hk peaks is typical of turbostratic structure and is a consequence of the shape of the related reciprocal nodes, which elongate for graphenes that behave as isolated ones, yet piled-up. (b) HRTEM image of a portion of a p-CNF. Modified from [51] (a and b). (c) Overall diffraction pattern of a bunch of ground p-CNFs such that in (b). (d) Detail of (c) showing the presence of 112 diffraction spots, revealing the graphitic (3D) structure of the p-CNFs. Reprinted with permission from [11] Copyright (2007) The Royal Society of Chemistry.

turbostratic) should be possible, as examples are still lacking in the literature. Because of the curvature of their wall, whatever the texture, the only possible stacking sequence for graphenes in MWCNT wall among those just listed is genuine turbostratic, as the other stacking sequences require the graphenes involved to be flat (of course, the local occurrence of a specific superimposition display is always locally possible, but not in a periodic manner over the whole graphene surface). Hence, they are naturally possible in textures in which graphenes are naturally flat, such as the platelet texture in CNFs [11] (Figures 1.11a and 1.14b). Another possibility is to make the graphenes become flat, by subjecting MWCNTs to thermal annealing at very high temperatures. This results in the migration of in-plane defects, the progressive healing of the distortions as seen in Figure 1.12b, and the faceting of the MWCNT cross-section (specifically for *c*-MWCNTs). Within each graphene stacks making the flat facets, which are separated from the neighbouring ones by grain boundaries in which defects have concentrated, the structure of genuine graphite (Bernal stacking) can develop. Finally, for very large radii of curvature (> ~ 500 nm), commensurate stacking such as the AB sequence and the related 3D hexagonal structure may occur locally over a sufficiently large area for the related diffraction spots to show up [50] (Figure 1.14c–d).

1.5 Electronic Structure

The electronic structure of SWCNTs is derived from that of a single graphene, however, first restricted to a ribbon and second curved into a cylinder. Starting from a single, flat, isometrical graphene which is a semi-conducting bi-dimensional crystal with a nil band gap, the limitation in one dimension restricts the number of allowed electron states in the circumferential direction (while the number of allowed electron states remains large in the axial direction) on the one hand, and the short radius of curvature distorts the electronic band structure with respect to that of pristine graphene on the other hand. Hence, considering this quantization of the electron states and the way the graphene is wrapped with respect to the lattice symmetry elements, the Fermi energy level (defined as the highest value of the last occupied energy measured at absolute zero) of a given (n,m) SWCNT (see Figure 1.2) will appear to fall either within an allowed, not fully occupied energy band, the so-called conduction band (thus making the SWCNT behave like a metal), or at the top of the last, fully occupied valence band, which is separated by a gap of forbidden energy states from the first conduction band (Figure 1.15). In graphene system, this gap is nil to relatively small only, making the related SWCNTs always behave like a semiconductor (meaning that electrons from the last valence band can be helped for jumping to the first conduction band through the forbidden band by providing energy from an external source such as photo-excitation or electric potential). In any case, no SWCNT can behave like an insulator.

Practically, all (n,m) SWCNTs with $n = m$ (armchair tubes) exhibit a metallic behaviour. Regarding the other SWCNTs, whether they are of zigzag ($m = 0$) or chiral ($n \neq m \neq 0$) type, and provided the curvature of the graphene is not too much pronounced (say, for radii of curvature beyond ~ 0.7 nm), they remain zero band gap semiconductors (hence assimilated to a metal) if $n - m = 3q$ (where q is an integer), and are large band gap semiconductors otherwise. In other words, one third of non-armchair tubes behave like

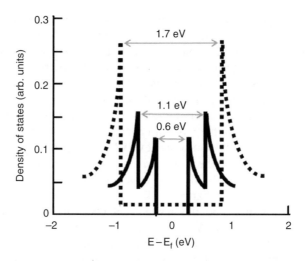

Figure 1.15 *Sketch of the density of states for two hypothetical SWCNTs, one being an armchair tube hence exhibiting a metallic behaviour (dotted line plot), the other exhibiting a semiconducting behaviour (heavy solid line plot). E_f is the Fermi energy, hence the Fermi level is located at $(E - E_f) = 0$. The metallic behaviour is ascertained by the occurrence of a finite density of states at the Fermi level for the dotted line plot, whereas there is no energy state allowed at the Fermi level for the solid line plot. The discretization of the possible energy states induced by the 1D morphology of the nanotubes generates the spiky features in the plots, which are known as van Hove singularities.*

metals whereas the other two thirds behave like semiconductors. In first approximation, this can be summarized by the well-known sketch displayed as Figure 1.16. However, for smaller diameters down to ~0.5 nm, this is no longer valid, and the strong deformation of the graphene makes all the chiral and the zigzag SWCNTs be semiconductors with energy gaps ranging from 0.5 to 1 eV [52]. For even shorter radii of curvature, the latter is again no longer true, and metal behaviour can be found again for supposedly semiconducting tubes such as the chiral (4,3) [52].

The picture, however, becomes rapidly complex as soon as bundled SWCNTs are considered. The intertube distance in a bundle of SWCNTs is short enough for a SWCNT to be significantly affected by the potential of the neighbouring one(s). This breaks the symmetry of the single, isolated SWCNT lattice. As a result, a bundle of identical armchair SWCNTs, although each would behave as metals if single and isolated, can exhibit a band gap making it a semiconducting object [53]. What could be the incidence of such interactions in real-life bundles where the SWCNTs involved exhibit a variety of helicities, cannot be predicted simply and requires calculations accounting for the variety of the interacting electronic structures and their relative positions in space.

As far as coaxial SWCNTs are concerned (i.e. *c*-MWCNTs), calculations have shown that each of the SWCNTs tends to maintain its own behaviour, yet tube interactions may occur that are likely to modify the electronic behaviours for specific relative positions of the superimposed graphene lattices [54,55]. However, this is specifically sensitive for

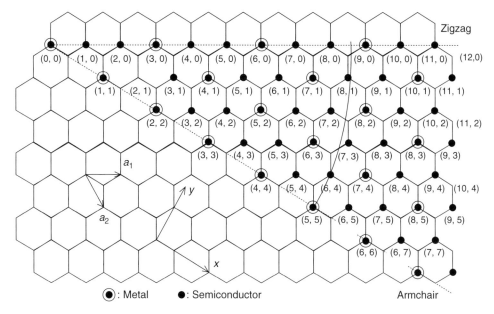

Figure 1.16 *Pairs of indices defining SWCNT helicity according to Figure 1.2 reported on a graphene lattice, along with their related electronic behaviour obtained from calculations. Of course, some of the tubes shown have no physical existence (e.g. (0,0) and (1,1)) or are purely hypothetical as they are highly unstable (typically, SWCNTs with very low n or m values). For instance, the smallest SWCNTs whose existence has been demonstrated so far exhibit a diameter of 0.4 nm. This corresponds to (n,m) = (3,3), (5,0), or (4,2). Reprinted from [1].*

tubes with relatively small diameters. As soon as radii of curvature are larger than 7 nm, which is quite frequent in MWCNTs, the band gap energy of the semiconducting SWCNTs is lower than the thermal energy at room temperature, meaning that all *c*-MWCNTs whose outer diameter is larger than 14 nm exhibit a metallic behaviour, at least at room conditions [1].

1.6 Some Properties of Carbon Nanotubes

It is of common knowledge that carbon nanotubes exhibit exceptional properties in many instances, including thermal, mechanical, and electrical. Table 1.3 illustrates this by gathering some values from the literature, which are all record values with respect to other solids. However, yet *h*-MWCNTs can have specific applicative interests, only *c*-MWCNTs with perfect nanotexture and SWCNTs are involved in such record values. This is due to a combination of their most specific features such as structural perfection, extreme anisotropy, the existence of delocalized electrons, the C-C atomic bond shorter, hence stronger, than in diamond, and so on.

Table 1.3   *Some amazing properties of carbon nanotubes. They all applied to SWCNTs, yet some are shared by high-nanotexture c-MWCNTs (e.g. aspect ratio and tensile modulus). Modified from [9].*

Properties	Values	Comments
Aspect ratio	~1 000 – 10 000	*Possibly higher*
Specific surface area	~2 780 m²/g	*When considering both surfaces of open SWCNTs*
Tensile strength	> 45 GPa	*Other values up to 100 GPa can be found in the literature*
Tensile modulus	1 to 1.3 TPa	*Independent on diameter when > 1nm*
Tensile strain	> 40%	*Provides toughness values higher than that of spider web*
Flexural modulus	1.2 TPa	
Thermal stability	> 3000°C	*In oxygen-free atmosphere*
Electrical conductivity	$10^4 - 10^7$ S/cm	*Better than copper*
Transport regime	Ballistic, up to superconductivity	*$T_c < 1 k$*
Thermal conductivity	~6 000 W/mk	*Better than diamond*
Electron emission	$10^6 - 10^9$ A/cm²	*Highest current density*

It would take books to describe how far and why carbon nanotubes are superior to other materials for so many properties. Figure 1.17 illustrates this for the *mechanical properties,* by gathering both the tensile strength-to-failure and Young's modulus of many kinds of processed fibres, including the best current ones which also are carbon-made. A first statement is that the best performances are by far achieved by carbon fibres. Another statement is that, if some pitch-based carbon fibres are able to exhibit Young's modulus values close to 1 TPa, the very best carbon fibre is limited to 7 GPa regarding tensile strength. A final statement is that it has been impossible so far to fabricate a material able to exhibit both the best tensile strength and the best tensile modulus, except SWCNTs. SWCNTs are able to exhibit a Young's modulus value better than 1 TPa but most of all is meanwhile able to exhibit a tensile strength-to-failure value that exceeds that of the best current carbon fibre by a factor of nine, or more.

Of course, a major issue is now to prepare micrometre-sized fibres able to be woven out of SWCNTs while maintaining enough of the superior behaviour. As fibres are dedicated to be incorporated into matrices, another issue will be to be able of an optimized interaction with the embedding material. If those two progresses could not be made, the practical use of SWCNTs for high mechanical property applications would be very limited.

Thanks to their exceptional *electrical conductivity* (Table 1.3), making an insulating material (typically a polymer) become electrically conductive is probably one of the most popular reasons for attempting to make composites loaded with carbon nanotubes. As opposed to composites dedicated to mechanical applications, nanotube-containing composites do not require paying much attention to the nanotube/matrix interaction, as what are supposed to be preferably handled by the nanotubes, charge carriers, do not have to transit via the matrix. Hence, it is enough for the nanotubes to build an interconnected

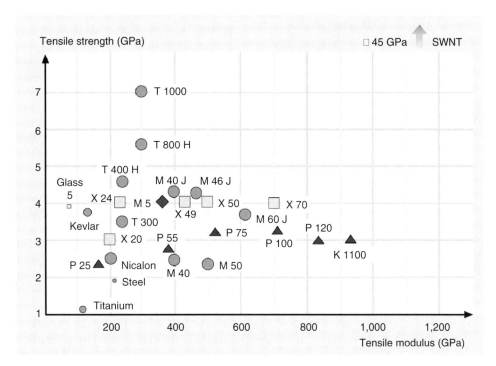

Figure 1.17 *Plots of the mechanical property values for many kinds of fibres including well-known materials (steel, Kevlar, glass), advanced carbon fibres, and SWCNTs. Triangles are pitch-based carbon fibres (the one with the highest modulus value, K1100, is from AMOCO). Solid circles are polyacrilonitrile-based carbon fibres (the one with the highest tensile strength value, T1000, is from TORAYCA). Modified with permission from [9] Copyright (2010) Springer Science + Business Media.*

network within the matrix for allowing the charge carriers to travel and jump from a nanotube to another until crossing through the whole composite material section. No need either to make fibres out of the nanotubes. The more the nanotubes are actually dispersed, the lower the proportion of the nanotubes in the matrix and the more the composite will maintain the optical properties (e.g. transparency) of the matrix alone. In this matter, the aspect ratio of the filler has a dramatic consequence (Figure 1.18). Indeed, conductive polymers are known to be obtained by loading them with carbon blacks for instance but the proportion of the latter has to be ~15 vol.% for a reasonable conductivity to be achieved. This makes any material totally dark. On the other hand, a good conductivity can be obtained with only 0.1–3 vol.% of nanotubes in an insulating polymer matrix, depending on the type of nanotubes (SWCNTs or MWCNTs) selected as filler, as a direct consequence of aspect ratios as high as 1000 and more. An important issue here is to be able to disperse as homogeneously as possible the nanotubes within the matrix, in a way that ideally leaves each nanotube individual, which is quite challenging as far as SWCNTs are concerned and still a matter of investigation.

A graphene surface is poorly reactive by nature, and graphene *reactivity* mostly comes from the presence of free edges at which edge carbon atoms exhibit dangling bonds.

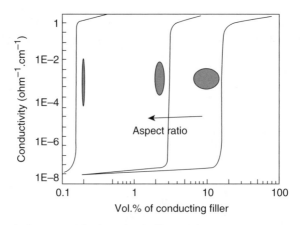

Figure 1.18 *Sketch illustrating the beneficial effect of the increasing aspect ratio of the material making the conducting filler on the filler concentration value at which the percolation threshold takes place.*

However, three features are able to increase the reactivity of graphene: (i) curvature, thanks to the partial sp^3 character subsequently induced to the carbon atom hybridization; (ii) the presence of defects of any kind, typically heterocycles and vacancies; (iii) the substitution of carbon atoms by others in the graphene lattice, allowing a specific chemistry to take place.

From the three features mentioned, the reactivity of carbon nanotubes and nanofibres can be qualitatively anticipated. Tube ends will be the most reactive parts, either because they are open, leaving free graphene edges, or because their closure requires the presence of heterocycles (six pentagons, for regular SWCNTs). From titration experiments, it can be deduced that the cylinder part of SWCNTs may contain many heterocycles, in the range of one Dienes defect (see the Figure 1.1 caption) every 5 nm for instance [56], possibly more abundant for arc-grown SWCNTs than for CCVD-grown SWCNTs, due to the much faster growth speed of the former which is likely to favour the occurrence of lattice defects [57]. Assuming that the same is true for c-MWCNTs with perfect nanotexture (as for arc-grown c-MWCNTs, for example, Figures 1.9a and 1.11c), which is not ascertained, the presence of heterocycles in the concentric graphenes would not affect the reactivity of c-MWCNTs as much as for SWCNTs, because the chemical reaction will not progress through the tube wall unless all the graphene cylinders making the c-MWCNT would present heterocycles at the same location, which is unlikely (except in the case of faceting, Figure 1.19a). Also, if the reactivity of SWCNTs (as well as their instability) increases as their radius of curvature decreases, the curvature effect is of lesser consequence on c-MWCNTs, as they most often exhibit much larger outer diameters than SWCNTs (i.e. much larger radii of curvature). In addition, the propensity of SWCNTs to self-assemble into bundles creates several favourable sites (surface, groove, channel and pore) for adsorption (Figure 1.20). All this makes that SWCNTs are much more reactive, relatively speaking, than c-MWCNTs with comparable graphene perfection. Of course, this is no longer true as soon c-MWCNTs with lower nanotexture quality are considered, and the lower the nanotexture (from right to left in Figure 1.12b), the higher the reactivity.

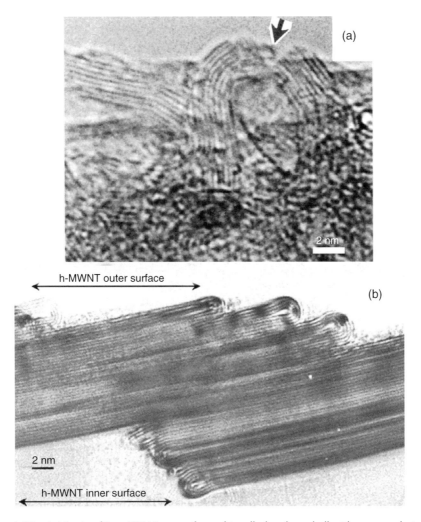

Figure 1.19 *(a) Lattice fringe TEM image of a multi-walled carbon shell with a somewhat polyhedral morphology subjected to chemical oxidation treatment by HNO$_3$. From the angle made by the wall remnants on each side of the opening resulting from the oxidation, it can be guessed that the oxidation has initiated where the graphenes were bent the most, possibly due to the presence of a pentagon line (i.e. every superimposed graphenes were wearing a pentagon at the same location, inducing a 113° angle Chinese hat deformation of the graphene stack) (Reprinted with permission from [58] Copyright (2001) Elsevier). (b) Lattice fringe TEM image of a h-MWCNT similar to that in Figure 1.12d, after it was subjected to a high temperature heat treatment (2400°C) in argon. The graphene-free edges have buckled locally, forming loops in which the graphene stacking distance is relaxed for minimizing the energetic cost (image supplied by H. Allouche, CEMES-CNRS).*

Finally, as edges are the most reactive part of graphene, all the texture types involving accessible graphene edges (p-CNFs, h-CNFs, h-MWCNTs) are likely to be highly reactive, starting at room temperature with easy physisorption of the surrounding moieties but also allowing chemical reactions and intercalation between graphenes. However, either deliberately

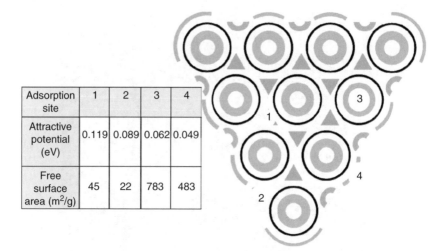

Adsorption site	1	2	3	4
Attractive potential (eV)	0.119	0.089	0.062	0.049
Free surface area (m²/g)	45	22	783	483

Figure 1.20 *Sketch showing the various adsorption sites in a bundle of SWCNTs seen in cross-section. Sites 1 to 4 correspond to so-called interstitial, groove, pore and surface sites respectively. Modified with permission from [9] Copyright (2010) Springer Science + Business Media.*

by applying thermal treatments or spontaneously during the growth process, the edge reactivity and intergraphene space accessibility can be minimized by the occurrence of local graphene closures (Figure 1.19b). HRTEM investigations are therefore always recommended.

1.7   Conclusion

There is no doubt that carbon nanotubes are extraordinary materials with respect to their amazing properties as well as their morphological, textural, nanotextural and structural versatility. However, this means that using them in the most optimized way appears not to be an easy task. As far as MWCNTs are concerned, the abundance of varieties requires a thorough investigation of the materials using advanced characterization techniques such as HRTEM. On the other hand, the most unique properties are encountered for SWCNTs whose variability is only structural. They however come with other challenging issues related to various needs such as sorting the metallic from the semiconducting ones, tailoring their surface interaction with a surrounding medium, preventing them from bundling, and so on. With all those issues, meta-nanotubes are likely to help. This is what is going to be described in the following chapters.

References

[1] M.S. Dresselhaus, G. Dresselhaus and P.C. Eklund, *Science of Fullerenes and Carbon Nanotubes*, Academic Press, San Diego (CA), USA (1996).
[2] R. Saito, G. Dresselhaus and M.S. Dresselhaus, *Physical Properties of Carbon Nanotubes*, Imperial College Press, London, UK (1998).
[3] S. Reich, C. Thomsen and J. Maultzsch, *Carbon Nanotubes: Basic Concepts and Physical Properties*, Wiley-VCH, Weinheim, Germany (2004).

[4] M.J. O'Connell (ed.), *Carbon Nanotubes: Properties and Applications*, CRC Press/Taylor & Francis, Boca Raton (FL), USA (2006)

[5] A. Loiseau, P. Launois, P Petit, S. Roche and J.-P. Salvetat (eds.), *Understanding Carbon Nanotubes: From Basics to Applications*, in *Lect. Notes Phys.*, **677**, Springer-Verlag, Heidelberg, Germany (2006).

[6] Y. Gogotsi (ed.), *Nanotubes and Nanofibers*, CRC Press, Boca Raton, FL, USA (2006).

[7] A. Jorio, M.S. Dresselhaus and G. Dresselhaus (eds.), *Carbon Nanotubes*, in *Topics Appl. Phys.* **111**, Springer-Verlag, Heidelberg, Germany, (2008).

[8] P.F.J. Harris, *Carbon Nanotube Science: Synthesis, Properties, and Applications*, Cambridge University Press, Cambridge, UK (2009).

[9] M. Monthioux, P. Serp, E. Flahaut, M. Razafinimanana, C. Laurent, A. Peigney, W. Bacsa and J.-M. Broto, Introduction to Carbon Nanotubes. In *Nanotechnology Handbook* (ed. B. Bhushan), 3rd Edition (revised), Springer-Verlag, Heidelberg, Germany, 47–118 (2010).

[10] D.M. Guldi and N. Martin (eds.), *Carbon Nanotubes and Related Structures: Synthesis, Characterization, Functionalization, and Applications*, Wiley-VCH, Weinheim, Germany (2010).

[11] M. Monthioux, L. Noé, L. Dussault, J.-C. Dupin, N. Latorre, T. Ubieto, E. Romeo, C. Royo, A. Monzón and C. Guimon, Texturising and structurising mechanisms of carbon nanofilament during growth, *J. Mater. Chem.*, **17**, 4611–4618 (2007).

[12] W. Krätschmer, L.D. Lamb, K. Fostiropoulos and D.R. Huffman, Solid C_{60}: A new form of carbon, *Nature*, **347**, 354–358 (1990).

[13] S. Iijima, Helical microtubules of graphite carbon, *Nature*, **354**, 56–58 (1991).

[14] C. Journet, W.K. Maser, P. Bernier, A. Loiseau, L.M. de la Chapelle, S. Lefrant, P. Deniard, R. Lee and J.E. Fischer, Large-scale production of single-walled carbon nanotubes by the electric-arc technique, *Nature*, **388**, 756–758 (1997).

[15] M. Monthioux, M. Pacheco, H. Allouche, M. Razafinimanana, N. Caprais, L. Donadieu and A. Gleizes, New data about the formation of SWNTs by the electrical arc method, In *Electronic Properties of Molecular Nanostructures* (eds. H. Kuzmany, J. Fink, M. Mehring and S. Roth), *Amer. Inst. Phys. Conf. Proc.*, **633**, 182–185 (2002).

[16] Y. Saito, M. Okuda, N. Fujimoto, T. Yoshikawa, M. Tomita and T. Hayashi, single-wall carbon nanotubes growing radially from Ni fine particles formed by arc evaporation, *Jpn. J. Appl. Phys.*, **33**, L526–L529 (1994).

[17] P.-X. Hou, C. Liu and H.-M. Cheng, Purification of carbon nanotubes, *Carbon*, **46**, 2003–2025 (2008).

[18] M. Monthioux, B.W. Smith, B. Burteaux, A. Claye, J.E. Fischer, D.E. Luzzi, Sensitivity of single-wall carbon nanotubes to chemical processing: an electron microscopy investigation, *Carbon*, **39**, 1251–1272 (2001).

[19] T. Kyotani, L.-F. Tsai and A. Tomita, Preparation of ultrafine carbon tubes in nanochannels of an anodic aluminum oxide film, *Chem. Mater.*, **8**, 2109–2113 (1996).

[20] X. Wang, Q. Li, J. Xie, Z. Jin, J. Wang, Y. Li, K. Jiang and S. Fan, Fabrication of ultralong and electrically uniform single-walled carbon nanotubes on clean substrates, *Nano Lett.*, **9**, 3137–3141 (2009).

[21] Wikipedia (2011) *Carbon Nanotube,* http://en.wikipedia.org/wiki/Carbon_nanotube. [accessed 18 July 2011].

[22] M.B. Nardelli, B.I. Yakobson and J. Bernholc, Mechanism of strain release in carbon nanotubes, *Phys. Rev. B*, **57**, R4277–R4280 (1998).

[23] A.J. Stone and D.J. Wales, Theoretical studies of icosahedral C_{60} and some related species, *Chem. Phys. Lett.*, **128**, 501–503 (1986).

[24] G.J. Dienes, Mechanism for self-diffusion in graphite, *J. Appl. Phys.*, **23**, 1194–1200 (1952).

[25] Z. Wang, S. Irle, G. Zheng and K. Morokuma, Analysis of the relationship between reaction energies of electrophilic SWNT additions and sidewall curvature: chiral nanotubes, *J. Phys. Chem. C*, **112**, 12697–12705 (2008).

[26] N. Hamada, S.I. Sawada and A. Oshiyama, New-one dimensional conductors, graphite microtubules, *Phys. Rev. Lett.*, **68**, 1579–1781 (1992).

[27] C. Noguez (2006) *Optical Properties of Nanostructures* http://www.fisica.unam.mx/cecilia/ [accessed 18 July 2011].

[28] J.C. Meyer, M. Paillet, T. Michel, A. Moréac, A. Neumann, G.S. Duesberg, S. Roth, J.-L. Sauvajol, Raman modes of index-identified freestanding single-walled carbon nanotubes, *Phys. Rev. Lett.*, **95**, 217401 (2005).

[29] H. Kataura, Y. Kumazawa, Y. Maniwa, I. Uemezu, S. Suzuki, Y. Ohtsuka and Y. Achiba, Optical properties of single-wall carbon nanotubes, *Synt. Met.*, **103**, 2555–2558 (1999).

[30] H. Telg, J. Maultzsch, S. Reich, F. Hennrich and C. Thomsen, Chirality distribution and transition energies of carbon nanotubes, *Phys. Rev. Lett.*, **93**, 177401 (2004).

[31] A. Merlen, N. Bendiab, P. Toulemonde, A. Aouizerat, A. San Miguel, J.L. Sauvajol, G. Montagnac, H. Cardon and P. Petit, Resonant Raman spectroscopy of single-wall carbon nanotubes under pressure, *Phys. Rev. B*, **72**, 035409 (2005).

[32] I.C. Gerber, P. Puech, A. Gannouni and W. Bacsa, Influence of nitrogen doping on the radial breathing mode in carbon nanotubes, *Phys. Rev. B*, **79**, 075423 (2009).

[33] V. Tishkova, P.-I. Raynal, P. Puech, A. Lonjon, M. Le Fournier, P. Demont, E. Flahaut and W. Bacsa, Electrical conductivity and Raman imaging of double wall carbon nanotubes in a polymer matrix, *Compos. Sci. Technol.*, in print (2011).

[34] P. Puech, F. Puccianti, R. Bacsa, C. Arrondo, V. Paillard, A. Bassil, M. Monthioux, E. Flahaut, F. Bardé and W. Bacsa, Ultraviolet photon absorption in single, double wall carbon nanotubes and peapods: heating-induced phonon line broadening, wall coupling, and transformation, *Phys. Rev. B*, **76**, 054118 (2007).

[35] S. Cui, I.A. Kinloch, R.J. Young, L. Noé and M. Monthioux, The effect of stress transfer within double-walled carbon nanotubes upon their ability to reinforce composites, *Adv. Mater.*, **21**, 3591–3595 (2009).

[36] S. Cui, I.A. Kinloch, R.J. Young and M. Monthioux, Response to "Comment on the effect of stress transfer within double-walled carbon nanotubes upon their ability to reinforce composites", *Adv. Mater.*, **22**, 1180–1181 (2010).

[37] F.F. Xu, Y. Bando and D. Golberg, The tubular conical helix of graphitic boron nitride, *New J. Phys.*, **5**, 118.1–118.16 (2003).

[38] J. Vera-Agullo, H. Varela-Rizo, J.A. Conesa, C. Almansa, C. Merino and I. Martin-Gullon, Evidence for growth mechanism and helix-spiral cone structure of stacked-cup carbon nanofibers, *Carbon*, **45**, 2751–2758 (2007).

[39] J.B. Bai, Growth of nanotube/nanofibre coils by CVD on an alumina substrate, *Mater. Lett.*, **57**, 2629–2633 (2003).

[40] X. Sun, R. Li, B. Stansfield, J.-P. Dodelet, G. Ménard and S. Désilets, Controlled synthesis of pointed carbon nanotubes, *Carbon*, **45**, 732–737 (2007).

[41] G. Meng, Y.J. Jung, A. Cao, R. Vajtai and P.M. Ajayan, Controlled fabrication of hierarchically branched nanopores, nanotubes, and nanowires, *Proc. Nat. Acad. Sci.*, **102**, 7074–7078 (2005).

[42] Q. Liu, W. Liu, Z.-M. Cui, W.-G. Song and L.-J. Wan, Synthesis and characterization of 3D double branched K junction carbon nanotubes and nanorods, *Carbon*, **45**, 268–273 (2007).

[43] X. Zhao, Y. Liu, S. Inoue, T. Suzuki, R.O. Jones and Y. Ando, Smallest carbon nanotube is 3Å in diameter, *Phys. Rev. Lett.*, **92**, 125502 (2004).

[44] A. Oberlin, J. Goma and J.-N. Rouzaud, Techniques d'étude des structures et textures (micro-textures) des matériaux carbonés, *J. Chem. Phys.*, **81**, 701–710 (1984).

[45] X. Bourrat, *Contribution à l'Etude de la Croissance du Carbone en Phase Vapeur*, 'Doctorat d'Etat', University of Pau, France (1987).

[46] M. Monthioux, Structures, textures, and thermal behaviour of polyaromatic solids. In *Carbon Molecules and Materials* (eds. R. Setton, P. Bernier and S. Lefrant), 127–177. Taylor & Francis, London (2002).

[47] F. Tuinstra and J.L. Koenig, Raman spectrum of graphite, *J. Chem. Phys.*, **53**, 1126–1131 (1970).

[48] A.C. Ferrari and J. Robertson, Interpretation of Raman spectra of disordered and amorphous carbon, *Phys. Rev. B*, **61**, 14095–14107 (2000).

[49] I. Calizo, A.A. Balandin, W. Bao, F. Miao and C.N. Lau, Temperature dependence of the Raman spectra of graphene and graphene multilayers, *Nano Lett.*, **7**, 2645–2649 (2007).

[50] H. Allouche and M. Monthioux, Chemical vapor deposition of pyrolytic carbon on carbon nanotubes. Part II: Texture and structure, *Carbon*, **43**, 1265–1278 (2005).

[51] A. Oberlin, J.-L. Boulmier and M. Villey, Electron microscopy study of kerogen microtexture, in *Kerogen* (ed. B. Durand), Technip, Paris, France, 191–241 (1980).

[52] V. Z'olyomi and J. Kürti, First-principles calculations for the electronic band structures of small diameter single-wall carbon nanotubes, *Phys. Rev. B,* **70**, 085403 (2004).

[53] P. Delaney, H.J. Choi, J. Ihm, S.G. Louie and M.L. Cohen, Broken symmetry and pseudogaps in ropes of carbon nanotubes, *Nature*, **391**, 466–468 (1998).

[54] P. Lambin, L. Philippe, J.-C. Charlier and J.-P. Michenaud, Electronic band structure of multi-layered carbon tubules, *Comput. Mater. Sci.*, **2**, 350–356 (1994).

[55] Y.-K. Kwon and D. Tománek, Electronic and structural properties of multiwall carbon nanotubes, *Phys. Rev. B,* **58**, R16001–R16004 (1998).

[56] M. Monthioux, Filling single-wall carbon nanotubes, *Carbon*, **40**, 1809–1823 (2002).

[57] M. Monthioux, E. Flahaut and L. Noé, unpublished data.

[58] M. Monthioux, B.W. Smith, B. Burteaux, A. Claye, J.E. Fischer and D.E. Luzzi, Sensitivity of single-wall carbon nanotubes to chemical processing: an electron microscopy investigation, *Carbon*, **39**, 1251–1272 (2001).

2
Doped Carbon Nanotubes: (X:CNTs)

Alain Pénicaud[1], Pierre Petit[2] and John E. Fischer[3]
[1]*CRPP, CNRS, Université Bordeaux I, France*
[2]*Institut Charles Sadron, CNRS, University of Strasbourg, France*
[3]*University of Pennsylvania, Philadelphia, USA*

2.1 Introduction

2.1.1 Scope of this Chapter

Before going any further it is necessary to define the term *doped* and the way in which it will be used throughout this chapter. If one searches for the verb 'to dope' in the Oxford American Dictionary, here is the definition that one will find: (i) 'administer drugs (to a racehorse, greyhound or an athlete) in order to inhibit or enhance sporting performance'. Outside the sporting field, the scientific concept of doping is likewise: add something to a pristine (starting) material to enhance performance. A few lines below, in the same dictionary, we find this more specialized definition: (ii) 'electronics: add an impurity (to a semiconductor) to produce a desired electrical characteristics'. This actually, will be one end of the doping spectrum we will consider. Indeed, this is true for many other compounds such as graphite intercalation compounds (GICs), fullerenes (C_{60}, C_{70}, etc.), bronzes, and so on. Doping in nanotubes span all the way from impurity-like random doping similar to what is encountered in the semiconductor technology all the way to stoichiometric compounds with ordered dopant superlattices like GICs.

Generally speaking, there are three ways one can 'dope' nanotubes: substitutional doping, endohedral doping and external doping. The same is true for fullerenes [1]. Substitutional doping means that some carbon atoms of the nanotube skeleton are replaced

Carbon Meta-Nanotubes: Synthesis, Properties and Applications, First Edition. Edited by Marc Monthioux.
© 2012 John Wiley & Sons, Ltd. Published 2012 by John Wiley & Sons, Ltd.

by nonisoelectronic heteroatoms or combinations thereof, the most widely studied being boron and nitrogen. This is the subject of Chapter 6 and will not be covered here. Endohedral doping covers situations where the dopant is inside the inner cavity of the nanotubes, by reference to the fullerene field where endohedral fullerenes refer to fullerenes with heteroatoms inside. This is the subject of Chapter 5 and will, in general, not be covered here. In a few occasions, though, we will make some intrusion into that field since the electronic consequences of endohedral doping are very similar to the external doping situation. Finally, external doping is the subject of this chapter, that is, the dopant is neither part of the skeleton, nor inside the nanotubes but outside the nanotubes.

This chapter deals exclusively with single-walled nanotubes (SWCNTs). There is, indeed, far less work on doping multi-walled nanotubes. n- and p-doping of double-walled nanotubes has not been covered here. Previous reviews have appeared on the subject of doped nanotubes, either as a whole or addressing part of the field. Early reviews include Duclaux's account of p- and n-doping of SWCNTs and MWNTs [2] and Fischer's account of SWCNT doping [3]. More specialized reviews have appeared on the structural aspects of doped SWCNTs [4] and on electrochemical doping of nanotubes [5–7]. Hole doping (p-doping) of SWCNTs has also been covered rather recently [8]. This book being materials oriented (rather than devices, for example), no attempt has been made to cover the transistors and sensors literature on nanotubes.

Why should one dope SWCNTs in the first place? For the same reasons as for graphite, fullerenes, or conjugated polymers. As will be described in this chapter, doping leads to an increase in conductivity, all tubes actually behaving as metals whether or not their structure, in the neutral state, corresponds to a semiconducting or a metallic tube. Additional incentives are the search for superconductivity, an unclear issue in the field of nanotubes probably due to disorder and/or heterogeneous samples, and the search for applications such as Li-batteries.

2.1.2 A Few Definitions

Exohedral doping of carbonaceous materials, as will be seen later, is analogous to graphite doping, leading to graphite intercalation compounds (GICs) and fullerene doping. Formally, IUPAC reserves the term intercalation to '…compounds resulting from reversible inclusion, without covalent bonding, of one kind of molecule in a solid matrix of another compound, which has a laminar structure.' [9]. Fullerenes and nanotubes do not have a laminar structure and thus, the compounds resulting from exohedral doping should not be, strictly speaking, called intercalation compounds but insertion compounds. This said, the term intercalation compounds might sometimes have escaped our attention in this chapter and indeed simply reflects analogous physics to the graphite world.

Species can be inserted within the nanotube bundles with or without charge transfer (doping). To the best of our knowledge, although examples of neutral guest molecules inserted within a fullerite lattice abound [10], such (exohedral) insertion compounds with neutral guests have not been reported with SWCNTs (yet it is likely that it may occur when attempting to fill the SWCNT tubular cavity – see Chapter 5 – with molecules whose dimensions fit that of the triangular channels in SWCNT bundles). Carbon nanotubes have been shown to be intrinsically *amphoteric*, that is, they can be doped with either acceptor (halogens and so on) or donor (alkali metals and so on) species, like graphite [11] and rather unlike fullerenes which are far easier to reduce than to oxidize [12]. The doping species may thus be n-doping (addition of electrons, i.e., reduction) or p-doping (removal of electrons, i.e., oxidation). After a rather general and historical introduction, the main body of this chapter will be divided into n-doped SWCNTs and p-doped SWCNTs.

It is well-known that when two phases are brought into contact, their previously different Fermi levels are brought into coincidence as thermodynamic equilibrium is achieved. This may be accompanied by charge transfer between the phases, depending on the spectrum of energy levels over which electrons are distributed. This is true whatever the physical states of these phases. In some works on nanotubes, the term 'work function' has been used which, of course, is nothing else than the energy difference between vacuum and the Fermi level. There are often discussions whether the Fermi level is dependent or independent of diameter. For sake of preciseness, one should recall that the value alluded to is then the Fermi level in the neutral state. Any decrease or increase of the Fermi level with respect to the neutral state is a result of p- or n-doping.

When speaking about charge transfer between two phases, especially when one of these phases is a solution containing ionized or neutral molecules or atoms, the terms 'Fermi level', 'electrochemical potential' and 'redox potential' are generally used depending on whether the speaker is a physicist or a chemist. We emphasize here that all these terms are equivalent: the Fermi level is the electrochemical potential and is identical to the redox potential (multiplied by the electron charge -|e|), with respect to a suitable reference level. So, the use of one of these terms in a solution is only a matter of taste, as it is shown in a simple didactic discussion reported by Reiss [13].

It is the electrochemical potential which determines when, and in what direction, a charged species will be transferred between two points in a system. When, for example, an electron can be transferred between two systems it will move from the system of higher electrochemical potential to that of lower electrochemical potential. Thus the system of higher electrochemical potential (Fermi level) 'reduces' that of lower potential. When the Fermi levels are equal, no electron transfer occurs. Furthermore, it follows that the difference in electrochemical potential between the two systems measures the reversible work which must be performed *on* or *by* the system (depending on the direction of transfer) in making the transfer.

2.1.3 Doped/Intercalated Carbon Allotropes – a Brief History

An important feature shared by nanotubes with other carbon polymorphs is the versatility with which structure and properties can be modified by adding heteroatoms or molecules to the all-carbon host [14]. This is dramatically true for graphite [11], solid C_{60} [15] and carbon nanotubes [16,17], and perhaps also for amorphous sp^2-bonded carbons [18]. Conjugated linear polymers, for example, polyacetylene $(CH)_x$ share this attribute. In general, dopant species occupy specific sites in the host lattice. Intercalation, a special case of doping, occurs when the sites form their own ordered sublattice, the most dramatic example being the stacking sequences, or 'stages' of n-graphene and 1-intercalant layers which form ordered 1D superlattices in 'stage n' graphite intercalation compounds, or GICs [11].

Doping concentrations in carbons are generally large compared to those encountered in classic doped inorganic semiconductors. In common with many of the latter, amphoteric charge transfer between dopant and host increases the delocalized electron or hole concentration to yield n- or p-type carbons, as discussed in detail below for nanotubes.

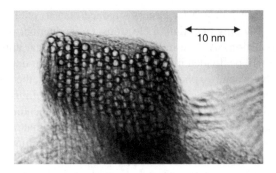

Figure 2.1 *HRTEM image of the cross-section of a semicrystalline rope of carbon nanotubes [19].*

Along with property modification, the partially ionic character of the doped/intercalated solid contributes importantly to its stability relative to the isolated constituents (i.e., the Madelung energy in the case of ordered ionic superlattices).

2.1.3.1 Structure and 'Staging'

Shortly after the discovery of single-wall carbon nanotubes, Smalley's group found a synthesis procedure which yielded semi-crystalline bundles, or 'ropes' (Figure 2.1), with well-defined interstitial sites suggesting the possibility of ordered doped phases, or intercalation compounds [19].

These would be natural extensions of the ordered 1D superlattices [20] observed in alkali-doped $(CH)_x$ [21], whose crystal chemistry and phase equilibria are in turn closely related to those of the quasi-2D GICs. The taxonomy of carbon-derived intercalation compounds would then be completed by including the 3D alkali-doped fullerides, cf. Figure 2.2.

At this time it was well-established that the ordered superlattices in GICs and doped polymers resulted from competition between the anisotropic dopant-dopant and guest-host interactions (enthalpies), entropy and volume [22]. Consequently, minimization of the Gibbs free energy versus chemical potential μ, temperature T and pressure P determined the equilibrium structures, in particular how they evolved with concentration. The top panel of Figure 2.3 shows the 2D, or layer intercalate case, for which the consequences of competing interactions on phase equilibria have been verified experimentally. The parameter U_o sets the energy scale of in-plane interactions among neighboring intercalates.

For staging to occur, the *net* in-plane and out-of-plane interactions must be attractive and repulsive respectively (clearly there must be more than just the electrostatic forces arising from guest-host charge exchange – see the following). For example, if $U_o > 300\,K$, staging transitions versus fractional site filling X will not occur at room temperature, an important attribute for battery electrodes due to reversible structural phase transitions which degrade the material upon multiple cycling. At high temperature, entropy dominates, the host lattice becomes enthalpically irrelevant, and for any site filling fraction X the intercalates are distributed randomly in all three dimensions corresponding to the paramagnetic state in the Ising magnetic analog. The statistical mechanics of staging superlattices in intercalated graphite is formally identical to the Ising model of spin-ordering in magnetic systems, where filled and empty galleries correspond to spin-up and spin-down magnetic ions. The high temperature

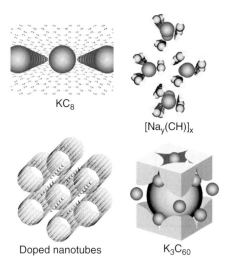

KC$_8$

[Na$_y$(CH)]$_x$

Doped nanotubes   K$_3$C$_{60}$

Figure 2.2 *Schematic structural models for alkali-intercalated carbonaceous host lattices. Clockwise from upper right: stage 1 potassium-intercalated graphite KC$_8$; saturated phase of Na-intercalated polyacetylene in which close-packed Na chains occupy all available interstitial channels; potassium fulleride with both tetrahedral and the single octahedral site in the fcc host are singly occupied; hypothetical saturation-intercalated SWCNT triangular lattice corresponding approximately to KC$_{13}$. All but the last one have been identified crystallographically. For a better understanding of the figure, please refer to the colour plate section. Adapted from [14] Copyright (2000) American Chemical Society.*

paramagnetic limit corresponds to random spins (no net moment) by analogy to random dilute filling of all interlayer spaces (dilute stage 1). Staging at lower temperatures (energy dominates entropy) corresponds to ordered magnetic states, viz. ferromagnetic, antiferromagnetic and so on. At low temperature, the dominant enthalpic term is the in-plane attraction among intercalates so the interlayer spaces are fully occupied (maximal areal density), in a layer sequence consistent with X (including phase separation if the 'stage index' 1/X is non-integral). When X is close to unity, the stage 1 sequence persists along with random vacancies, increasing in concentration with increasing temperature. At sufficiently high temperature, the 2D intercalate layer 'melts' to again yield the 'dilute stage 1' lattice gas. Ordered stages or stage mixtures constitute an array of ferro – or antiferromagnetic states in the Ising model.

Figure 2.4 provides a rationale for the occurrence of staging transitions near 300 K in intercalated graphite and their absence in layer chalcogenides [24]. At low concentration, local host layer distortions around well-separated intercalated alkali ions create elastic dipoles [22], the magnitude of which depends on the strain gradient, which in turn determines the decay length. Neighboring dipoles in the same interlayer gallery (i.e., the interlayer empty space) attract each other. Due to strong covalent bonds linking Ti and S atoms comprising the very rigid unitary trilayer, distortions in TiS$_2$ decay slowly with distance from the ion, the strain gradient (layer curvature) is small, and the dipoles are weak. Consequently U$_o$ is small, T/U$_o$ is large enough for 300 K to be above the ferromagnetic limit, and the only stable structure is stage 1 with variable 'coverage' in all galleries. All of the above is reversed for floppy graphene layers, which is why GICs are the prototypical multistaged layer

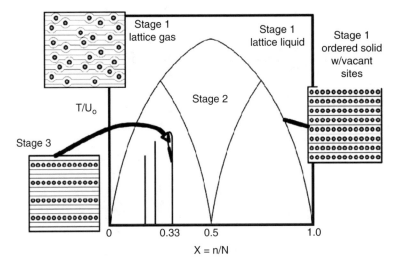

Staged superlattices in conjugated polymers

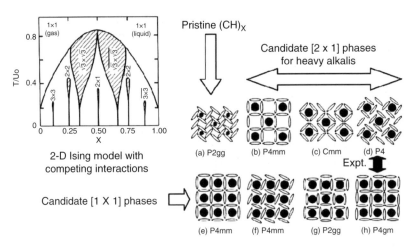

Figure 2.3 *Phase diagrams for 1D (**top**) and 2D (**bottom**) Ising-like intercalation compounds where spin down/up is the conceptual equivalent to site filled/empty. Adapted with permission from [20] Copyright (1988) Elsevier Ltd (top) and [23] Copyright (1991) American Physical Society (bottom).*

intercalates. Structural phase transitions versus charge/discharge present challenges for achieving long cycle life in Li ion batteries with well-ordered graphitic anodes [18,25].

As a host lattice for intercalation, polyacetylene can be considered a 2D analog[1] of 1D graphite, cf. Figure 2.2, upper right. A new feature here is that conjugated polymer chains are ribbon-like, which can rotate axially to optimize the intercalation channel (Figure 2.3, bottom panel). The mean field theory neglecting these rotations predicts a phase diagram similar to that of GICs [20] while concentration-dependent in situ XX-ray diffraction shows a sequence of phases involving both selective channel filling (staging) and different 2D lattice symmetries

---

[1] The ordering is between and among filled interchain spacings, thus the superlattice is indeed 2D, not 1D.

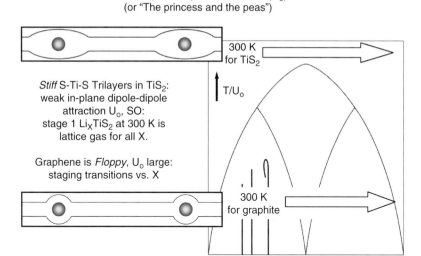

Elastic dipoles, layer stiffness and $|U_o|$
(or "The princess and the peas")

Stiff S-Ti-S Trilayers in TiS$_2$: weak in-plane dipole-dipole attraction U_o, SO: stage 1 Li$_X$TiS$_2$ at 300 K is lattice gas for all X.

Graphene is *Floppy*, U_o large: staging transitions vs. X

300 K for TiS$_2$

T/U_o

300 K for graphite

Figure 2.4 *Schematic representation of elastic dipoles, that is, the origin of in-plane attraction. These produce weak forces in stiff hosts and vice versa, thus graphite exhibits staging whereas TiS$_2$ does not. Adapted with permission from [24] Copyright (1987) American Physical Society.*

depending on the rotation motif of the chains [21,23]. Here the competing interactions which define the intercalation superlattices are chain rotations, chemical potential and Coulomb repulsion between ionized dopants, typically alkali metals or halogens.

For completeness, consider the 3D face-centered cubic (fcc) fullerene molecular crystal [15,26], see Figure 2.2 (lower right). At low concentration, alkali ions preferentially and randomly occupy the larger octahedral sites up to 1:1 stoichiometry, accompanied by slight lattice *contraction* due to ionic interactions because even Cs$^+$ is too small to 'fill' the sites. Approaching A$_3$C$_{60}$ (where A = alkali metal ion) we find either lattice gas or phase separation as the much smaller tetrahedral sites (two per molecule) begin to fill, depending on the choice of A (or A' = alkaline earth, or alloys of both). Beyond A$_3$C$_{60}$, the host lattice transforms to tetragonal, which accommodates four intercalates per fullerene since the latter are no longer close-packed. Finally, the body-centered tetragonal (bct) structure gives way to the body-centered cubic (bcc) structure with a limiting composition A$_6$C$_{60}$. Lattice gas behavior, that is, random occupancy of a subset of lattice sites with no correlations among filled sites, is not observed beyond the fcc limit, but variable charge per fullerene can be realized by mixed alkali/alkaline earth site occupancy, a feature which was exploited to establish the correct physics of fulleride superconductivity [27].

2.1.3.2 *Electronic Properties*
How does chemical doping affect the electronic properties of our prototypical carbon host lattices, and what new physics can be thereby probed? Starting with GICs we find a textbook embodiment of simple Fermi liquid behavior [11]. All the classic metal physics concepts apply in their simplest one-electron garb: Matthiessen's rule for resistivity ρ versus temperature [28], Drude-like IR-VIS optical properties uncomplicated by overlapping

interband transitions [29], Pauli susceptibility with minimal many-body corrections, [30,31] etc. Some GICs superconduct at very low temperatures and/or at high pressure. The high quality of samples based on highly oriented pyrolytic graphite and natural crystals makes it possible to tune and study the effects of anisotropy – the electron-conducting alkali GICs are more 3D than the parent graphite due to partial hybridization of alkali's and graphitic π orbitals, while hole-conducting acceptor compounds (protonic acids, superacids, halogens, metal halides, etc.) are more nearly 2D because the large c-axis expansion and absence of interlayer hybridization effectively decouples the adjacent conducting sheets from each other.

Charge transfer or redox doping has a more dramatic effect on the electronic properties of conjugated polymers, exemplified by $(CH)_x$. The pristine host is an intrinsic semiconductor with a very high electrical resistivity, typical of polymers. Very dilute donor or acceptor doping decreases the resistivity to the metallic range, but without the emergence of Pauli spins, up to a threshold of ~6% in the case of iodine [32]. This unexpected behavior led to the concept of spinless charge carriers called solitons, and their intimate connection to conjugation defects in the *trans* isomer. The surprising new physics embodied by the soliton bears a strong family resemblance with superconductivity [33].

Donor-doped solid C_{60} yields superconducting phases over a range of molecular valences bracketing the optimal value 3 (i.e., A_3C_{60}), with surprisingly high transition temperatures T_c exceeded only by the cuprates [26,34]. Weak intermolecular overlap in the fcc solid leads to narrow bands and large density of states at the Fermi energy when optimally doped. High frequency intramolecular modes contribute importantly to electron-phonon coupling, so the high T_c can be understood in the framework of the Bardeen-Cooper-Schrieffer (BCS) theory albeit with all the relevant parameters pushed to their extremes to explain the high T_c [34]. Unique among carbon hosts, C_{60} is difficult to oxidize, either in solution or solid state [12].

Carbon nanotube aggregates respond to chemical doping much like graphite. The ~2/3 fraction of semiconducting tubes becomes metallic, electrical conductivity is enhanced for donor or acceptor doping, Pauli spins are detected, alkali-doped samples change color, and so on. Doping of individual tubes has also been carried out. All of this is treated in detail in the rest of this chapter.

2.1.4   What Happens upon Doping SWCNTs?

Synthesizing doped SWCNTs means performing redox reactions between the dopant species and SWCNT bundles or ropes, which is achievable by numerous synthetic routes. Thus, if the Fermi levels of the dopant species, whether n or p-type, and SWCNT bundles are different, charge transfer will occur between the two systems. This, along with changes in atomic structure, imply changes in the electronic structure of the tubes, leading to altered physical (and chemical) properties, in particular their optical, vibrational and electronic transport properties.

2.1.4.1   *Lattice Structure*
In light of Section 2.1.3, one might expect doped nanotubes to reflect some aspects of intercalated conjugated polymers. Superlattices appropriate to the circular cross-section have been proposed [16] then optimized theoretically [35] and the related band structures have been calculated [36]. Figure 2.5 shows examples in which the C:A (C=carbon, A=alkali) ratio evolves from ~40 to ~13 to 8 from top to bottom, assuming close-packed 1D alkali metal chains and (10,10) nanotubes.

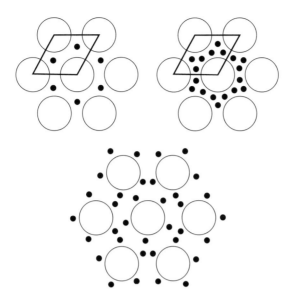

Figure 2.5   *Hypothetical 2D intercalation lattices for nanotube 'ropes'. (**Top left**) Two intercalate chains (black dots) per tube (open circles) corresponds to ~MC_{40} for close-packed 1D chains of M in a lattice of (10,10) tubes. (**Top right**) smaller M, or lattice expansion, permits six chains per tube, or ~KC_{13}. (**Bottom**) Epitaxial wrapping of 2×2 M superlattice on the surface of each tube, in direct analogy to MC_8. Reprinted with permission from [3] Copyright (2002) American Chemical Society.*

However, structural data on doped SWCNT ropes is sparse and incomplete, as discussed next. In particular, diffraction patterns for lattices shown above, which can be readily simulated [35,37], have not been observed, thus there is no evidence for ordered intercalate superlattices. Perhaps due in part to the lack of well-ordered nanotube crystals, one could argue that there is insufficient driving force for staging even in a perfect 3D ordered nanotube lattice (which would of course be limited to modulo six tube indices compatible with the triangular packing). First of all, the projected mass density of a tube is a circle, ruling out the analog of chain rotations in conjugated polymers. Furthermore, a tube is stiffer than a sheet, so local tube distortions around isolated intercalate ions would be minimal.

2.1.4.2   Electronic Structure

As described in Chapter 1, the helicity of a nanotube leads to a very peculiar electronic structure related to the quantization of the electron wave function around the circumference of the tube, and to the corresponding energy dependence of the electronic density of states (DOS) [38]. Transferring electrons to (from) SWCNTs populates (depopulates) the initially empty (filled) electronic states in the DOS. Thus, their Fermi level may be upshifted (downshifted) into the conduction (valence) band (Figure 2.6). The first obvious consequence is that, if the Fermi level of a semiconducting tube is shifted above (below) the first van Hove singularity, the density of states at the Fermi level is non-zero and the tube becomes metallic with an electron (hole) conductivity. This consequence on the electrical transport properties is detailed for n-doping in Section 2.2.4.

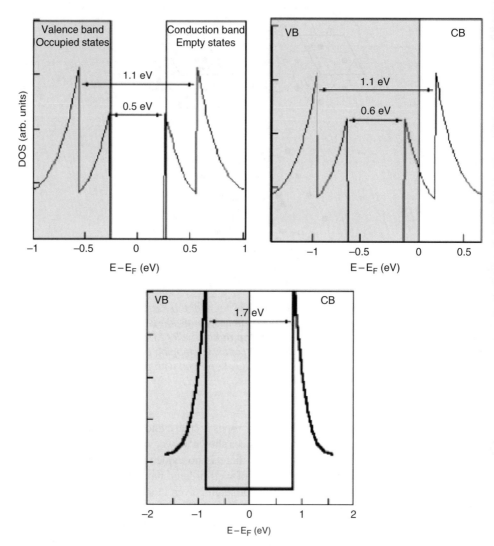

Figure 2.6 *(Top left)* Schematic representation of the density of states of an undoped semiconducting SWCNT. *(Top right)* shifting E_F above the first van Hove singularity transforms the undoped semiconducting SWCNT to a n-doped metallic one with $n(E_F) \neq 0$. Shifting E_F below the first van Hove singularity (not shown) results in a p-doped metallic tube. *(Bottom)* Schematic representation of the density of states of a neutral metallic SWCNT. The energies between van Hove singularities correspond to tubes having a diameter of about 1.4 nm.

2.1.4.3 Optical Properties

As described in Chapter 1, the energy gaps in the electronic structure of SWCNTs lie in the UV-VIS-NIR range allowing the study of this electronic structure by optical absorption experiments [39]. Because of selection rules, and particularly the change in the parity of the electron wave function, optical transitions are allowed only between mirror spikes in

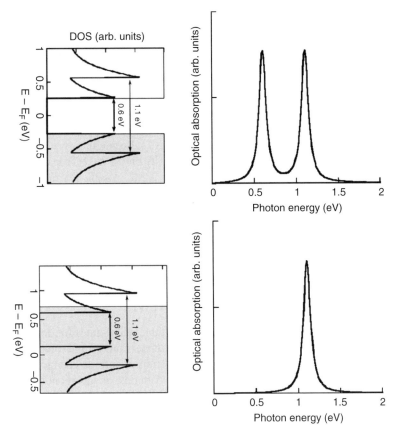

Figure 2.7 *(Top)* DOS of a neutral semiconducting SWCNT and the corresponding simulated optical absorption bands induced by transitions between van Hove singularities. *(Bottom-left)* DOS of the same semiconducting SWCNT after n-doping. The Fermi level being shifted above the first van Hove singularity, the transition of lowest energy is suppressed. *(Bottom-right)* Corresponding simulated optical absorption spectrum.

the DOS, in other words from an occupied state to the corresponding symmetric unoccupied one (Figure 2.6). Consequently, in the energy range considered, a semiconducting tube with a diameter of 1.4 nm will be characterized by two absorption bands, sometimes referred to as S11 and S22, at 0.6 (S11) and 1.1 eV (S22), a metallic tube by one absorption band at 1.7 eV (M11). Thus, shifting the Fermi level of a semiconducting tube between the two first singularities, for example (Figure 2.7, bottom), leads to the suppression of the transition between the first mirror peaks in the DOS, leaving the one at 1.1 eV unchanged. The shift of E_F above the second singularity results in the suppression of the two first optical transitions. The same holds of course for metallic tubes [40]. These changes in the optical properties of SWCNTs are illustrated in Section 2.2.3 for n-doped nanotubes.

2.1.4.4 Vibrational Properties

In the field of doped/intercalated carbon allotropes, Raman spectroscopy is specifically powerful for characterizing SWCNTs thanks to their resonant character, which allows very small amounts of material to be studied. Moreover, thanks to their peculiar electronic structure, the response of semiconducting tubes can be enhanced with respect to metallic tubes (and vice versa) by tuning the wavelength of the exciting light (see Figure 2.26 later on) [41,42]. Upon doping, as the optical transitions are suppressed, the resonant character of the Raman response is suppressed, making Raman spectroscopy an indirect probe of the electronic structure and electrical behavior of SWCNTs. Beside this very peculiar feature, it is expected that the increase of electron (hole) density onto SWCNTs leads to a softening (stiffening) of their vibrational modes. In Section 2.2.6, in which a rapid overview of Raman spectroscopy of n-doped SWCNTs and its correlation with optical and electrical properties is presented, it will be shown that the simplistic view of the softening/stiffening of the vibrational modes upon n-doping is more complicated than first expected.

2.2 n-Doping of Nanotubes

2.2.1 Synthetic Routes for Preparing Doped SWCNTs

Similar to conducting polymers, GICs, and C_{60}, the three main routes for doping SWCNTs are exposure to a vapor phase, electrochemistry, and immersion in ionic solutions of definite redox potentials. The common goal is to obtain materials of definite chemical composition that could be used, for example, for energy storage, batteries, or superconducting devices.

2.2.1.1 Vapor Phase Synthesis

This kind of sample preparation is performed using a quartz tube sealed under high vacuum, preferably 10^{-5}–10^{-6} mbars, in which the sample to be doped and the dopant are placed at each extremity (Figure 2.8). The tube is then heated in a temperature gradient in order to keep the sample at a temperature higher than the dopant, typically 25°C. The dopant is thus evaporated and diffuses into the sample. Finally, the sample is annealed for two days to obtain a homogeneously doped sample [16].

This method is principally used for doping SWCNTs with alkali metals but has been also used to adsorb donor molecules such as butylamine and propylamine for example [43].

2.2.1.2 Electrochemical Doping

Electrochemistry is particularly interesting for doping carbon materials because it allows one to continuously shift the Fermi level by changing the cell potential. Concomitantly, it affords precise control of guest concentration allowing the determination of the chemical composition of the samples, facilitates in situ experiments on air-sensitive materials, and provides a controlled method to test reversibility. This method, used primarily for alkali ions because their redox potential is quite high (close to –3V for all alkali metals allowing a large potential window using aprotic electrolyte solutions [5,44–47], leads to the co-insertion of the solvent which solvates the cations [48].

A typical setup for studying electrochemical insertion of lithium into SWCNT samples by galvanostatic charge-discharge and cyclic voltammetry experiments has been reported by Claye et al. [45], (Figure 2.9). Basically, Li metal, used as both counter and reference

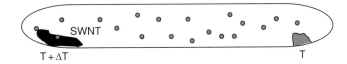

Figure 2.8 *Schematic of a simple device used for doping by vapor phase.*

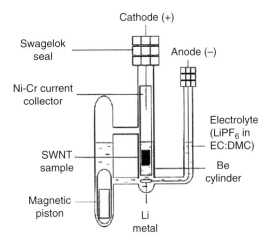

Figure 2.9 *Electrochemical cell used for in situ X-ray diffraction. The cell is shown in the configuration for galvanostatic charge-discharge. For X-ray measurements, the piston is raised such that the electrolyte falls into the reservoir. Reprinted with permission from [45] Copyright (2000) Electrochemical Society.*

electrodes, and a SWCNT working electrode, are immersed in an organic liquid electrolyte allowing the migration of Li$^+$ ions from the anode (Li electrode) to the cathode (SWCNT electrode). To avoid oxygen contamination, air-tight screw-in electrochemical cells are used.

2.2.1.3 Doping by Solutions of Definite Redox Potential

This method consists of immersing SWCNTs into a radical anion salt solution of known redox potential [40,49]. The apparatus consists of a U-shaped glass tube where the sample to be doped is placed in one branch, the second containing neutral organic molecules in solution in pure THF with a piece of alkali metal. The whole apparatus is sealed under high vacuum. Electron transfer from, for example, lithium, to organic molecules results in the radical-anion form of the molecule with Li$^+$ as the counter ion. This reaction is easily checked by the change in color of the solution and the appearance of the characteristic absorption spectrum of the radical-anion. This experimental arrangement, developed initially for n-doping of polyacetylene [50], allows the doping of the sample by bringing the solution into contact with it. Once the doping is achieved, the excess of doping solution is removed, the sample rinsed by internal distillation of the solvent and then dried by cooling the ampoule at $-70°C$. The chemical composition of the doped sample is given by the change in the amplitudes of the absorption spectra of the doping solutions before and after the doping process, proportional to the number of electrons transferred onto the tubes, thus to the number of inserted Li$^+$ ensuring the neutrality of the system. Similar to the electrochemical doping, this method leads to the co-insertion of solvent and alkali ions [51]. Using redox solutions, only few molecules are available leading to a discrete set of redox potential values whose highest value is slightly lower than those of alkali ions ($-2.5\,V$ using the anion of naphthalene [52] compared to $-3\,V$ using alkali electrochemical doping).

2.2.1.4 Other Methods

Adhesion of inorganic and organic molecules onto single isolated carbon nanotubes has also revealed the extreme sensitivity of the electrical properties of SWCNTs to molecular species [53,54]. Here, rather than a complete ionization of each donor or acceptor as in the former methods, partial electron donating or accepting by adsorbed molecular species is exploited to dope SWCNTs. Examples of this will be given in Chapters 3 and 4, dedicated to the functionalization and decoration of carbon nanotubes respectively.

2.2.2 Crystalline Structure and Chemical Composition of n-Doped Nanotubes

Dopant intercalation within the bundles of SWCNTs has been established experimentally by the analysis of transmission electron microscopy (TEM) images [55], electron diffraction and electron energy-loss spectroscopy (EELS) [56,57], X-ray diffraction [2,45], and neutron diffraction [51]. Structural studies are presented below according to the sample preparations.

2.2.2.1 Samples Doped by Alkali Vapors

X-ray diffraction patterns of SWCNTs doped to saturation by K, Rb and Cs vapors show that the first order *10* peak characteristic of the 2D triangular lattice within bundles shifts to lower q-values, the shift increasing with the atomic number of the alkali [58] (Figure 2.10). This observation implies the lattice expansion of the SWCNT bundles, and also reveals that they remain crystalline upon doping. The chemical composition of the three compounds determined by weight uptake corresponds to one alkali atom for eight carbon atoms. However, in this experiment the starting material results from a high temperature annealing of SWCNT pristine sample which leads to an increase of SWCNT diameters and a concomitant decrease of the lattice parameters [58]. Thus, the observed shift of the *10* peak between the starting material and the saturated doped materials does not correspond to the real lattice expansion and cannot be compared to analogous experiments performed on GICs.

More relevant is the fact that, despite the relatively high concentration of inserted alkali atoms, the observed diffraction patterns do not exhibit new lattice symmetries and cannot be simulated by ordered structures as shown in Section 2.1.4.1, [35,37], thus there is no evidence for ordered intercalate superlattices.

In situ electron diffraction experiments on *pristine* SWCNTs [56] show that, during the intercalation in vapor phase, the *10* peak progressively shifts as the K/C ratio increases, indicating an intercalation of alkali ions in between the tubes (Figure 2.11). Consistent with X-ray diffraction, there is no evidence for new lattice symmetries. The progressive evolution (no staging) is consistent with a residue compound, that is, the dopant does not form stages: Compared to graphite, the stiff nanotubes give elastic distortions too weak for 'in-channel' attraction, hence no well defined compound forms, unlike graphite intercalation compounds. The relative intensity of the K 2p and C 1s core excitation signals obtained from electron energy loss spectroscopy (EELS) allows the A/C ratio during the doping process to be estimated (Figure 2.11). The chemical composition of the saturated sample found in this work is KC_7.

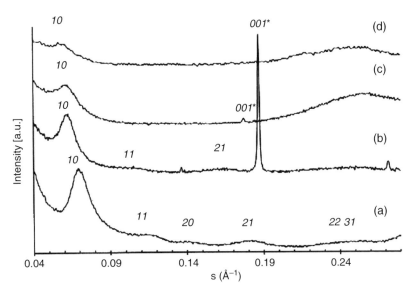

Figure 2.10 *X-ray diffraction patterns of (a) starting material, and SWCNTs doped to saturation by (b) K; (c) Rb; or (d) Cs. Reprinted with permission from [58] Copyright (2003) Elsevier Ltd.*

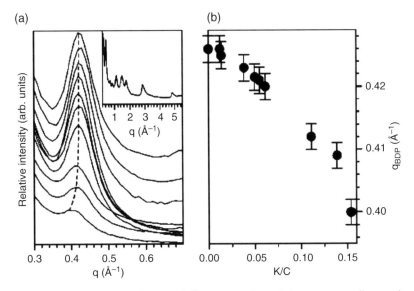

Figure 2.11 *(a) First-order 10 electron diffraction peak, and (b) corresponding peak value of SWCNT bundles progressively doped in situ with potassium (derived from the zero loss dispersion in an EELS experiment). In (a), K/C increases from top to bottom and the downshift of q(max) is qualitatively consistent with Figure 2.10. The inset in (a) shows the diffraction pattern of pristine SWCNTs in a wide q region. Reprinted with permission from [56] Copyright (2003) American Physical Society.*

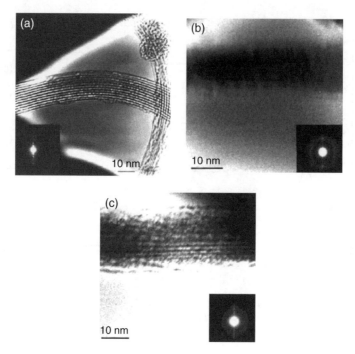

Figure 2.12 *TEM images of (a) a pristine SWCNT bundle; (b) a Cs-doped and (c) dedoped bundle. Reprinted with permission from [55] Copyright (1988) Springer Science + Business Media.*

The vapor method for synthesizing doped samples is reversible in the sense that the *pristine* sample can be recovered after exposing the sample to air, which is known to cause de-intercalation in alkali-metal-intercalated carbon structures [59]. This was observed by electron diffraction and TEM imaging by Bower and coworkers [55] (Figure 2.12). The characteristic fringes of SWCNT bundles disappear upon doping, revealing that alkali atoms intercalate but no new fringe spacings are observed indicating the lack of intercalate superlattices. The reappearance of lattice fringes and diffraction spots after air exposure shows that alkali intercalation with SWCNT bundles is reversible. Although the bundle is less ordered than the unreacted material, that is, the nanotubes at the outermost surface appear to bend away from the bundle presumably due to de-intercalation, none of the individual SWCNTs observed shows signs of structural damages or defects.

These works on vapor-doped samples show evidence for uniform lattice expansion and that doped samples remain crystalline, however, no signature of intercalate superlattices have been observed. Moreover, the relative intensities are still inconsistent with appreciably filled channels in the doped state and a convincing fit of scattering data to structural model has yet to be achieved.

2.2.2.2 Samples Prepared by Electrochemical Doping

An attempt to solve structural issues has been done using two half-cells consisting of *purified* SWCNTs and Li as electrodes [45]. In this work, nanotubes exhibited reversible capacities in the range of 460 mAh g^{-1}, corresponding to a chemical composition of Li$_{1.23}$C$_6$

(a)

(b)

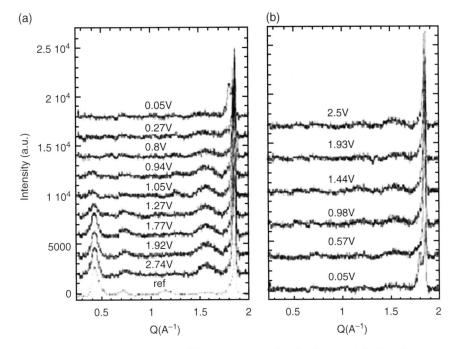

Figure 2.13 *In situ powder X-ray profiles measured at fixed cell potentials (**panel a,** **bottom** to **top**) during Li insertion and (**panel b, bottom** to **top**) de-insertion. As insertion progresses, the first-order 10 peak at Q=0.43 A⁻¹ decreases in intensity without shifting or broadening. The sharp peak at high Q is the 002 peak from residual graphite impurity, the evolution of which upon doping/dedoping shows the proper operation of the cell. Reprinted with permission from [3] Copyright (2002) American Chemical Society.*

using $LiPF_6$ in dimethylcarbonate. Upon doping, a progressive decrease in bundle intensities is observed, but with no change in position contrary to vapor doping results. In common with all other alkali doping experiments, there is no evidence for structural ordering of the dopant. Similar results are obtained for K doping using 1 M KCN in anhydrous $(C_2H_5)_3B/$ THF [48]. The fact that the *10* diffraction peak intensity decreases but does not shift with increasing A/C (A=Li, K) was interpreted in terms of a two-phase model: the continuous but incomplete intensity loss suggests that some regions are fully doped and noncrystalline, while some others are not doped and retain the pristine crystal structure (Figure 2.13).

However, reversing the current (dedoping) does not lead to the recovery of pristine bundles. Evidently, electrochemical doping 'frays' the bundles by progressive solvent co-intercalation rather than exfoliating them, since crystallinity can be subsequently restored upon removal of the electrolyte, washing, and annealing.

2.2.2.3   *Samples Prepared by Doping with Solutions of Definite Redox Potential*
As mentioned above, this method allows synthesizing samples of well-defined chemical composition. Using the radical-anions of naphthalene, fluorenone and anthraquinone, with Li^+ and K^+ as counter ions, solutions of redox potentials of 2.5, 1.3 and 0.85 eV have been placed in contact with *purified* SWCNTs from the same batch as was used for the

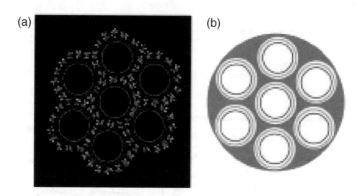

Figure 2.14   *Schematic representations of the structure of a LiC6-THF bundle of seven tubes (a) atomically resolved and (b) approximated by homogeneous densities of scatterers. Reprinted with permission from [51] Copyright (2005) American Physical Society.*

electrochemical studies above. At thermodynamic equilibrium, samples of chemical composition AC_{25}, AC_{10} and AC_6 (A=K, Li) were obtained [49]. Their structures were determined by neutron diffraction experiments [51]. In the diffraction patterns, a high incoherent flat intensity over the whole Q range, unambiguously related to the presence of a large amount of hydrogen in the sample, indicates that, as for the samples prepared by electrochemistry, the samples are ternary compounds $A(THF)_nC_x$ due to the solvation of alkali ions by THF. The interpretation of these experiments is shown in Figure 2.14. The alkali ions form a shell of homogeneous density of scatterers surrounding the SWCNTs, and the THF molecules decorate these objects. Lattice expansion of the SWCNT bundles is estimated to be 0.85 nm and 1.06 nm for $Li(THF)_{0.8}C_6$ and $K(THF)_{0.8}C_6$ respectively, very much larger than was found for samples prepared by vapor phase. Simulations of the neutron diffraction profiles give a lower bound n=0.8 for the number of THF solvating each cation. Interestingly, the triangular lattice of the SWCNTs in the bundle is preserved, albeit strongly dilated, in spite of the intertube interactions being strongly modified by the intercalated objects.

Hence, it appears that doping SWCNTs is a complicated issue. A proof is given by the large variety of chemical composition reported for the saturated samples by different groups, for example, $KC_{6.5}$ [56] to KC_{24} [55], CsC_8 [2,55] to CsC_{24} [55] using vapor doping, $Li_{1.23}C_6$ [45] to $Li_{2.7}C_6$ [60] by electrochemical doping, $Li(THF)C_6$ [49] to $Li(THF)C_{10}$ [61] using ionic solutions. These discrepancies may find a lot of explanations, such as the air sensitivity of alkali metals, the presence of defects in purified samples, the differences in crystallinity from sample to sample, the difference in the techniques of measurement, etc. However, from all structural studies, the main conclusion that can be drawn is that there is no signature of any ordered intercalate superlattices at any intercalate concentration and regardless of the SWCNT synthesis route (laser ablation, arc-discharge, purified or pristine), as opposed to what is found in other prototypical carbon host. Why such a difference? Some hints have been given in the introduction of this chapter: Lack of ordered nanotube crystals, insufficient driving force for staging even in a perfect 3D ordered nanotube lattice, curvature of nanotubes, stiffness... Anyway, the experimental evidence of the lack of intercalate superlattices rules out all the theoretical models relying on 2D intercalation lattices that have been used for the simulations of X-ray patterns of vapor doped samples (see Section 2.1.4.1) and is in favor of a random insertion of dopant species, supporting the use of homogeneous densities of scattering centers for simulating experimental data [51].

2.2.3 Modification of the Electronic Structure of SWCNTs upon Doping

In terms of the rigid band model, shifting the Fermi level of a SWCNT results in doping it, provided that this shifts exceeds half the gap between its first pair of van Hove singularities for semiconducting tubes, while the band structure remains intact. Much work has been performed on the gating of individual isolated SWCNTs [62]. For example, by applying a voltage between an individual SWCNT and a reference electrode in an electrolyte [63], significant changes in the electronic transport properties and phonon spectra of nanotubes are observed. Sharp increases in current through metallic nanotubes gated electrochemically indicate that the Fermi energy reaches valence or conduction band. Other examples showing the evolution of SWCNTs from p-type to n-type by irreversible adsorption of polymers have also been reported [43].

Recently, it has been shown that the Fermi levels of SWCNTs in the neutral state are not identical but scale linearly with the gap energy of the tubes [64]. However, as far as solid samples are concerned, the Fermi levels, or redox-potentials, of all the tubes in the contacted bundles are aligned in the middle of the gap and all the accessible electronic states in the DOS are filled below this energy level. Upon doping by alkali vapors, electrochemistry or redox reactions with solutions of known redox potential, the absolute value of the redox potential of a doped sample is that of the alkali metal, the cell potential or the redox potential of the solution.

The relationship between the electronic structure of SWCNTs and optical absorption has been presented in Section 2.1.4.3. However, for a real sample, the optical absorption spectrum is the superposition of the response of each tube. Since macroscopic SWCNT samples have different diameters and helicities which define the energy gaps in the DOS, the overall spectra display broadened absorption bands.

Figure 2.15 is an illustration of the progressive suppression of the optical transitions with increasing Fermi level of SWCNTs. In this experiment [40], thin films were exposed to solutions of radical-anions of naphthalene, fluorenone and anthraquinone with redox potentials of -2.5, -1.3 and -0.85 V respectively (the counter ion in the present experiment is Li^+, but the same results have been obtained with K^+) [65].

For the starting material (curve a), the first two features at 0.6 and 1.2 eV originate from bandgap transitions in semiconducting tubes, whereas the third at 1.7 eV originates from metallic tubes (see Section 2.1.4.2).

Exposure of the SWCNTs to the Anthraquinone$^-$-Li$^+$ solution leads to the disappearance of the band at 0.6 eV, the other optical transitions being unaffected. This shows that the Fermi level of the doped sample set at 0.85 eV lies above the first singularity in the conduction band, and thus all the tubes of the sample are metallic. Doping the sample with the Fluorenone-Li$^+$ solution also completely removes the feature at 0.6 eV and the low-energy shoulder of the second absorption band (inset) while the feature at 1.3 eV and the band corresponding to metallic tubes are unaffected. Thus, the effect of setting the Fermi level of SWCNTs to 1.3 eV is to fill up the first and second peaks in the DOS of semiconducting SWCNTs characterized by an energy gap lower than 1.3 eV. Finally, all the features

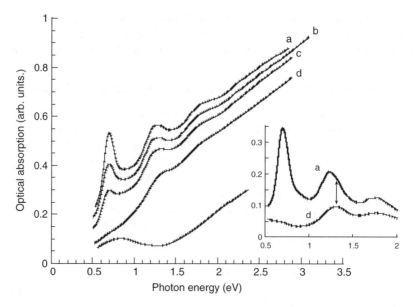

Figure 2.15 *Optical absorption spectra of n-doped SWCNT thin films. The linear increase of the background is generated by π-plasmon absorption. **(Curve a)** Pristine sample; **(Curves b** and **c)** Disappearance of the absorption band at 0.6 eV upon doping with anthraquinone radical anion. **(Curve d** and **inset)** effect of doping with fluorenone radical anion on the optical absorption spectrum of SWCNTs. **(Bottom left curve)** effect of doping with naphthalene radical anion; the occurrence of an absorption band at 0.8 eV is probably due to the modification of π-plasmon background upon doping. Adapted with permission from [40] Copyright (1999) Elsevier Ltd.*

characteristic of the pristine material are removed after exposure of the sample to a solution of Naphthalene-Li⁺. These experiments allowed to estimate the absolute value of the redox potential of SWCNTs to −0.5 V, comparable to that reported for C_{60} [66].

Many authors have used optical absorption to check the effect of doping, particularly upon electrochemical doping as this doping process allows tuning continuously and reversibly the cell potential, and thus the Fermi level of the tubes (Figure 2.16) [46,67].

Anodic polarization shifts the Fermi level below its standard value, depleting progressively the electronic states of the DOS, and forbidding the transitions between the singularities. Analogously, cathodic polarization leads to sequential filling of the initially empty states of the DOS with the same effect on the transitions, in agreement with the amphoteric nature of SWCNTs. In both cases, the optical absorption bands disappear in the same sequence, that is, in the series of increasing gap energies. The suppression of the first transition can be detected even in an aqueous electrolyte solution, but higher gap energies are poorly accessible owing to the limited potential window. Further depletion/filling of electronic states is possible only in a nonaqueous medium [46–48,68].

Within the rigid band model, these works show that the initially semiconducting tubes become metallic while the initially metallic ones simply become better metals with higher electronic density of states at the Fermi level.

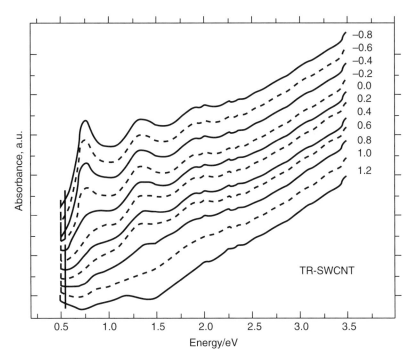

Figure 2.16 *Vis-NIR spectra of ITO-supported nanotube film in 0.2 M LiClO₄/acetonitrile. The applied potential (in V versus Ag-reference electrode) is labeled on each curve. Reprinted with permission from [46] Copyright (2001) American Chemical Society.*

2.2.4 Electrical Transport in Doped SWCNTs

SWCNT networks or materials constitute mixtures of semiconducting and metallic tubes that coexist in bundles with weak electronic coupling [69]. Electrical transport in such samples is expected to be dominated by the contained metallic tubes, especially at low temperatures where the semiconductor contribution should be frozen out. In practice, however, the samples normally show metallic behavior only at higher temperatures, whereas below a certain critical temperature (T*) an increase of the resistance with decreasing temperature is observed. Values between 90 and 350 K have been observed for T*, depending on the pretreatment(s) such as annealing of the sample [70]. The negative temperature coefficient of resistance below T* has been ascribed to several different mechanisms, the most important of which are weak localization [71,72], Kondo effect associated with the presence of magnetic impurities [73], 2D or 3D variable range hopping [74], as well as tunneling between metallic regions connected in series [75]. Among these, the latter accounts most appropriately for the experimental data in several cases [76,77]. Within this model, the barriers that separate the metallic regions may be structural defects or shape deformations within single tubes, intertube contacts within the bundles, or the contacts between the bundles. In contrast to these model-dependent conclusions derived more or less directly from transport, electron spin resonance and thermoelectric power measurements have shown that despite the negative slope of the resistance, SWCNT materials are metallic at every temperature [70,78] and behave as homogeneous materials as the resistance is insensitive to the frequency of the applied electrical field. This would of course be inconsistent with the earlier-mentioned transport models which either depend on some inhomogeneity of multiphase behavior or are intrinsically frequency-dependent. Moreover, the temperature dependence of resistance is qualitatively similar for single ropes and more complex 'mats' from 13 to 200 K [79], ruling out tunneling between ropes as the limiting factor in the undoped material.

2.2.4.1 SWCNTs Doped with Alkali Vapors

The effect of n-doping on the electrical conductivity was first reported by Lee et al. [16]. For saturation doped samples, the important observations are twofold [79] (Figure 2.17). First, saturation doping reduces the resistance by a factor 40 at 300 K, second dR/dT is positive at all temperatures, and data fits the classical theory for metals: $R = A + BT^n$, where A represents extrinsic elastic scattering of electrons (or holes) by impurities, and so on, while the power law term comes from electron-phonon scattering. Comparable results are obtained on doped single ropes [79] indicating that the enhancement of the conduction reflects charge transfer rather than improvement of rope-rope contacts, or other morphological effects. Finally, the suppression of gate modulation of the rope conductance after doping confirms that all tubes become metallic after doping.

Since then, many groups have explored the effect of doping on the electronic transport properties of SWCNTs. All experiments, whatever the doping method and the starting material, show a comparable drop of the resistance for samples doped at saturation [45,49,60,80–82], comparable to GICs but less dramatic than fullerides or conjugated polymers.

Doping SWCNTs with alkali vapors does not allow an estimate of the Fermi level of the material except for saturation doped samples for which it reaches that of the alkali metal. Most of the reported chemical compositions for samples doped at saturation are MC_8 [2,16,80,83], where M = K, Rb, Cs, associated with a change in the material resistance of more than one order of magnitude, the absolute value of the ratio R(pristine)/R(doped) being slightly different depending on the authors, probably due to the difference in the quality of the pristine material. However, all experiments show that most of the resistance decrease occurs at the very beginning of the doping process, as shown on Figure 2.18.

Since the '8' in MC_8 GICs is imposed on large alkali ions by the host structure, its recurrence in doped SWCNTs must have more to do with the Fermi level shift, that is, electrochemical potential differences, rather than steric considerations.

2.2.4.2 SWCNTs Doped by Electrochemical Method and by Ionic Solutions

Figures 2.19 and 2.20 show the variation of the resistance of two samples from the same batch, doped by ionic solutions and electrochemical charge injection respectively. These two experiments can be directly compared because resistance measurements are performed at thermodynamic equilibrium as a function of the redox potential of the reducing medium and the starting mats are identical.

First of all, both experiments show a rapid decrease of the resistance for small changes of the Fermi level of the solutions followed by a smooth monotonic decrease. In terms of conductivity, it corresponds to a linear variation with the Fermi level [3,49]. The slope of this linear variation is exactly the same in both experiments (Figure 2.19 in this chapter and Figure 2.7 in [3]). The increase of conductivity is linear in electrochemical potential but highly nonlinear in charge carrier concentration per carbon atom (x in the formula Li_xC),

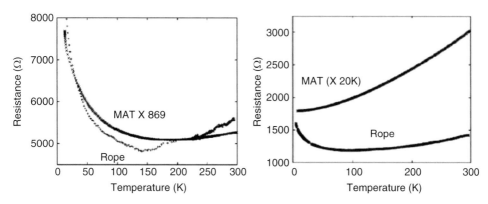

Figure 2.17 **(Left)** *Four-point resistance versus temperature for a pristine rope and for a mat of similar material normalized at 200 K and plotted on an expanded R scale to emphasize the similar T dependence above 13 K.* **(Right)** *Four-point resistance versus temperature for the rope and mat samples after doping with potassium (reprinted with permission from [79] Copyright (2000) American Physical Society.*

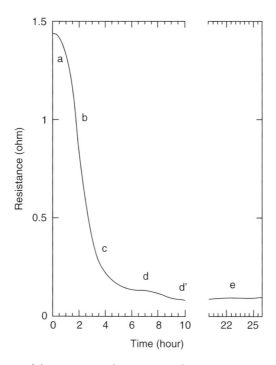

Figure 2.18 *Variation of the resistance during in situ doping experiment with Rb vapor. Reprinted with permission from [82] Copyright (2001) American Physical Society.*

since most of the Li uptake occurs at potentials below 0.5 eV as shown on Figure 2.21: the charge carrier concentration added to the tubes first increases rapidly at low energy, then increases almost linearly with the energy. The observed monotonous increase shows a one-to-one map between energy and charge concentration, indicating that there is no

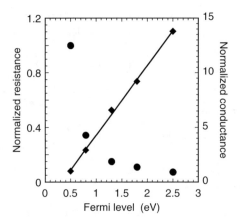

Figure 2.19 *Variation of the normalized resistance (solid circles) and normalized conductance (solid diamonds) of SWCNT mats doped with ionic solutions as a function of the Fermi level of the mat. 0.5 eV corresponds to the estimated Fermi level of the undoped material.*

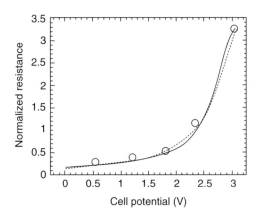

Figure 2.20 *Variation of the normalized resistance (solid line: upon doping; dashed line: upon dedoping) of SWCNT mat electrochemically doped as a function of the cell potential – the 3 V potential corresponds to the undoped mat. Open circles are resistance data from Figure 2.19 normalized to the electrochemical parameters. Adapted from [14].*

staging of GIC-like intercalated structures in SWCNT materials. Similar results are found for electrochemical doping; the electrochemical potential varies continuously and monotonously with the Li concentration [45]. Conductance enhancement therefore occurs in the early stages of doping, as predicted by theoretical calculations [84].

Moreover, in these experiments, the absence of anomalies or nonmonotonic behavior, which might be expected if the Fermi level were to encounter an electronic singularity as x is varied, clearly demonstrates that van Hove singularities characteristic of 1D density of electronic states are washed out in doped samples. A possible interpretation would be a crossover between a 1D to a 3D regime upon doping. However, the x(E) data on Figure 2.21

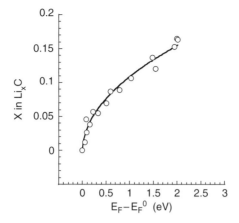

Figure 2.21 *Variation of the charge carrier concentration added to SWCNT material upon doping with ionic solutions. Open circles correspond to the value obtained at saturation by electrochemical doping with Li. Solid line is a $E^{1/2}$ fit of the data.*

can be surprisingly well fitted by a $E^{1/2}$ law [85]. Within the rigid band picture, this would lead to an $E^{-1/2}$ variation of the density of states, as, assuming that the Fermi surface of SWCNTs is not modified upon doping, charge carrier concentration is the integral of the density of states over the Fermi surface. Thus, doped SWCNTs would behave as a one-dimensional free electron gas. In that case, the smearing of van Hove singularity in the DOS could be attributed to the loss of long range order in doped SWCNTs due to the random potential induced by Li^+ ions probed by conduction electrons.

The rigid band approach to doping is supported by the fact that doped SWCNTs behave as normal metals. Indeed, using high resolution photoemission spectroscopy, analysis of the spectral shape near the Fermi level reveals a transition from Tomonaga-Luttinger liquid behavior in pristine samples [86] to a normal Fermi liquid behavior at a doping concentration high enough to fill the conduction bands of the semiconducting tubes [87].

2.2.4.3 *Reversibility of the Change in Electrical Transport*
Doping with alkali vapors is only partially reversible by heating the samples in vacuum, as shown in Figure 2.22, while it is fully reversible by electrochemistry (Figure 2.20) or by ionic solutions (Figure 2.23). Having in mind that the crystallinity of the bundles is not recovered upon dedoping, at least in electrochemical experiments, these results indicate that the conductivity of SWCNTs is not sensitive to structural disorder.

2.2.5 Spectroscopic Evidence for n-Doping

2.2.5.1 *Electron Spin Resonance*
Electron spin resonance (ESR) and nuclear magnetic resonance (NMR) are very powerful techniques for the study of metals since they provide a direct measurement of the density of electronic states at the Fermi energy. However, their great success in the study of GICs, alkali fullerides and doped conjugated polymers is tempered by the presence of ferromagnetic catalytic particles in SWCNT materials. Both the incomplete removal of magnetic

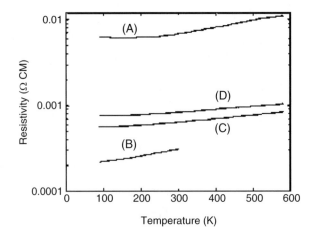

Figure 2.22 *Resistivity versus temperature for a bulk SWCNT sample. (A) Pristine material; (B) After doping with potassium at 473 K; (C) After heating in the cryostat vacuum to 580 K overnight; (D) After 3 days at 580 K. Reprinted with permission from [16] Copyright (1997) Macmillan Publishing Ltd.*

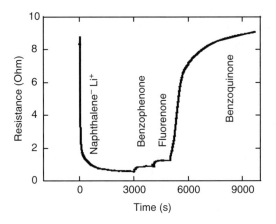

Figure 2.23 *Variation of the resistance of SWCNTs exposed successively to the radical-anion of naphthalene and to neutral solutions of benzophenone, fluorenone and benzoquinone. In this experiment, dedoping occurs by charge transfer from doped SWCNTs to the neutral aromatic molecules to form their ionic species. Benzoquinone was chosen because its redox potential of −0.4 V is closed to that of pristine SWCNTs. Reprinted with permission from [49] Copyright (2000) Elsevier Ltd.*

impurities and their generally unknown concentration are probably responsible for the diverse ESR and NMR literature results.

Conduction electron spin resonance (CESR) of pristine SWCNTs is generally undetectable, which is quite surprising since statistically 1/3 of the tubes in an ensemble should be metallic. Upon doping with K or Rb, an asymmetric Dysonian signal appears at the free electron value [2,58,88]. Such asymmetric lines are observed for metals when the skin depth is smaller than the dimension of the sample, leading to an inhomogeneous phase of the exciting field and a mixture of absorption and dispersion signals [89,90]. The CESR

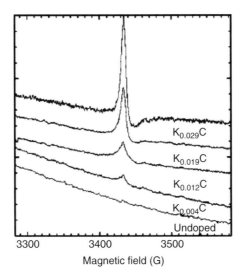

Figure 2.24 *Evolution of the CESR signal of electrochemically doped SWCNTs versus doping. Reprinted with permission from [48] Copyright (2000) American Physical Society.*

linewidth, proportional to the spin relaxation rate, is independent of doping concentration but increases with the atomic number of the alkali metal [2] due to spin-orbit coupling [91], as found for GICs [92,93], alkali fullerides [94], and doped polyacetylene [95].

A detailed analysis of ESR on K-doped SWCNTs (K:SWCNTs) using in situ electro-chemical method is given by Claye et al. [48], allowing to determine the evolution of both spin susceptibility and microwave conductivity upon doping. In these experiments, the saturation value is found to be K/C=1/24 rather than 1/8 with vapor doping, due to the co-insertion of THF solvent with K ions. The spin susceptibility of the saturated phase corresponds to a density of states $n(E_F)=0.015$ state per eV per spin per C atom, which is about five times smaller than a theoretical estimate [96]. More interesting is the evolution of the CESR signal upon doping. The intensity of the CESR signal increases continuously with increasing K/C (Figure 2.24). Figure 2.25 shows that the deduced spin susceptibility and microwave conductivity increase with K/C and are fully reversible upon dedoping.

As for samples doped with alkali vapors, the CESR linewidth is found to be independent of the K/C ratio, which is not consistent with the standard picture of ESR for uniform Pauli spins in a metal. An interpretation of these ESR observations consistent with the X-ray results would be a two-domain picture of K-saturated domains embedded in undoped regions [48].

2.2.5.2 Raman Spectroscopy

Raman spectroscopy is one of the most efficient tools to investigate the vibrational properties of SWCNTs in relation with their structural and electronic properties [41,42]. Raman scattering from SWCNT is a resonant process, as illustrated by dramatic changes of the tangential mode (TM) (1400–1700 cm^{-1}) in the Raman spectrum with laser excitation energy. Figure 2.26a displays the Raman results for a SWCNT sample with an average

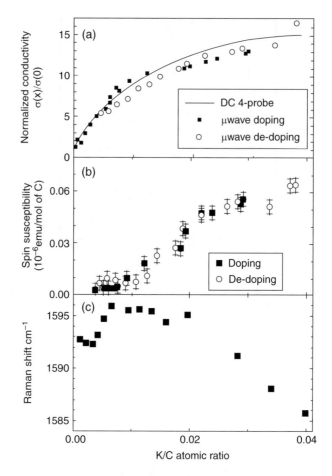

Figure 2.25 *In situ measurements of (a) electrical conductivity, (b) spin susceptibility, and (c) shift of the Raman-active tangential G-band as functions of K/C ratio during galvanostatically controlled insertion and de-insertion. Raman data are discussed in Section 2.2.6. Reprinted with permission from [3] Copyright (2002) American Chemical Society.*

diameter of 1.4 nm and a dispersion of ±0.2 nm. A broad and asymmetric band around 1540 cm^{-1} and a sharp peak at 1580 cm^{-1} are observed with the 1.92 eV excitation energy (Figure 2.26a, bottom) while several symmetric lines with a dominant peak around 1590 cm^{-1} and two other strong features around 1560 and 1550 cm^{-1} appear using 2.41 eV excitation energy (Figure 2.26a, top). Other intense lines observed in the range 100–200 cm^{-1} (not shown) are associated with the radial breathing modes (RBM).

Many experiments and calculations have shown that the RBM frequencies are related to the tube diameters [97,98]. Despite these well established results, the mechanism giving rise to the first order spectrum is still subject to debate between single-resonance and double-resonance scattering. Single-resonance scattering is associated with allowed optical transitions between spikes in the 1D electronic density of states, which fall in the visible and near-infrared ranges for pristine samples. The allowed optical transition energies

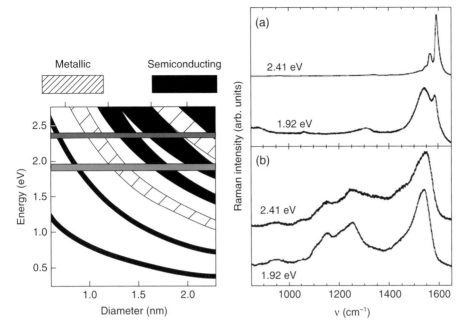

Figure 2.26 *(Left) Allowed optical transitions as a function of the diameter of SWCNTs.*
(Right) Raman spectra in the TM range of (a) pristine SWCNTs and (b) saturated Rb:SWCNT
bundles at laser excitation energies 2.41 eV (top) and 1.92 eV (bottom). For a better
understanding of the figure, please refer to the colour plate section. Adapted with permission
from [65] Copyright (2003) Elsevier Ltd.

depend both on the diameter and metallic or semiconducting character of the tubes, such
that when the excitation energy matches a particular, allowed optical transition, the Raman
response of the corresponding tube is enhanced. For a tube of diameter 1.4 nm, the resonance process predicts an enhanced response of metallic (semiconducting) tubes for
the 1.92 eV (2.41 eV) excitation respectively. From the profile of the main peak of the
Raman response of metallic tubes excited at 1.92 eV (Figure 2.26a, bottom), it was found
that the shape of the lowest frequency dominant component of TM region is well described
using a Breit–Wigner–Fano (BWF) resonance which involves the coupling between a
phonon and an electronic continuum (plasmon) [39,99–101], that has recently been reinter-
preted as due to electron-phonon coupling [102–104].

Because Raman vibrational features are sensitive to both intercalation and charge transfer, this spectroscopy allows the evolution of the structural and electronic properties of
SWCNTs upon doping to be investigated. The first Raman evidence for charge transfer in
doped carbon nanotube bundles has been given in the pioneering work by Rao et al. on
saturated K- and Rb:SWCNT bundles (also on iodine- and bromine-doped SWCNTs) [17].
The Raman spectrum of *saturated* alkali-doped SWCNTs presents the following features:
(i) the high frequency tangential modes shift substantially to lower frequencies due to electron transfer onto SWCNTs; (ii) the lineshape of the most intense component is well
described by a BWF resonance as all tubes became metallic; (iii) the intensities of both TM
and RBM vanish because of the loss of resonance.

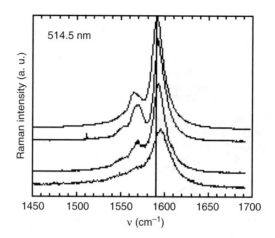

Figure 2.27 *Evolution of the Raman spectrum of the TM for different doping levels. From top to bottom: LiC_{25}, LiC_{10}, LiC_7 and LiC_6. Adapted with permission from [65] Copyright (2003) Elsevier Ltd.*

Raman studies performed on samples doped with different alkali metals (Li, K, Rb, Cs) and from different routes reveal a universal behavior versus concentration, independent of the nature of the alkali metal [48,65,82,105–106]. All these studies show that for all alkali metals, the high frequency tangential modes first upshift at very low concentration, then monotonically downshift with further doping to saturation. In a first step, the profile of TM region progressively changes upon doping from a multi-peak structure to a single upshifted line (Figure 2.27). This behavior is related to a concomitant monotonic decrease of the resistance (Figure 2.18) and a progressive vanishing of peaks associated with van Hove singularities in optical absorption spectra (Figure 2.15).

In a second step, a drastic change of the Raman profile occurs when the resistance reaches its final stable value. Two distinct Raman responses have been identified in relation with two plateaus in the resistance curve (Figure 2.18) and are assigned to two different doped phases: *upshifted symmetric single line* associated to phase I (first plateau (point d) in the resistance curve in Figure 2.18, spectrum I in Figure 2.28) and *downshifted asymmetric BWF component* associated to phase II (second plateau (point e) in the resistance curve in Figure 2.18, spectrum II in Figure 2.28).

With regards to the significant difference in the size of the dopant species, the strong similarity between the behaviors measured on the different doped samples must be pointed out. This similarity suggests that, whatever the doping species, for a given stoichiometry, intercalated SWCNT materials have similar structure.

Some analogies with the behavior of the Raman response upon doping in graphite intercalation compounds can be found [11]: (i) upshifted component at low doping level (stages 2 and 3 in GICs); (ii) downshifted BWF component at the highest doping level (stage 1 GICs). At low doping level, the spectra for alkali-doped graphite exhibit a doublet structure. The intensity of the component close to the frequency of the E_{2g} Raman-active mode for pristine graphite decreases with increasing doping level, while the intensity of the upshifted line increases with increasing doping level. With regards to these behaviors, the lower and upper frequency components are, respectively, identified with vibrations in

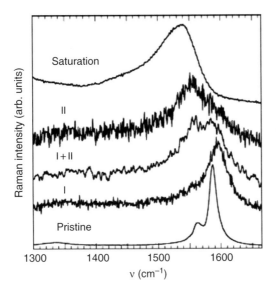

Figure 2.28 *Evolution of the Raman response at high doping levels. Spectra were recorded during combined in situ resistivity and Raman experiments. Spectra labeled I and II correspond to different doping phases associated to two plateaus in the resistivity measurements (adapted with permission from [65] Copyright (2003) Elsevier Ltd).* **(Top curve)** *Raman spectrum of an ex situ saturated sample [107].*

planes not adjacent to and planes adjacent to an intercalate layer plane. However, in alkali-doped SWCNTs, an upshift of the highest-frequency component of the TM group is clearly evidenced and no doublet structure is observed. This shift is weaker than the doublet splitting for GICs, suggesting weaker interactions between the tubes and the alkali atoms in SWCNTs than in graphite. A possible interpretation could then be [65]: (i) in a first step, a progressive and homogeneous intercalation of the bundles by alkali atoms occurs, the hollow channels formed by three tubes being the most convenient sites that allow a monotonic intercalation of alkali atoms inside the bundles. This gives rise to only slight lattice expansion and therefore to a slight shift of TM spectra; (ii) in a second step, the sites between two tubes can be occupied, these latter sites being comparable to those involved in graphite intercalation compounds (intercalation between two planes). Within this interpretation, and by analogy with graphite, the hardening of the TM component could be the consequence of the progressive insertion of alkali atoms inside channels between the tubes, structure I corresponding to a saturated phase for alkali atoms localized inside these channels. We emphasize, however, that there is no direct structural evidence for this scenario.

In summary, the agreement between the Raman data obtained from a variety of doped samples prepared by different routes, and the correlation of doping-induced evolution of Raman response with resistivity and optical absorption support the consistency of results and unambiguously establishes the universal character of the stoichiometry dependence of the Raman spectrum in alkali-doped SWCNT bundles.

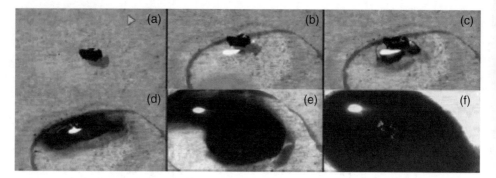

Figure 2.29   *Sequential photographs of the spontaneous dissolution of a sodium salt of nanotubes. Reprinted with permission from [61] Copyright (2005) American Chemical Society.*

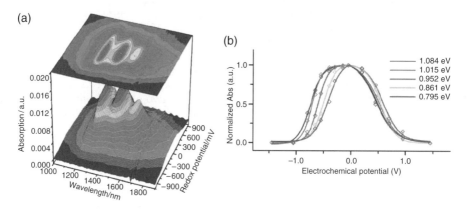

Figure 2.30   *(a) Smoothed surface plot of optical absorption intensity as a function of wavelength and electrode potential in the S11 region for a solution of a potassium salt of HiPco SWCNTs in DMSO. (b) Intensity of each of the five bands selected in the spectra above after normalization as a function of potential. The fitting curves (full lines) were calculated by assuming that only the neutral tubes absorb, and by using a simple Beer-Lambert law wherein the concentrations of reduced/neutral and neutral/oxidized tubes where expressed as a function of electrode potential, using Nernst equation [114]. This way, oxidation and reduction potentials could be determined (corresponding to the inflection points in each sigmoidal curve). In all plots, raw electrochemical data, that is, uncorrected for ohmic drop, are referenced to standard Calomel electrode. For a better understanding of the figure, please refer to the colour plate section. Reprinted with permission from [6] Copyright (2008) Royal Society of Chemistry.*

2.2.6   Solutions of Reduced Nanotubes

Unexpectedly, reduced (n-doped) nanotubes have been found useful for processing SWCNTs. By analogy to soluble, rigid, inorganic polymers such as $Mo_6Se_6^{2-}$ [108], alkali metal salts of nanotubes, that is, reduced SWCNTs, spontaneously dissolve (Figure 2.29) in polar organic solvents such as DMSO without the need for added energy such as the ubiquitous sonication step when dispersing SWCNTs with surfactants [61]. This mild

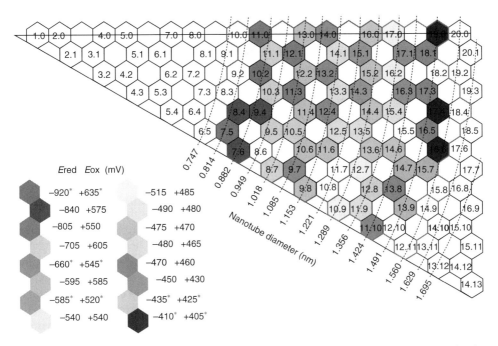

Figure 2.31 *Chirality map displaying the average standard potentials associated to each of the SWCNT structures identified in this work. HiPco SWCNTs are located inside the red line, while arc-discharge SWCNTs are inside the blue line. For a better understanding of the figure, please refer to the colour plate section. Reprinted with permission from [114] Copyright (2008) American Chemical Society.*

dissolution route yields concentrated (ca. 2 mg ml^{-1}) solutions with the characteristic signatures of reduced nanotubes in Raman spectroscopy, that is, loss of intensity, upshift and broadening of the tangential signal around 1600 cm^{-1}, and disappearance of the radial breathing mode at low wavenumbers [109].

AFM and X-ray analyses show full exfoliation of bundles, in the case of SWCNTs [110] yielding concentrated solutions of individualized tubes and disentanglement in the case of multi-walled nanotubes. Solutions of nanotube salts have been used for functionalization of carbon nanotubes [111–113]. Spectro-electrochemical analysis of these solutions allowed the determination of reduction and oxidation potentials as a function of tube diameter (Figure 2.30 and Figure 2.31). Furthermore, by comparison between the resulting electrochemical gap and the optical gap, exciton binding energies could be experimentally derived [6,114].

2.3 p-Doping of Carbon Nanotubes

Upon p-doping, electrons are removed from the valence band of nanotubes which are, thus, oxidized. On the other hand, carbon nanotubes can be oxidized by acids such as boiling HNO$_3$, leading to nanotubes bearing oxygen groups. It is often assumed that an oxidized nanotube is always a nanotube bearing oxygen-containing groups. This is not true. Indeed, nanotubes can be oxidized simply by removal of electrons without adding any functional group, the same way reduced nanotubes (n-doped), seen in the previous section, are nanotubes bearing added electrons. Since this chapter is dealing with doped nanotubes, that is, nanotubes with excess or deficit of electrons, we will be only considering oxidation from the point of view of electron removal without disruption of the carbon skeleton. Functional oxidation is discussed in Chapter 3 (functionalized nanotubes). For sake of clarity, we will be talking of p-doping rather than oxidation.

The structure of carbon nanotubes situates them right in between graphite and fullerenes. Hence, the chemistry that was tried on them was often inspired from either of those two fields. We saw in previous section that alkali metals have been widely and preferentially used to reduce nanotubes. From oxidation point of view, the first p-doping of single-walled nanotubes was performed with halogens [16,17]. This will be described in Section 2.3.1. Carbon nanotubes have been subsequently p-doped with $FeCl_3$, acceptor molecules, thionyl chloride, acids, and superacids. This will be the subject of Sections 2.3.2–2.3.6. As far as this book is material oriented, the literature about the analytical electrochemistry on p-doped nanotubes will not be covered here, unless necessary. Reviews on the subject have recently appeared [5,7]. Although SWCNT p-doping should be a quite symmetrical concept versus the previously treated n-doping, the structures of the two sections are entirely different. Indeed, because n-dopants are mainly alkali metals for which several synthesis routes have been developed, it was thus convenient to describe separately the different synthesis techniques and the properties, whereas p-dopants are quite numerous and different in nature. Hence, the authors preferred to treat each of the dopants as a specific case and describe it in extenso from the synthesis to the properties. For each of the following subsections, it has therefore been attempted to describe the synthesis, the mechanism of the SWCNT/dopant interaction, and the structure (when known), properties and physical characterization of the resulting material. When applications are known or expected, they will be briefly mentioned.

2.3.1 p-Doping of SWCNTs with Halogens

Along with alkali metal doping, bundles of SWCNTs were first doped with halogens in 1997. In two simultaneous papers [16,17], samples of SWCNTs were exposed to Br_2 and I_2 vapors in vacuumed ampoules. It was shown that the resistivity of a SWCNT mat decreases from $0.016\,\Omega$ cm to $0.001\,\Omega$ cm in a matter of minutes when exposed to bromine vapor (Figure 2.32) [16]. It was also found that the T* temperature, the temperature for minimum resistivity associated with disorder [115], was significantly lowered upon Br doping.

In the companion paper, Eklund and coworkers discussed Raman changes upon doping [17]. In particular, it was seen that Br_2 produces a rather large upshift of the tangential mode ($+24\,cm^{-1}$). Noteworthy is the fact that the RBM mode does not disappear upon doping, as it was observed with alkali doping (see the related sections earlier in this chapter). Preliminary work with iodine doping caused much less spectral changes which lead the authors to question the occurrence of an actual doping of the SWCNTs by iodine. As we will see below, later work on iodine doping showed that doping depends on the reaction conditions. Materials formed with bromine or iodine are sufficiently different to justify seperate sections for Br- or I-doped carbon nanotubes. As far as we can tell, there is no report in the literature on chlorine p-doping of carbon nanotubes.

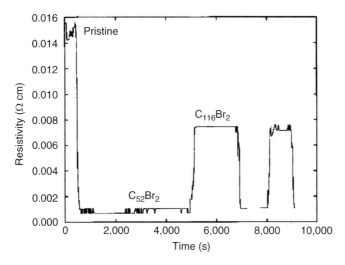

Figure 2.32 *Directionally averaged resistivity (ca. 300 K) of a bulk SWCNT sample versus time of exposure to Br_2 vapor. Within 3 min, ρ has dropped from 0.015 to 0.001 Ω cm, with weight uptake (determined in a seperate experiment) corresponding approximately to $C_{52}Br_2$. After 3000 s, the apparatus was opened to the air, with no change in ρ. After 5000 s, the sample was heated in air with a heat gun for ca. 200 s, which increased ρ to 0.008 Ω cm and decreased the Br_2 content as shown. Re-exposing the same sample to Br_2 vapor at 7000 s returned ρ to its previous minimum value. One additional heating/redoping cycle is included in the figure. Reprinted with permission [16] Copyright (1997) Macmillan Publishing Ltd.*

2.3.1.1 p-Doping with Bromine

After the initial work by Lee et al. [16] and Rao et al. [17], most of the work performed on halogen-doped nanotubes dealt with iodine doping. A few authors, nevertheless, studied Br doped nanotubes, whose work is presented below. Interestingly enough, reversible Br-doping has been used as a mean of purification (Section 2.3.1.1.4) and sorting (Section 2.3.1.1.5).

2.3.1.1.1 Synthesis Several techniques have been used to dope SWCNTs with bromine. Haddon et al. [116] doped CS_2 solutions of s-SWCNTs (for soluble (end-functionalized) SWCNTs) with bromine or iodine and recorded changes on the UV-vis spectrum (see the following). All the other reports, starting with Rao et al. [17] and Lee et al. [16] are variations of exposure of solid nanotubes (soot) to bromine vapor in a vacuumed ampoule. Kataura et al. doped raw nanotubes with bromine vapor in an evacuated ampoule and quantified the results with Raman spectroscopy [117]. Minami et al. and Jacquemin et al., studied the effect of doping on thin films of SWCNTs, sprayed on surfaces from ethanol suspension [44,118]. The resulting films were doped by exposure to dopant vapor in a two-zone vacuumed quartz tube. Doping was monitored as a function of dopant concentration by optical absorption, and dc conductivity. Additionally, Sundqvist et al. performed bromine doping by exposing SWCNTs to bromine vapor in an evacuated ampoule [119]. It should be pointed out that doping with bromine is rather fast, in the range of few minutes at room temperature [17].

2.3.1.1.2 Structure No work has been performed on Br-doped nanotube structure. However, several groups have performed calculations on bromine doped nanotubes [120–122]. They found substantial charge transfer from the nanotubes to the bromine molecule and point out preferential stoichiometries Br_3 or Br_5 when the dopant is inside the nanotube [122]. Independently, from high pressure Raman studies of bromine doped SWCNTs, it has been suggested that bromine is present under the form of solid Br_2, at the surface of the tubes, analogous to bromine intercalated graphite [123] and polybromine chains [119].

2.3.1.1.3 Characterization Strong evidence for p-doping in bromine-treated nanotubes comes from absorption and Raman spectroscopy, and conductivity measurements. The resistivity of a SWCNT mat at room temperature decreased by more than one order of magnitude upon exposure to bromine (Figure 2.32) [16]. Temperature dependent measurements showed a decrease of resistivity with temperature with suppression of the minimum resistivity found for mats of undoped tubes, hence showing an increased metallic character for the Br-doped mats compared to the undoped ones.

As explained in Section 2.1.4 (*What Happens upon Doping SWCNTs*), optical absorption spectroscopy show three distinct absorption bands, labelled S11, S22 and M11 that correspond to transitions between symmetrical van Hove singularities. The two with the lowest energy are for semiconducting tubes while the third one corresponds to metallic tubes. Bromine doping leads to the disappearance of all three bands [40,44,118], showing that Br_2 is a sufficiently strong oxidizer to remove electrons from all three van Hove singularities of the valence band. Upon saturation doping, new, weak, bands appear at ca. 1.07 eV in the case of bromine [44]. Similar effect had been observed for n-doping. [40] Although the matter has not been conclusively settled, Minami et al. propose that depletion of the M11 singularity allows new optical transitions from deep lying valence bands to that singularity. They further point out that the new bands obtained at high n-doping (1.3 eV) or p-doping (1.07 eV) do not have the same energy, and thus, should this hypothesis be valid, they represent a signature of the asymmetry of the nanotube band structure.

The first Raman spectrum of Br doped samples (on laser-grown SWCNTs) showed an upshift of the tangential mode (TM) signal from 1593 cm^{-1} (undoped tubes, 514 nm (green) laser) to 1617 cm^{-1} [17]. Later work always observed such an upshift, although the exact values themselves may vary from sample to sample or from lab to lab. For example, Sundqvist et al. reported an upshift from 1594 to 1603 cm^{-1} (for 12 h doping of arc-synthetized tubes) [119], Wilson et al. reported an upshift from 1598 to 1610 cm^{-1} (red laser, 633 nm) in their purification procedure [124] that they ascribed to partial doping by comparing to the 24 cm^{-1} upshift of Rao et al. Indeed, although their doping procedure consists in immersing SWCNTs in liquid bromine, this is performed at room temperature and may lead to milder doping level. Contrary to the alkali metal doping, there are observable peaks in the radial breathing mode (RBM) region, at low wavenumbers. However, it is not clear whether those are still RBM signals from the nanotubes or interhalogen bands [119].

2.3.1.1.4 Reversible p-Doping as a Mean of Purification HiPco SWCNTs have been purified by immersing the raw nanotubes in air- and moisture-free liquid Br_2. The metal present (Fe) is quickly oxidized to its bromide salt and removed by rinsing with water or diluted acid. The amount of metal catalyst in the samples has been reduced from ca. 27% to ca. 3% by this procedure. The bromine treated samples showed evidence of partial p-doping: upshift of the tangential mode from 1598–1610 cm^{-1} and vanishing of the RBM mode [124]. The doping was fully reversible and the SWCNTs were returned to the neutral state by heating the sample at 400°C under dry N$_2$ gas for one hour.

2.3.1.1.5 *Enrichment in Metallic or Semiconducting Nanotubes*   If SWCNTs are to have a future in electronics, the issue of sorting semiconducting tubes from the metallic ones must be resolved. The first reports on nanotube separation according to their electronic character appeared in 2003 [125–128].The most efficient demonstration of sorting, as of today, has been obtained by density gradient ultracentrifugation [129]. One of the 2003 reports [128] relied on charge transfer between nanotubes and bromine. Aqueous bromine solution was added to a surfactant (Triton X100) stabilized nanotube aqueous dispersion. The resulting mixture was centrifuged (12 h at 24000 g) and the supernatant and sediment separately analyzed. Absorption spectrum was used to assess the degree of enrichment. Since absorption and scattering cross sections of semiconducting and metallic tubes respectively are unknown, the authors normalized their spectra with respect to the metallic band M11 and compared semiconducting band (S11 or S22) intensities. The semiconducting fraction in the supernatant was greatly enhanced compared with the sediment. At first, the authors attributed this effect to preferential interaction of the metallic tubes with bromine, but further work showed that it is more likely that surfactant destabilization by bromine complexation with the tubes leaves the metallic tubes without their surfactant envelope [130]. It has been shown that surfactant covered nanotubes have a density close to 1 g cc^{-1}, [131] compared to nanotubes themselves which have a density close to 1.4. Hence in a mixture, naked SWCNTs will sediment preferentially to surfactant-coated SWCNTs.

2.3.1.2 *p-Doping with Iodine*
Iodine behaves differently from bromine when doping bundles of SWCNTs, from the synthesis all the way to the structural characterization. In particular, polyiodine anions I$_n^-$ get into the inner cores of the nanotubes and thus form a special class of hybrid nanotubes (I@SWCNTs) with charge transfer. Hence, although such materials belong to the Chapter of this book dealing with filled nanotubes (Chapter 5), we present here a brief description of I:SWCNTs for sake of better, more accurate comparison with the other p-doped SWCNTs.

2.3.1.2.1 *Synthesis*   Iodine, the heavier of the stable halogen elements differs from its lighter fellows by a smaller electron affinity. Iodine is a weaker oxidant than bromine, with an oxidation potential +0.6 V versus H$_2$/H$^+$ compared to +1.08 V for bromine, and is unique in its ability to form polyatomic anions I$_n^-$ with n = 3, 5,…[132]. In the first trials to dope SWCNTs with iodine [17], samples were placed in an evacuated ampoule with the iodine on the opposite side of the ampoule. The reaction was carried out at room temperature for several hours. Slight changes in the Raman spectra were observed, far less spectacular than with bromine or alkali metal doping. Specifically, an upshift from 186 to 188 cm^{-1} of the radial breathing mode and a downshift from 1593 to 1590 cm^{-1} of the tangential mode were observed. The authors ascribed this to either very weak charge transfer or simple physisorption of I$_2$ molecules on the SWCNT surface.

It was found later on that, upon contact with molten iodine (compared to iodine vapor in the previous case), a clearer doping of the nanotube occurred [133]. Samples were prepared by immersing SWCNTs in molten iodine at 140°C in evacuated quartz tubes for several hours. Post-treatment consisted in keeping the sample at 60–80°C while maintaining the other end of the ampoule in liquid nitrogen. This allowed for annealing of the doped sample and removal of excess iodine. Weight uptake and thermogravimetric analysis both suggested an average composition of IC_{12}. Note that this would correspond to one net charge per 60 carbon atoms if iodine is under the form of I_5^- anions, hence a rather low doping level.

Upon the subsequent evidence that polyiodide chains can form within the tubes instead of between them in bundles [134] despite their not being subjected to any opening process such as acid reflux ([135,136] and Chapter 5 in this book), it was hypothesized that highly oxidizing I^+ species, originating from I_2 dissociation into I^+ and I_3^- in the molten state [137] are responsible for the tube opening.

2.3.1.2.2 Structure Within a close-packed nanotube bundle forming a hexagonal array, there exists only one true intercalation site, namely the triangular void between three adjacent tubes (Figure 2.5). Additionally, as we have seen in the case of alkali metal doping, the dopant can fix on the outer surface of the bundles, either on the surface of a tube or in the grooves between two tubes. This picture is valid for closed tubes. If the tubes are open, then the inside becomes an additional possible insertion site.

A first evidence of iodine localization in iodine-doped samples was given by Z-contrast scanning transmission electron microscopy (Z-contrast STEM) [133]. Z contrast, that is, the atomic number contrast, scales with the square of the atomic number [138]. Hence, iodine ($Z=53$) is highly visible within a carbon environment ($Z=6$). Iodine atoms were found to form linear chains with an average spacing between chains of ca. 21 Å (this particular sample was described as 'moderately intercalated' by the authors).

Much higher quality images were obtained later on and, together with theoretical calculations, allowed describing much more precisely the structure of I:SWCNT samples: helical, linear chains of iodine atoms have been observed [134] in the inside of the tubes, with up to three chains spiralling together within the tubes [139]. Extensive X-ray and neutron diffraction have been performed on iodine-doped nanotube bundles by Bendiab et al. [140]. A strong shift of the *10* reflection of the two-dimensional bundle lattice is observed by X-ray diffraction, towards lower q. Such a shift has often been taken as evidence of intercalation of the rope triangular lattice accompanying a small lattice expansion. Intuitively indeed, intercalation should lead to the lattice expansion of the bundles, hence larger lattice parameter and lower q value as it has been observed for example in alkali-doped fullerenes. However, neutron diffraction shows little if no shift for the *10* reflection (Figure 2.33). This inconsistency has not been explained yet.

Filling of the inside cavity of the nanotubes creates an accidental extinction phenomenon [141]. The authors conclude that I_n^- chains (n=3 or 5 and comes from Raman experiments) filling the inside of the tubes and a small proportion being into the interstitial site. Indeed, if the tubes are open, inner channels present a much larger volume than interstitial channels. Further evidence from X-ray absorption fine structure (EXAFS) confirms the presence of I_5^- chains inside the SWCNTs but rules out the presence of I_3^- species [142]. It should be noted however that other authors performed X-ray absorption near edge structure

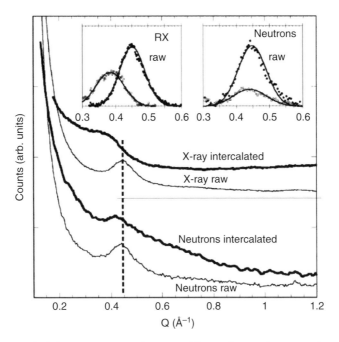

Figure 2.33 *Detail of the 10 Q range. The vertical bar indicates the 10 peak position for the raw sample. The Q shift is 3% for neutrons and 15% for X-ray. (**Inset**) 10 peak after background subtraction. Reprinted with permission from [140] Copyright (2004) American Physical Society.*

(XANES) analysis and found no evidence for any p-doping of the nanotubes with iodine and actually claim small n-doping with formation of I^+ species [143]. However, this latter work involves vapor phase synthesis instead of the immersion in liquid iodine, as a possible hint to explain the discrepancy observed.

2.3.1.2.3 Characterization When it comes to I:SWCNTs, the main question, apart from the structure issue, is the amount of charge transfer if any. Since neutral tubes present a series of well-defined van Hove singularities in their band structure, leading to specific optical transitions, the unambiguous way to measure doping is by absorption spectroscopy. When doping a CS_2 solution of end-functionalized SWCNTs with iodine, Haddon et al. observed the full disappearance of the 0.67 eV feature in the absorption spectrum (5386 cm^{-1} in Ref. [116]), unambiguously showing that the first van Hove singularity of those electric-arc-prepared nanotubes (Carbolex) had been depleted [116]. The same result, that is, disappearance of only the lowest energy transition, was obtained when exposing a thin film of SWCNTs to iodine vapor [40]. Hence iodine is a strong enough oxidant to empty the first van Hove transition of the semiconducting tubes. However, two papers exist about iodine intercalation where the authors observed no charge transfer [144], and even reverse (n-doping) charge transfer [143]. No absorption spectrum is shown, though, in these two reports.

Raman spectroscopy, otherwise acknowledged as an invaluable tool to study carbon nanotubes, may be ambiguous, in the case of halogen doping. The first problem comes from the RBM modes that occur in the same energy region as interhalogen vibrations. Indeed, the modes observed between 100 and $180\,cm^{-1}$ have been attributed to I_n^- species. Regarding the tangential mode region, around $1600\,cm^{-1}$, the first report on halogen-doping showed that, whereas bromine-doping led to an upshift from 1593 to $1617\,cm^{-1}$, iodine-doping led to a downshift to $1587\,cm^{-1}$. Considering that there is a neat relationship between charge and TM displacement in fullerenes [1], this could be taken as an indication of n-doping. However, future work never showed such a clean, linear, relationship between charge and vibration energy in the case of the nanotubes. Furthermore, a very recent work showed that apparent TM shift were due to respective importance of the G+ and G− components of the TM band rather than doping level [145].

2.3.2 p-Doping with Acceptor Molecules

Statistically, two thirds of a macroscopic ensemble of nanotubes are semiconducting molecular wires. Research on nanotube-based electronic devices has thus been very active in the last decade, both on assemblies of nanotubes and single-tube devices. Originally, it was found that semiconducting nanotubes were p-doped. It was soon recognized though, that charge transfer from ambient molecular oxygen was responsible for the apparently intrinsic p-doped character of nanotube devices [54]. Enforced p-doping of nanotubes has been obtained with NO_2 [146] and tetrafluorotetracyanoquinodimethane ($TCNQF_4$) [68,147,148]. Amphoteric doping has been demonstrated by Takenobu et al. by filling the inside of opened SWCNTs (see Chapter 5) with donor and acceptor molecules [149], leading to air-stable transistors [150,151].

2.3.2.1 Oxygen Sensitivity

Cycling between vacuum and ambient air exposure has a clear effect on the resistivity of SWCNTs. Four-probe measurements on thin films of SWCNTs [54] show a resistivity change of 10 to 15 % (Figure 2.34). Likewise, sign and magnitude of the thermoelectric power changed. It thus appears that 'neutral' nanotubes are slightly n-doped rather than p-doped. However, besides oxygen, it has been hypothesized by several groups that difference between the electrodes and nanotube work function could also create apparently intrinsic doping [54,152–155]. It was found that oxygen doping could be only partially reversed, by annealing at high temperature (110 to 150°C in vacuum). Calculations have been performed on O_2 adsorption. Adsorption energy of 0.25 eV for the binding of O_2 to the surface of a SWCNT was calculated and it was found that semiconducting tubes were rendered metallic due to p-doping by oxygen with an estimated 0.1 electron transfered from the SWCNTs to each O_2 molecule [156]. Contrary to the above, it has also been reported that the apparent O_2 sensitivity of SWCNTs was due to contaminants, coming from the purification process. Removal of the contaminants would render SWCNTs insensitive to O_2, CO_2, H_2O and N_2, while a strong sensitivity to toxic gases such as NO_2, SO_2 and NH_3 was observed (Figure 2.35) [157]. In general, the interaction of gases with SWCNTs probably deserves a word of caution since it has also been observed that gas sensitivity is dependent on temperature history of the sample [158]. Likewise, N_2 and He collisions with SWCNTs have a measurable influence on the thermoelectric power (Figure 2.36) [159].

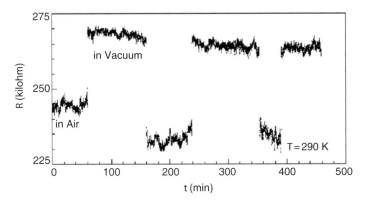

Figure 2.34 *Sensitivity of the electrical resistance R of SWCNT films to gas exposure at a temperature (T) of 290 K. The nanotube resistance switched by 10 to 15% as the chamber surrounding the sample was alternatively flooded with air or evacuated. Identical results were obtained if pure, dry oxygen was used instead of air, indicating that oxygen was the source of the effect. Likewise, switching the chamber purge gas from oxygen to any inert gas resulted in similar stepwise changes in R. Reprinted with permission from [54] Copyright (2000) AAAS.*

2.3.2.2 NO_2 Sensing

NO_2 sensing was observed on single tube devices made from a metal(s)-SWCNT/metal junction [53] (where 's-SWCNT' stands for 'semiconducting carbon nanotube'). The SWCNTs were grown by chemical vapor deposition (CVD) on patterned catalyst islands. Conductance of such devices changed dramatically upon exposure to NO_2 gas (Figure 2.37). The same effect was observed with mats rather than individual tubes but with a much lesser amplitude. Slow diffusion of the reactant and unavailability of the inner tubes within bundles were proposed as reasons for this smaller effect. Based on density functional theory calculation, charge transfer was found to be responsible for the change in conductivity [157].

2.3.2.3 TCNQ and $TCNQF_4$ Doping

Thin films of SWCNTs were prepared by spraying an ethanol dispersion onto quartz substrates or semitransparent platinum films. These films were immersed in CS_2 solutions to which a variety of organic dopants were added: benzoquinone, bromanil, TCNQ, $TCNQF_4$, as electron acceptors (oxidants) hydroquinone, bis-ethylenedithiotetrathiafulvalene (BEDT-TTF), tetramethyltetrathiafulvalene (TMTTF) as electron donors (reducing agents). Out of those reagents, TCNQ and $TCNQF_4$ lead to considerable changes in the absorption spectrum of the films, showing close to full disappearance of the S11 absorption. [68,147]. Field effect transistors were prepared with controlled carrier density by doping with $TCNQF_4$ in solution [150].

2.3.2.4 Amphoteric Doping Leading to Hybrid SWCNTs

Takenobu and coworkers have doped carbon nanotubes with a variety of organic molecules, both acceptors and donors, thus inducing amphoteric doping [149]. The starting SWCNTs were laser-grown and purified with H_2O_2, HCl, and NaOH. Apparently, the purification

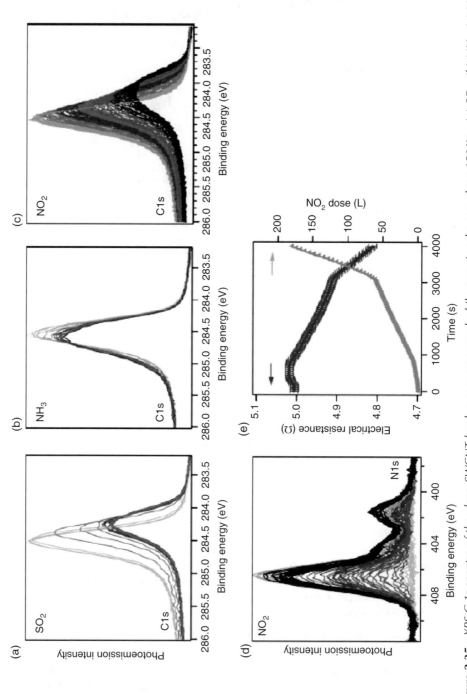

Figure 2.35 XPS C 1s spectra of the clean SWCNT buckypaper measured while exposing the sample at 150 K to (a) SO_2; (b) NH_3; (c) NO_2; (d) The evolution of the N 1s spectra in the case of NO_2; (e) SWCNT buckypaper resistance (dots) as at partial pressures of about 10^{-8} mbar; (d) The evolution of the N 1s spectra in the case of NO_2; (e) SWCNT buckypaper resistance (dots) as

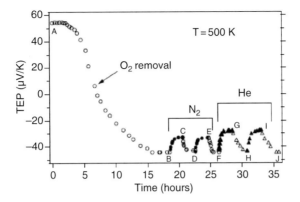

Figure 2.36 *Thermopower S versus time for a mat of SWCNTs at T = 500 K initially saturated with air at ambient conditions. The sample is under dynamic vacuum when open symbols are used, and the dark symbols represent intervals when N₂ and He are present. N₂ is introduced at B and D; He is introduced at F and H. Vacuum pumping is applied at A, C, E, G, and I. Reprinted with permission from [159] Copyright (2000) American Physical Society.*

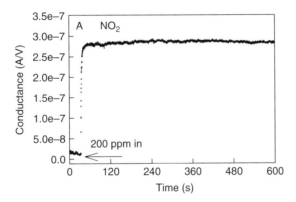

Figure 2.37 *Electrical response of a semiconducting SWCNT to NO₂ molecules. Conductance (under Vg = +4 V, in an initial insulating state) versus time in a 200 ppm NO₂ flow. Reprinted with permission from [53] Copyright (2000) AAAS.*

procedure opened up in the SWCNT wall holes with sufficient width for molecules to enter [160]. After doping with organic donors or acceptors, it was found that the dopant preferentially bound in the inner cavity of the tubes, thus forming hybrid nanotubes (see also Chapter 5). The amount of doping is controlled by the ionization energy (n-doping) or electron affinity (p-doping) of the encapsulated molecule (Figure 2.38). Varied doping amount was checked by optical absorption (Figure 2.38) and Raman (Figure 2.39). Subsequent calculations confirmed the experimental results [161]. In the case of TCNQ,

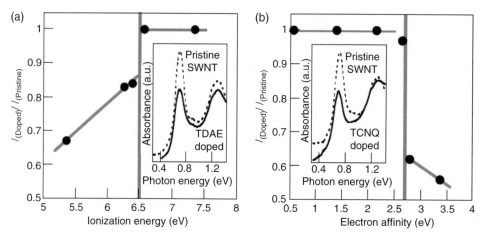

Figure 2.38 *Relationship between the absorption intensity and ionization energy or electron affinity of molecules in Table 2.1. (a) Intensity ratio versus ionization energy (note that the data for TCNQ was out of range.). (b) Intensity ratio versus electron affinity. The* **insets** *display absorption spectra of undoped SWCNT (dashed lines), TDAE-doped SWCNT, and TCNQ-doped SWCNT (solid lines). Adapted with permission from [149] Copyright (2003) Macmillan Publishers Ltd.*

Table 2.1 *Electron affinity and ionization energy for molecules used in Ref. [149]. TDAE=tetrakis(dimethylamino)ethylene, TMTSF=tetramethyltetraselenafulvalene, TCNQ=tetracyanoquinodimethane.*

Molecule	Electron affinity (eV)	Ionization energy (eV)
TDAE		5.36
TMTSF		6.27
Tetrathiafulvalene		6.40
pentacene	1.392	6.58
anthracene	0.56	7.36
3,5-dinitrobenzonitrile	2.16	
C_{60}	2.65	
TCNQ	2.80	9.50
Tetrafluoro-TCNQ	3.38	

XPS measurements also confirmed the charge transfer and showed that TCNQ was in the TNCQ^{-1} state [148] while transistors were actually fabricated and tested [151].

2.3.3 p-Doping of SWCNTs with FeCl$_3$

p-doping of SWCNTs has been performed by FeCl$_3$ and the resulting material analyzed by X-ray diffraction, absorption and Raman spectroscopies and electron energy loss spectroscopy (EELS) showing reversible p-doping.

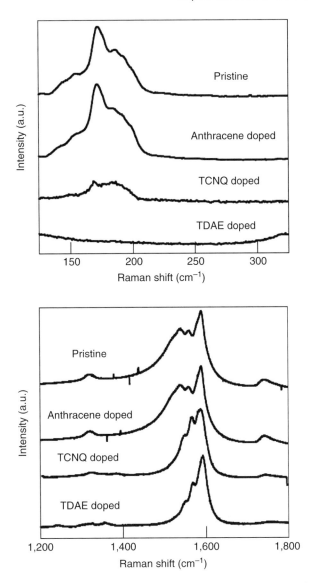

Figure 2.39 *Raman spectra of undoped SWCNT, anthracene-, TCNQ-, and TDAE-doped SWCNTs. (a) Radial-breathing mode of undoped and doped SWCNTs. The excitation wavelength is 632.8 nm. (b) G-band of undoped and doped SWCNTs. Adapted with permission from [149] Copyright (2003) Macmillan Publishers Ltd.*

2.3.3.1 Synthesis

Doping with $FeCl_3$ has been performed on laser grown tubes (mean diameter = 1.3(1) nm) and HiPCO tubes (mean diameter = 1.0(2) nm) [162]. Doping was performed in a two-zone sealed tube under vacuum. Dopant and nanotubes were maintained at sublimation temperature and 574 K respectively [163]. Doping was monitored by Raman spectroscopy

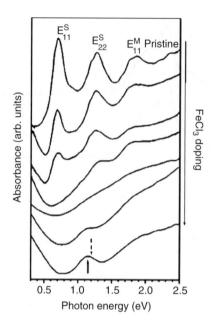

Figure 2.40   *The evolution of the optical absorption spectrum with increasing FeCl₃ doping. From top to bottom the doping level increases. The arrow in the bottom spectrum indicates the doping-induced optical peak. The dashed arrow points out the plasmon energy. Reprinted with permission from [164] Copyright (2004) American Physical Society.*

and could take several hours to reach maximum doping level. Upon annealing to 500°C in vacuum, doping was found to be reversible [164].

2.3.3.2   Characterization

Electron diffraction upon doping showed a loss of the *10* peak bundle signal around 0.4Å^{-1} typical of bundles. This loss was attributed to lattice expansion and induced disorder of the bundles [163]. Raman spectroscopy of the doped material [162,163] showed the usual loss of intensity of the TM and RBM peaks. No significant shift of the TM band was observed. For HiPco tubes, RBM bands had different behavior as a function of doping (see the following). Core level excitation at the C_{1s} excitation-edge showed an additional feature when compared to undoped tube, due to empty states in the valence band. Absorption spectroscopy showed the gradual disappearance of the S11, then S22, then M11 optical transitions, corresponding to depletion of the respective van Hove singularities (or filling in case of n-doping) (Figure 2.40). At high doping level, new, weak bands appear that we already mentioned, at ca. 1.1 eV. Contrarily to Minami et al. [44], and as it has been interpreted for n-doping [40], the authors attributed it to a collective excitation (plasmon) of the holes introduced by p-doping. Whatever its origin, it is worth noting that this band has been observed by a large number of teams, both for n- and p-doping.

It is worth noting that, when FeCl₃ doping was performed on HiPco tubes, the decrease in the RBM mode was observed to happen preferentially for low frequency peaks (i.e., large diameter tubes) [163] and attributed to diameter selective doping. A discussion on this and other related works is presented in forthcoming Section 2.3.8.

2.3.4 p-Doping of SWCNTs with $SOCl_2$

There have been several reports on the reaction of SWCNTs with $SOCl_2$. Although the chemistry actually involved is not clear, there are unambiguous experimental evidences for p-doping such as enhanced conductivity, optical absorption decrease, and doping reversibility.

2.3.4.1 Synthesis

$SOCl_2$ doping of nanotubes was performed on HiPco, laser ablation and electric arc nanotubes [165,166]. Samples were prepared in two ways: buckypaper (obtained by filtering nanotube suspension) was immersed in thionyl chloride ($SOCl_2$), and then dried in air. Alternatively, SWCNTs were stirred in $SOCl_2$ at 45°C for ca. 24h and then used to prepare buckypaper or composites [167]. As pointed out by the authors, $SOCl_2$ is a corrosive liquid that reacts with water and moisture forming SO_2 and HCl. From organic chemistry textbooks, it is also known that $SOCl_2$ will react with alcohols (R-OH) and carboxylic acids (R-COOH) to yield respectively RCl or RCOCl plus SO_2 and HCl. R-OH and RCOOH are often encountered on SWCNTs due to the oxidative purification treatments. In the absence of moisture, when $SOCl_2$ is used as an oxidant, the reaction products are $S+SO_2+Cl^-$. In all these cases, SO_2 is one of the decomposition products of $SOCl_2$. SO_2 has been shown to p-dope nanotubes [157]. Hence, p-doping by SO_2 is a plausible hypothesis (also tentatively made by Detllaff-Weglikowska et al. [165]) behind the p-doping observed when treating SWCNTs with $SOCl_2$.

By placing $SOCl_2$ treated mats in water and monitoring conductivity as a function of water immersion time, it was found that the originally enhanced conductivity goes back to the conductivity of the undoped mats [166].

2.3.4.2 Characterization

The features of doped-nanotubes are observed after treatment with $SOCl_2$: room temperature conductivity is enhanced by a factor of up to 5 while the temperature dependent resistance shows a shallow minimum around 150 K instead of the increasing resistance with decreasing temperature that the authors observed with buckypaper made with undoped-tubes (Figure 2.41).

The doped buckypaper shows enhanced mechanical properties. Elemental analysis (XPS and EDX) show the presence of S, Cl and O, the latter being already present albeit in a lesser amount, in the starting samples [165]. Optical absorption shows the disappearance of the S11 transition, together with a shift in the minimum of absorbance towards higher energy [166]. Rather large downshifts (up to $20 \, cm^{-1}$) are observed for the Raman signals, both RBM and TM [166].

2.3.4.3 Conducting and Transparent SWCNT/Polymer Composites [167]

Thin films of polymethylmetacrylate (PMMA)/doped nanotubes were prepared by mixing solutions of $SOCl_2$-modified SWCNTs with PMMA in $CHCl_3$. The resulting films (ca. 20 micrometer thick) show transmittance of 90% at 0.1% SWCNT mass fraction with a four-probe conductivity of $0.4 \, 10^{-2} \, S \, cm^{-1}$. Additionally, it was observed that even such low doping levels could increase the mechanical properties of the composites [168].

2.3.5 p-Doping of SWCNTs with Acids

From the bulk production [6,169] of SWCNTs onwards, purifying the raw nanotubes was attempted by applying oxidative and/or acid treatments, among others. The question thus arises of the influence of these treatments upon SWCNTs, in particular with respect to

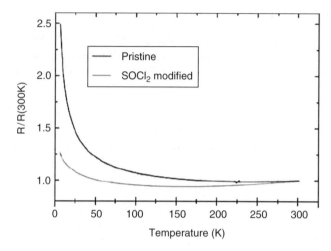

Figure 2.41 *Temperature dependence of the resistance of undoped and SOCl$_2$ modified HiPco buckypaper. Reprinted with permission from [165] Copyright (2005) American Chemical Society.*

p-doping. Indeed, acids can be classified as oxidizing acids, such as H$_2$SO$_4$ or HNO$_3$, and nonoxidizing acids, such as HCl. Oxidizing acids dissociate into H$^+$ (H$_3$O$^+$) and oxidizing anions able to oxidize the substrate. Hence purification might already p-dope the tubes even if not always recognized. Inversely, it has been suggested that protonation rather than oxidation might be occurring when reacting Brønsted acids (proton releasing acids) with carbon nanotubes [8]. The fact is, although these acid-doped materials are pretty well characterized in terms of physical properties, their chemical characterization is far from being unambiguous. This situation might change in the future with carbon nanotube samples getting more homogeneous with time, which in turn, should allow better analysis of the changes occurring during a chemical treatment. Most of the p-doping work has been done with nitric acid HNO$_3$ and sulfuric acid H$_2$SO$_4$, and compared to what had been observed before in GICs.

The amount of p-doping varies considerably with reaction conditions (diluted or concentrated acids, exposure time, sonication, reflux...), hence this section is not separated between synthesis, structure properties, and so on, rather, it attempts to sum up what happens as a function of dopant and doping conditions. Nanotubes used are from electric arc or pulse laser vaporization origin (quite similar diameter distribution D = 1.3 +/– 0.1 nm) or from the HiPCO process (D = 1.0 +/– 0.3 nm [170]). A word of caution is probably necessary about HNO$_3$ doping. Indeed nitric acid is widely used for purification treatment. Although work on p-doping (see following) shows in general good reversibility, HNO$_3$ is known to promote functionalization through the oxidation of the carbon network [171] (see Chapter 3 on functionalized nanotubes).

Electrochemical oxidation of a SWCNT mat has been performed in sulfuric acid with a standard three-electrode configuration and analysed using in situ Raman spectroscopy and mass uptake [172]. Before the application of any electrochemical potential, a spontaneous charge transfer reaction is observed unlike graphite where charge transfer happens only when electrochemically forced [173–175]. This spontaneous p-doping is followed by a reversible

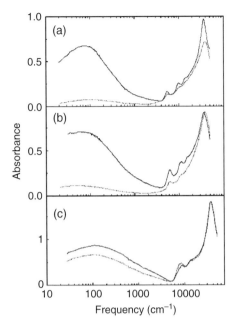

Figure 2.42 *Absorbance spectra of purified (solid line) and as-prepared (dotted line) thin films of the SWCNT material produced by (a) the electric arc technique; (b) the laser ablation technique; (c) the HiPco process. Reprinted with permission [178] Copyright (2002) American Chemical Society.*

doping upon injected charge followed by an irreversible doping for high transferred charge. Dopant was assumed to be a mixture of neutral H_2SO_4 and charged HSO_4^- species.

Upon immersion into nitric acid (70%) for two hours or less, X-ray diffraction showed a displacement of the *10* lattice reflection from 0.39 Å^{-1} to 0.36 Å^{-1}. This shift, attributed to HNO_3 intercalation between SWCNTs within bundles, could be reversed by heating at 500 K in vacuum. Further reaction time lead to irreversible changes including the alteration of SWCNT bundles into carbon onion-like particles [176]. Similarly, it has been shown that exposure to 2 M nitric acid for 48 h produces bleaching of the S11 and S22 interband transitions. This bleaching is reversible upon heating under vacuum or upon laser irradiation (due to local heating) [177]. The amount of bleaching varies between different reports, probably simply reflecting the doping amplitude, itself related to acid concentration, exposure time, experimental conditions, and so on. Transmission studies on a wide energy scale (from 10 to 50 000 cm^{-1}, ca. 1 meV to 6 eV) have shown that, concomitantly with the S11 and S22 bleaching, was a dramatic increase of a wide band centered around 100 cm^{-1} (ca. 12 meV) that the authors attributed to free-carrier intraband transitions near the Fermi level of p-doped semiconducting tubes [178] (Figure 2.42).

Comparative measurements have been performed on SWCNTs treated with HNO_3 and H_2SO_4 (and HCl [178]). Raman spectroscopy [180,181], XPS [179], X-ray diffraction (XRD), conductivity, thermopower and reflectivity [181] have been measured and compared for HNO_3 and H_2SO_4 doping. Raman spectroscopy, as has been seen earlier, can be used to selectively probe resonant tubes by choosing the adequate laser energy. Hence, for 1.3 nm diameter tubes (middle of the diameter distribution for electric arc and pulsed-laser vaporized

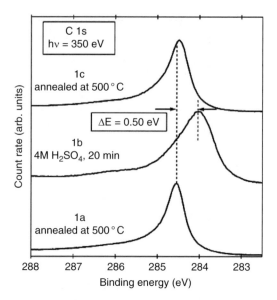

Figure 2.43 *(1a) C 1s core level spectra of the SWCNT buckypaper recorded after an initial annealing step; (1b) after 20 min in 4 M H_2SO_4; (1c) after additional annealing at 500°C. Reprinted with permission from [179] Copyright (2003) PCCP Owner Societies.*

nanotubes), the red laser (632 nm) probes the metallic tubes. Metallic tubes, in the neutral state, show a large profile [17] that used to be associated to a Breit-Wigner-Fano coupling [101,182] but has recently been associated to electron-phonon coupling [102,103]. Upon doping with HNO_3 and H_2SO_4, this large profile diminishes (HNO_3) or disappear (H_2SO_4). The neutral profile can be restored either by local heating, increasing the laser power or by thermal treatment [180,181]. For all kinds of nanotubes and for all physical measurements, the same trend has been observed, that is, the doping effect is more pronounced for H_2SO_4 than for HNO_3 [179–181]. For both acids and chlorhydric acid, XPS study of the doped nanotubes reveals a downshift of the C 1s core level, consistent with p-doping of the tubes and shift of the Fermi energy within the valence band [179]. The measured C1s core level shift is 0.1 eV (HCl), 0.2 eV (HNO_3) and 0.5 eV (H_2SO_4). Only in the case of H_2SO_4 is this shift stable in time (Figure 2.43), whereas with HCl and HNO_3, it diminishes in a matter of days. Peaks corresponding to sulfur, nitrogen and chlorine atoms have also been observed. In the case of H_2SO_4, two sulfur peaks are present, the major component being in the VI+ state, either H_2SO_4 or HSO_4^-, by analogy with graphite intercalation compounds [11,183]. In the case of HCl, two peaks are also observed for Cl, the minor one attributed to chloride anion, the major one attributed to chlorine atoms covalently bound to carbon. This latter fact probably indicates chlorination of the carbon skeleton of the SWCNTs. The picture is more complicated with nitrogen, with some nitrogen already present after the HNO_3 purifying procedure. The possibility of substitutional doping with N that would have occurred during SWCNT synthesis was also mentioned by the authors [179].

Conductivity measurements show that acid p-doping reduces resistivity at all temperatures and leads to weaker (resistive) temperature dependence. Sulfuric acid doping leads to metallic behavior $d\rho/dT > 0$, above 100 K [181] (Figure 2.44).

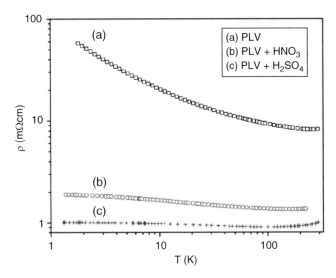

Figure 2.44 *The doping effect on resistivity and its temperature dependence.*
Doping reduces $\rho_{(300K)}$, approximately twice as much for sulfuric than nitric acid.
The temperature dependence becomes weaker after doping, more dramatically for sulfuric
than nitric acid. PLV stands for nanotubes synthetized by the pulse laser vaporization
method. Reprinted with permission from [181] Copyright (2005) American Physical Society.

It is interesting to compare the doping-induced Fermi level downshift on different samples and by different techniques. From thermopower and reflectivity measurements, values from 0.32 to 0.5 eV were obtained for HNO_3 doping both for HiPco and pulsed laser vaporization samples, whereas H_2SO_4 doping lead to downshifts of 0.46 to 0.75 eV [181]. These values compare well with those obtained by XPS on PLV samples: −0.2 and −0.5 eV for HNO_3 and H_2SO_4 respectively [179].

Finally, it is worth noting that, although HNO_3 doping effect has been observed to diminish with time [179], H_2SO_4-doped nanotubes show no change under air exposure [181], unlike the alkali metal-doped (n-doped) nanotubes.

2.3.6   p-Doping of SWCNTs with Superacids

In the first articles on the exposure of SWCNTs to superacids, the term protonation was employed and a mechanism was proposed [184]. However, subsequent characterizations (see the following section on SWCNT fibers) are only understandable if the tubes have been oxidized, that is, p-doped [185]. It should be stressed again that oxidation does not necessarily means the addition of oxygen-containing groups but simply the loss of one or several electrons without any change to the skeleton of a molecule. Although the debate between protonation and p-doping of SWCNTs as the effect of superacids is probably not over as of today, it certainly deserves to be kept as an open question because of the importance of superacid solutions of carbon nanotubes. Whatever the case, it should be pointed out that superacids play a double role in nanotube chemistry, acting both as the agent performing protonation or p-doping, and as the solvent that dissolves the p-doped or protonated tubes. This is sensibly different from the situation for C_{60} where Brønsted acidity and oxidizing capacity have been separated by using the superacid $H(CB_{11}H_6X_6)$ (X = Cl or Br) stoichiometrically as reagent in a solvent [186].

Indeed, it is worth reminding that substances that we know as acids can be classified in many ways, for example, Brønsted and Lewis acids. Brønsted acids are what is generally meant by an acid, that is, a protonating species AH which, upon dissociation, leads to H^+, generally present as H_3O^+ in aqueous media. While the concept of acidity was first defined in water, it can be extended to other solvents, in which AH can dissociate to form the species A^- and a protonated solvent molecule [187]. Superacids [188,189] are nothing more than very powerful acids. In terms of Brønsted acidity, a superacid was defined by Gillespie as 'an acid stronger than 100% sulfuric acid' [190], and superacids based on Lewis acidity have been defined similarly.

Dissolution: SWCNTs were found to be soluble in superacids such as the strictly anhydrous 100% sulfuric acid, or oleum (up to 20% free SO_3), trifluoromethanesulfonic acid and chlorosulfonic acid [184]. Solubilities are significant, probably the highest found for unfunctionalized nanotubes, up to 45 mg ml^{-1} in chlorosulfonic acid. It is noteworthy that in the case of chlorosulfonic acid, no sonication was needed, as was the case for alkali metal salts solutions [61]. Solutions ranged in color from golden yellow to pale brown when diluted, and did not exhibit optical transitions between van Hove singularities of the neutral suspensions but rather a broad signal around 900 nm (Figure 2.45). Disappearance, or quenching, of the van Hove transitions are expected both for p- (or n-) doping and for functionalization. The origin of the broad signal is ascribed by the authors to a charge-transfer process [184].

The phase diagram of these solutions as a function of concentration has been studied, going from diluted to semidiluted to isotropic concentrated phase and finally, above 4%vol to a biphasic system leading to a liquid crystalline phase [192]. Interestingly enough, above 0.01%vol, the nanotubes arrange in a *spaghetti* phase of entangled SWCNTs with no visible ends [184] (Figure 2.46).

Starting from concentrated solutions, different kind of materials have been prepared from superacid SWCNT solutions: (i) liquid crystalline phase; (ii) SWCNT *alewives*; (iii) SWCNT fibers.

Liquid crystalline phase: It is well known from Onsager theory that concentrated solutions of rod-like particles should lead to a nematic phase. The critical concentration for reaching that phase is dependent on the L/D (length over diameter) ratio of the rods and their polydispersity. Concentrated superacid solutions of SWCNTs indeed show nematic order [193] (Figure 2.47).

SWCNT alewives: Superacid SWCNT solutions are highly moisture-sensitive and should be handled in rigorously dry conditions. Further evidence for the aligned nematic phase comes from the fact that elongated particles (ca. 10 μm long) precipitate from this nematic phase when slowly exposed to moisture. They were dubbed *alewives* by the authors, by analogy to the alewife fish [184] (Figure 2.48).

SWCNT fibers: Wet spinning of the concentrated SWCNT superacid solutions in varied coagulating baths, analogous to rod-like polymers and SWCNT surfactant suspensions [194], yields SWCNT fibers (Figure 2.49) [195]. Water and residual acid are removed through annealing at 850°C in a flow of H_2/Ar. Mechanical measurements show a Young's

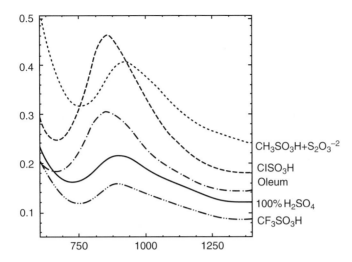

Figure 2.45 *Vis-NIR absorption spectra (in nm) of dilute solutions of SWCNTs in various superacids. Reprinted with permission from [184] Copyright (2004) American Chemical Society.*

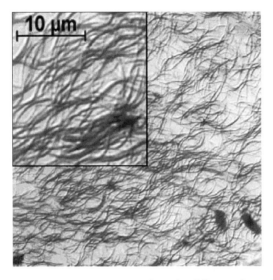

Figure 2.46 *The transmission optical micrograph in the differential interference contrast mode of a 0.25 wt.% solution of SWCNTs in 100% sulfuric acid. Inset shows a magnified view of the same sample. Reprinted with permission from [184] Copyright (2004) American Chemical Society.*

modulus of 120 ± 10 Gpa and a tensile strength of 116 ± 10 MPa [195]. Note that after annealing, the SWCNTs are no longer p-doped but neutral. Extensive physical measurements have been performed on the fibers, before and after annealing [185]. Conductivity measurements show that, at 300 K, annealed fibers are 10 times more resistive than

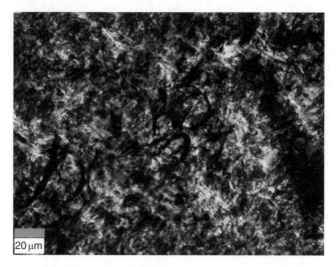

Figure 2.47   *The bottom phase of a centrifugated sample of a SWCNT solution in 123% sulfuric acid seen via crossed-polarizer optical microscopy shows a birefringent polydomain structure. Reprinted with permission from [193] Copyright (2005) American Chemical Society.*

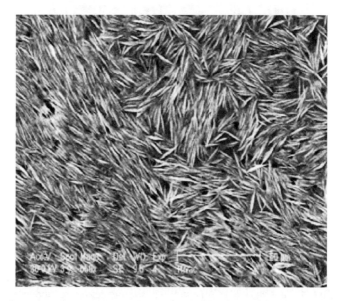

Figure 2.48   *Scanning electron microscopy image of SWCNT alewives, precipitated from a 4 wt.% solution of SWCNTs in oleum (20% SO_3) by slow exposure to moisture. Scale bar is 50 μm. Reprinted with permission from [184] Copyright (2004) American Chemical Society.*

as-extruded fibers [185], that is, a fact that strongly supports the p-doping hypothesis. Four-point resistivity measurements actually show a metallic behavior above 200 K (Figure 2.49). Noteworthy is the fact that this is obtained on air exposed - that is, air stable - as-extruded fibers.

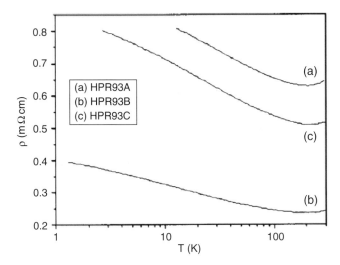

Figure 2.49 *Four-point resistivity versus temperature for three as-extruded fibers. Nonmetallic behavior at low T levels off as T → 0 (nondivergent behavior) while metallic behavior is observed above 200 K. HPR93 designates batch number, while A, B and C denote SWCNT concentrations and orifice diameters of 8 wt.%/500 mm; 6 wt.%/250 mm and 6 wt.%/125 mm respectively. Reprinted with permission from [185] Copyright (2004) American Institute of Physics.*

Structural studies have been performed on annealed fibers, swollen in sulfuric acid [196,197]. H_2SO_4 molecules are found to form a partly ordered structure around the nanotubes. Upon cooling, full crystallization of the surrounding sulfuric acid layer is templated by the SWCNT surface, showing directional interactions between the SWCNT surface and the H_2SO_4 molecules, taken as evidence for protonation [196]. An update has recently appeared on the dissolution of carbon nanotubes into chlorosulfonic acid and the corresponding phase diagram [198].

Superacid SWCNT solutions as reaction intermediates: Superacids can also be successfully used for subsequent functionalization of SWCNTs. Although this specific part may pertain more to Chapter 3 of this book, we include it here for sake of completeness. SWCNTs have been aryl-functionalized by dissolution in oleum and reaction with sodium nitrite in presence of substituted anilines, taking advantage of the exfoliation power of superacids on SWCNT bundles. AFM statistics of the reaction products show mostly exfoliated material [199] (Figure 2.50). The same reaction has been performed in concentrated sulfuric acid (96%) rather than oleum [200] and in a mixture of oleum and nitric acids where the nitric acid is found to cut the SWCNTs yielding functionalized, short, and soluble SWCNTs [189,191].

2.3.7 p-Doping with other Oxidizing Agents

2.3.7.1 Doping with K_2IrCl_6

Zheng and Diner have studied the reaction of DNA-wrapped nanotubes with the inorganic oxidizing agent K_2IrCl_6 [201]. Their starting material was a highly purified sample [126,202], containing a majority of (6,5) tubes. Redox titration of (6,5) enriched CoMoCAT

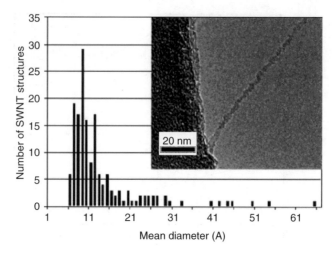

Figure 2.50 *AFM-derived histogram (mica) of tube/bundle mean diameters present in a typical sample of functionalized SWCNTs. Bundles begin to appear at 15 Å. Over 160 structures were sampled. The nanotubes were generally short, exhibiting a mean length of 100 nm (inset). TEM of a functionalized SWCNTs suspended from the lacey carbon TEM grid (bar = 20 nm). Reprinted with permission from [199] Copyright (2004) American Chemical Society.*

SWCNTs [203] with aqueous solutions of K_2IrCl_6 showed full disappearance of the (6,5) absorption bands (Figure 2.51) and allowed a reduction potential of ca. 800 mV versus the standard hydrogen electrode (SHE) to be estimated.

2.3.7.2 Doping with AuCl₃
Another demonstration of the progressive unfilling of the van Hove singularities (Figure 2.52) was made by Kim et al. when doping thin films of arc-discharge SWCNTs with increasing amounts of $AuCl_3$ [204]. The disappearance of the S11 transition is first observed, followed by S22 and finally M11. Concomitantly, a new band appears at 1.12 eV, that we already mentioned and that these authors attributed to an asymmetric transition from deep levels in the valence band to depleted van Hove singularities.

2.3.8 Diameter Selective Doping

There have been several reports on diameter selective doping of carbon nanotubes. Li and Nicholas have p-doped HiPco nanotubes with an oxidant whose oxidizing potential was pH dependent [205]. The experiment was performed on aqueous suspensions of individualized SWCNTs (according to O'Connell et al. procedure [132]) dispersed with SDS with hexacyanoferrate(III) as the oxidant. By varying the pH, they observed a progressive vanishing of the E11 transitions, starting with those of the larger diameter tubes all the way to the smaller diameter tubes (see zone between 1100 and 1300 cm⁻¹ in Figure 2.53). Bandgap energies, that is, the energy separation between the van Hove singularities being inversely proportional to diameter [39], each of the absorption bands in Figure 2.53 corresponds to a group of nanotubes of similar diameter (see Figure 2.31). From Figure 2.53, one infers that only large semiconducting tubes are oxidized at high pH values by

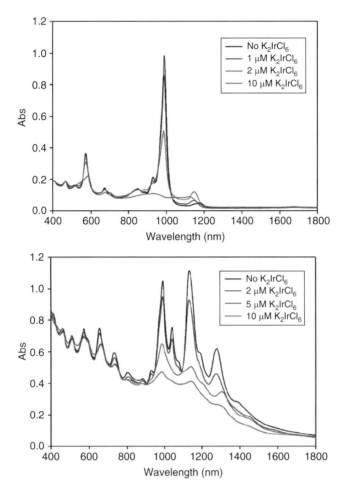

Figure 2.51 *(Top) Redox titration of (6,5) tubes. A (6,5)-enriched CNT solution was centrifugated using a Microcon spin filter YM100 (Millipore) and was resuspended in D_2O. Freshly made K_2IrCl_6 solution in D_2O was added to 100 ml of the CNT solution to the indicated final concentration. The spectrum was recorded after 10 min of incubation. (**Bottom**) Redox titration of unfractionated HiPco tubes under similar experimental conditions. For a better understanding of the figure, please refer to the colour plate section. Reprinted with permission from [201] Copyright (2004) American Chemical Society.*

$Fe(CN)_6^{3-}$. When lowering the pH, the reduction potential of $Fe(CN)_6^{3-}$ becomes more negative and smaller tubes are also oxidized.

The effect of $FeCl_3$ (p-doping) and potassium (n-doping) as dopant on HiPCO tubes have been investigated [163]. At low doping levels, a selective loss of Raman resonance was observed which was attributed to selective charge transfer and further observed by electrochemical doping of the same nanotubes [206]. However this selective loss of resonance was later studied by p-doping SWCNTs with nitric and sulfuric acids with very different conclusions [181]. It is noteworthy that all three papers, while describing very different experimental conditions (n- or p-doping, acid doping, electrochemical doping in

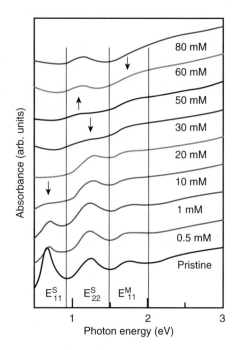

Figure 2.52 *UV-vis-NIR optical absorption spectra with increasing AuCl₃ concentration. The arrows indicate the disappearance of van Hove singularity transitions of p-doped SWCNTs and the appearance of the new peak induced at highly p-type doping. E_{ii}^S/E_{ii}^M indicates the van Hove singularity transitions between the ith energy levels of the semiconducting/metallic SWCNT. Reprinted with permission from [204] Copyright (2008) American Chemical Society.*

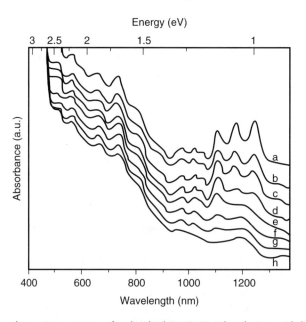

Figure 2.53 *The absorption spectra of individual SWCNT-rich solution with hexacyanoferrate (III) doping at various pH values of (a) 11.9; (b) 11.1; (c) 10.6; (d) 9.5; (e) 8.7; (f) 4.0; (g) 3.3; (h) 2.9. Reprinted with permission from [205] Copyright (2004) Institute of Physics.*

ionic liquids…) report the same observation: Upon doping, the large diameter tubes are first affected, as can be seen by the loss of intensity of their RBM signals, followed by the small diameter SWCNTs while the mid-size ones remain much less affected.

It has now been agreed [106,181] that this apparent diameter selective effect is merely a resonance condition effect: Tubes which are in good resonance conditions show strong intensity changes as a function of doping, because of the loss of resonance associated to doping, whereas tubes only weakly resonating have a far less changing Raman RBM intensity, which could be taken, misleadingly, as low- or no-doping.

2.4 Practical Applications of Doped Nanotubes

Several applications of doped carbon nanotubes have been considered, in particular for different forms of energy storage. The enhanced electrical conductivity observed after p- or n-doping suggests their generic application as synthetic metals, although they suffer the same drawback as GICs, namely sensitivity to oxygen and moisture. Redox doping has been demonstrated as a novel approach to enhance the device characteristics of SWCNT field effect transistors.

Hydrogen offers an alternative sustainable fuel source to replace petroleum. A major obstacle is the achievement of realistic hydrogen storage capacity onboard a fuel cell powered vehicle. Carbon nanotube materials are among the obvious candidates [14]. Theoretically they can meet the capacity requirements, but the van der Waals interaction between essentially sp^2 carbon and molecular hydrogen remains too weak for reversible adsorption/desorption at realistic temperatures and pressures. One approach to solving this limitation borrows from what we know about alkali graphite compounds, namely the partially covalent metal-carbon bonding which leaves unfilled orbitals to hybridize with hydrogen molecular orbitals. This would result in higher enthalpies of adsorption, intermediate between van der Waals and complete covalent bonding. A first attempt to demonstrate enhanced storage in alkali-doped nanotubes, claiming 20 wt.% storage [207], suffered from uncontrolled exposure to moisture [208]. More careful experiments are in progress in several labs around the world.

A more promising embodiment of the same general idea is to decorate the SWCNT surface with light transition metals, for example, titanium, in essence an 'inorganic functionalization'. Here the Ti d-orbital/nanotube π^* orbital hybridization creates a robust 'bond' which can then bind multiple H_2 molecules as the band fills to completion [209]. The same principles apply to fullerenes and perhaps disordered porous carbons with much larger surface area available for Ti attachment and subsequent H_2 binding.

Rechargeable battery technology is another field in which doped nanotubes can play a role. Recent developments in rechargeable lithium-ion battery technology have been dominated by efforts to replace metallic lithium electrodes with lithium–carbon guest–host compounds for improved safety and cyclability [18]. 'Rocking chair' or 'shuttlecock' batteries shuttle Li^+ between two guest–host solids serving as cathode and anode, preventing the reduction of Li^+ to dangerous metallic lithium at any point in the charge–discharge cycle. A major goal is to achieve higher lithium capacities in the carbon anode to partly offset the weight penalty of the carbon.

A prerequisite for rechargeable battery anodes is reversible insertion and extraction of lithium in the carbon host. This has been demonstrated for SWCNT materials by several groups. Early attempts using unpurified material were unsuccessful; they showed large irreversible capacity and little or no reversible capacity. Much better results were obtained with purified material [45], with reversible lithium capacities 25% greater than that of graphite. A team at the University of North Carolina, Chapel Hill, has achieved better success by studying the effects of ball milling after a simpler purification procedure [60]. They were able to produce material with about twice the lithium capacity of graphite. These two results are promising, but the details of the charge–discharge behavior tell us that a lot more needs to be done for these materials to be of practical interest. Furthermore, as in the hydrogen storage problem, there is no hard evidence to convince us that we are in fact exploiting the interior and exterior sites implied by the idealized models. SWCNTs maintain high lithium capacity at high current densities [45], a property that will be important if the lithium-ion battery market is extended to high-power applications such as electric vehicles and load-leveling devices.

The concept of redox doping also applies to individual nanotubes. A charge transfer p-doping scheme has been reported [210] which utilizes one-electron oxidizing molecules to obtain stable, unipolar carbon nanotube transistors. The p-doping improves carrier injection and allows tuning of the threshold voltage. In particular, as-prepared devices which are unipolar are converted to ambipolar, that is, they conduct in the on-state by either electrons or holes depending on the sign of the gate voltage. This allows for greater flexibility in device design, and simplifies the task of assembling different devices into specific logic gates. Additional benefits of these doped 'tubeFETs' are a 2–3 decade increase in saturation current and significant suppression of the off-state current. The mechanism behind all these improvements is believed to involve doping-induced modification of the Schottky barrier heights at source and drain contacts. The same authors studied charge transfer interactions between as-prepared tube-FETs and several amine-containing molecules including hydrazine, polyaniline, and aminobutyl phosphonic acid [210]. Hydrazine converts p-channel to n-channel behavior, while reversible switching between p- and n-channel is achieved by accessing different oxidation.

2.5 Conclusions, Perspectives

As with graphite and then the fullerenes, doping has brought a wealth of new properties to nanotube soot. However, superconductivity, observed for alkali metal doping of both graphite [211] and C_{60} [212] has been scarcely observed with nanotubes for which only proximity-induced superconductivity and intrinsic superconductivity below 1 K have been reported [213]. A good reason for this near absence of superconductivity might well be the inherent disorder and inhomogeneity of nanotube samples. The question remains whether doping of a crystalline, homogeneous (in diameter and chirality), bundle of carbon nanotubes would exhibit superconductivity and at what temperature [214,215]. From a wider point of view, the recent progress on sorting nanotubes [216–218] and controlled synthesis of specific chiralities [219] should bring new or better properties when doping homogeneous nanotube samples. In particular, far-reaching advances can be reasonably expected for devices and sensors, once they are produced with a single chirality of nanotubes. Doping of nanotubes, as fullerenes before, was greatly inspired and aided by the wealth of knowledge on graphite intercalation compounds. With the field of graphene acquiring full speed and somewhat building on the knowledge developed on nanotubes, the loop is now closing onto itself. New forms of ordered carbons will probably appear in the future, schwartzite for example [220] where, again, the knowledge developed on previous forms of carbon, including nanotubes will help to work on these novel materials, including doping them.

References

[1] M.S. Dresselhaus, G. Dresselhaus and P.C. Eklund, *Science of Fullerenes and Carbon Nanotubes*, Academic Press, San Diego (1996).

[2] L. Duclaux, Review of the doping of carbon nanotubes (multiwalled and single-walled), *Carbon*, **40**, 1751–1764 (2002).

[3] J.E. Fischer, Chemical doping of single-wall carbon nanotubes, *Acc. Chem. Res.*, **35**, 1079–1086 (2002).

[4] L. Duclaux, J.-L. Bantignies, L. Alvarez, R. Almairac and J-L. Sauvajol, The structure of doped single wall carbon nanotubes, in *Annual Review of Nanoresearch Volume 1*, G. Cao and C. Jeffrey Brinker (eds.), World Scientific Publishing, Chapter 5, 215–250 (2006).

[5] L. Kavan and L. Dunsch, Spectroelectrochemistry of carbon nanostructures, *Chem. Phys. Chem.*, **8**, 974–998 (2007).

[6] M. Iurlo, D. Paolucci, M. Marcaccio and F. Paolucci, Electron transfer in pristine and function-alised single-walled carbon nanotubes, *Chem. Commun.*, 4867–4874 (2008).

[7] M. Marcaccio, D. Paolucci and F. Paolucci, Molecular devices based on fullerenes and carbon nanotubes, in *Electrochemistry of Functional Supramolecular Systems*, M. Venturi, A. Credi and P. Ceroni Editors, John Wiley & Sons, Inc. (2009).

[8] D. Claves, Hole doping of fullerenes and carbon nanotubes by way of intercalation chemistry, *J. Nanosci. Nanotechnol.*, **7**, 1221–1238 (2007).

[9] IUPAC GoldBook, http://goldbook.iupac.org/I03076.html [accessed 21/11/2008] (2010).

[10] See for example: A. Pénicaud, Building solids with Buckminsterfullerenes, *Fullerene Sci. Technol.*, **6**, 731–741 (1998).

[11] M.S. Dresselhaus, and G. Dresselhaus, Intercalation compounds of graphite, *Adv. Phys.*, **30**, 139–326 (1981).

[12] C.A. Reed and R.D. Bolskar, Discrete fulleride anions and fullerenium cations, *Chem. Rev.*, **100**, 1075–1120 (2000).

[13] H. Reiss, The Fermi level and the redox potential, *J. Phys. Chem.* **89**, 3783 (1985).

[14] J.E. Fischer, Carbon nanotubes: a nanostructured material for energy storage, *Chem. Innov.*, **30**, 21–27 (2000).

[15] D.W. Murphy, M.J. Rosseinsky, R.M. Fleming, R. Tycko, A.P. Ramirez, R.C. Haddon, T. Siegrist, G. Dabbagh, J.C. Tully and R.E. Walstedt, Synthesis and characterization of alkali metal fullerides: A_xC_{60}, *Carbon*, **30**, 151–162 (1993).

[16] R.S. Lee, H.J. Kim, J.E. Fischer, A. Thess and R.E. Smalley, Conductivity enhancement in K- and Br-doped single-wall carbon nanotube bundles, *Nature*, **388**, 255–257 (1997).

[17] A.M. Rao, P.C. Eklund, S. Bandow, A. Thess and R.E. Smalley, Evidence for charge transfer in doped carbon nanotube bundles from Raman scattering, *Nature*, **388**, 257–259 (1997).

[18] J.R. Dahn, T. Zheng, Y. Liu and J.S. Xue, Mechanisms for lithium insertion in carbonaceous materials, *Science*, **272**, 590–593 (1995).

[19] A. Thess, R. Lee, P. Nikolaev, H. Dai, P. Petit, J. Robert, C. Xu, Y.H. Lee, S.G. Kim, A.G. Rinzler, D.T. Colbert, G.E. Scuseria, D. Tománek, J.E. Fischer and R.E. Smalley, Crystalline ropes of metallic carbon nanotubes, *Science*, **273**, 483–487 (1996).

[20] J. Ma, H.-Y. Choi, E.J. Mele and J.E. Fischer, Staging in doped polymers, *Synth. Met.* **27**, A75–A81 (1988).

[21] L.W. Shacklette, N.S. Murthy and R.H. Baughman, New structural phases of polymer battery anode materials – alkali-metal doped polyacetylene and polyphenelene, *Mol. Cryst. Liq. Cryst.*, **121**, 201–209 (1985).

[22] S.A. Safran and D.R. Hamman, Long-range elastic interactions and staging in graphite intercalation compounds, *Phys. Rev. Lett.*, **42**, 1410–1413 (1979).

[23] P.A. Heiney, J.E. Fischer, D. Djurado, J. Ma, D. Chen, M.J. Winokur, N. Coustel, P. Bernier and F.E. Karasz, Channel structures in alkali-doped conjugated polymers: Broken-symmetry intercalation lattices, *Phys. Rev. B*, **42**, 2507–2512 (1991).

[24] J.E. Fischer and H.J. Kim, Elastic effects in intercalation compounds: Comparison of lithium in graphite and TiS_2, *Phys. Rev. B*, **35**, 3295–3299 (1987).

[25] J.R. Dahn, D.C. Dahn and R.R. Haering, Elastic energy and staging in intercalation compounds, *Solid State Commun.*, **42**, 179–183 (1982).

[26] T. Yildirim, O.Z. Zhou and J.E. Fischer, Intercalation compounds of fullerenes II: Structure and superconductivity of alkali metal fullerides, in *The Physics of Fullerene-Based and Fullerene-Related Materials*, W. Andreoni (ed.), Kluwer, Amsterdam (2000).

[27] T. Yildirim, L. Barbedette, J.E. Fischer, C.L. Lin, J. Robert, P. Petit and T.T.M. Palstra, T_c vs carrier concentration in cubic fulleride superconductors, *Phys. Rev. Lett.*, **77**, 167–170 (1996).

[28] D.G. Onn, G.M.T. Foley and J.E. Fischer, Electronic properties, resistive anomalies and phase transitions in the graphite intercalation compounds with K, Rb and Cs, *Phys. Rev. B*, **19**, 6474–6479 (1979).

[29] J.E. Fischer, J.M. Bloch, C.C. Shieh, M.E. Preil and K. Jelley, Reflectivity spectra and dielectric function of stage 1 donor intercalation compounds of graphite, *Phys. Rev. B*, **31**, 4773–4779 (1985).

[30] F.J. DiSalvo and J.E. Fischer, Anisotropic magnetic susceptibility and phase transitions in intercalated graphite: KC_{24}, *Solid State Commun.*, **28**, 71–75 (1978).

[31] B.R. Weinberger, J. Kaufer, A.J. Heeger, J.E. Fischer, M.J. Moran and N.A.W. Holzwarth, Magnetic spin susceptibility of AsF_5-intercalated graphite: Determination of the density of states at the fermi energy, *Phys. Rev. Lett.*, **41**, 1417–1420 (1978).

[32] C.K. Chiang, C.R. Fincher, Y.W. Park, A.J. Heeger, Electrical conductivity in doped polyacetylene, *Phys. Rev. Lett.*, **39**, 1098–1101 (1977).

[33] A.J. Heeger, S. Kivelson, J.R. Schrieffer and W.-P. Su, Solitons in conducting polymers, *Rev. Modern Phys.*, **60**, 781–852 (1988).

[34] O. Gunnarsson, *Alkali-Doped Fullerides: Narrow-Band Solids with Unusual Properties*, World Scientific, Singapore (2004).

[35] G. Gao, T. Çağin and W.A. Goddard III, Position of K atoms in doped single-walled carbon nanotube crystals, *Phys. Rev. Lett.*, **80**, 5556–5559 (1998).

[36] J.C. Charlier, P. Lambin and T.W. Ebbesen, Electronic properties of carbon nanotubes with polygonized cross sections, *Phys. Rev. B*, **54**, R8377–R8380 (1996).

[37] J.E. Fischer, A. Claye and R.S. Lee, Crystal chemistry of nanotube lattices, *Mol. Cryst. Liq. Cryst.*, **340**, 737–742 (2000).

[38] P. Lambin, Electronic structure of carbon nanotubes, *C. R. Physique*, **4**, 1009 (2003).

[39] H. Kataura, Y. Kumazawa, Y. Maniwa, I. Umezu, S. Suzuki, Y. Ohtsuka and Y. Achiba, Optical properties of single-wall carbon nanotubes, *Synth. Metals*, **103**, 2555–2558 (1999).

[40] P. Petit, C. Mathis, C. Journet and P. Bernier, Tuning and monitoring the electronic structure of carbon nanotubes, *Chem. Phys. Lett.*, **305**, 370–374 (1999).

[41] M.S. Dresselhaus and P.C. Eklund, Phonons in carbon nanotubes, *Adv. in Phys.*, **49**, 705 (2000).

[42] J.-L. Sauvajol, E. Anglaret, S. Rols and L. Alvarez, Phonons in single wall carbon nanotube bundles, *Carbon*, **40**, 1697–1714 (2001).

[43] J. Kong and H. Dai, Full and modulated chemical gating of individual carbon nanotubes by organic amine compounds, *J. Phys. Chem. B*, **105**, 2890–2893 (2001).

[44] N. Minami, S. Kazaoui, R. Jaquemin, H. Yamawaki, K. Aoki, H. Kataura and Y. Achiba, Optical properties of semiconducting and metallic single wall carbon nanotubes: effect of doping and high pressure, *Synth. Met.*, **116**, 405–409 (2001).

[45] A.S. Claye, J.E. Fischer, C.B. Huffman, A.G. Rinzler and R.E. Smalley, Solid-state electrochemistry of the Li single wall carbon nanotube system, *J. Electrochem. Soc.*, **147**, 2845–2852 (2000).

[46] L. Kavan, P. Rapte, L. Dunsch, M.J. Bronikowski, P. Willis and R.E. Smalley, Electrochemical tuning of electronic structure of single-walled carbon nanotubes: in-situ Raman and vis-NIR study, *J. Phys.Chem. B*, **105**, 10764–10771 (2001).

[47] P. Corio, P.S. Santos, V.W. Brar, G.G. Samsonidze, S.G. Chou and M.S. Dresselhaus, Potential dependent surface Raman spectroscopy of single wall carbon nanotube films on platinum electrodes, *Chem. Phys. Lett.*, **370**, 675–682 (2003).

[48] A.S. Claye, N.M. Nemes, A. Janossy and J.E. Fischer, Structure and electronic properties of potassium-doped single-wall carbon nanotubes, *Phys. Rev. B*, **62**, R4845–R4848 (2000).

[49] E. Jouguelet, C. Mathis and P. Petit, Controlling the electronic properties of single-wall carbon nanotubes by chemical doping, *Chem. Phys. Lett.*, **318**, 561–564 (2000).

[50] C. Mathis and B. François, Experimental studies of n-doped $(CH)_x$: d.c. electrical conductivity, *Synth. Met.*, **9**, 347–354 (1984).

[51] J. Cambedouzou, S. Rols, N. Bendiab, R. Almairac, J.-L. Sauvajol, P. Petit, C. Mathis, I. Mirebeau and M. Johnson, Tunable intertube spacing in single-walled carbon nanotube bundles, *Phys. Rev. B*, **72**, 041404/1–4 (2005).

[52] M. Swarcz, in *Carbanions, Living Polymers and Electron Transfer Processes*, John Wiley & Sons, Inc., New York (1968).

[53] J. Kong, N.R. Franklin, C. Zhou, M. G. Chapline, S. Peng, K. Cho and H. Dai, Nanotube molecular wires as chemical sensors, *Science*, **287**, 622–625 (2000).

[54] P.G. Collins, K. Bradley, M. Ishigami and A. Zettl, Extreme oxygen sensitivity of electronic properties of carbon nanotubes, *Science*, **287**, 1801–1804 (2000).

[55] C. Bower, S. Suzuki, K. Tanigaki and O. Zhou, Synthesis and structure of pristine and alkali-metal-intercalated single-walled carbon nanotubes, *Appl. Phys. A*, **67**, 47–52 (1998).

[56] X. Liu, T. Pichler, M. Knupfer and J. Fink, Electronic and optical properties of alkali-metal-intercalated single-wall carbon nanotubes, *Phys. Rev. B*, **67**, 125403/1–8 (2003).

[57] T. Pichler, M. Sing, M. Knupfer, M.S. Golden and J. Fink, Potassium intercalated bundles of single-wall carbon nanotubes: electronic structure and optical properties, *Solid State Commun.*, **109**, 721–726 (1999).

[58] L. Duclaux, J.-P. Salvetat, P. Lauginie, T. Cacciaguera, A.M. Faugère, C. Goze-Bac and P. Bernier, Synthesis and characterization of SWNT-heavy alkali metal intercalation compounds, effect of host SWNTs materials, *J. Phys. Chem. Solids*, **64**, 571–581 (2003).

[59] O. Zhou, R.M. Fleming, D.W. Murphy, C.H. Chen, R.C. Haddon, A.P. Ramirez and S.H. Glarum, Defects in carbon nanostructures, *Science*, **263**, 1744–1747 (1994).

[60] B. Gao, A. Kleinhammes, X.P. Tang, C. Bower, L. Fleming, Y. Wu and O. Zhou, Electrochemical intercalation of single-walled carbon nanotubes with lithium, *Chem. Phys. Lett.*, **307**, 153–157 (1999).

[61] A. Pénicaud, P. Poulin, A. Derré, E. Anglaret and P. Petit, Spontaneous dissolution of a single wall carbon nanotube salt, *J. Am. Chem. Soc.*, **127**, 8–9 (2005).

[62] M. Burghard, Electronic and vibrational properties of chemically modified single-wall carbon nanotubes, *Surf. Sci. Rep.*, **58**, 1–109 (2005) and references therein.

[63] S.B. Cronin, R. Barnett, M. Tinkham, S.G. Chou, O. Rabin, M.S. Dresselhaus, A.K. Swam, M.S. Ünlü and B.B. Goldberg, Electrochemical gating of individual single-wall carbon nanotubes observed by electron transport measurements and resonant Raman spectroscopy, *Appl. Phys. Lett.*, **84**, 2052–2054 (2004).

[64] M.J. O'Connell, E.E. Eibergen and S.K. Doorn, Chiral selectivity in the charge-transfer bleaching of single-walled carbon-nanotube spectra, *Nature Mater.*, **4**, 412–418 (2005).

[65] J.-L. Sauvajol, N. Bendiab, E. Anglaret and P. Petit, Phonons in alkali-doped single-wall carbon nanotube bundles, *C. R. Phys.*, **4**, 1035–1045 (2003).

[66] P.M. Allemand, A. Koch, F. Wudl, Y. Rubin, F. Diederich, M.M. Alvarez, S.J. Anz and R.L. Whetten, Two different fullerenes have the same cyclic voltammetry, *J. Am. Chem. Soc.*, **113**, 1050–1051 (1991).

[67] S. Kazaoui, N. Minami, N. Matsuda, H. Kataura and Y. Achiba, Electrochemical tuning of electronic states in single-wall carbon nanotubes studied by *in situ* absorption spectroscopy and ac resistance, *Appl. Phys. Lett.*, **78**, 3433–3435 (2001).

[68] S. Kazaoui, N. Minami, H. Kataura, Y. Achiba, Absorption spectroscopy of single-wall carbon nanotubes: effects of chemical and electrochemical doping, *Synth. Met.*, **121**, 1201–1202 (2001).

[69] Y. Otsuka, Y. Naitoh, T. Matsumoto and T. Kawai, Point-contact current-imaging atomic force microscopy: Measurement of contact resistance between single-walled carbon nanotubes in a bundle, *Appl. Phys. Lett.*, **82**, 1944–1946 (2003).

[70] J. Hone, M.C. Llaguno, A.T. Johnson, J.E. Fischer, D.A. Walters, M.J. Casavant, J. Schmidt and R.E. Smalley, Electrical and thermal transport properties of magnetically aligned single wall carbon nanotube films, *Appl. Phys. Lett.*, **77**, 666–668 (2000).

[71] J.E. Fischer, H. Dai, A. Thess, R. Lee, N.M. Hanjani, D.L. Dehaas, and R.E. Smalley, Metallic resistivity in crystalline ropes of single-wall carbon nanotubes, *Phys. Rev. B*, **55**, R4921–R4924 (1997).

[72] J. Hone, I. Ellwood, M. Muno, A. Mizel, M.L. Cohen and A. Zettl, Thermoelectric power of single-walled carbon nanotubes, *Phys. Rev. Lett.*, **80**, 1042–1045 (1998).

[73] L. Grigorian, G.U. Sumanasekera, A.L. Loper, S.L. Fang, J.L. Allen and P.C. Eklund, Giant thermopower in carbon nanotubes: A one-dimensional Kondo system, *Phys. Rev. B*, **60**, R11309–R11312 (1999).

[74] B. Liu, B. Sundqvist, D. Li and G. Zou, Resistivity and fractal structure in carbon nanotube networks, *J. Phys. Cond. Matter*, **14**, 11125–11129 (2002).

[75] A.B. Kaiser, G. Duesberg and S. Roth, Heterogeneous model for conduction in carbon nanotubes, *Phys. Rev. B*, **57**, 1418–1421 (1998).

[76] M.S. Fuhrer, M.L. Cohen, A. Zettl and V. Crespi, Localization in single-walled carbon nanotubes, *Solid State Commun.*, **109**, 105–109 (1999).

[77] G.T. Kim, S.H. Jhang, J.G. Park, Y.W. Park and S. Roth, Non-ohmic current–voltage characteristics in single-wall carbon nanotube network, *Synth. Met.*, **117**, 123–126 (2001).

[78] P. Petit, E. Jouguelet, J.E. Fischer, A.G. Rinzler and R.E. Smalley, Electron spin resonance and microwave resistivity of single-wall carbon nanotubes, *Phys. Rev. B*, **56**, 9275–9278 (1997).

[79] R.S. Lee, H.J. Kim, J.E. Fischer, J. Lefebvre, M. Radosavljević, J. Hone, and A.T. Johnson, Transport properties of a potassium-doped single-wall carbon nanotube rope, *Phys. Rev. B*, **61**, R4526–R4529 (2000).

[80] L. Grigorian, G.U. Sumansekera, A.L. Loper, S. Fang, J.L. Allen and P.C. Ecklund, Transport properties of alkali-metal-doped single-wall carbon nanotubes, *Phys. Rev. B*, **58**, R4195–R4198 (1998).

[81] S. Kazaoui, N. Minami, R. Jacquemin, H. Kataura and Y. Achiba, Amphoteric doping of single-wall carbon-nanotube thin films as probed by optical absorption spectroscopy, *Phys. Rev. B*, **60**, 13339–13342 (1999).

[82] N. Bendiab, L. Spina, A. Zahab, P. Poncharal, C. Marlière, J.L. Bantignies, E. Anglaret and J.L. Sauvajol, Combined in situ conductivity and Raman studies of rubidium doping of single-wall carbon nanotubes, *Phys. Rev. B*, **63**, 153407/1–4 (2001).

[83] S. Suzuki, C. Bower and O. Zhou, In-situ TEM and EELS studies of alkali–metal intercalation with single-walled carbon nanotubes, *Chem. Phys. Lett.*, **285**, 230–234 (1998).

[84] F. Leonard and J. Tersoff, Novel length scales in nanotube devices, *Phys. Rev. Lett.*, **83**, 5174–5177 (1999).

[85] P. Petit, unpublished.

[86] M. Bockrath, D.H. Cobden, J. Lu, A.G. Rinzler, R.E. Smalley, L. Balents and P.L. McEuen, Luttinger-liquid behaviour in carbon nanotubes, *Nature*, **397**, 598–601 (1999).

[87] H. Rauf, T. Pichler, M. Knupfer, J. Fink, and H. Kataura, Transition from a Tomonaga-Luttinger liquid to a Fermi liquid in potassium-intercalated bundles of single-wall carbon nanotubes, *Phys. Rev. Lett.*, **93**, 096805/1–4 (2004).

[88] S. Bandow, M. Yudasaka, R. Yamada, S. Iijma, F. Kokai and K. Takahashi, Electron spin resonance of K-doped single-wall carbon nanohorns and single-wall carbon nanotubes, *Mol. Cryst. Liq. Cryst.*, **340**, 749–756 (2000).

[89] G. Feher and A.F. Kip, Electron spin resonance absorption in metals. I. Experimental, *Phys. Rev.*, **98**, 337–348 (1955).

[90] F.J. Dyson, Electron spin resonance absorption in metals. II. Theory of electron diffusion and the skin effect, *Phys. Rev.*, **98**, 349–359 (1955).

[91] R.J. Elliott, Theory of the effect of spin-orbit coupling on magnetic resonance in some semiconductors, *Phys. Rev.*, **96**, 266–279 (1954).

[92] K.A. Müller and R. Kleiner, Conduction carrier spin resonance in the alkalimetal-graphites C_8Me and $C_{24}Me$, *Phys. Lett.*, **1**, 98–100 (1962).

[93] S.K. Khanna, E.R. Falardeau, A.J. Heeger and J.E. Fischer, Conduction electron spin resonance in acceptor-type graphite intercalation compounds, *Sol. State Comm.*, **25**, 1059–1065 (1978).

[94] P. Petit, J. Robert, T. Yildirim and J.E. Fischer, ESR evidence for phonon-mediated resistivity in alkali-metal-doped fullerides, *Phys. Rev. B*, **54**, R3764–R3767 (1996).

[95] F. Rachdi and P. Bernier, ESR study of the temperature dependence of metallic complexes of alkali-metal-doped polyacetylene, *Phys. Rev. B*, **33**, 7817–7819 (1986).

[96] A. Maarouf, C.L. Kane and E.J. Mele, Electronic structure of carbon nanotube ropes, *Phys. Rev. B*, **61**, 11156–11165 (2000).

[97] S. Rols, A. Righi, L. Alvarez, E. Anglaret, R. Almairac, C. Journet, P. Bernier, J.-L. Sauvajol, A.M. Benito, W.K. Maser, E. Muñoz, M.T. Martinez, G.F. de la Fuente, A. Girard and J.-C. Ameline, Diameter distribution of single wall carbon nanotubes in nanobundles, *Eur. Phys. J. B*, **18**, 201–205 (2000).

[98] L. Henrard, V.N. Popov and A. Rubio, Influence of packing on the vibrational properties of infinite and finite bundles of carbon nanotubes, *Phys. Rev. B*, **64**, 205403/1–10 (2001).

[99] M.A. Pimenta, A. Marucci, S.A. Empedocles, M.G. Bawendi, E.B. Hanlon, A.M. Rao, P.C. Eklund, R.E. Smalley, G. Dresselhaus and M.S. Dresselhaus, Raman modes of metallic carbon nanotubes, *Phys. Rev. B*, **58**, R16016–R16019 (1998).

[100] L. Alvarez, A. Righi, T. Guillard, S. Rols, E. Anglaret, D. Laplaze and J.-L. Sauvajol, Resonant Raman study of the structure and electronic properties of single-wall carbon nanotubes, *Chem. Phys. Lett.*, **316**, 186–190 (2000).

[101] S.D.M. Brown, A. Jorio, P. Corio, M.S. Dresselhaus, G. Dresselhaus, R. Saito and K. Kneipp, Origin of the Breit-Wigner-Fano lineshape of the tangential G-band feature of metallic carbon nanotubes, *Phys. Rev. B*, **63**, 155414/1–8 (2001).

[102] M. Oron-Carl, F. Hennrich, M.M. Kappes, H.V. Lohneysen and R. Krupke, On the electron-phonon coupling of individual single-walled carbon nanotubes, *Nano Lett.*, **5**, 1761–1767 (2005).

[103] M. Lazzeri, S. Piscanec, F. Mauri, A.C. Ferrari, J. Robertson, Phonon linewidths and electron–phonon coupling in graphite and nanotubes, *Phys. Rev. B*, **73**, 155426 (2006).

[104] S. Piscanec, M. Lazzeri, J. Robertson, A.C. Ferrari, F. Mauri, Optical phonons in carbon nanotubes: Kohn anomalies, Peierls distortions, and dynamic effects, *Phys. Rev. B*, **75**, 035427/1–22 (2007).

[105] N. Bendiab, E. Anglaret, J.-L. Bantignies, A. Zahab, J.-L. Sauvajol, P. Petit, C. Mathis and S. Lefrant, Stoichiometry dependence of the Raman spectrum of alkali-doped single-wall carbon nanotubes, *Phys. Rev. B*, **64**, 245424/1–6 (2001).

[106] L. Kavan, M. Kalbac, M. Zukalova and L. Dunsch, Electrochemical doping of chirality-resolved carbon nanotubes, *J. Phys. Chem. B*, **109**, 19613–19619 (2005).

[107] N. Bendiab, A. Righi, E. Anglaret, J.-L. Sauvajol, L. Duclaux and F. Beguin, Low-frequency Raman modes in Cs- and Rb-doped single wall carbon nanotubes, *Chem. Phys. Lett.*, **339**, 305–310 (2000).

[108] J.M. Tarascon, F.J. Di Salvo, C.H. Chen, P.J. Carroll, M. Walsh and L.J. Rupp, First example of monodispersed $(Mo_3Se_3)^-_\infty$ clusters, *J. Solid State Chem.*, **58**, 290–300 (1985).

[109] E. Anglaret, F. Dragin, R. Martel and A. Pénicaud, Raman studies of solutions of single-wall carbon nanotube salts, *J. Phys. Chem. B*, **110**, 3949–3954 (2006).

[110] A. Pénicaud, F. Dragin, E. Anglaret and R. Almairac, unpublished data.

[111] O. Roubeau, A. Lucas, A. Pénicaud and A. Derré, Covalent functionalization of carbon nanotubes through organometallic reduction and electrophilic attack, *J. Nanosc. Nanotechnol.*, **7**, 1–5 (2007).

[112] D. Voiry, O. Roubeau and A. Pénicaud, Stoichiometric control of single walled carbon nanotubes functionalization, *J. Mat. Chem.*, **20**, 4385–4391 (2010).

[113] D. Voiry, C. Vallés, O. Roubeau and A. Pénicaud, Dissolution and alkylation of industrially produced multi-walled carbon nanotubes, *Carbon*, **49**, 170–175 (2010).

[114] D. Paolucci, M. Melle Franco, M. Iurlo, M. Marcaccio, M. Prato, F. Zerbetto, A. Pénicaud and F. Paolucci, Singling out the electrochemistry of individual single-walled carbon nanotubes in solution, *J. Am. Chem. Soc.*, **130**, 7393–7399 (2008).

[115] J.E. Fischer, H. Dai, A. Thess, R. Lee, N.M. Hanjani, D.L. Dehaas and R.E. Smalley, Metallic resistivity in crystalline ropes of single-wall carbon nanotubes, *Phys. Rev. B*, **55**, R4921–R4924 (1997).

[116] J. Chen, M.A. Hamon, H. Hu, Y. Chen, A.M. Rao, P.C. Eklund and R.C. Haddon, Solution properties of single-walled carbon nanotubes, *Science*, **282**, 95–98 (1998).

[117] H. Kataura, Y. Kumazawa, N. Kojima, Y. Maniwa, I. Umezu, S. Masabuchi, S. Kazama, Y. Ohtsuka, S. Suzuki and Y. Achiba, Resonance Raman scattering of Br_2 doped single-walled carbon nanotube bundles, *Mol. Cryst. Liq. Cryst.*, **340**, 757–762 (2000).

[118] R. Jacquemin, S. Kazaoui, D. Yu, A. Hassanien, N. Minami, H. Kataura and Y. Achiba, Doping mechanism in single-wall carbon nanotubes studied by optical absorption, *Synth. Met.*, **115**, 283–287 (2000).

[119] B. Liu, Q. Cui, M. Yu, G. Zou, J. Carlsten, T. Wagberg and B. Sundqvist, Raman study of bromine-doped single-walled carbon nanotubes under high pressure, *J. Phys. Condens. Matter*, **14**, 11255–11259 (2002).

[120] S.-H. Jhi, S.G. Louie and M.L. Cohen, Electronic properties of bromine-doped carbon nanotubes, *Solid State Commun.*, **123**, 495–499 (2002).

[121] N. Park, D. Sung, S. Hong, D. Kang and W. Park, Metallization of the semiconducting carbon nanotube by encapsulated bromine molecules, *Physica E*, **29**, 693–697 (2005).

[122] D. Sung, N. Park, W. Park and S. Hong, Formation of polybromine anions and concurrent heavy hole doping in carbon nanotubes, *Appl. Phys. Lett.*, **90**, 093502/1–3 (2007).

[123] S. Matsuzaki, T. Kyoda, T. Ando, M. Sano, Raman spectra of graphite-bromine intercalation compounds at high pressures, *Sol. State Comm.*, **67**, 505–507 (1998).

[124] Y. Mackeyev, S. Bachilo, K.B. Hartman, L.J. Wilson, The purification of HiPco SWCNTs with liquid bromine at room temperature, *Carbon*, **45**, 1013–1017 (2007).

[125] D. Chattopadhyay, I. Galeska, F. Papadimitrakopoulos, A route for bulk separation of semiconducting from metallic single-wall carbon nanotubes, *J. Am. Chem. Soc.*, **125**, 3370–3375 (2003).

[126] M. Zheng, A. Jagota, M.S. Strano, A.P. Santos, P. Barone, S.G. Chou, B.A. Diner, M.S. Dresselhaus, R.S. McLean, G.B. Onoa, G.G. Samsonidze, E.D. Semke, M. Usrey and D.J. Walls, Structure-based carbon nanotube sorting by sequence-dependent DNA assembly, *Science*, **302**, 1545–1548 (2003).

[127] R. Krupke, F. Hennrich, H.V. Löhneysen, M.M. Kappes, Separation of metallic from semiconducting single-walled carbon nanotubes, *Science*, **301**, 344–347 (2003).

[128] Z. Chen, X. Du, M.-H. Du, C.D. Rancken, H.-P. Cheng and A.G. Rinzler, Bulk separative enrichment in metallic or semiconducting single wall carbon nanotubes, *Nanolett.*, **3**, 1245–1249 (2003).

[129] M.S. Arnold, A.A. Green, J.F. Hulvat, S.I. Stupp and M.C. Hersam, Sorting carbon nanotubes by electronic structure using density differentiation, *Nature Nanotechnol.*, **1**, 60–65 (2006).

[130] Z. Chen, Z. Wu, J. Sippel and A.G. Rinzler, Metallic/semiconducting nanotube separation and ultra-thin, transparent nanotube films, in *XVII International Winterschool/Euroconference on Electronic Properties of Novel Materials*. AIP Conference Proceedings, **723**, 69–74 (2004).

[131] M.J. O'Connell, S.M. Bachilo, C.B. Huffman, V.C. Moore, M.S. Strano, E.H. Haroz, K.L. Rialon, P.J. Boul, W.H. Noon, C. Kittrell, J. Ma, R.H. Hauge, R.B. Weisman and R.E. Smalley, Band gap fluorescence from individual single-walled carbon nanotubes, *Science*, **297**, 593–596 (2002).

[132] F.A. Cotton and G. Wilkinson, *Advanced Inorganic Chemistry*, John Wiley & Sons, Inc. New York, 577 (1988).

[133] L. Grigorian, K.A. Williams, S. Fang, G.U. Sumanasekera, A.L. Loper, E.C. Dickey, S.J. Pennycook and P.C. Eklund, Reversible intercalation of charged iodine chains into carbon nanotube ropes, *Phys. Rev. Lett.*, **80**, 5560–5563 (1998).

[134] X. Fan, E.C. Dickey, P.C. Eklund, K.A. Williams, L. Grigorian, R. Buczko, S.T. Pantelides, S.J. Pennycook, Atomic arrangement of iodine atoms inside single-walled carbon nanotubes, *Phys. Rev. Lett.*, **84**, 4621–4624 (2000).

[135] J. Liu, A.G. Rinzler, H. Dai, J.H. Hafner, R.K. Bradley, P.J. Boul, A. Lu, T. Iverson, K. Shelimov, C.B. Huffman, F. Rodriguez-Macias, Y.-S. Shon, T.R. Lee, D.T. Colbert and R.E. Smalley, Fullerene pipes, *Science*, **280**, 1253–1256 (1998).

[136] J. Sloan, J. Hammer, M. Zwiefka-Sibley and M.L.H. Green, The opening and filling of single walled carbon nanotubes (SWTs), *Chem. Commun.*, 347–348 (1998).

[137] S.A. Jenekhe, S.T. Wellinghoff and J.F. Reed, Synthesis of highly conducting heterocyclic polycarbazoles by simultaneous polymerization and doping in liquid iodine, *Mol. Cryst. Liq. Cryst.*, **105**, 175–189 (1984).

[138] S.J. Pennycook and D.E. Jesson, Atomic resolution Z-contrast imaging of interfaces, *Acta Metallurg. Mater.*, **40**, S149–S159 (1992); D.E. Jesson and S.J. Pennycook, *Proc. R. Soc. London A*, **449**, 273 (1995).

[139] L. Guan, K. Suenaga, Z. Shi, Z. Gu and S. Iijima, Polymorphic structures of iodine and their phase transition in confined nanospace, *Nano Lett.*, **7**, 1532–1535 (2007).

[140] N. Bendiab, R. Almairac, S. Rols, R. Aznar, J.-L. Sauvajol and I. Mirebeau, Structural determination of iodine localization in single-walled carbon nanotube bundles by diffraction methods, *Phys. Rev. B*, **69**, 195415/1–8 (2004).

[141] J. Cambedouzou, V. Pichot, S. Rols, R. Klement, P. Launois, P. Petit, H. Kataura and R. Almairac, On the diffraction pattern of C_{60} peapods, *Eur. Phys. J. B*, **42**, 31–45 (2004).

[142] T. Michel, L. Alvarez, J.-L. Sauvajol, R. Almairac, R. Aznar, J.-L. Bantignies, and O. Mathon, EXAFS investigations of iodine-doped carbon nanotubes, *Phys. Rev. B*, **73**, 195419/1–5 (2006).

[143] Y. Jung, S.-J. Hwang and S.-J. Kim, Spectroscopic evidence on weak electron transfer from intercalated iodine molecules to single-walled carbon nanotubes, *J. Phys. Chem. C*, **111**, 10181–10184 (2007).

[144] K.R. Kissel, K.B. Hartman, P.A.W. Van der Heide and L.J. Wilson, Preparation of I_2@ SWCNTs: Synthesis and spectroscopic characterization of I_2-loaded SWCNTs, *J. Phys. Chem. B*, **110**, 17425–17429 (2006).

[145] H. Farhat, H. Son, G.G. Samsonidze, S. Reich, M.S. Dresselhaus and J. Kong, Phonon softening in individual metallic carbon nanotubes due to the Kohn anomaly, *Phys. Rev. Lett.*, **99**, 145506/1–4 (2007).

[146] J. Kong, N.R. Franklin, C. Zhou, M.G. Chapline, S. Peng, K. Cho and H. Dai, Nanotube molecular wires as chemical sensors, *Science*, **287**, 622–625 (2000).

[147] S. Kazaoui, Y. Guo, W. Zhu, Y. Kim and N. Minami, Optical absorption spectroscopy of single-wall carbon nanotubes doped with a TCNQ derivative, *Synth. Met.*, **135–136**, 753–754 (2003).

[148] M. Shiraishi, S. Swaraj, T. Takenobu, Y. Iwasa, M. Ata, W.E.S. Unger, Spectroscopic characterization of single-walled carbon nanotubes carrier-doped by encapsulation of TCNQ, *Phys. Rev. B*, **71**, 125419 (2005).

[149] T. Takenobu, T. Takano, M. Shiraishi, Y. Murakami, M. Ata, H. Kataura, Y. Achiba, Y. Iwasa, Stable and controlled amphoteric doping by encapsulation of organic molecules inside carbon nanotubes, *Nat. Mater.*, **2**, 683–688 (2003).

[150] T. Takenobu, T. Kanbara, N. Akima, T. Takahashi, M. Shiraishi, K. Tsukagoshi, H. Kataura, Y. Aoyagi and Y. Iwasa, Control of carrier density by a solution method in carbon nanotube devices, *Adv. Mat.*, **17**, 2430–2434 (2005).

[151] M. Shiraishi, S. Nakamura, T. Fukao, T. Takenobu, H. Kataura and Y. Iwasa, Control of injected carriers in tetracyano-p-quinodimethane encapsulated carbon nanotube transistors, *Appl. Phys. Lett.*, **87**, 093107/1–3 (2005).

[152] C. Zhou, J. Kong, E. Yenilmez and H. Dai, Modulated chemical doping of individual carbon nanotubes, *Science*, **290**, 1552–1555 (2000).

[153] S. Tans, A. Verschueren and C. Dekker, Room-temperature transistor based on a single carbon nanotube, *Nature*, **393**, 49–52 (1998).

[154] R. Martel, T. Schmidt, H.R. Shea, T. Hertel and P. Avouris, Single- and multi-wall carbon nanotube field-effect transistors, *Appl. Phys. Lett.*, **73**, 2447/1–3 (1998).

[155] C. Zhou, J. Kong and H. Dai, Electrical measurements of individual semiconducting single-walled carbon nanotubes of various diameters, *Appl. Phys. Lett.*, **76**, 1597–1599 (1999).

[156] S.-H. Jhi, S.G. Louie, M.L. Cohen, Electronic properties of oxidized carbon nanotubes, *Phys. Rev. Lett.*, **85**, 1710–1713 (2000).

[157] A. Goldoni, R. Larciprete, L. Petaccia and S. Lizzit, Single-wall carbon nanotube interaction with gases: Sample contaminants and environmental monitoring, *J. Am. Chem. Soc.*, **125**, 11329–11333 (2003).

[158] K. Bradley, S.-H. Jhi, P.G. Collins, J. Hone, M.L. Cohen, S.G. Louie and A. Zettl, Is the intrinsic thermoelectric power of carbon nanotubes positive? *Phys. Rev. Lett.*, **85**, 4361–4364 (2000).

[159] G.U. Sumanasekera, C.K.W. Adu, S. Fang and P.C. Eklund, Effects of gas adsorption and collisions on electrical transport in single-walled carbon nanotubes, *Phys. Rev. Lett.*, **85**, 1096–1099 (2000).

[160] Y. Maniwa, Y. Kumazawa, Y. Saito, H. Tou, H. Kataura, H. Ishii, S. Suzuki, Y. Achiba, A. Fujiwara and H. Suematsu, Anomaly of X-ray diffraction profile in single-walled carbon nanotubes, *Jpn. J. Appl. Phys.*, **38**, L668–L670 (1999).

[161] J. Lu, S. Nagase, D. Yu, H. Ye, R. Han, Z. Gao, S. Zhang and L. Peng, Amphoteric and controllable doping of carbon nanotubes by encapsulation of organic and organometallic molecules, *Phys. Rev. Lett.*, **93**, 116804/1–4 (2004).

[162] A. Kukovecz, T. Pichler, R. Pfeiffer, C. Kramberger and H. Kuzmany, Diameter selective doping of single wall carbon nanotubes, *Phys. Chem. Chem. Phys.*, **5**, 582–587 (2003).

[163] A. Kukovecz, T. Pichler, R. Pfeiffer and H. Kuzmany, Diameter selective charge transfer in p-and n-doped single wall carbon nanotubes synthesized by the HiPCO method, *Chem. Commun.*, 1730–1731 (2002).

[164] X. Liu, T. Pichler, M. Knupfer, J. Fink and H. Kataura, Electronic properties of $FeCl_3$-intercalated single-wall carbon nanotubes, *Phys. Rev. B*, **70**, 205405/1–5 (2004).

[165] U. Dettlaff-Weglikowska, V. Skalalova, R. Graupner, S.H. Jhang, B.H. Kim, H. J. Lee, L. Ley, Y.W. Park, S. Berber, D. Tomanek and S. Roth, Effect of $SOCl_2$ treatment on electrical and mechanical properties of single-wall carbon nanotube networks, *J. Am. Chem. Soc.*, **127**, 5125–5131 (2005).

[166] V. Skalalova, A.B. Kaiser, U. Dettlaff-Weglikowska, K. Hrncarikova and S. Roth, Effect of chemical treatment on electrical conductivity, infrared absorption, and Raman spectra of single-walled carbon nanotubes, *J. Phys. Chem. B*, **109**, 7174–7181 (2005).

[167] U. Dettlaff-Weglikowska, M. Kaempgen, B. Hornbostel, V. Skalalova, J. Wang, J. Liang and S. Roth, Conducting and transparent SWCNT/polymer composites, *Phys. Stat. Sol. (b)*, **243**, 3440–3444 (2006).

[168] V. Skalalova, U. Dettlaff-Weglikowska, S.Roth, Electrical and mechanical properties of nanocomposites of single wall carbon nanotubes with PMMA, *Synth. Met.*, **152**, 349–352 (2005).

[169] C. Journet, W.K. Maser, P. Bernier, A. Loiseau, M. Lamy de la Chapelle, S. Lefrant, P. Deniard, R. Lee and J.E. Fischer, Large-scale production of single-walled carbon nanotubes by the electric-arc technique, *Nature*, **388**, 756–758 (1997).

[170] I.W. Chiang, B.E. Brinson, A.Y. Huang, P.A. Willis, M.J. Bronikowski, J.L. Margrave, R.E. Smalley and R.H. Hauge, Purification and characterization of single-wall carbon nanotubes (SWCNTs) obtained from the gas-phase decomposition of CO (HiPco Process), *J. Phys. Chem. B*, **105**, 8297–8301 (2001).

[171] M.N. Tchoul, W.T. Ford, G. Lolli, D.E. Resasco and S. Arepalli, Effect of mild nitric acid oxidation on dispersability, size, and structure of single-walled carbon nanotubes, *Chem. Mater.*, **19**, 5765–5772 (2007).

[172] G.U. Sumanasekera, J.L. Allen, S.L. Fang, A.L. Loper, A.M. Rao and P.C. Eklund, Electrochemical oxidation of single wall carbon nanotube bundles in sulfuric acid, *J. Phys. Chem. B*, **103**, 4292–4297 (1999).

[173] W. Rüdorff and U. Hoffmann, Über Graphitsalze, *Z. Anorg. Allg. Chem.*, **238**, 1–50 (1938).

[174] W. Rüdorff, Kristallstruktur der Säureverbindungen des graphits, *Z. Phys. Chem. B*, **45**, 42–68 (1940).

[175] A. Metrot and J.E. Fischer, Charge transfer reactions during anodic oxidation of graphite in H_2SO_4, *Synth. Met.*, **3**, 201–207 (1981).

[176] C. Bower, A. Kleinhammes, Y. Wu and O. Zhou, Intercalation and partial exfoliation of single-walled carbon nanotubes by nitric acid, *Chem. Phys. Lett.*, **288**, 481–486 (1998).

[177] F. Hennrich, R. Wellmann, S. Malik, S. Lebedkin and M.M. Kappes, Reversible modification of the absorption properties of single-walled carbon nanotube thin films via nitric acid exposure, *Phys. Chem. Chem. Phys.*, **5**, 178–183 (2003).

[178] M.E. Itkis, S. Niyogi, M.E. Meng, M.A. Hamon, H. Hu and R.C. Haddon, Spectroscopic study of the Fermi level electronic structure of single-walled carbon nanotubes, *Nano Lett.*, **2**, 155–159 (2002).

[179] R. Graupner, J. Abraham, A. Vencelova, T. Seyller, F. Hennrich, M.M. Kappes, A. Hirsch and L. Ley, Doping of single-walled carbon nanotube bundles by Brønsted acids, *Phys. Chem. Chem. Phys.*, **5**, 5472–5476 (2003).

[180] Z. Yu and L.E. Brus, Reversible oxidation effect in Raman scattering from metallic single-wall carbon nanotubes, *J. Phys. Chem. A*, **104**, 10995–10999 (2000).

[181] W. Zhou, J. Vavro, N.M. Nemes, J.E. Fischer, F. Borondics, K. Kamarás and D.B. Tanner, Charge transfer and Fermi level shift in p-doped single-walled carbon nanotubes, *Phys. Rev. B*, **71**, 205423–7 (2005).

[182] K. Kempa, Gapless plasmons in carbon nanotubes and their interactions with phonons, *Phys. Rev. B*, **66**, 195406/1–5 (2002).

[183] W.R. Salaneck, C.F. Brucker, J.E. Fischer and A. Metrot, X-ray electron spectroscopy of graphite intercalated with H_2SO_4, *Phys. Rev. B*, **24**, 5037–5046 (1981).

[184] S. Ramesh, L.M. Ericson, V.A. Davis, R.K. Saini, C. Kittrell, M. Pasquali, W.E. Billups, W. Wade Adams, R.H. Hauge and R.E. Smalley, Dissolution of pristine single walled carbon nanotubes in superacids by direct protonation, *J. Phys. Chem. B*, **108**, 8794–8798 (2004).

[185] W. Zhou, J. Vavro, C. Guthy, K.I. Winey, J.E. Fischer, L.M. Ericson, S. Ramesh, R. Saini, V.A. Davis, C. Kittrell, M. Pasquali, R.H. Hauge and R.E. Smalley, Single-walled carbon nanotube fibers extruded from super-acid suspensions: preferred orientation, electrical and thermal transport, *J. Appl. Phys.*, **95**, 649–655 (2004).

[186] C.A. Reed, K.-C. Kim, R.D. Bolskar and L.J. Mueller, Taming superacids: Stabilization of the fullerene cations HC_{60}^+ and C_{60}^+, *Science*, **289**, 101–104 (2000).

[187] G.A. Olah, G.K.S. Prakash and J. Sommer, *Superacids*, John Wiley & Sons, Inc., New-York (1985).

[188] N.F. Hall and J.B. Conant, A study of superacids solutions. I. The use of the chloranil electrode in glacial acetic acid and the strength of certain weak bases, *J. Am. Chem. Soc.*, **49**, 3047–3061 (1927).

[189] It is worth quoting here James Tour's warning [191] regarding experimental conditions.

CAUTION: Oleum and 70% nitric acid are extremely hazardous materials that should be handled with utmost care. Users should wear an acid-proof apron, heavy gloves, safety glasses, and a face shield, and all operations shouls be done in a fume hood when handling these reagents.

[190] R.J. Gillespie, Fluorosulfuric acid and related superacid media, *Acc. Chem. Res.*, **1**, 202–209 (1968).

[191] Z. Chen, K. Kobashi, U. Rauwald, R. Booker, H. Fan, W.F. Hwang and J.M. Tour, Soluble ultra-short single-walled carbon nanotubes, *J. Am. Chem. Soc.*, **128**, 10568–10571 (2006).

[192] V.A. Davis, L.M. Ericson, A.N.G. Parra-Vasquez, H. Fan, Y. Wang, V. Prieto, J.A. Longoria, S. Ramesh, R.K. Saini, C. Kittrell, W.E. Billups, W.W. Adams, R.H. Hauge, R.E. Smalley and M. Pasquali, Phase behaviour and rheology of SWCNTs in superacids, *Macromolecules*, **37**, 154–160 (2004).

[193] P.K. Rai, R.A. Pinnick, A.N.G. Parra-Vasquez, V.A. Davis, H.K. Schmidt, R.H. Hauge, R.E. Smalley and M. Pasquali, Isotropic-nematic phase transition of single-walled carbon nanotubes in strong acids, *J. Am. Chem. Soc.*, **128**, 591–595 (2005).

[194] B. Vigolo, A. Penicaud, C. Coulon, C. Sauder, R. Pailler, C. Journet, P. Bernier and P. Poulin, Macroscopic fibers and ribbons of oriented carbon nanotubes, *Science*, **290,** 1331–1334 (2000).

[195] L.M. Ericson, H. Fan, H. Peng, V.A. Davis, W. Zhou, J. Sulpizio, Y. Wang, R. Booker, J. Vavro, C. Guthy, A.N.G. Parra-Vasquez, M.J. Kim, S. Ramesh, R.K. Saini, C. Kittrell, G. Lavin,

H. Schmidt, W.W. Adams, W.E. Billups, M. Pasquali, W.-F. Hwang, R.H. Hauge, J.E. Fischer and R.E. Smalley, Macroscopic, neat, single-walled carbon nanotube fibers, *Science*, **305**, 1447–1450 (2004).

[196] W. Zhou, P.A. Heiney, H. Fan, R.E. Smalley and J.E. Fischer, Single walled carbon nanotube-templated crystallization of H_2SO_4: Direct evidence for protonation, *J. Am. Chem. Soc.*, **127**, 1640–1641 (2005).

[197] W. Zhou, J.E. Fischer, P.A. Heiney, H. Fan, V.A. Davis, M. Pasquali and R.E. Smalley, Single walled carbon nanotube in superacid: X-ray and calorimetric evidence for partly ordered H_2SO_4, *Phys. Rev. B*, **72**, 045440/1–5 (2005).

[198] V.A. Davis, A.N.G. Parra-Vasquez, M.J. Green, P.K. Rai, N. Behabtu, V. Prieto, R.D. Booker, J. Schmidt, E. Kesselman, W. Zhou, H. Fan, W.W. Adams, R.H. Hauge, J.E. Fischer, Y. Cohen, Y. Talmon, R.E. Smalley and M. Pasquali, True solutions of single-walled carbon nanotubes for assembly into macroscopic materials, *Nature Nanotechnol.*, **4**, 830–834 (2009).

[199] J.L. Hudson, M.J. Casavant and J.M. Tour, Water soluble, exfoliated, nonroping single-wall carbon nanotubes, *J. Am. Chem. Soc.*, **126**, 11158–11159 (2004).

[200] J.J. Stephenson, J.L. Hudson, S. Azad and J.M. Tour, Individualized single walled carbon nanotubes from bulk material using 96% sulfuric acid as solvent, *Chem. Mater.*, **18**, 374–377 (2006).

[201] M. Zheng and B.A. Diner, Solution redox chemistry of carbon nanotubes. *J. Am. Chem. Soc.*, **126**, 15490–15494 (2004).

[202] M. Zheng, A. Jagota, E.D. Semke, B.A. Diner, R.S. McLean, S.R. Lustig, R.E. Richardson and N.G. Tassi, DNA-assisted dispersion and separation of carbon nanotubes, *Nature Mater.*, **2**, 338–342 (2003).

[203] B. Kityanan, W.E. Alvarez, J.H. Harwell and D.E. Resasco, Controlled production of single-wall carbon nanotubes by catalytic decomposition of CO on bimetallic Co–Mo catalysts, *Chem. Phys. Lett.*, **317**, 497–503 (2000).

[204] K.K. Kim, J.J. Bae, H.K. Park, S.M. Kim, H.-Z. Geng, K.A. Park, H.J. Shin, S.M. Yoon, A. Benayad, J.-Y. Choi and Y.H. Lee, Fermi level engineering of single-walled carbon nanotubes by $AuCl_3$ doping, *J. Am. Chem. Soc.*, **130**, 12757–12761 (2008).

[205] L.-J. Li and R.J. Nicholas, bandgap-selective chemical doping of semiconducting single-walled carbon nanotubes, *Nanotechnol.*, **15**, 1844–1847 (2004).

[206] L. Kavan and L. Dunsch, diameter-selective electrochemical doping of HiPco single-walled carbon nanotubes, *Nanolett.*, **3**, 969–972 (2003).

[207] P. Chen, X. Wu, J. Lin and K.L. Tan, High H_2 uptake by alkali-doped carbon nanotubes under ambient pressure and moderate temperatures, *Science*, **285**, 91–94 (1999).

[208] R.T. Yang, Hydrogen storage by alkali-doped carbon nanotubes-revisited, *Carbon*, **38**, 623–641 (2000).

[209] S. Dag, Y. Ozturk, Y.S. Ciraci and T. Yildirim, Adsorption and dissociation of hydrogen molecules on bare and functionalized carbon nanotubes, *Phys. Rev. B*, **72**, 155404/1–8 (2005).

[210] P. Avouris, M. Freitag and V. Perebeinos, Carbon-nanotube optoelectronics, *Topics Appl. Phys.*, **111**, 423–454 (2008).

[211] N. Emery, C. Hérold, M. d'Astuto, V. Garcia, C. Belin, J.-F. Marêché, P. Lagrange, G. Loupias, Superconductivity of Bulk CaC_6, *Phys. Rev. Lett.*, **95**, 087003/1–4 (2005).

[212] A.F. Hebard, M.J. Rosseinsky, R.C. Haddon, D.W. Murphy, S.H. Glarum, T.T.M. Palstra, A.P. Ramirez and A.R. Kortan, Superconductivity at 18 K in potassium-doped C_{60}, *Nature*, **350**, 600–601 (1991).

[213] A. Kasumov, M. Kociak, M. Ferrier, R. Deblock, S. Guéron, B. Reulet, I. Khodos, O. Stéphan and H. Bouchiat, Quantum transport through carbon nanotubes: Proximity-induced and intrinsic superconductivity, *Phys. Rev. B*, **68**, 214521/1–16 (2003).

[214] L. Simonelli, M. Fratini, V. Palmisano and A. Bianconi, Possible clean superconductivity in doped nanotube crystals, *J. Phys. Chem. Solids*, **67**, 2187–2191 (2006).

[215] J. Gonzalez, Superconductivity in carbon nanotube ropes, *Phys. Rev. B*, **67**, 014528/1–12 (2003).

[216] M.S. Arnold, A.A. Green, J.F. Hulvat, S.I. Stupp and M.C. Hersam, Sorting carbon nanotubes by electronic structure using density differentiation. *Nature Nanotechnol.*, **1**, 60–65 (2006).

[217] R. Martel, Sorting carbon nanotubes for electronics, *ACS Nano*, **2**, 2195–2199 (2008).

[218] S. Fogden, C.A. Howard, N.T. Skipper and M. Shaffer, Separation of single-walled carbon nanotubes, *ChemOnTubes-2010, Proc.*, Arcachon, France, April 11–15 (2010).

[219] G. Lolli, L.A. Zhang, L. Balzano, N. Sakulchaicharoen, Y.Q. Tan and D.E. Resasco, Tailoring (n,m) structure of single-walled carbon nanotubes by modifying reaction conditions and the nature of the support of CoMo catalysts, *J. Phys. Chem. B.*, **110**, 2108–2115 (2006).

[220] A.L. Mackay and H. Terrones, Diamond from graphite, *Nature*, **352**, 762 (1991).

3

Functionalized Carbon Nanotubes: (X-CNTs)

Stéphane Campidelli[1], *Stanislaus S. Wong*[2,3] *and Maurizio Prato*[4]

[1]*Laboratoire d'Electronique Moléculaire, DSM/IRAMIS/SPEC, CEA Saclay, France*

[2]*Department: of Chemistry, State University of New York at Stony Brook, USA*

[3]*Cond. Matt. Phys. and Mater. Sci. Dept, Brookhaven National Laboratory, Upton, USA*

[4]*Dipartimento di Scienze Farmaceutiche, Universita' di Trieste, Italy*

3.1 Introduction

Functionalization of carbon nanotubes (CNT) has been pursued by several groups and has led to a completely new class of meta-CNTs, commonly called functionalized carbon nanotubes (X-CNT). X-CNTs offer the invaluable opportunity of combine the outstanding properties of CNT with those of other classes of materials. This chapter deals with a basic introduction to CNT functionalization, followed by an extensive description of X-CNT applications.

3.2 Functionalization Routes

Many different synthesis protocols have been optimized for the functionalization of CNTs. Basically, either a noncovalent or a covalent approach can be used. The two methodologies differ in that the supramolecular chemistry does not interfere with the extended π-electron

Carbon Meta-Nanotubes: Synthesis, Properties and Applications, First Edition. Edited by Marc Monthioux.
© 2012 John Wiley & Sons, Ltd. Published 2012 by John Wiley & Sons, Ltd.

system of CNTs. Instead, an extensive organic functionalization transforms sp^2 carbons in sp^3, thereby interrupting conjugation. This can cause problems at the level of applications when the intact electronic properties of CNTs are necessary. For this reason, when the covalent approach is used, a moderate level of functionalization is preferred, where the electronic properties are retained.

3.2.1   Noncovalent Sidewall Functionalization of SWCNTs

This is one of the more facile applications for the surface modification of carbon nanotubes. Theoretically, noncovalent attachment of aromatic and other organic molecules, such as benzene and cyclohexane, has been studied using first principles calculations [1]. Experimentally, the physisorption and subsequent organization of octadecylamine along nanotube sidewalls have been reported as a means for isolating semiconducting tubes in the supernatant versus the precipitate within a stable dispersion of single-wall carbon nanotubes (SWCNTs) [2]. Moreover, the sidewalls of purified SWCNTs have been non-covalently functionalized [3] with a bifunctional molecule, namely 1-pyrenebutanoic acid succinimidyl ester, which bonds via π stacking with the graphene sidewall surface. Amine groups on proteins then react with the anchored ester to form amide bonds for protein immobilization. In an analogous way [4], ferritin, cytochrome *c*, glucose oxidase, streptavidin, and biotin have been similarly attached to nanotube surfaces with high specificity and efficiency. In a similar experiment, SWCNTs have also been functionalized with bovine serum albumin (BSA) via a diimide-activated amidation under ambient conditions. Based on data such as atomic force microscopy (AFM), thermogravimetric analysis (TGA), Raman spectroscopy, and gel electrophoresis, the resultant nanotube-BSA conjugates are structurally interlinked and bioactive, forming dark-colored aqueous solutions [5].

It turns out that π-π interactions also play a role in the noncovalent immobilization [6,7] of conjugated tetra-*tert*-butylphthalocyanines and porphyrins onto nanotube surfaces. Interestingly, optically active porphyrins have been found to bind with different affinities to the left- and right-handed helical nanotube isomers to form complexes with unequal stabilities that can then be readily separated. Moreover, single-stranded DNA fragments (such as alternating guanine and thymidine amino acid sequences) can bind to nanotubes through π-stacking, resulting in helical wrapping to the surface, and enabling DNA-assisted separation of SWCNTs into fractions with different electronic structures and different diameter distributions upon processing with ion-exchange chromatography [8–10].

3.2.2   Covalent Functionalization of SWCNTs

3.2.2.1   *Fluorination*

A lot of effort, including density functional tight binding calculations [11], has been dedicated to probing the fluorination of tubes. SWCNTs treated at temperatures ranging from 150 to 325°C survive the fluorination process. Two major diagnostic features [12] in the Raman spectra of the fluorinated SWCNTs are (i) a decrease in the intensity of the radial breathing mode near 200 cm^{-1} and (ii) an increase in the intensity of the disorder D band at ~1320 cm^{-1}, the latter of which can be ascribed to the destruction of the intrinsic sp^2-hybridized structure as C-F bonds are created and sidewall carbons become sp^3-hybridized. Scanning tunneling microscopy (STM) images of fluorinated tubes reveal a dramatic banded structure indicating broad continuous regions of fluorination terminating abruptly in bands orthogonal to the tube axis [13]. These results suggest that fluorine adds more favorably around the circumference of the tube (preferably to tube tip), an assertion backed up by semi-empirical AM1 calculations, and that the fluorine atoms tend to bond next to each other [14]. A review article [15] suggests that fluorinated nanotubes exhibit altered conductive and optical properties and that they are also useful precursors for various subsequent reactions.

Hence, sidewall-fluorinated nanotubes can be further modified by either reacting them with alkyl magnesium bromides in a Grignard synthesis or by reaction with alkyllithium precursors to yield sidewall-alkylated nanotubes, via a concerted, allylic displacement mechanism [16]. Covalent attachment to the sidewalls was confirmed by UV-visible-near-IR spectroscopy [17], wherein optical features originating from transitions between the van Hove singularities in the 1-D electron density of states of the tube disappear as the electronic structure becomes perturbed. Likewise, fluorinated SWCNTs can also undergo Diels-Alder addition reactions, as proven by consistent IR, Raman, AFM, and ^{13}C NMR measurements [18]. Moreover, upon reaction of fluorinated SWCNTs with organic peroxides including benzoyl and lauroyl peroxide, phenyl and undecyl sidewall functionalized tubes can be correspondingly produced, though the reactions occur faster with C_{60} [19]. Characterization of these particular functionalized derivatives has been accomplished using Raman, FTIR, and UV-visible-near IR spectra as well as TGA/MS, TGA/FTIR, and TEM data [19]. In addition, heating fluorinated SWCNTs at 70 to 170°C in the presence of terminal alkylidene diamines leads to sidewall functionalization at levels as high as 1 over 8–12 carbon atoms [20].

Oxidizing these alkylated nanotubes in air allows for recovery of pristine nanotubes. As for fluorinated tubes, it was found that hydrazine could be used to regenerate the unfluorinated starting material [21]; a procedure was also established to 'tune'' the fluorine content of the tubes by first fluorinating heterogeneously, solvating in alcohol, and then defluorinating with substoichiometric quantities of hydrazine so as to generate tubes with different fluorine contents [22]. Heating fluorinated tubes at 100°C in the presence of noble gases can also lead to recovery of pristine tubes, as determined by a decrease in the IR intensity of C-F stretch bands at 1207 cm^{-1}, as well as a reduced resistance of the resulting nanotube samples [23]. More recently, fluorinated SWCNTs have been reacted with diols and glycerol in the presence of alkali metals or alkali hydroxides to yield –OH terminated moieties on nanotube sidewalls [24]. In addition, amino-sidewall functionalized tubes have been obtained at room temperature by fluorine displacement reactions of fluorinated SWCNTs with amines [25]. Moreover, thiol- and thiophene-functionalized SWCNTs have been prepared via the reaction of fluoronanotubes with appropriately derivatized amines [26].

3.2.2.2 Organic Functionalization
The reactivity of SWCNTs towards sidewall addition depends on the diameter of the tubes as well as on the pyramidalization angle of the carbon atom from the tube surface where the cycloaddition reaction takes place. Indeed, the convex surface [27] is more reactive than the concave one and the difference in reactivity increases with increasing pyramidalization angle and decreasing tube diameter. Moreover, the reactivity of the convex surface is enhanced by the pronounced exposure of the hybrid π orbitals from the exterior surface which favors orbital overlap with incoming addends [28].

(a)

(b)

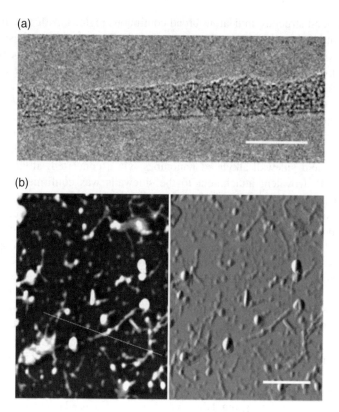

Figure 3.1 *(a) TEM image (scale bar = 10 nm) and (b) typical AFM topography (left) and amplitude (right) images of a derivatized polyimide with pendant hydroxyl groups (PI$_{OH}$).- SWCNT sample (scale bar = 1 μm). Reprinted with permission from [36] Copyright (2005) American Chemical Society.*

SWCNTs have been chemically functionalized in situ by a thermally induced reaction with diazonium compounds, generated by the action of isoamyl nitrite on aniline derivatives [29]. While the reaction was first done electrochemically, it was later shown that the diazotization functionalization could also be achieved at high temperatures by refluxing the reaction mixture [30,31]. This procedure allows for a very high degree of functionalization (estimated to be 8–12% from TGA experiments) to be achieved. A solvent-free analog of this system was reported, wherein this potentially large-scale reaction proceeds by mechanical stirring to yield highly functionalized nanotubes [32,33]. The derivatization reaction with arenediazonium in particular was studied mechanistically using GC-MS and XPS [34]. In addition, the reaction with 4-chlorobenzenediazonium was proposed as a model electron-transfer reaction that is selective for metallic versus semiconducting tubes via a charge-transfer stabilization of complexes at the surfaces of the former [35].

The Tour group, which pioneered diazotization, later developed a means for functionalizing SWCNTs in fuming sulfuric acid to afford unbundled individual SWCNT-arylsulfonic acids that did not require either surfactant or polymer-based prewrapping (Figure 3.1 and 3.2) [36,37], centrifugation, or sonication [38]. Moreover, they also proposed an environmentally-friendly methodology [39] for exfoliating SWCNTs from bundles and

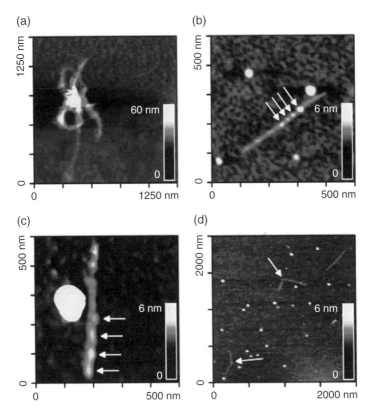

Figure 3.2 *Atomic force microscopy images of the SWCNT dispersions deposited onto mica. The dispersing agents are (a) Poly-L-Lys·HBr; (b,d) Poly-L-(Trp,Lys·HBr); and (c) Poly-L-(Tyr,Lys·HBr). The arrows indicate adsorbed polypeptide in (b) and (c), and end-to-end linking of SWCNTs in (d). Z-scales are given for all figures. Reprinted with permission from [37] Copyright (2007) American Chemical Society.*

functionalizing them as individuals by grinding them for minutes at room temperature with aryldiazonium salts in the presence of ionic liquids and K_2CO_3.

Direct addition to the unsaturated π-electron systems of SWCNTs can also occur through a number of different reactions (Figure 3.3) [40–43], including (i) a [2+1] cycloaddition of nitrenes, where, in the presence of alkyl azidoformates, alkoxycarbonylaziridino-SWCNTs can be formed, as deduced by XRD, XPS, and Raman spectroscopy [44,45]; (ii) a [4+2] Diels-Alder cycloaddition of *o*-quinodimethane theoretically predicted to be viable due to aromaticity stabilization [46] at the corresponding transition states and products, and later experimentally verified under microwave irradiation conditions [47]; (iii) the addition of nucleophilic carbenes [48], which have been theoretically probed using a two-layered ONIOM approach [49,50]; and (iv) a cyclopropanation process accomplished under Bingel reaction conditions [51].

For the nucleophilic Bingel reaction in particular, it has been suggested that, after an incubation time of several hours, this reaction induces highly regular, long-distance patterns in the nanotube that can be visualized using the STM [52]. Moreover, the Bingel reaction [51] allows

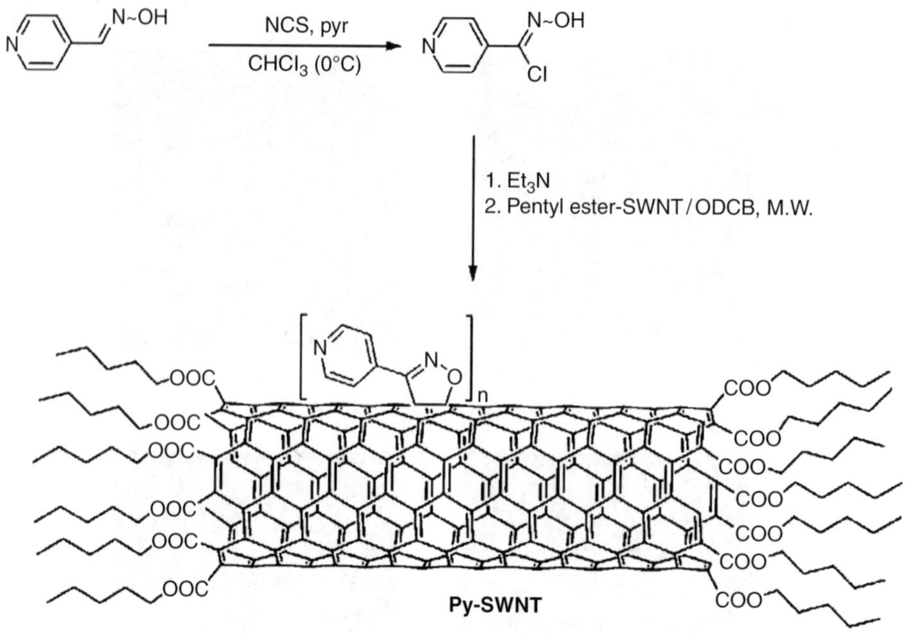

Figure 3.3 *Synthetic route for the preparation of 4-pyridyl isoxazolino-SWCNT (Py-SWCNT).*

for the introduction of 'chemical tags' such as Au colloids onto SWCNT surfaces, thereby enabling a means of spatially locating functional groups and determining their distribution.

Reductive alkylation of SWCNTs using lithium and alkyl halides in liquid ammonia yields sidewall functionalized tubes, such as dodecylated SWCNTs, in which alkyl radicals are involved as intermediates [53]. Alkyl groups can also be introduced through the nucleophilic addition of *t*-BuLi to the sidewalls of the tubes and subsequent reoxidation of the intermediates [54]. An aryl analog to this type of reaction has recently been reported [55]. Moreover, polyacylation under Friedel-Crafts conditions is an excellent way of purifying and decorating nanotubes with a variety of alkyl, aryl, and fluoryl groups [56]. Though SWCNTs can be functionalized under mild conditions by ball milling at room temperature in the presence of vapors of alkyl halides [57], free alkyl radicals generated by the decomposition of benzoyl peroxide in the presence of alkyl iodides have been more effective at derivatizing the sidewalls of small-diameter SWCNTs with an efficiency reported to be as high as 20% [58].

In terms of other significant sidewall reactions, derivatization based on 1,3-dipolar cycloaddition of azomethine ylides, generated by the condensation of an α-amino acid and an aldehyde, has been utilized to purify commercial HiPCO SWCNTs [41,59]. This covalent functionalization of SWCNTs with pyrrolidine rings allows for the introduction of a wide number of other functionalities, such as ferrocene, to be immobilized onto the SWCNT surface [60]. An analogous reaction has been reported with a microwave-assisted cycloaddition involving aziridines [61].

Rational functionalization in solution of the sidewalls of SWCNTs has also been accomplished by a relatively facile ozonolysis process, which was determined to be a

1,3-dipolar cycloaddition process that was selective for smaller diameter tubes. Ozonolysis [62–65] not only conserved the structural integrity of the tube bundles but also purified them, removing a large number of metal impurities and amorphous carbon entities. More importantly, the protocol presented a nondestructive, low-temperature method of introducing oxygenated functionalities directly onto the sidewalls and not simply at the end caps of SWCNTs. Thus, molecular species that can be currently tethered to the ends of the nanotubes could now be chemically dispersed along the sidewalls as well, with implications for the design of nanotube-based optoelectronic devices.

Additionally, nanotubes have even been hydrogenated via a modified Birch reduction method [66]. Nitrile ylides and nitrile imines have been theoretically predicted [67] and experimentally shown [68] to be excellent candidates to undergo 1,3-dipolar cycloaddition onto SWCNT sidewalls; in particular, available data suggest that electron transfer occurs from the electron rich substituents of the pyrazoline ring to the electron acceptor termini such as the nanotube sidewalls.

Many of these reactions are useful in nanotube manipulation. That is, tubes functionalized in these various ways enable a large variety of terminal functional groups to be introduced onto nanotube surfaces with implications for nanocomposite preparation and biomedical applications. For instance, reaction of SWCNTs with either succinic or glutaric acid acyl peroxides in *o*-dichlorobenzene at 80–90°C results in sidewall derivatization with pendant carboxylic groups that can be subsequently converted into a wide range of active, functional moieties [69]. In addition, a detailed investigation of the sidewall functionalization of nanotubes using a family of (R-) oxycarbonyl nitrenes, with the R group encompassing alkyl chains, aromatic groups, dendrimers, crown ethers, and oligoethylene glycol units, has recently been reported. The degree of functionalization, though, was <5% in that set of reactions [70].

3.2.2.3 *Inorganic Functionalization*
The reactivity of SWCNTs towards metal complexes is substantially different from that of fullerenes, even though both possess a non-planar sp^2 configuration. Curvature effects and different degrees of local strain between the two structures likely account for the observed differences in behavior. Specifically, the presence of five-membered cyclopentadienyl-like rings in C_{60} substantially enhances the affinity of (6,6) bonds in coordinating metal complexes and allows for strong back donation in the resulting adducts, thereby promoting their stability. However, theoretical calculations on molecular fragments replicating SWCNT surface curvature indicate that, unlike the case of fullerenes, five-membered rings are not present to stabilize π^* ligand orbitals and hence, back-bonding interactions are much weaker in these systems [71]. In fact, the elimination of pyramidalization of carbon atoms in fullerenes as well as the release of local and global strain on η^2 coordination have been the driving forces for metal-complexation reactions of fullerenes [72]. The reactivity of SWCNTs with respect to metal coordination compounds has not been as straightforward as well, though we have found that coordination at oxygenated sites appears to be a favorable mode of bonding.

In fact, it has been demonstrated that one can coordinate Vaska's compound [73], [Ir(CO) Cl(PPh$_3$)$_2$], onto the ends and sidewalls of SWCNTs. The compound complexes to raw nanotubes by η^2 coordination across the graphene double bonds. Indeed, binding the ligand is a quasi-associative process with the formation of a π complex and is expected to be similar to that observed [74,75] in either (TCNE)IrBr(CO)(PPh$_3$)$_2$ or the fullerene adduct [76]. This type of reaction, in which the ligand acts as a weak base towards the metal complex, has also been referred to as a 'reductive addition' (with respect to the metal) [77]. For the oxidized carbon nanotubes, coordination through the oxygen atoms seems a more likely possibility. In other words, there is an oxidative addition process that results in a hexa-coordinated structure. Oxidative additions to form 18e$^-$ 'coordinatively saturated' compounds are well known for Vaska's compound and other related d^8 complexes, and have been extensively studied [78,79]. Thus, it is postulated that the oxidized nanotubes coordinate to the compound via oxidative addition to form an Ir(III) complex. Additional theoretical studies [80] using hybrid quantum mechanics/molecular mechanics calculations suggest that the formation of a stronger adduct occurs when the transition metal center coordinates to carbon atoms belonging to pentagonal rings, as in defects or end caps.

Functionalizing oxidized nanotubes with Vaska's complex [73] renders them soluble and stable in organic solution, originating from exfoliation of large crystalline ropes to smaller bundles and individual tubes and thereby enabling further exploitation of their wet chemistry. Moreover, as the tubes are easily recoverable from solution, this finding has significant scientific and economic implications for nanotubes as reusable catalyst supports, particularly for expensive catalyst materials. Similar behavior has also been demonstrated for Wilkinson's compound [81], which is a Rh-based system, useful in the recyclable catalytic hydrogenation of alkenes, such as cyclohexene to cyclohexane. Vanadyl salen complexes covalently anchored to SWCNTs have been used as heterogeneous catalysts for the cyanosilylation of aldehydes with trimethylsilylcyanide [82].

Ruthenium-based compounds [83] as well as rigid dendritic ruthenium complexes [84] have been utilized to generate interconnects and assemblies of nanotubes. Moreover, covalent attachment [85] of ruthenium(II)-tris(2,2'-bipyridine) to SWCNTs rendered them sensitive to light that was absorbed by the ruthenium complex and made the nanotubes persistently photoconductive with a quantum yield of 0.55.

Furthermore, in an initial study, oxidized, cut SWCNTs were reacted with lanthanide salts containing Eu, La, and Tb. These studies found that the lanthanide ions [86] likely coordinate to SWCNTs through a disruption of hydrogen bonding in bundles of oxidized SWCNTs. Adducts were analyzed using FT-IR, Raman, and photoluminescence spectroscopies and were structurally characterized using AFM and TEM, along with energy-dispersive X-ray spectroscopy (EDS). In these systems, SWCNTs can serve as electron acceptors [87–89], and hence, can act as an energy sink in an excited-state energy transfer mechanism. Indeed, upon coordination of these lanthanide ions to oxidized SWCNTs and subsequently, upon relaxation after excitation, some degree of energy transfer is expected from the metal to the low-lying excited states in the visible and near-IR regions of the SWCNTs, which thereby resulted in a decreased intensity in the intrinsic emission of the lanthanide ions as a function of increased reaction time and concentration. Nonetheless, as confirmed by a more recent work [90] involving the complexation of a specific Eu complex, the presence of a SWCNT framework does not noticeably affect the basic lanthanide-centered luminescence stemming from characteristic electronic transitions within the 4f shell of the Eu^{III} ions. All of these results have implications for potential sensor and imaging applications.

The interaction of solution phase OsO$_4$ with SWCNTs [91] (Figure 3.4) in the presence of UV irradiation demonstrates its chemical specificity toward the nanotube electronic

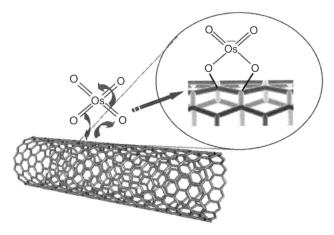

Figure 3.4 *Reaction scheme illustrating the cycloaddition of osmium tetroxide (OsO₄) to the sidewall of a single-walled carbon nanotube (SWCNT) leading to an osmyl ester of a five-membered ring structure (curved arrows indicate the flow of electrons during bond formation). Reprinted with permission from [91] Copyright (2003) American Chemical Society.*

structure. Nanotube osmylation appears to be chemically selective in differentiating between the reactivity of metallic versus semiconducting nanotubes [92]. That is, metallic nanotubes, with a larger electron density near the Fermi level, are better able to stabilize an intermediate charge transfer complex with an excited OsO_4 species. When the charge transfer complex dissociates, the net results are (i) covalent sidewall functionalization of these nanotubes through disruption of the conjugated π-electron structure, as well as (ii) reduction of the osmium tetroxide species to OsO_2 nanoparticles, which are then templated onto the sidewall surface. A systematic Raman study of our nanotube samples at three different excitation wavelengths, probing different electronic populations of tubes, provides for strong evidence of the higher reactivity of metallic tubes with respect to osmylation. These observations bear out a theoretical prediction of covalent sidewall osmylation (through the mediation of osmate ester adducts) [93] of nanotubes. In fact, the templating of OsO_2 on metallic nanotube surfaces may serve as the basis of achieving chiral separation of these nanotubes through centrifugation based on density differences.

With respect to Si-containing complexes, there was a recent report [94] on the effective silylation of raw, pristine SWCNTs. Specifically, SWCNTs were functionalized at their ends and sidewalls (i) with trimethoxysilane, and (ii), in a separate experiment, with hexaphenyldisilane. Raman analyses demonstrated selective reactivity of predominantly smaller-diameter semiconducting nanotubes. Whereas microscopy results clearly showed that the functionalization reaction was structurally non-destructive to the tube structure, spectroscopy data provided evidence for chemical attachment of organosilanes onto the carbon nanotube surface. UV-visible data also yielded evidence for selectivity and functionalization. Furthermore, control over the actual thickness of a SiO_2 coating on SWCNT surfaces was accomplished soon afterwards using two protocols based on an electrochemical sol-gel process [95]. In one of them, a SWCNT mat was used as a working electrode for the direct deposition of silica. In the other, nanotubes were dispersed in solution and silica was deposited onto these solubilized nanotubes in the presence of a platinum working electrode. The thickness of the silica coating could be controllably altered by varying the potential of the working electrode as well as the concentration of the sol solution. Both of these methodologies possessed several advantages, including, ease of use, environmental-friendliness, and utilization of relatively mild reaction conditions. Appropriately functionalized SWCNTs could therefore become an important component, for example, of individualized gate dielectric materials in field effect transistors.

Finally, SWCNTs have also been used as substrates for the deposition of metals by electron-beam evaporation, resulting, for the cases of Au, Al, Pb, and Fe, in the formation of discrete particles on nanotubes due to a weak interaction between metals and nanotubes [96]. Acid groups on treated SWCNTs can act as nucleation sites for a well-dispersed deposition of Pt clusters [97]. With Ti, Ni, and Pd, quasi-continuous coatings are possible, resulting in the formation of nanotube-supported metal nanowire structures [98]. This aspect is specifically detailed in Chapter 4.

3.2.2.4 Functionalization of Cut-CNT Ends

The purification of SWCNTs can be carefully monitored not only by microscopy but also quantitatively by near-IR spectroscopy [99]. It is not surprising that reactions primarily aiming at purification can not only open [100] carbon nanotubes but also functionalize them as well. For instance, thermal oxidation in air [101], nitric acid treatment [102], 'piranha' etching [103], ultrasonication [104], UV ozonation [105], soaking in potassium permanganate solution [106], as well as acid modification under supercritical water conditions [107] have all been proposed and/or used as purification methods, and can all be used to modify and generate chemical functionalities onto nanotube ends, sidewalls, and defect sites. These functional groups include aldehyde, carboxyl, ketone, ether, and hydroxyl moieties, and can be quantified using XPS [108] and acid-base titration [109]. What is clear is that acidic sites can be considered as key reactive 'flash points', for example, for the subsequent derivatization of either thiol groups [110,111] or alkyl substituents [112] with implications for self-assembly. End-to-end heterojunctions can be formed, for instance, by reacting chloride terminated nanotubes with aliphatic diamines [113]. Ester linkages, created by these processes, can be used to attach either viologens (Figure 3.5) [114], phthalocyanine chromophores [115], pyrenes [116,117], tetrathiafulvenes (Figure 3.6) [118], or napthalimide fluorescent moieties [119], useful for developing solar, photochromic and photocatalytic devices with associated control over charge transfer. Dendrimers have been linked directly to SWCNTs using a similar strategy without provoking significant damage in the intrinsic conjugated π-system [120]. A few papers have reported on asymmetric end-functionalization of carbon nanotubes with the interesting idea of eventually generating donors and acceptors attached to opposite ends of a nanotube respectively [121,122], analogous to SWCNT/pyrene/porphyrin nanohybrids [123] previously reported.

3.2.2.5 Additional Chemical Functionalization Strategies

SWCNTs can also be modified by a number of other methodologies. For instance, using radio frequency glow-discharge plasma activation [124], amino-dextran chains have been immobilized onto acetaldehyde-treated aligned SWCNTs through a Schiff-base formation and reductive stabilization with sodium cyanoborohydride, while periodate-oxidized dextran-FITC chains have been chemically grafted onto ethylenediamine-derivatized SWCNTs via the same reaction. Similarly, SWCNTs have been

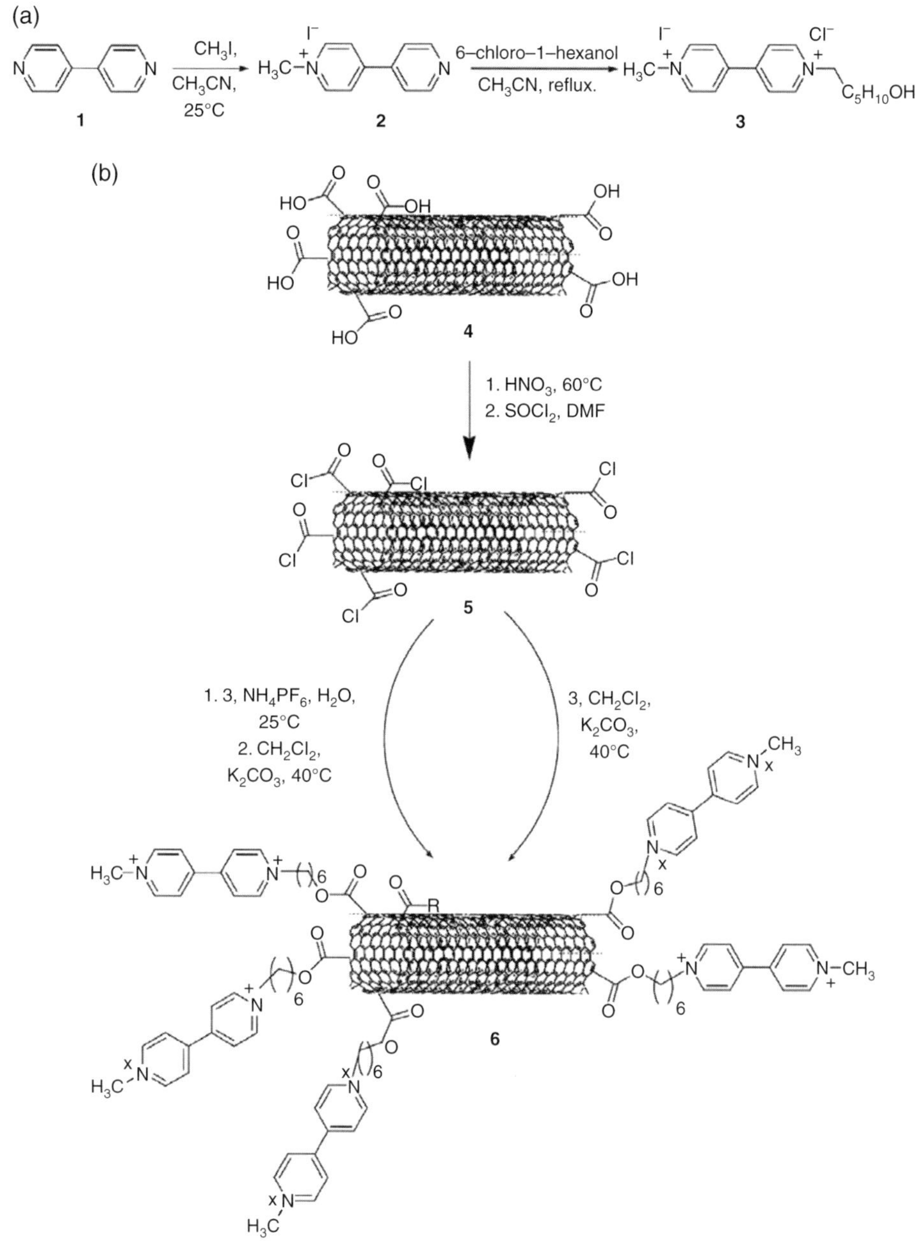

Figure 3.5 *(a) Synthesis of asymmetrically substituted viologen 3 and (b) Purification and chlorination of SWCNT and covalent attachment of 3 to SWCNT. Reprinted with permission from [114] Copyright (2005) American Chemical Society.*

subjected to microwave-generated N_2 plasma (at intermediate distances, typically 2.5 cm) for incorporating nitrogen functional groups into SWCNTs via sidewall attachment. The incorporation of nitrogen and oxygen was observed, as expected [125]. SWCNTs can also be derivatized by bombardment [126,127] with CH_3 and CF_3 radicals impacting

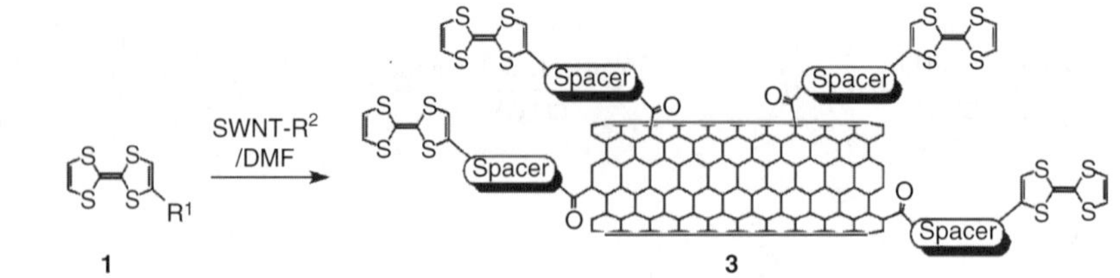

1 3

a: R^1 = –CH_2–OH; R^2 = –COOH; Spacer = –CH_2–O–
b: R^1 = –CH_2–O–CH_2–C_6H_4–COOH; R^2 = –CO–NH–CH_2–CH_2–NH_2; Spacer = –CH_2–O–CH_2–C_6H_4–CO–NH–CH_2–CH_2–NH–

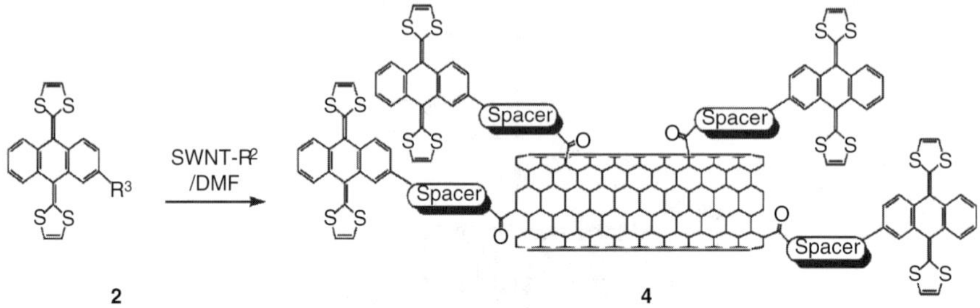

2 4

a: R^3 = –NH_2; R^2 = –COCl; Spacer = –NH–
b: R^3 = –CH_2–OH; R^2 = –COCl; Spacer = –CH_2–O–
c: R^3 = –CH_2–O–CH_2–C_6H_4–COOH; R^2 = –CO–NH–CH_2–CH_2–NH_2; Spacer = –CH_2–O–CH_2–C_6H_4–CO–NH–CH_2–CH_2–NH–

Figure 3.6 *Synthesis of tetrathiafulvene (TTF)- and π-extended tetrathiafulvalene (exTTF)-functionalized SWCNTs. Reprinted with permission from [118] Copyright (2006) John Wiley and Sons.*

with incident energies of 10, 45, and 80 eV, as was confirmed by X-ray photoelectron spectroscopy and scanning electron microscopy.

Chemical functionalization of SWCNTs attached to conventional atomic force microscopy (AFM) probes has also been demonstrated as a methodology of yielding high-resolution, chemically-sensitive images on samples containing multiple chemical domains [128–130]. Specifically, samples prepared by microcontact printing were imaged in ethanol and as a function of pH, using functionalized carbon nanotube tips. Lateral chemical resolution of about 3 nm and less was obtained on mixed bilayer samples [129]. An alternative approach for assessing the functionality at a specific nanotube tip end has been to measure the adhesion force between the tip and a surface that terminates in a known chemical functionality (i.e., chemical force microscopy).

Functionalized nanotube tips can be created by shortening the tubes in an oxidizing environment and then derivatizing them by means of standard peptide coupling chemistry. By this means, hydrophobic and basic amine-terminated probes have been formed [129]. The nature of the tip functionality could then be assessed by means of force measurements. Biological probes [128] were created in the same manner through the use of biotin cadaverine as the amine. With this strategy, specific biotin-streptavidin interactions, an example of ligand-receptor interactions, were determined to be ca. 200 pN per biotin-streptavidin pair. A new way of derivatizing nanotube tips was proposed by shortening and sharpening them in different gaseous environments [131]. In nitrogen, basic tips can be created whereas in hydrogen, nanotube tips terminated by a hydrophobic functionality can be formed.

3.3   Properties and Applications

3.3.1   Electron Transfer Properties and Photovoltaic Applications

The first step toward the realization of energy conversion devices is the design of structures in which electron donor and electron acceptor moieties are in close proximity and can interact with each other. The electrical conductivity, morphology, and good chemical stability of SWCNTs are promising features that stimulate their integration into photovoltaic systems. It has been demonstrated that carbon nanotubes readily accept charges, which can be transported under nearly ideal conditions along the tubular SWCNT axis [132,133]. In these systems, carbon nanotubes act as electron acceptors and many electron donors (including discrete molecules like porphyrins, phthalocyanines, tetrathiafulvalenes (TTF) or conjugated polymers like polythiophenes or poly-*meta*-phenylenevinylenes) have been linked to nanotubes either by covalent or noncovalent coupling.

3.3.1.1   Covalent Approach

Carbon nanotubes can be produced by several methods. A constant amongst the various synthesis routes is that the resulting raw materials usually contain carbon nanotubes mixed with amorphous carbon and/or catalytic metal particles as impurities. Consequently, an effective purification of the nanotubes is required before their further processing. The treatment of the raw materials under strong acidic and oxidative conditions – typically sonicating in a mixture of concentrated nitric and sulfuric acid, or heating in hydrogen peroxide or in a H_2O_2/H_2SO_4 mixture – results in shortened, opened tubes [134,135]. The grafting of carboxyl groups on the nanotube sidewalls by a series of oxidative treatments represents a powerful pathway for further modifications of the nanotubes since the acid functionalities can react with alcohols or amines to give ester or amide derivatives, as mentioned in the previous section.

The covalent linkage of porphyrins to SWCNT was achieved through an esterification reaction between the carboxylic groups of SWCNTs and porphyrins containing hydroxyl groups (Figure 3.7) [136]. Photophysical investigations of the SWCNT-H_2P conjugates showed that the quenching of the fluorescence of the porphyrins was strongly dependent on the length of the spacer between the nanotube and the chromophore. In case **1** of Figure 3.7 where the spacer is a CH_2 moiety, no fluorescence quenching was observed while in case **2** where the spacer contains a six-carbon chain, the fluorescence of the porphyrin was approximately 70% of that of the reference compound. This result could be attributed to the flexibility of the alkyl chain which allowed for a better interaction between the nanotube and the porphyrin.

In the same way, the photophysical properties of a family of SWCNT–TTF derivatives were investigated as a function of the linker between the SWCNTs and the TTF moieties, and also as a function of the nature of the tetrathiafulvalene unit (simple TTF or π-extended TTF) (Figure 3.5 and 3.7) [118]. The SWCNT-TTF conjugates gave rise to photoinduced electron transfer, while the lifetime of the charge separated states (typically on the order of several hundreds of nanoseconds) depended on the length of the spacer between the two electroactive moieties (longer lifetimes when the length increases) and on the nature of the TTF (longer lifetimes with π-extended TTF probably due to a better charge delocalization).

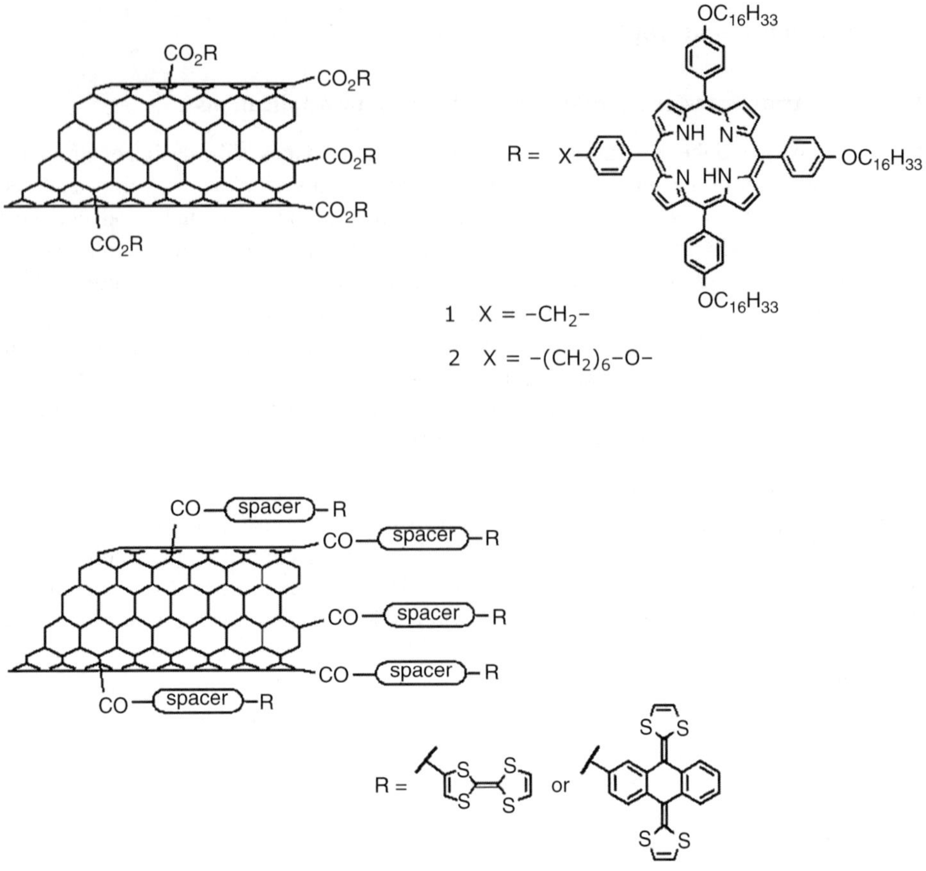

Figure 3.7 *Example of SWCNT functionalized with electro-active species (i.e., porphyrin and tetrathiafulvalene (TTF) moieties).*

More recently, photoactive films made of porphyrin-SWCNT were prepared and used to fabricate photoelectrochemical devices on nanostructured SnO_2 electrodes [137]. The films exhibited a photon-to-photocurrent efficiency of about 4.9% under bias voltage of 0.08 V. According to the authors, the electron injection from the excited states of the SWCNTs to the conduction band of the SnO_2 electrode is responsible for the photocurrent generation. Despite the efficient quenching of the excited singlet state of the porphyrin by the SWCNTs in the porphyrin-linked SWCNTs, the photocurrent action spectra revealed that the excitation of the porphyrin moieties makes no contribution to the photocurrent generation. The evolution of an exciplex between the excited singlet state of the porphyrin and the SWCNTs, and the subsequent rapid decay to the ground state without generating the charge-separated state is proposed to explain the unusual photoelectrochemical behavior.

In 2005, the modification of a self-assembled monolayer of cysteamine/2-thioethanesulfonic acid on gold surface with short oxidized carbon nanotubes was reported. SWCNTs were linked to the amino surface by one of their extremities while CdS nanoparticles were

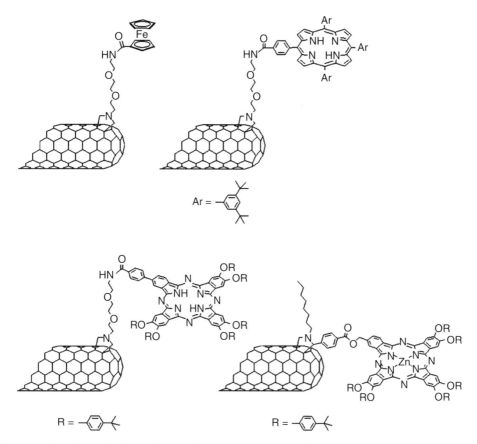

Figure 3.8 *Example of functionalized SWCNT containing ferrocene, porphyrin or phthalocyanine derivatives.*

attached to the other extremity [138]. The photocurrent generated by the functionalized electrode in the presence of triethanolamine as a sacrificial electron donor was found to be as high as 830 nA under photon irradiation at $\lambda = 390$ nm. The authors estimated a quantum efficiency for the photon-to-electron conversion of about 25%.

In an earlier work, the covalent attachment of ferrocene (Fc) onto nanotube sidewalls was achieved (Figure 3.8) [139]. The photoexcitation of Fc-SWCNT with visible light leads to electron transfer that yields a long-lived $Fc^{•+}$-SWCNT$^{•-}$ species. This result suggests that nanotube-based covalent nanohybrids are suitable for solar energy conversion.

Based on the same approach, SWCNTs grafted with phthalocyanine and porphyrin derivatives were obtained (Figure 3.8) [115,140]. For all nanoconjugates, a strong communication (i.e., photoinduced electron transfer) between the nanotube and the macro-cycle subunits has been observed. Phthalocyanines are of particular interest since they are synthetic porphyrin analogs, exhibiting particularly intense absorption characteristics in the red spectral region, where porphyrins fail to absorb appreciably.

This approach has a severe limitation in that a low number of photoactive groups can be introduced on the nanotube sidewalls. Indeed it is generally well admitted that the extensive covalent functionalization of SWCNT sidewalls disrupts the conjugated π-system of the tubes and subsequently affects their optical and electronic properties [141]. In this context, the functionalization of nanotubes with polymers or dendrimers represents a particularly promising strategy. Dendrimers are regular hyperbranched macromolecules. At high generation, they possess a globular structure with a large density of functional groups at the periphery [142–146]. In order to increase significantly the number of light-harvesting chromophores on the nanotubes, a polyamidoamine (PAMAM) dendrimer [147] was built on the nanotube sidewalls. The grafting site of the dendrimer on the functionalized nanotubes is the amine group. The nanotubes bearing dendrons were further functionalized with tetraphenylporphyrins and it was estimated that, on the average, only two porphyrins were grafted on each dendron (Figure 3.9) [120]. In response to visible light irradiation, SWCNT-(H_2P)x nanoconjugates gave rise to fast charge separation evolving from the photoexcited H_2P chromophores. The oxidized H_2P chromophore was identified through its fingerprint absorption in the 550–800 nm wavelength range, while the signature of the reduced SWCNT appeared in the 850–1400 nm wavelength range.

Another limitation toward the fabrication of nanotube-based functional materials comes from the incorporation of the chromophores onto the nanotube sidewalls. The development of methods allowing for an efficient grafting of molecules on the nanotubes is, therefore, highly desirable. The emerging field of 'click chemistry' has the potential to provide an elegant protocol to prepare carbon nanotube-based functional materials. The term 'click chemistry' defines a series of chemical reactions which are versatile, specific, easy to realize, assoiated with a simple purification process (absence of by-products) and compatible with aqueous media (green chemistry). This new concept has been introduced by Sharpless in 2001 [148]. Among the large collection of organic reactions, the Cu(I) catalyzed variation [149,150] of the *Huisgen* 1,3-dipolar cycloaddition [151,152] represents the most effective reaction in 'click chemistry'. This reaction involves a 1,3-dipolar cycloaddition between azide and acetylene derivatives in the presence of Cu(I) catalyst giving rise to 1,2,3-triazole rings.

Recently, several examples of the functionalization of carbon nanotubes via the Cu(I)-catalyzed azide-alkyne 3+2 cycloaddition (CuAAC) reaction have been reported [153]. In particular, CuAAC has permitted researchers to incorporate very efficiently zinc-phthalo-cyanines (ZnPc) and zinc-porphyins (ZnP) onto SWCNTs (Figure 3.10) [154,155]. The nanotube-based nanoconjugates were investigated with a series of steady-state and time resolved spectroscopy experiments and photoinduced electron transfers from the photoactive molecules (i.e., phthalocyanines and porphyrins) to the nanotubes have been identified in both cases. SWCNT-ZnPc conjugate was integrated into photoactive electrodes (i.e., photoanodes), using ITO as a transparent semiconductor and tested in a photoelectrochemical cell configuration [154]. Upon illumination, stable and reproducible photocurrents were observed. For the SWCNT-ZnP conjugates, the originality of the work resides in the structures attached on the nanotubes which permit a fine tuning of the absorption cross-section throughout the visible part of the solar spectrum [155].

The previous examples highlighted the use of dendritic structures to increase the number of functional groups on the nanotube sidewalls. Recently, dendrimers attached onto nanotubes served as templates for the synthesis of metallic or semiconducting nanoparticles. The

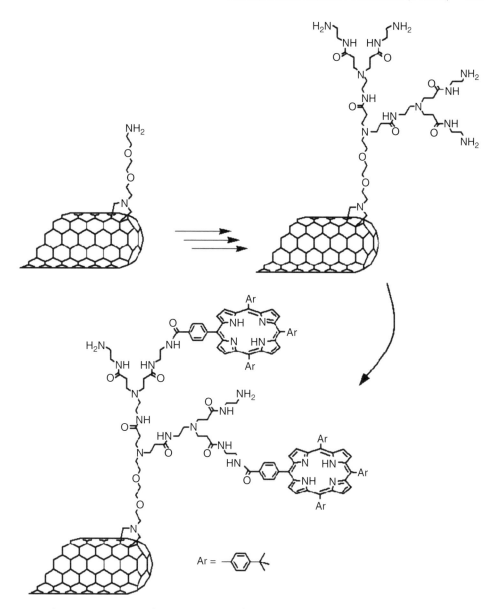

Figure 3.9 *SWCNT bearing PAMAM dendrimers functionalized with porphyrins.*

synthesis of Ag particles attached to PAMAM dendrimers bonded to MWCNTs was reported [156], while in another contribution, the formation of CdS nanoparticles promoted by their attachment to the dendrimers was described [157]. These examples demonstrate that dendrimers can play at least two roles: (i) amplifier of existing functional groups present on nanotube surface and (ii) template promoter for the growth of nanoparticles subsequently attached to the dendrimers. The latter would be a very elegant way to combine the properties

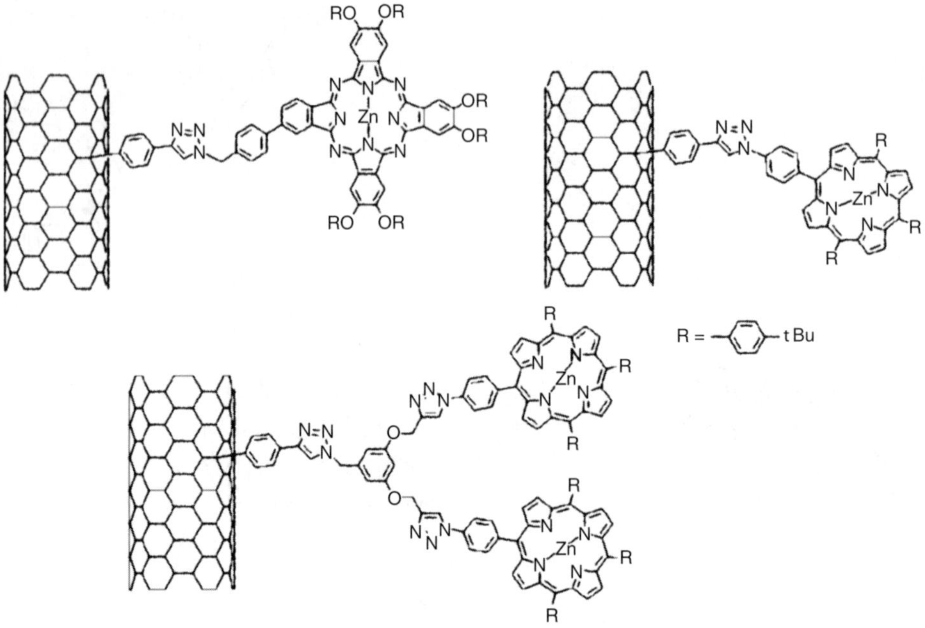

Figure 3.10 *Example of SWCNT functionalization with phthalocyanine and porphyrins via 'click chemistry'.*

of nanotubes (electrical conductivity, aspect ratio, etc.) with those of nanoparticles (catalysis, luminescence, etc.).

In general, covalent linkage of photo/electro-active groups to the SWCNT sidewalls has allowed model compounds to be obtained for studying electron and/or energy transfer processes in solution. However, such nanoconjugates remain difficult to synthesize and the degradation of the nanotube properties due to the oxidative treatments or to the insertion of sp^3 carbon in the conjugated π-system could be a serious drawback for the device efficiency.

3.3.1.2 Hybrid Covalent/Supramolecular Approach

The functionalization of carbon nanotubes with isoxazolines containing pyridyl pendant groups was reported recently [158]. In this construction, pyridyl moieties form axial complexes with zinc porphyrins. Upon photoexcitation, energy transfer between the singlet excited state of the porphyrin and the nanotube was observed in the complexes.

As discussed earlier, the functionalization of SWCNTs with polymers can be of particular interest to introduce a large number of repetitive units on nanotubes without inducing the transformation ($sp^2 \rightarrow sp^3$) of too many carbon atoms of the framework. Several studies have been reported, including the functionalization of SWCNTs with poly(sodium 4-styrenesulfonate) to form PSS^{n-}-SWCNT [159]. The negative charges on the polymer were used to form an electrostatic complex with a positively charged porphyrin (H_2P^{8+}) (Figure 3.11) [160,161]. In another example, polyvinylpyridines were attached to SWCNTs and zinc porphyrins were complexed axially to the pyridine moieties (Figure 3.11) [162].

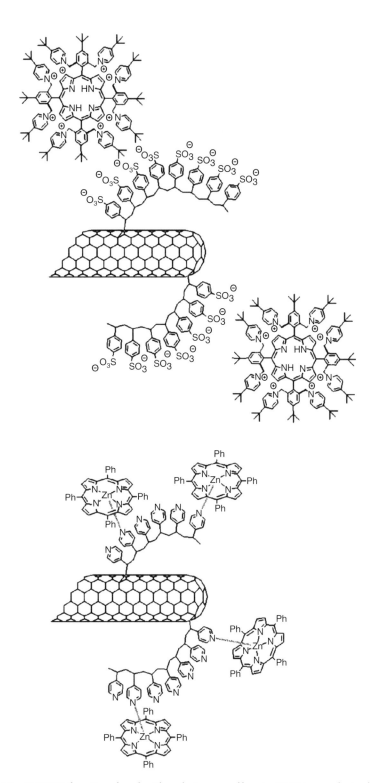

Figure 3.11 *SWCNT functionalized with polystyrenesulfonate (PSS^n-) or polyvinylpyridine (PVP) complexed with porphyrin derivatives.*

The polymers covalently linked to the sidewalls ensure the solubility of the nanotubes and the interaction with the chromophores. Since only a limited number of polymer chains are attached on the nanotubes, the absorption spectra showed that the electronic fine structures of the SWCNT are maintained in the vis-NIR region. The complexation of SWCNT-polymer conjugates with porphyrin derivatives was followed by absorption and fluorescence spectroscopy. In the complexes, the photoexcitation of the porphyrin chromophore led to a rapid and efficient intrahybrid charge transfer. The lifetime of the charge-separated radical ion pair was found to be in the range of several microseconds ($11 \mu s$ for PSS^{n-}/H_2P^{8+}-SWCNT and $3.8 \mu s$ for PVP/ZnP-SWCNT).

3.3.1.3 Supramolecular Approach

For applications where the electronic properties of SWCNTs are important, the most promising approach remains the pure supramolecular functionalization (i.e., noncovalent association of nanotubes with electron donors). The interaction of nanotubes with functionalized surfactants is so far the easiest method to disperse nanotubes in water or in organic solvents and many examples have been reported during the last five years [163–167].

3.3.1.3.1 Wrapping Molecular Entities around a SWCNT

The dispersion of SWCNTs by wrapping with a photoactive polymer was reported [168]. The polymer is in fact a copolymer of methyl methacrylate and porphyrin-modified methacrylic acid (Figure 3.12). The photophysical properties of the wrapped SWCNTs were determined by means of transient absorption spectroscopy. The complex gave rise to a photoinduced electron transfer from the porphyrin moiety to the nanotube. A similar approach was described for the complexation of SWCNTs with porphyrin-rich units [169]. Here, a macromolecule containing 16 porphyrin pendant groups was used to disperse the nanotubes. The authors suggested that a diameter-selective dispersion was accomplished through noncovalent complexation of the nanotubes with the flexible porphyrin polypeptide. In addition, photoexcitation of the supramolecular complex affords the long-lived charge-separated species. An interesting feature of these approaches is that the control of the chromophore quantity incorporated in the copolymer can allow for a fine tuning of the properties of the resulting nanohybrid.

Photovoltaic devices were fabricated very simply by mixing SWCNTs with poly-3-octylthiophene (P3OT) in chloroform. The photoactive films were deposited by drop or spin casting from a solution on indium-tin oxide (ITO) on quartz substrates followed by evaporation of aluminum [170]. Diodes consisting of Al/SWCNT-P3OT/ITO with a low nanotube concentration (<1%) showed photovoltaic behavior, with an open circuit voltage of 0.7–0.9 V. The short circuit current was increased by two orders of magnitude as compared with the pristine polymer diodes. It was proposed that the main reason for this increase was the photoinduced electron transfer at the polymer/nanotube interface. SWCNT/conductive polymer composites may represent an alternative class of organic semiconducting materials, promising for organic photovoltaic cells with improved performance.

3.3.1.3.2 π-Stacking

Large π-conjugated systems exhibit strong interaction with the nanotube sidewalls. In a family of aromatic compounds containing a polar head, polyaromatic compounds such as anthracene or, even better, pyrene, were shown to be able to give stable suspensions of SWCNTs [167]. Aromatic macrocycles such as porphyrins or

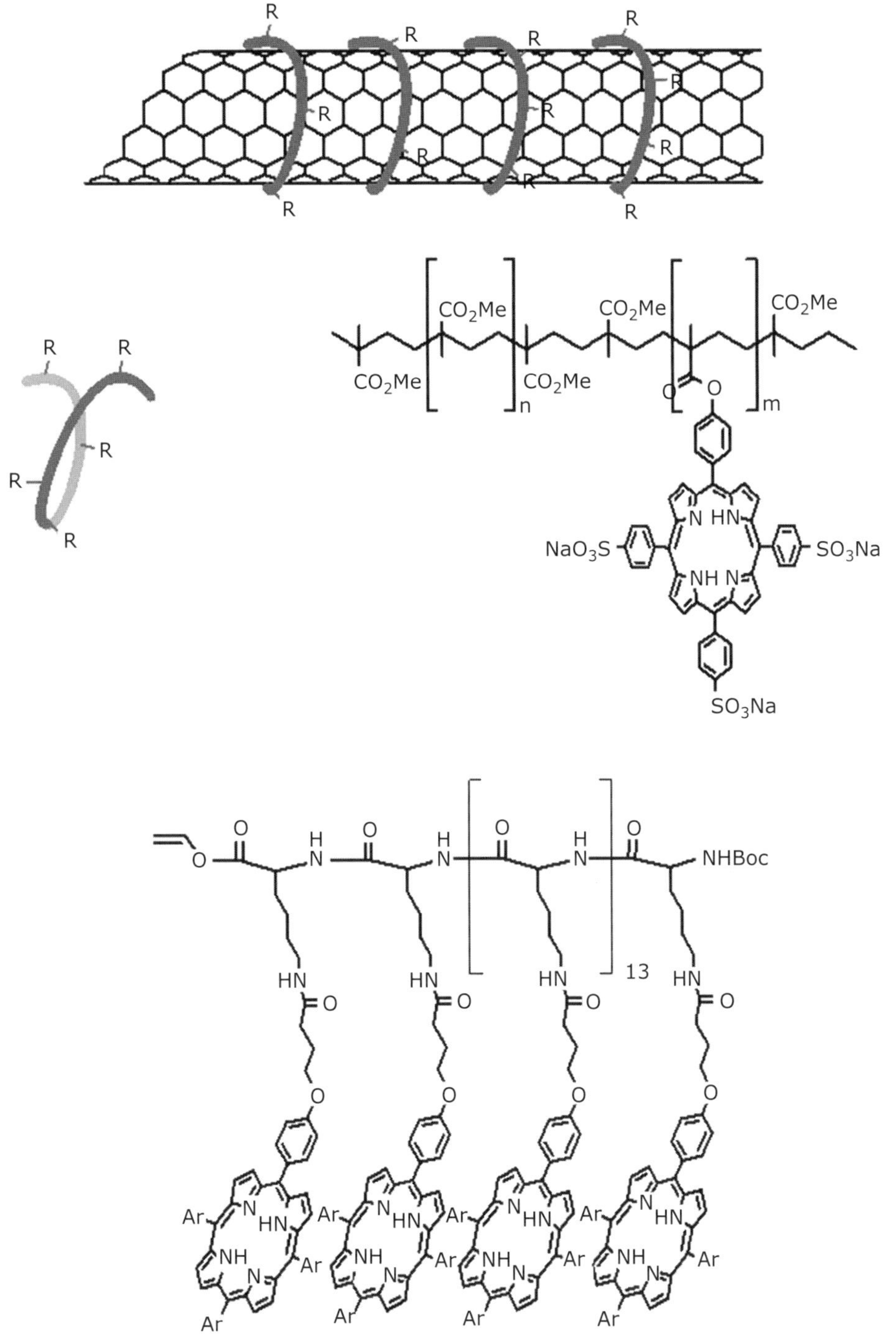

Figure 3.12 Schematic representation of an oligomer or a polymer wrapped around a SWCNT.

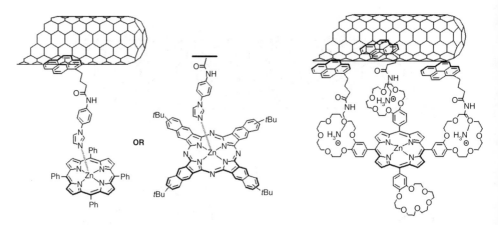

Figure 3.13 *Pyrene-SWCNT supramolecular assemblies complexed with porphyrins or naphthalocyanines.*

phthalocyanines are also suitable for nanotube dispersion. In particular, several groups demonstrated that monomeric [7,171–173] or polymeric porphyrins [174] as well as fused porphyrins [175] could stick to the nanotube surface by π-stacking interactions. However, despite the simplicity and the versatility of these systems, their photovoltaic properties have not been tested yet.

The dispersion of SWCNTs with a pyrene derivative bearing an imidazol ring was achieved. The imidazol moiety was used for axial complexation of zinc porphyrin (ZnP) and naphthalocyanine (ZnNc) derivatives (Figure 3.13) [176]. Photophysical measurements showed efficient fluorescence quenching of the donor ZnP and ZnNc entities in the nanohybrids and revealed that the photoexcitation of the chromophores results in the one-electron oxidation of the donor unit with a simultaneous one-electron reduction of the SWCNTs. The experiments were also conducted in the presence of electron and hole mediators (hexyl-viologen di-cation and 1-benzyl-1,4-dihydronicotinamide respectively). Accumulation of the radical cation (HV$^{•+}$) was observed in high yields, which provided additional proof for the occurrence of photoinduced charge separation. The same type of experiments were also carried out with porphyrins bearing 18-crown-6 substituents and complexed to nanotubes via pyrene ammonium salt anchors [177].

Moreover, pyrene-containing TTF and π-extended TTF were synthesized and combined with nanotubes via π-stacking interactions (Figure 3.14) [178,179]. Photoexcitation of the SWCNT-exTTF nanohybrid allowed, for the first time, a complete characterization of the radical ion pair state, especially in light of injecting electrons into the conduction band of SWCNTs. These electrons, injected from photoexcited exTTF, shift the transitions that are associated with the van Hove singularities to lower energies.

Amphiphilic pyrene derivatives are known to disperse carbon nanotubes through π-π interactions [166]. Using 1-(trimethylammonium acetyl)pyrene bromide (pyrene$^+$), CNTs were dispersed in aqueous media and were combined with negatively charged chromophores [180–183]. The interactions between pyrenes and nanotubes were investigated by absorption and emission spectroscopy. The 'immobilization' of pyrene on SWCNTs causes a slight red-shift (i.e., 1–2 nm) of the π–π* transitions of pyrene$^+$ indicating electronic

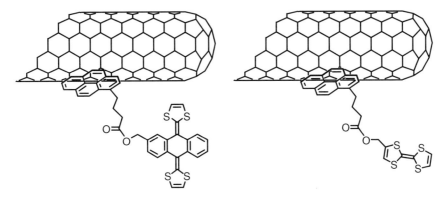

Figure 3.14 *SWCNT functionalized by π-stacking with pyrene-TTF derivatives.*

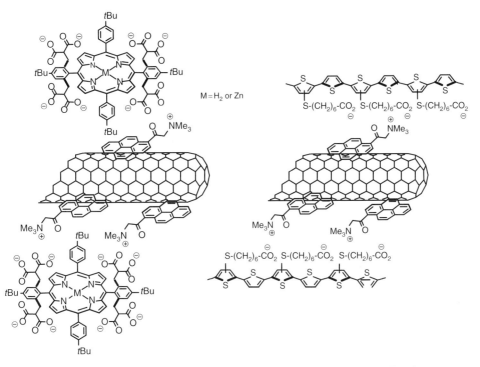

Figure 3.15 *Example of supramolecular donor/acceptor assemblies formed by electrostatic interactions.*

communication between the two components of the system. To prepare donor-acceptor complexes, the trimethylammonium group of pyrene$^+$ was used as an electrostatic anchor to bind anionic porphyrin (H$_2$P^{8-} and ZnP^{8-}) or polythiophene derivatives (Figure 3.15). In the SWCNT/pyrene/porphyrin composite systems, fluorescence and transient absorption studies in solutions showed rapid intrahybrid electron transfer, creating intrinsically

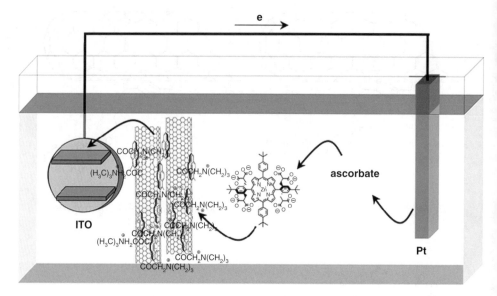

Figure 3.16 *Schematic representation of an electrochemical photovoltaic cell containing carbon nanotubes, positively charged pyrenes and negatively charged porphyrins.*

long-lived radical ion pairs. Following the initial charge separation event, the spectroscopic features of the oxidized donors disappear with time. Through analysis at several wavelengths, it was possible to obtain lifetimes for the newly formed ion pair state of about $0.65\,\mu s$ and $0.4\,\mu s$ for H_2P^{8-} and ZnP^{8-} respectively [123].

The favorable charge separation features that result from the combination of SWCNT with porphyrins in SWCNT/pyrene$^+$/MP^{8-} (M = H$_2$ or Zn) are promising for the construction of photoactive electrode surfaces. Using electrostatically driven layer-by-layer assembly technique, semitransparent ITO electrodes have been realized from SWCNT/pyrene$^+$/MP^{8-} and -SWCNT/pyrene$^+$/polythiophene^{n-}. The ITO electrodes were first coated by poly(diallyl dimethylammonium) chloride (PDDA^{n+}) and sodium poly(styrene-4-sulfonate) (PSS^{n-}). The hydrophobic interactions between the surface and the polymer chains ensure the stability of the modified electrode on which the nanotubes will be deposited. After deposition of a layer of pyrene$^+$-SWCNTs on ITO coated with PDDA^{n+}/PSS^{n-}, the layer of negatively charged porphyrin or polythiophene was deposited. The process was repeated to obtain electrodes containing up to 15 layers of SWCNT/pyrene$^+$/MP^{8-} (or polythiophene^{n-} instead of MP^{8-}) [161,182,183].

The photoelectrochemical cells were finally constructed using a Pt electrode connected to the modified ITO electrode. An example of a cell is given in Figure 3.16. Upon illumination, electron transfer from the porphyrins to the nanotubes occurs. The electrons are then injected into the ITO layer and then travel to the Pt electrode. The oxidized porphyrins are converted to their ground state through reduction via sodium ascorbate, which serves as a sacrificial electron donor. These systems gave rise to promising monochromatic internal photoconversion efficiencies of up to 8.5% [161].

A similar approach has been recently described, in which ITO electrodes were modified with carbon nanotubes and connected to ruthenium complexes via viologen derivatives [184]. The formation of the donor-acceptor hybrids (SWCNT/viologen/Ru complexes) was ensured by electrostatic interactions, typically the interaction of carboxylate groups on SWCNTs and of Ru complexes with the ammonium groups of the viologen derivative. Photovoltaic properties of the ITO-modified electrode were measured in a three-electrode glass cell containing a reference electrode (Ag/AgCl) and a platinum counter electrode in a solution of I^-/I_3^- in acetonitrile. The device showed a photocurrent of $10.3 \, nA \, cm^2$ under white light illumination ($100 \, mW \, cm^2$).

3.3.2 Chemical Sensors (FET-Based)

As seen in Chapter 1, the electronic properties of SWCNTs depend on their helicity, as the only material known to date to have this unique property [185]. The use of carbon nanotubes for producing field effect transistors (FETs) has been extensively studied but for such applications, only semiconducting tubes are suitable [186].

The first examples of carbon nanotube field effect transistors (CNT-FETs) (Figure 3.17) were reported in 1998 simultaneously by Avouris [187] and Dekker [188]. The transistors were made from laser-ablation nanotubes dispersed by sonication and deposited on a Si/SiO_2 surface patterned with noble metal electrodes. Since SWCNTs are made from rolled up graphene, all the atoms constituting the nanotubes are located on the surface and there are no constituent atoms inside, as it is the case for bulk materials or MWCNTs. All the atoms of a SWCNT are in close contact with the environment, making nanotubes highly sensitive materials.

Recently, Blanchet *et al.* demonstrated on CNT-FETs that the covalent functionalization of SWCNTs with fluorinated olefins can drastically increase the on/off ratio (and so improve the transistor characteristics) without affecting too much the carrier mobility (at least for low functionalization degrees) [189]. As it has been already observed for diazonium additions to SWCNTs [190], the fluorinated olefins reacted preferentially on metallic SWCNTs via a [2+2] cycloaddition. For low concentrations of olefin (less than $0.01 \, M$ of olefin per M of SWCNTs), mainly metallic nanotubes were functionalized while the semiconducting ones were unaffected. For high concentrations of olefin, the semiconducting tubes reacted as well and the properties of the devices were, in this case, affected. Cabana and Martel also explored the influence of covalent functionalization on the properties of CNT-FET [191]. In particular, they studied the reversibility of the diazonium addition reaction by cycling functionalization and defunctionalization processes. They demonstrated that the reaction was not fully reversible and led to an accumulation of defects onto the SWCNT sidewall during the reaction/annealing cycles which degrade the properties of the devices.

In 2000, chemical sensors based on individual single-walled carbon nanotubes FET were described [192]. The electrical characteristic of the transistors changed upon exposure to gaseous molecules such as NO_2 or NH_3 in argon flow. The nanotube sensors exhibited a fast response and a high sensitivity for $1\% \, NH_3$ and $200 \, ppm \, NO_2$. The reversibility was achieved by slow recovery under ambient conditions or by heating to high temperatures. At the same time, the effect of oxygen on CNT-FETs was demonstrated [193].

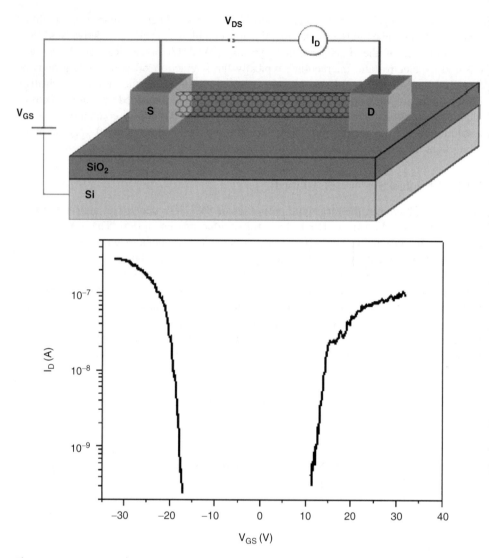

Figure 3.17 *(Top) Schematic representation of a carbon nanotube field effect transistor (CNT-FET). (Bottom) Representation of the ambipolar electrical characteristics of a CNT-FET.*

By comparing the characteristics of CNT-FETs in vacuo and in the presence of oxygen, it was observed that exposure to air or oxygen influences the electrical resistance of the SWCNT; the resistance decreased when the devices were exposed to oxygen. In these two examples, one can see that adsorbed gas strongly influences the behavior of the CNT-FETs, thus demonstrating that CNTs are extremely sensitive to their chemical environment. However, this extreme sensitivity is countered by a lack of selectivity. In 2003, the fabrication of CNT-FET arrays was reported for the detection of gas molecules [194]. The functionalization of the nanotubes with polymers was used to impart high sensitivity and selectivity to the sensors. Polyethyleneimine coatings afforded n-type nanotube devices capable of detecting NO_2 at less than 1 ppb concentrations while being insensitive to NH_3. Coating nanotubes with Nafion (a sulfonated tetrafluorethylene copolymer) allowed for the selective detection of NH_3. CNT-FETs functionalized noncovalently with peptide-modified polymers were also tested for the selective detection of heavy metals [195]. They allowed for the detection of metal ions with concentrations in the pico- to micromolar range using several different peptides.

Another way was explored for ammonia detection [196,197]. The approach was based on covalent functionalization of SWCNTs with poly(m-aminobenzenesulfonic acid) (PABS). The polyaniline-based polymer linked to nanotubes was found to be sensitive to NH_3 by deprotonation of the sulfonic groups of the PABS side chains. The devices fabricated with PABS-SWCNTs showed an increase of resistance during exposure to ammonia compared to pristine nanotubes. The PABS-SWCNT sensors rapidly recovered their resistance when NH_3 was replaced with nitrogen and this system allowed detection of NH_3 at concentrations as low as 5 ppm.

The use of carbon nanotubes as gas sensors went beyond the research step. Nanomix Co (www.nano.com) commercialized a detection platform based on carbon nanotube networks. The principle is based, as in the other cases, on the changes in the electronic characteristics of the device as it interacts with the analyte. The carbon nanotube networks are functionalized with different recognition agents to induce the proper performance characteristics such as specificity, sensitivity, and so on. [198–202]. Notably, the coating of CNT networks by metal nanoparticles for gas detection was described [202]. In this work, the differences in catalytic activity of eighteen metals for detection of H_2, CH_4, CO and H_2S gas were examined. The electronic response of metal-decorated CNT-FET devices to all four combustible gases was similar in that the presence of a combustible gas resulted in a decrease in the device conductance. Furthermore, a sensor array was fabricated by site-selective electroplating of Pd, Pt, Rh, and Au metals on isolated SWCNT networks located on a single chip (Figure 3.18). The resulting electronic sensor array was exposed to a randomized series of toxic/combustible gas and the electronic responses of all sensor elements were recorded and analyzed using statistical analysis tools allowing the determination of the specific response of each element.

In the previous series of examples, it has been possible to highlight the high sensitivity of CNT towards the detection of chemicals through the direct interaction between the analyte and the nanotube or through the interaction between the analyte and a receptor in close interaction with the nanotube. If the receptor is light sensitive, its light-induced transformation can change the environment of the nanotube which can detect the event electrically.

3.3.3 Opto-Electronic Devices (FET-Based)

Recently, the use of functionalized CNT-FETs as light detectors was reported [203]. The nanotube transistors were functionalized noncovalently with a zinc porphyrin derivative by drop casting of a solution of porphyrin onto the SWCNT network. Upon illumination, the response of the device was a shift of the threshold voltage toward positive voltages, indicating hole doping of the SWCNTs. The direction of the threshold voltage shift indicates a photoinduced electron transfer from the nanotubes to the porphyrins. In the

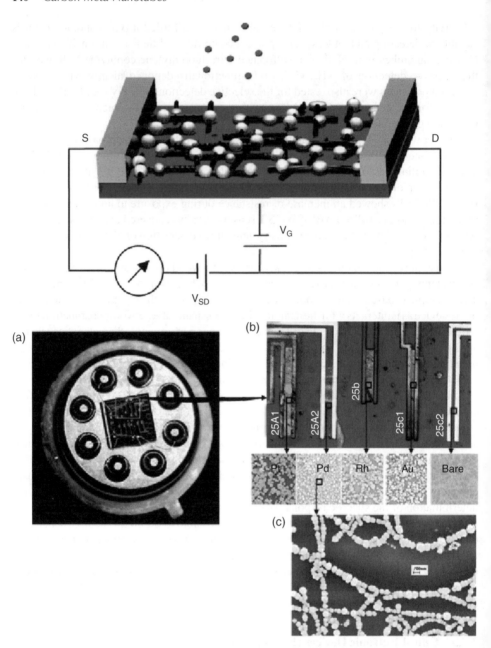

Figure 3.18 *Gas sensors based on CNT-FETs functionalized with metal nanoparticles. Reprinted with permission from [202] Copyright (2006) American Chemical Society.*

same way, Zhou *et al.* fabricated a nanoscale color detector based on a single-walled carbon nanotube functionalized with azobenzene chromophores [204]. The subunits were assembled through pi-stacking interactions between the nanotube and the pyrene moieties of the chromophores. By synthesizing azobenzene derivatives with specific absorption

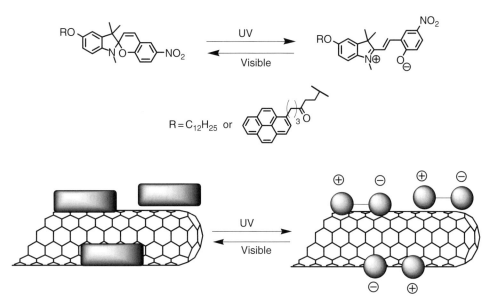

Figure 3.19 *SWCNT functionalized by π-stacking with photoactive switches.*

windows, they demonstrated the controlled detection of visible light of low intensity in narrow ranges of wavelengths. The effect was attributed to the *cis/trans* isomerization of the azobenzene moieties which creates a large change in the dipole moment, thereby altering the electrostatic environment of the nanotube.

In a similar approach, photochromic molecules such as spiropyrans [205] were used to switch the conductance of a single-walled carbon nanotube transistor. Spiropyrans are well known photoswitchable molecules. The spiro form (colorless), in which the two aromatic parts of the molecule are separated by a spiro sp^3 carbon, can open under UV irradiation leading to a completely conjugated zwitterionic molecule (called merocyanine). CNT-FETs were functionalized with spiropyran molecules and the influence of the irradiation on the transistor characteristics was studied [206]. These authors used an alkyl chain or a pyrene moiety as an anchor to hold the photoswitchable spiropyrans in proximity to the tube surface (Figure 3.19). Under UV irradiation, the conductance of the photosensitive device decreased significantly while the threshold voltage did not change appreciably, and then, after irradiation with visible light, the initial characteristics were restored. Therefore, the decrease of conductivity is due to the isomerization of the spiropyrans and the authors explained this by the fact that the charge-separated state of the merocyanine introduces scattering sites for the carriers by creating localized dipole fields around the tubes. These sites then scatter charges when they flow in the nearby SWCNT channel and thereby lower the mobility in the devices. Another possibility may be that the nearby phenoxide ion quenches the *p*-type carriers in the tubes and behaves like a charge trap.

Light-induced isomerization of azobenzene-based chromophores was also used to control the CNT-FET characteristics [207]. First the nanotube-based transistors were fabricated and then the SWCNTs were functionalized with an azobenzene derivative bearing a pyrene subunit using π-stacking interactions. Upon UV illumination, the conjugated chromophore gave rise to *cis-trans* isomerization leading to charge redistribution near the nanotube. This charge redistribution changed the local electrostatic environment, shifting the threshold voltage and increasing the conductivity of the nanotube transistor. The conductance change was reversible and repeatable over long periods of time.

In 2004, a CNT-FET was combined with a photosensitive polymer to fabricate optoelectronic memory devices [208]. They demonstrated that the electrical characteristics of CNT-FETs coated with poly{(*m*-phenylenevinylene)-*co*-[(2,5-dioctyloxy-*p*-phenylene)vinylene]} (PmPV) or poly-3-octylthiophene (P3OT) changed upon illumination, and required a long time to recover their original conductance after the illumination was stopped. However, the origin of this effect remained mainly unexplained until 2006, when the group of Bourgoin fabricated the same kind of CNT-FETs based on one or a few nanotubes and covered them with P3OT (Figure 3.20) [209]. This group proposed a mechanism for the memory effect based on the trapping of photogenerated electrons at the nanotube/gate dielectric interface. The optical gating mechanism can be understood under the assumption that, when the photo-excitation is turned on, a large quantity of electron-hole pairs are generated throughout the whole polymer film. Figure 3.20b shows that the CNT-FET gets settled in its 'ON'-state regardless of V_{GS}. The *p*-type transistor induces the depletion of holes and accumulation of electrons at the polymer/SiO$_2$ interface. Because of their density and proximity, these trapped electrons define the nanotube conductance more efficiently than the back gate. When the illumination is stopped and due to the positive gate voltage, the electrons remain trapped and keep the transistor 'ON'. The device is brought back to its initial state by applying a short negative V_{GS} pulse (-4 V).

3.3.3.1   *SWCNT Electrodes*

In 2005, a method to connect molecules across gaps between cut nanotubes and to study the conductance through the junction was described [210]. To this aim, gaps between carbon nanotubes grown on a surface were first opened via oxidative cutting. Then the molecules under study were introduced by covalent coupling on the carboxylic functions decorating the mouths of both cut parts of the tube. This technique gave rise to SWCNT electrodes separated by gaps of ≤ 10 nm bridged with a series of molecules. It is important to note that the reconnected SWCNTs recovered their original general electrical behavior (either metallic or semiconducting). The nanotube gaps were functionalized with several π-conjugated molecules like benzoxazole derivatives, oligothiophenylethynylenes, terpyridine complexes or oligoanilines (Figure 3.21a). A series of protonation and deprotonation experiments were performed on the oligoaniline-based devices and the result was that the protonated form was more conductive than the neutral form. These devices provided a local probe for monitoring pH on the basis of one or only a few molecules. Molecular switches based on photoisomerizable diarylethene derivatives were also introduced in the nanotube gaps (Figure 3.21b) [211]. As in the case of spiropyran-merocyanine systems, diarylethene derivatives [212] possess both open and closed forms. However, in the case of diarylethene derivatives, the open form is nonconjugated while the closed form is conjugated. It was found that under UV irradiation (ring closure), the conductance of the devices increased up to 25-fold when thiophene-based switches were tested. However, these structures

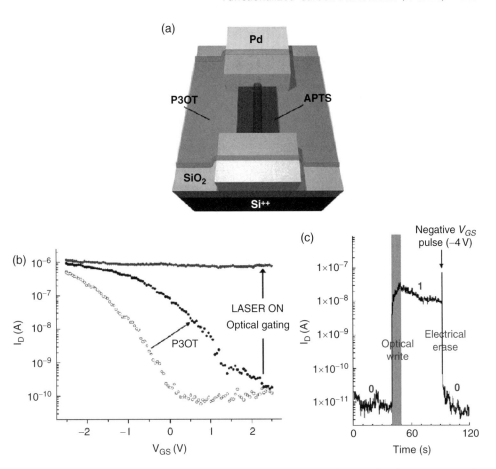

Figure 3.20 *(a) Representation of the optical gated CNT-FET; (b) Transfer characteristics of the naked transistor in the dark (open circles), coated with P3OT in the dark (filled black circles), and upon illumination (λ = 457 nm, grey circles); (c) Principle of the writing and erasing of the memory device. The band in grey represents the light pulse used to write the electric information. Reprinted with permission from [209] Copyright (2006) John Wiley and Sons.*

did not permit one to demonstrate the reversibility of the isomerization, which was demonstrated using pyrrole-based switches instead.

In another study, cut SWCNTs were first functionalized with diaminofluorenone derivatives and then biotin probes were attached. These probes were linked to gold nanoparticles coated with streptavidin through formation of noncovalent complexes (Figure 3.21c) [213]. Each step of the chemical functionalization and biological assembly was detected electrically at the single event level. The formation of the biotinylated derivatives gave rise to a decrease in the 'ON'-state resistance and threshold voltage of the devices. When coupling with streptavidin gold nanoparticles, large changes in the resistance were noticed. Because these devices are able to sense individual binding events, this approach makes possible the formation of ultrasensitive and real-time measurements of individual binding events.

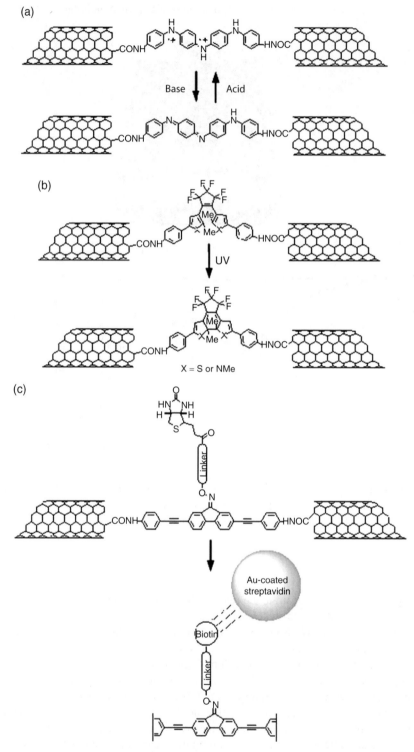

Figure 3.21 *Representation of nanotube gaps functionalized with active molecules.*

Very recently, the same group attached several DNA sequences between single-walled carbon nanotube electrodes and measured their electrical properties [214]. Well-matched duplex DNA in the gap between the electrodes exhibited a resistance on the order of a few MΩ. However, a single mismatch in the DNA sequence led to a dramatic increase of the resistance of the devices.

3.3.4 Biosensors

Electronic detection of biomolecules using carbon nanotubes appears to be a field of intense research and several reviews have recently listed the principal efforts made in this direction [215–218]. In this section, we will mainly focus on sensors based on nanotube field effect transistors but we will also give the principles underlying and a few examples of nanotube-based electrochemical biosensors.

3.3.4.1 CNT-FET-Based Biosensors

In 2002, the interactions were investigated between streptavidin (SA) and SWCNTs, which demonstrated that proteins could link through nonspecific binding with nanotubes via hydrophobic interactions [219]. To prevent such nonspecific binding, nanotubes were functionalized by the co-adsorption of triton and poly(ethylene glycol) on the sidewalls. This polymer coating did not allow for the fixation of streptavidin. On the other hand, specific binding of SA onto SWCNTs was achieved by the cofunctionalization of nanotubes with biotin and protein-resistant polymers. A similar approach was used to fabricate nanotube-based biosensors capable of the selective detection of biological objects in solution [220]. In this work, several biological targets like biotin, staphylococcal protein A (SpA) and U1A antigen (a 33 kDa protein) were covalently attached to Tween 20 surfactant. These assemblies were 'immobilized' on the nanotube surface for specific recognition respectively with streptavidin, immunoglobulin G (IgG) and 10E3 monoclonal antibodies (Figure 3.22a-c). The binding process of the biological objects with their respective targets immobilized on CNTs was followed by quartz crystal microbalance (QCM) analysis and CNT-FET electrical resistance measurements. For the microbalance measurements, the immunosensing system was assembled on a QCM crystal, whereas the electrical measurements were conducted on the CNTs bridging two microelectrodes. For example in the case of SpA-IgG system, the QCM frequency decreased upon the specific binding of the IgG whereas little perturbation was observed for proteins that did not interact specifically with SpA (Figure 3.22d). In separate experiments, it was demonstrated that CNT-FET coated with the SpA-Tween conjugate exhibited specific detection with an appreciable conductance change upon exposure to IgG but not to unrelated proteins (Figure 3.22e). Thus, specific interactions between antibodies and antigens can be probed by using nanotubes directly as electronic transducers.

Another example of Tween-functionalized CNT-FET was reported recently [221]. The nanotube-based sensor was designed to selectively detect thrombin (a coagulation protein) through selective interaction of the protein with a particular DNA sequence. The fabrication of the sensor was based on the modification of activated Tween 20 adsorbed on the side wall of the carbon nanotube transistor with a 3'-amino-modified single stranded DNA. The electrical transfer characteristics of the CNT-FET were measured at each process stage. The immobilization of the single stranded DNA caused a rightward shift in the gate-threshold voltage, presumably due to the negatively charged DNA backbone. Upon addition of thrombin, a sharp decrease in conductance of the device was observed. The

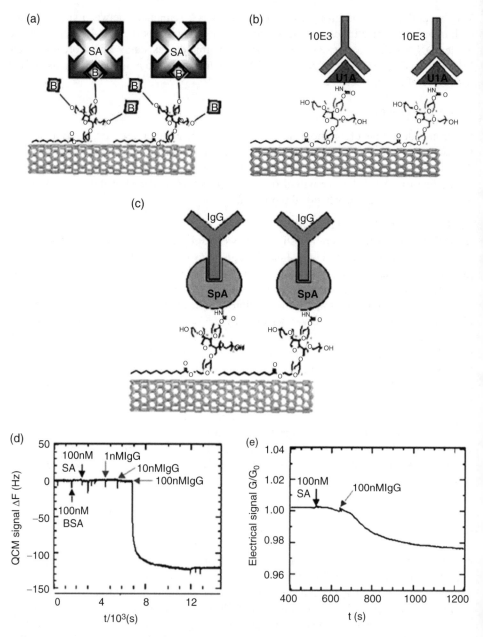

Figure 3.22 *(a–c) Schematic representation of SWCNTs non-covalently functionalized with bioactive species (biotins, U1A antigen and staphylococcal protein A). The bioactive moieties attached to Tween 20 surfactant ensured the recognition properties of the system and allowed for specific detection; (d–e) Examples of QCM and electrical characteristic curves are given after introduction of immunoglobulin G (IgG). Reprinted with permission from [220] Copyright (2003) National Academy of Sciences, USA.*

sensitivity of the device increased strongly up to a protein concentration of about 100 nM and then became saturated at around a concentration of 300 nM. The lowest detection limit of the sensor reported in this work was around 10 nM.

Steptavidin exhibits a strong affinity for biotin; the dissociation constant of the complex is on the order of $\sim10^{-15}$, ranking among one of the strongest noncovalent interactions known. This explains why the biotin-streptavidin complex has been extensively used in nanotechnology. The fabrication of a CNT-FET sensitive to SA using a biotin functionalized carbon nanotube bridging two microelectrodes was reported [198]. The CNT-FET was coated with a mixture of polyethyleneimine/polyethyleneglycol (PEI/PEG) and the amino groups present on PEI were allowed to react with biotin *N*-hydroxy-succinimidyl ester to permit the specific binding of SA on the device. The characteristics of the transistor showed significant changes upon streptavidin exposure. The control experiment, carried out on a coated CNT-FET but in the absence of biotin on PEI, permitted researchers to demonstrate that SA did not bind to the polymer layer. AFM was used to show the SA binding: the biotinylated device was exposed to streptavidin labeled with gold nanoparticles. The presence of the nanoparticle on the images confirmed the presence of SA.

The realization of individual CNT-FET functionalized with glucose oxidase (GOx) bearing pyrene moieties through π-π interactions was described [222]. Controlled immobilization of GOx onto the sidewall of a semiconducting SWCNT resulted in the decrease of the nanotube conductance. The conductivity of the GOx-coated SWCNTs exhibited a strong dependence on pH and showed an increase in conductance upon addition of glucose, suggesting their potential use as a sensor for enzymatic activity.

In 2005, the selective detection of prostate specific antigen (PSA), which is an oncological marker for the presence of prostate cancer, was reported using both n-type In_2O_3 nanowire and p-type carbon nanotube transistors [223]. The originality of this approach is the complementary detection of PSA using n- and p-type devices. To ensure the selective binding of PSA, the devices were functionalized with an anti-PSA monoclonal antibody. In the case of CNT-FETs, the SWCNT surface was first functionalized with 1-pyrenebutanoic acid succinimidyl ester followed by treatment with the PSA antibody solution. For the In_2O_3 nanowires, 3-phosphonopropionic acid was anchored on the surface after which the antibodies were introduced after the activation of the carboxylic groups. Upon addition of PSA, an enhancement of the conductance for nanowire devices and a reduction of conductance for CNT-FET were observed. The gate dependence of both the nanowire and the SWCNT devices also changed and in both cases, the threshold voltage of the nanowire device was shifted toward a more negative value. The influence of PSA resulted in an n-doping of the devices. The real-time detection measurements showed that upon exposure to PSA, the nanowire device showed an increase in conductance for protein concentration of 0.14 nM while the SWCNT device exhibited a decrease in conductance for protein concentrations of 1.4 nM. Therefore, the sensitivity of the In_2O_3-based sensor was found to be better than the one for a carbon nanotube-based sensor.

3.3.4.2 *Electrochemical Sensors*

Carbon materials have been widely used as components in electrochemical biosensors for decades and notably in these last years, carbon nanotubes have attracted particular attention from the scientific community [217,224,225]. The outstanding ability of CNTs to accept and transport charges makes them very promising materials for their incorporation in

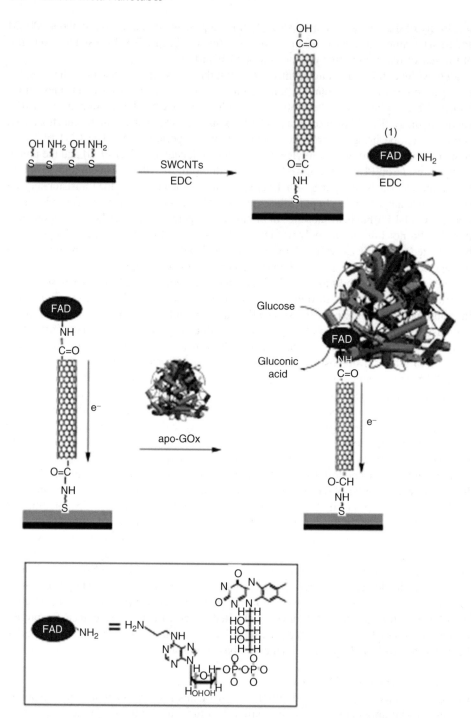

Figure 3.23 *Electrochemical sensor obtained by immobilization of glucose oxidase on a SWCNT vertically aligned on a gold electrode. Reprinted with permission from [229] Copyright (2004) John Wiley and Sons.*

electrochemical sensing devices. The modification of glassy carbon or metal electrodes with nonfunctionalized carbon nanotubes has been reported to improve the characteristics of the electrodes [225]. However, to improve the solubility and the specificity of the nanotubes in the device, their chemical functionalization seems much more interesting. For example, in order to detect DNA hybridization, amino terminated single strand DNA were covalently linked to oxidized MWCNTs on a gold substrate [226] or on a glassy carbon electrode [227,228]. In general, the fabrication of electrochemical sensors requires the use of MWCNTs because these nanotubes exhibit a metallic character and they are easier to manipulate as compared to SWCNTs. Only a few examples have reported the use of modified SWCNTs for electrochemical sensing applications.

The realization of an electrochemical glucose sensor was reported, based on the association of GOx on carbon nanotube modified electrodes [229]. SWCNT arrays were fabricated on a gold electrode by covalent linkage of shortened carbon nanotubes on a cystamine/thioethanol mixed self-assembled monolayer. An amino derivative of the flavine adenine dinucleotide cofactor (FAD) was attached on the second extremity of the nanotubes and then GOx was reconstituted on the surface of the SWCNT array. The nanotubes perpendicularly oriented to the surface played the role of electron acceptor and charge carrier from the reactive center to the electrode (Figure 3.23). Quartz crystal microbalance experiments were performed to estimate the surface coverage of the SWCNTs and cyclic voltammetry of FAD modified nanotube electrode was used to prove that the FAD units were electrically connected with the surface. Coulometric assay of the FAD redox wave and microgravimetric QCM experiments indicated an average surface coverage of about $1.5 \cdot 10^{-10}$ mol·cm^{-2}. Finally, the binding of GOx on the FAD cofactor was supported by AFM. Upon addition of glucose, an increase of the electrocatalytic anodic current was observed as the concentration of glucose increased and it was shown that the electron transfer rate was strongly dependent on the nanotube lengths. For example, very short nanotube wires on surface (i.e., 25 nm) improved the electrical communication between proteins and electrode.

In a similar way, SWCNTs functionalized with ferrocene were tested as an amperometric glucose sensor [230]. For this purpose, a Fc-SWCNT derivative (see Figure 3.8) was coimmobilized with glucose oxidase within a thin polypyrrole film adsorbed onto the glassy carbon electrode surface. The Fc-SWCNT/GOx/polypyrrole films were examined for their catalytic properties with respect to glucose oxidation. For the detection, the modified electrode potential was held at 0.5 V which corresponds to oxidation of ferrocenyl moieties, and the anodic current was monitored while adding subsequent amounts of glucose. After each addition, an anodic current step was observed, reaching a stationary value within 10 s; the glucose sensitivity of the composite film was found to be about 0.3 mA·M^{-1}·cm^{-2}. In contrast, no response was obtained with pure Fc-SWCNT/polypyrrole films, that is, in the absence of the redox protein.

3.4 Conclusion

The already rich variety of CNT applications can be further improved when these carbon cylinders are functionalized. The main reason is that *f*-CNT become more versatile while the new molecular pieces can be combined with the CNT properties. The result is a wide variety of interesting hybrid materials that can be used in a high number of applications. This field is in current expansion, as new methodologies for the functionalization of CNT are produced continuously and new applications are reported with improved performance.

References

[1] J. Zhao, J.P. Lu, J. Han and C.K. Yang, Noncovalent functionalization of carbon nanotubes by aromatic organic molecules, *Appl. Phys. Lett.*, **82**, 3746–3748 (2003).

[2] J. Chattopadhyay, I. Galeska and F. Papadimitrakopoulos, A route for bulk separation of semiconducting from metallic single-wall carbon nanotubes, *J. Am. Chem. Soc.*, **125**, 3370–3375 (2003).

[3] R.J. Chen, Y. Zhang, D. Wang and H. Dai, Noncovalent sidewall functionalization of single-walled carbon nanotubes for protein immobilization, *J. Am. Chem. Soc.*, **123**, 3838–3839 (2001).

[4] B.R. Azamian, J.J. Davis, K.S. Coleman, C.B. Bagshaw and M.L.H. Green, Bioelectrochemical single-walled carbon nanotubes, *J. Am. Chem. Soc.*, **124**, 12664–12665 (2002).

[5] W. Huang, S. Taylor, K. Fu, D. Zhang, T.W. Hanks, A.M. Rao and Y.-P. Sun, Attaching proteins to carbon nanotubes via diimide-activated amidation, *Nano Lett.*, **2**, 311–314 (2002).

[6] X. Wang, Y. Liu, W. Qiu and D. Zhu, Immobilization of tetra-*tert*-butylphthalocyanines on carbon nanotubes: a first step towards the development of new nanomaterials, *J. Mater. Chem.*, **12**, 1636–1639 (2002).

[7] H. Murakami, T. Nomura and N. Nakashima, Noncovalent porphyrin-functionalized single-walled carbon nanotubes in solution and the formation of porphyrin–nanotube nanocomposites, *Chem. Phys. Lett.*, **378**, 481–485 (2003).

[8] M. Zheng, A. Jagota, E.D. Semke, B.A. Diner, R.S. McLean, S.R. Lustig, R.E. Richardson and N.G. Tassi, DNA-assisted dispersion and separation of carbon nanotubes, *Nature Mater.*, **2**, 338–342 (2003).

[9] M.S. Strano, M. Zheng, A. Jagota, G.B. Onoa, D.A. Heller, P.W. Barone and M.L. Usrey, Understanding the nature of the DNA-assisted separation of single-walled carbon nanotubes using fluorescence and Raman spectroscopy, *Nano Lett.*, **4**, 543–550 (2004).

[10] M. Zheng, A. Jagota, M.S. Strano, A.P. Santos, P.W. Barone, S.G. Chou, B.A. Diner, M.S. Dresselhaus, R.S. McLean, G.B. Onoa, G.G. Samsonidze, E.D. Semke, M.L. Usrey and D.J. Walls, Structure-based carbon nanotube sorting by sequence-dependent DNA assembly, *Science*, **302**, 1545–1548 (2003).

[11] G. Seifert, T. Koler and T. Frauenheim, Molecular wires, solenoids, and capacitors by sidewall functionalization of carbon nanotubes, *Appl. Phys. Lett.*, **77**, 1313–1315 (2000).

[12] P.R. Marcoux, J. Schreiber, P. Batail, S. Lefrant, J. Renouard, G. Jacob, D. Albertini and J.-V. Mevellec, A spectroscopic study of the fluorination and defluorination reactions on single-walled carbon nanotubes, *Phys. Chem. Chem. Phys.*, **4**, 2278–2285 (2002).

[13] K.F. Kelly, I.W. Chiang, E.T. Mickelson, R.H. Hauge, J.L. Margrave, X. Wang, G.E. Scuseria, C. Radloff and N.J. Halas, Insight into the mechanism of sidewall functionalization of single-walled nanotubes: an STM study, *Chem. Phys. Lett.*, **313**, 445–450 (1999).

[14] C.W. Bauschlicher, Hydrogen and fluorine binding to the sidewalls of a (10,0) carbon nanotube, *Chem. Phys. Lett.*, **322**, 237–241 (2000).

[15] V.N. Khabashesku, W.E. Billups and J.L. Margrave, Functionalized carbon nanotubes: properties and applications, *Acc. Chem. Res.*, **35**, 1096–1104 (2002).

[16] P.J. Boul, J. Liu, E.T. Mickelson, C.B. Huffman, L.M. Ericson, I.W. Chiang, K.A. Smith, D.T. Colbert, J.L. Margrave and R.E. Smalley, Reversible sidewall functionalization of buckytubes, *Chem. Phys. Lett.*, **310**, 367–372 (1999).

[17] R.K. Saini, I.W. Chiang, H.P. Peng, R.E. Smalley, W.E. Billups, R.H. Hauge and J.L. Margrave, Covalent sidewall functionalization of single wall carbon nanotubes, *J. Am. Chem. Soc.*, **125**, 3617–3621 (2003).

[18] L. Zhang, J.Z. Yang, C.L. Edwards, L.B. Alemany, V.N. Khabashesku and A.R. Barron, Diels-Alder addition to fluorinated single walled carbon nanotubes, *Chem. Commun.* 3265–3267 (2005).

[19] H. Peng, P. Reverby, V.N. Khabashesku and J.L. Margrave, Sidewall functionalization of single-walled carbon nanotubes with organic peroxides, *Chem. Commun.* 362–363 (2003).

[20] J.L. Stevens, A.Y. Huang, H. Peng, I.W. Chiang, V.N. Khabashesku and J.L. Margrave, Sidewall amino-functionalization of single-walled carbon nanotubes through fluorination and subsequent reactions with terminal diamines, *Nano Lett.*, **3**, 331–336 (2003).

[21] E.T. Mickelson, C. Huffman, A.G. Rinzler, R.E. Smalley, R.H. Hauge and J.L. Margrave, Fluorination of single-wall carbon nanotubes, *Chem. Phys. Lett.*, **296**, 188–194 (1998).

[22] E.T. Mickelson, I.W. Chiang, J.L. Zimmerman, P.J. Boul, J. Lozano, J. Liu, R.E. Smalley, R.H. Hauge and J.L. Margrave, Solvation of fluorinated single-wall carbon nanotubes in alcohol solvents, *J. Phys. Chem. B*, **103**, 4318–4322 (1999).

[23] L. Zhao, V.U. Kiny, B. Zheng, J. Liu and T. Viswanathan, Thermal recovery behavior of fluorinated single-walled carbon nanotubes, *J. Phys. Chem. B*, **106**, 293–296 (2002).

[24] L. Zhang, V.U. Kiny, H. Peng, J. Zhu, R.F.M. Lobo, J.L. Margrave and V.N. Khabashesku, Sidewall functionalization of single-walled carbon nanotubes with hydroxyl group-terminated moieties, *Chem. Mater.*, **16**, 2055–2061 (2004).

[25] L. Valentini, D. Puglia, I. Armentano and J.M. Kenny, Sidewall functionalization of single-walled carbon nanotubes through CF_4 plasma treatment and subsequent reaction with aliphatic amines, *Chem. Phys. Lett.*, **403**, 385–389 (2005).

[26] L. Zhang, J. Zhang, N. Schmandt, J. Cratty, V.N. Khabashesku, K.F. Kelly and A.R. Barron, AFM and STM characterization of thiol and thiophene functionalized SWCNTs: pitfalls in the use of chemical markers to determine the extent of sidewall functionalization in SWCNTs, *Chem. Commun.* 5429–5431 (2005).

[27] H. Mauser, A. Hirsch, N.J.R. van Eikema Hommes and T. Clark, Chemistry of convex versus concave carbon: The reactive exterior and the inert interior of C-60, *J. Mol. Model.*, **3**, 415–422 (1997).

[28] Z. Chen, W. Thiel and A. Hirsch, Reactivity of the convex and concave surfaces of single-walled carbon nanotubes (SWCNTs) towards addition reactions: Dependence on the carbon-atom pyramidalization, *Chem. Phys. Chem.*, **4**, 93–97 (2003).

[29] J. Bahr, J. Yang, D.V. Kosynkin, M.J. Bronikowski, R.E. Smalley and J.M. Tour, Functionalization of carbon nanotubes by electrochemical reduction of aryl diazonium salts: a bucky paper electrode, *J. Am. Chem. Soc.*, **123**, 6536–6542 (2001).

[30] J. Bahr and J.M. Tour, Highly functionalized carbon nanotubes using in situ generated diazonium compounds, *Chem. Mater.*, **13**, 3823–3824 (2001).

[31] C.A. Dyke and J.M. Tour, Unbundled and highly functionalized carbon nanotubes from aqueous reactions, *Nano Lett.*, **3**, 1215–1218 (2003).

[32] C.A. Dyke and J.M. Tour, Solvent-free functionalization of carbon nanotubes, *J. Am. Chem. Soc.*, **125**, 1156–1157 (2003).

[33] C.A. Dyke and J.M. Tour, Overcoming the insolubility of carbon nanotubes through high degrees of sidewall functionalization, *Chem. Eur. J.*, **10**, 812–817 (2004).

[34] C.A. Dyke, M.P. Stewart, F. Maya and J.M. Tour, Diazonium-based functionalization of carbon nanotubes: XPS and GC–MS analysis and mechanistic implications, *Synth. Lett.* 155–160 (2004).

[35] M.S. Strano, Probing chiral selective reactions using a revised Kataura plot for the interpretation of single-walled carbon nanotube spectroscopy, *J. Am. Chem. Soc.*, **125**, 16148–16153 (2003).

[36] D. Hill, Y. Lin, L. Qu, A. Kitaygorodskiy, J.W. Connell, L.F. Allard and Y.-P. Sun, Functionalization of carbon nanotubes with derivatized polyimide, *Macromolecules*, **38**, 7670–7675 (2005).

[37] C.G. Salzmann, M.A.H. Ward, R.M. J. Jacobs, G. Tobias and M.L.H. Green, Interaction of tyrosine-, tryptophan-, and lysine-containing polypeptides with single-wall carbon nanotubes and its relevance for the rational design of dispersing agents, *J. Phys. Chem. C*, **111**, 18520–18524 (2007).

[38] J.L. Hudson, M.J. Casavant and J.M. Tour, Water-soluble, exfoliated, nonroping single-wall carbon nanotubes, *J. Am. Chem. Soc.*, **126**, 11158–11159 (2004).

[39] B.K. Price, J.L. Hudson and J.M. Tour, Green chemical functionalization of single-walled carbon nanotubes in ionic liquids, *J. Am. Chem. Soc.*, **127**, 14867–14870 (2005).

[40] A. Hirsch, Functionalization of single-walled carbon nanotubes, *Angew. Chem., Int. Ed.*, **41**, 1853–1859 (2002).

[41] D. Tasis, N. Tagmatarchis, V. Georgakilas and M. Prato, Soluble carbon nanotubes, *Chem. Eur. J.*, **9**, 4000–4008 (2003).

[42] S. Niyogi, M.A. Hamon, H. Hu, B. Zhao, P. Bhowmik, R. Sen, M.E. Itkis and R.C. Haddon, Chemistry of single-walled carbon nanotubes, *Acc. Chem. Res.*, **35**, 1105–1113 (2002).

[43] H.G. Chae, T.V. Sreekumar, T. Uchida and S. Kumar, A comparison of reinforcement efficiency of various types of carbon nanotubes in poly acrylonitrile fiber, *Polymer*, **46**, 10925–10935 (2005).

[44] M. Holzinger, J. Steinmetz, D. Samaille, M. Glerup, M. Paillet, P. Bernier, L. Ley and R. Graupner, [2+1] cycloaddition for cross-linking SWCNTs, *Carbon*, **42**, 941–947 (2004).

[45] M. Holzinger, O. Vostrowsky, A. Hirsch, F. Hennrich, M. Kappes, R. Weis and F. Jellen, Sidewall functionalization of carbon nanotubes, *Angew. Chem., Int. Ed.*, **40**, 4002–4005 (2001).

[46] X. Lu, F. Tian, N. Wang and Q. Zhang, Organic functionalization of the sidewalls of carbon nanotubes by Diels-Alder reactions: A theoretical prediction, *Org. Lett.*, **4**, 4313–4315 (2002).

[47] J.L. Delgado, P. de la Cruz, F. Langa, A. Urbina, J. Casado and J.T. López Navarrete, Microwave-assisted sidewall functionalization of single-wall carbon nanotubes by Diels–Alder cycloaddition, *Chem. Commun.*, 1734–1735 (2004).

[48] H. Hu, B. Zhao, M.A. Hamon, K. Kamaras, M.E. Itkis and R.C. Haddon, Sidewall functionalization of single-walled carbon nanotubes by addition of dichlorocarbene, *J. Am. Chem. Soc.*, **125**, 14893–14900 (2003).

[49] Y.-Y. Chu and M.-D. Su, Theoretical study of addition reactions of carbene, silylene, and germylene to carbon nanotubes, *Chem. Phys. Lett.*, **394**, 231–237 (2004).

[50] X. Lu, F. Tian and Q. Zhang, The [2+1] Cycloadditions of dichlorocarbene, silylene, germylene, and oxycarbonylnitrene onto the sidewall of armchair (5,5) single-wall carbon nanotube, *J. Phys. Chem. B*, **107**, 8388–8391 (2003).

[51] K.S. Coleman, S.R. Bailey, S. Fogden and M.L.H. Green, Functionalization of single-walled carbon nanotubes via the Bingel reaction, *J. Am. Chem. Soc.*, **125**, 8722–8723 (2003).

[52] K.A. Worsley, K.R. Moonoosawny and P. Kruse, Long-range periodicity in carbon nanotube sidewall functionalization, *Nano Lett.*, **4**, 1541–1546 (2004).

[53] F. Liang, A.K. Sadana, A. Peera, J. Chattopadhyay, Z. Gu, R.H. Hauge and W.E. Billups, A convenient route to functionalized carbon nanotubes, *Nano Lett.*, **4**, 1257–1260 (2004).

[54] R. Graupner, J. Abraham, D. Wunderlich, A. Vencelova, P. Lauffer, J. Rohrl, M. Hundhausen, L. Ley and A. Hirsch, Nucleophilic-alkylation-reoxidation: A functionalization sequence for single-wall carbon nanotubes, *J. Am. Chem. Soc.*, **128**, 6683–6689 (2006).

[55] J. Chattopadhyay, A.K. Sadana, F. Liang, J.M. Beach, Y.X. Xiao, R.H. Hauge and W.E. Billups, Carbon nanotube salts. Arylation of single-wall carbon nanotubes, *Org. Lett.*, **7**, 4067–4069 (2005).

[56] T.S. Balaban, M.C. Balaban, N. Malik, F. Heinrich, R. Fischer, H. Rosner and M. Kappes, Polyacylation of single-walled carbon nanotubes under Friedel-Crafts conditions: An efficient method for functionalizing, purifying, decorating, and linking carbon allotropes, *Adv. Mater.*, **18**, 2763–2767 (2006).

[57] R. Barthos, D. Mehn, A. Demortier, N. Pierard, Y. Morciaux, G. Demortier, A. Fonseca and J.B. Nagy, Functionalization of single-walled carbon nanotubes by using alkyl-halides, *Carbon*, **43**, 321–325 (2005).

[58] Y. Ying, R.K. Saini, F. Liang, A.K. Sadana and W.E. Billups, Functionalization of carbon nanotubes by free radicals, *Org. Lett.*, **5**, 1471–1473 (2003).

[59] V. Georgakilas, K. Kordatos, M. Prato, D.M. Guldi, M. Holzinger and A. Hirsch, Organic functionalization of carbon nanotubes, *J. Am. Chem. Soc.*, **124**, 760–761 (2002).

[60] N. Tagmatarchis and M. Prato, Functionalization of carbon nanotubes via 1,3-dipolar cycloadditions, *J. Mater. Chem.*, **14**, 437–439 (2004).

[61] F.G. Brunetti, M.A. Herrero, J.D. M. Munoz, S. Giordani, A. Diaz-Orthiz, S. Filippone, G. Ruaro, M. Meneghetti, M. Prato and E. Vázquez, Reversible microwave-assisted cycloaddition of aziridines to carbon nanotubes, *J. Am. Chem. Soc.*, **129**, 14580–14581 (2007).

[62] S. Banerjee and S.S. Wong, Rational sidewall functionalization and purification of single-walled carbon nanotubes by solution-phase ozonolysis, *J. Phys. Chem. B*, **106**, 12144–12151 (2002).

[63] S. Banerjee and S.S. Wong, Demonstration of diameter-selective reactivity in the sidewall ozonation of SWCNTs by resonance Raman spectroscopy, *Nano Lett.*, **4**, 1445–1450 (2004).

[64] L.C. Teague, S. Banerjee, S.S. Wong, C. A. Richter, B. Varughese and J.D. Batteas, Effects of ozonolysis and subsequent growth of quantum dots on the electrical properties of freestanding single-walled carbon nanotube films, *Chem. Phys. Lett.*, **442**, 354–359 (2007).

[65] T. Hemraj-Benny, T.J. Bandosz and S.S. Wong, Effect of ozonolysis on the pore structure, surface chemistry, and bundling of single-walled carbon nanotubes, *J. Coll. Interf. Sci.*, **317**, 375–382 (2008).

[66] S. Pekker, J.-P. Salvetat, E. Jakab, J.-M. Bonard and L. Forro, Hydrogenation of carbon nanotubes and graphite in liquid ammonia, *J. Phys. Chem. B*, **105**, 7938–7943 (2001).

[67] X. Lu, F. Tian, X. Xu, N. Wang and Q. Zhang, A theoretical exploration of the 1,3-dipolar cycloadditions onto the sidewalls of (n,n) armchair single-wall carbon nanotubes, *J. Am.Chem. Soc.*, **125**, 10459–10464 (2003).

[68] M. Alvaro, P. Atienzar, P. de la Cruz, J.L. Delgado, H. Garcia and F. Langa, Sidewall functionalization of single-walled carbon nanotubes with nitrile imines. Electron transfer from the substituent to the carbon nanotube, *J. Phys. Chem. B*, **108**, 12691–12697 (2004).

[69] H. Peng, L.B. Alemany, J.L. Margrave and V.N. Khabashesku, Sidewall carboxylic acid functionalization of single-walled carbon nanotubes, *J. Am. Chem. Soc.*, **125**, 15174–15182 (2003).

[70] M. Holzinger, J. Abraham, P. Whelan, R. Graupner, L. Ley, F. Hennrich, M. Kappes and A. Hirsch, Functionalization of single-walled carbon nanotubes with (R-)Oxycarbonyl Nitrenes, *J. Am. Chem. Soc.*, **125**, 8566–8580 (2003).

[71] F. Nunzi, F. Mercuri, A. Sgamellotti and N. Re, The coordination chemistry of carbon nanotubes: a density functional study through a cluster model approach, *J. Phys. Chem. B*, **106**, 10622–10633 (2002).

[72] R.C. Haddon, Chemistry of fullerenes – the manifestation of strain in a class of continuous aromatic molecules, *Science*, **261**, 1545–1550 (1993).

[73] S. Banerjee and S.S. Wong, Functionalization of carbon nanotubes with a metal-containing molecular complex, *Nano Lett.*, **2**, 49–53 (2002).

[74] F. McGinnety and J.A. Ibers, Structure and bonding in $[Ir(O_2)(Ph_2PCH_2 \cdot CH_2 \cdot PPh_2)_2][PF_6]$ and in $IrBr(CO)(PPh_3)_2C_2(CN)_4$, *J. Chem. Soc., Chem. Commun.*, 235–237 (1968).

[75] L. Manojlovic-Muir, K. Muir and J.A. Ibers, Geometries of transition metal complexes containing simple alkenes, *Discuss. Faraday Soc.*, **47**, 84–92 (1969).

[76] A.L. Balch, V.J. Catalano, D.A. Costa, W.R. Fawcett, M. Federco, A.S. Ginwalla, J.W. Lee, M.M. Olmstead, B.C. Noll and K. Winkler, Transition metal fullerene chemistry: The route from structural studies of exohedral adducts to the formation of redox active films, *J. Phys. Chem. Solids*, **58**, 1633–1643 (1997).

[77] L. Vaska, Reversible activation of covalent molecules by transition-metal complexes. The role of the covalent molecule, *Acc. Chem. Res.*, **1**, 335–344 (1968).

[78] F.A. Cotton, G. Wilkinson, C.A. Murillo, M. Bochmann, *Advanced Inorganic Chemistry, 6th Edition*, John Wiley & Sons, Inc., New York (1999).

[79] C.A. McAuliffe, W. Levason, *Phosphine, Arsine, and Stilbine Complexes of the Transition Elements*; Elsevier Scientific: New York (1979).

[80] F. Mercuri and A. Sgamellotti, Functionalization of carbon nanotubes with Vaska's Complex: A theoretical approach, *J. Phys. Chem. B*, **110**, 15291–15294 (2006).

[81] S. Banerjee and S.S. Wong, Structural characterization, optical properties, and improved solubility of carbon nanotubes functionalized with Wilkinson's catalyst, *J. Am. Chem. Soc.*, **124**, 8940–8948 (2002).

[82] C. Baleizao, B. Gigante, H. Garcia and A. Corma, Vanadyl salen complexes covalently anchored to single-wall carbon nanotubes as heterogeneous catalysts for the cyanosilylation of aldehydes, *J. Catal.*, **221**, 77–84 (2004).

[83] F. Frehill, J.G. Vos, S. Benrezzak, A.A. Koós, Z. Kónya, M.G. Rüther, W.J. Blau, A. Fonseca, J.B. Nagy, L.P. Biró, A.I. Minett and M. in het Panhuis, Interconnecting carbon nanotubes with an inorganic metal complex, *J. Am. Chem. Soc.*, **121**, 13694–13695 (2002).

[84] H. Chaturvedi and J.C. Poler, Binding of rigid dendritic ruthenium complexes to carbon nanotubes, *J. Phys. Chem. B*, **110**, 22387–22393 (2002).

[85] R.F. Khairoutdinov, L.V. Doubova, R.C. Haddon and L. Saraf, Persistent photoconductivity in chemically modified single-wall carbon nanotubes, *J. Phys. Chem. B*, **108**, 19976–19981 (2004).

[86] T. Hemraj-Benny, S. Banerjee and S.S. Wong, Interactions of lanthanide complexes with oxidized single-walled carbon nanotubes, *Chem. Mater.*, **16**, 1855–1863 (2004).

[87] H. Ago, M.S.P. Shaffer, D.S. Ginger, A.H. Windle and R.H. Friend, Electronic interaction between photoexcited poly(p-phenylene vinylene) and carbon nanotubes, *Phys. Rev. B*, **61**, 2286–2290 (2000).

[88] L. Cao, H.Z. Chen, H.B. Zhou, L. Zhu, J.Z. Sun, X.B. Zhang, J.M. Xu and M. Wang, Carbon-nanotube-templated assembly of rare-earth phthalocyanine nanowires, *Adv. Mater.*, **15**, 909–913 (2003).

[89] D.B. Romero, M. Carrard and L.A. Zuppiroli, A carbon nanotube organic semiconducting polymer heterojunction, *Adv. Mater.*, **8**, 899–902 (1996).

[90] G. Accorsi, N. Armaroli, A. Parisini, M. Meneghetti, R. Marega, M. Prato and D. Bonifazi, Wet adsorption of a luminescent EuIII complex on carbon nanotubes sidewalls, *Adv. Funct. Mater.*, **17**, 2975–2982 (2007).

[91] J. Cui, M. Burghard and K. Kern, Reversible sidewall osmylation of individual carbon nanotubes, *Nano Lett.*, **3**, 613–615 (2003).

[92] S. Banerjee and S.S. Wong, Selective metallic tube reactivity in the solution-phase osmylation of single-walled carbon nanotubes, *J. Am. Chem. Soc.*, **126**, 2073–2081 (2004).

[93] X. Lu, F. Tian, Y. Feng, X. Xu, N. Wang and Q. Zhang, Sidewall oxidation and complexation of carbon nanotubes by base-catalyzed cycloaddition of transition metal oxide: A theoretical prediction, *Nano Lett.*, **2**, 1325–1327 (2002).

[94] T. Hemraj-Benny and S.S. Wong, Silylation of single-walled carbon nanotubes, *Chem. Mater.*, **18**, 4827–4839 (2006).

[95] M. Kanungo, H.S. Isaacs and S.S. Wong, Quantitative control over electrodeposition of silica films onto single-walled carbon nanotube surfaces, *J. Phys. Chem. C*, **111**, 17730–17742 (2007).

[96] Y. Zhang and H. Dai, Formation of metal nanowires on suspended single-walled carbon nanotubes, *Appl. Phys. Lett.*, **77**, 3015–3017 (2000).

[97] R. Yu, L. Chen, Q. Liu, J. Lin, K.-L. Tan, S.C. Ng, H.S.O. Chan, G.-Q. XU and T.S.A. Hor, Platinum deposition on carbon nanotubes via chemical modification, *Chem. Mater.*, **10**, 718–722 (1998).

[98] Y. Zhang, N.W. Franklin, R.J. Chen and H. Dai, Metal coating on suspended carbon nanotubes and its implication to metal–tube interaction, *Chem. Phys. Lett.*, **331**, 35–41 (2000).

[99] R. Sen, S.M. Richard, M.E. Itkis and R.C. Haddon, Controlled purification of single-walled carbon nanotube films by use of selective oxidation and near-IR spectroscopy, *Chem. Mater.*, **15**, 4273–4279 (2003).

[100] B.W. Clare and D.L. Kepert, Opening of carbon nanotubes by addition of oxygen, *Inorg. Chim. Acta*, **343**, 1–17 (2003).

[101] L. Quingwen, Y. Hao, Y. Yinchun, Z. Jin and L. Zhongfan, Defect location of individual single-walled carbon nanotubes with a thermal oxidation strategy, *J. Phys. Chem. B*, **106**, 11085–11088 (2002).

[102] A. Kukovecz, C. Kramberger, M. Holzinger, H. Kuzmany, J. Schalko, M. Mannsberger and A. Hirsch, On the stacking behavior of functionalized single-wall carbon nanotubes, *J. Phys. Chem. B*, **106**, 6374–6380 (2002).

[103] K.J. Ziegler, Z.N. Gu, J. Shaver, Z.Y. Chen, E.L. Flor, D.J. Schmidt, C. Chan, R.H. Hauge and R.E. Smalley, Cutting single-walled carbon nanotubes, *Nanotechnology*, **16**, S539–S544 (2005).

[104] A. Koshio, M. Yudasaka, M. Zhang and S. Iijima, A Simple way to chemically react single-wall carbon nanotubes with organic materials using ultrasonication, *Nano Lett.*, **1**, 361–363 (2001).

[105] M.-L. Sham and J.-K. Kim, Surface functionalities of multi-wall carbon nanotubes after UV/Ozone and TETA treatments, *Carbon*, **44**, 768–777 (2006).

[106] T.J. Aitchison, M. Ginic-Markovic, J.G. Matisons, G.P. Simon and P.M. Fredericks, Purification, cutting, and sidewall functionalization of multiwalled carbon nanotubes using potassium permanganate solutions, *J. Phys. Chem. C*, **111**, 2440–2446 (2007).

[107] K.C. Park, H. Hayashi, H. Tomiyasu, M. Endo and M.S. Dresselhaus, Progressive and invasive functionalization of carbon nanotube sidewalls by diluted nitric acid under supercritical conditions, *J. Mater. Chem.*, **15**, 407–411 (2005).

[108] W.H. Lee, S.J. Kim, W.J. Lee, J.G. Lee, R.C. Haddon and P.J. Reucroft, X-ray photoelectron spectroscopic studies of surface modified single-walled carbon nanotube material, *Appl. Surf. Sci.*, **181**, 121–127 (2001).

[109] H. Hu, P. Bhowmik, B. Zhao, M.A. Hamon, M.E. Itkis and R.C. Haddon, Determination of the acidic sites of purified single-walled carbon nanotubes by acid–base titration, *Chem. Phys. Lett.*, **345**, 25–28 (2001).

[110] N.O.V. Plank, R. Cheung and R.J. Andrews, Thiolation of single-wall carbon nanotubes and their self-assembly, *Appl. Phys. Lett.*, **85**, 3229–3231 (2004).

[111] J.K. Lim, W.S. Yun, M.-H. Yoon, S.K. Lee, C.H. Kim, K. Kim and S.K. Kim, Selective thiolation of single-walled carbon nanotubes, *Synth. Met.*, **139**, 521–527 (2003).

[112] Y. Qin, J. Shi, W. Wu, X. Li, Z.-X. Guo and D. Zhu, Concise route to functionalized carbon nanotubes, *J. Phys. Chem. B*, **107**, 12899–12901 (2003).

[113] P. W. Chiu, G.S. Duesberg, U. Dettlaff-Weglikowska and S. Roth, Interconnection of carbon nanotubes by chemical functionalization, *Appl. Phys. Lett.*, **80**, 3811–3813 (2001).

[114] M. Alvaro, C. Aprile, P. Atienzar and H. Garcia, Preparation and photochemistry of single wall carbon nanotubes having covalently anchored viologen units, *J. Phys. Chem. B*, **109**, 7692–7697 (2005).

[115] B. Ballesteros, S. Campidelli, G. de la Torre, C. Ehli, D.M. Guldi, M. Prato and T. Torres, Synthesis, characterization and photophysical properties of a SWCNT-phthalocyanine hybrid, *Chem. Commun.*, 2950–2952 (2007).

[116] R.B. Martin, L.W. Qu, Y. Lin, B.A. Harruff, C.E. Bunker, J.R. Gord, L.F. Allard and Y.-P. Sun, Functionalized carbon nanotubes with tethered pyrenes: Synthes is and photophysical properties, *J. Phys. Chem. B*, **108**, 11447–11453 (2004).

[117] H. Li, M.E. Kose, L. W. Qu, Y. Lin, R.B. Martin, B. Zhou, B.A. Harruff, L.F. Allard and Y.-P. Sun, Excited-state energy transfers in single-walled carbon nanotubes functionalized with tethered pyrenes, *J. Photochem. Photobiol. A*, **185**, 94–100 (2007).

[118] M.A. Herranz, N. Martín, S. Campidelli, M. Prato, G. Brehm and D.M. Guldi, Control over electron transfer in tetrathiafulvalene-modified single-walled carbon nanotubes, *Angew. Chem., Int. Ed.*, **45**, 4478–4482 (2006).

[119] W. Zhu, N. Minami, S. Kazaoui and Y. Kim, Fluorescent chromophore functionalized single-wall carbon nanotubes with minimal alteration to their characteristic one-dimensional electronic states, *J. Mater. Chem.*, **13**, 2196–2201 (2003).

[120] S. Campidelli, C. Sooambar, E. Lozano-Diz, C. Ehli, D.M. Guldi and M. Prato, Dendrimer-functionalized single-wall carbon nanotubes: Synthesis, characterization, and photoinduced electron transfer, *J. Am. Chem. Soc.*, **128**, 12544–12552 (2006).

[121] M. Burghard, Asymmetric end-functionatization of carbon nanotubes, *Small*, **1**, 1148–1150 (2005).

[122] K.M. Lee, L. Li and L. Dai, Asymmetric end-functionalization of multi-walled carbon nanotubes, *J. Am. Chem. Soc.*, **127**, 4122–4123 (2005).

[123] C. Ehli, G.M.A. Rahman, N. Jux, D. Balbinot, D.M. Guldi, F. Paolucci, M. Marcaccio, D. Paolucci, M. Melle-Franco, F. Zerbetto, S. Campidelli and M. Prato, Interactions in single wall carbon nanotubes/pyrene/porphyrin nanohybrids, *J. Am. Chem. Soc.*, **128**, 11222–11231 (2006).

[124] Q. Chen, L. Dai, M. Gao, S. Huang and A. Mau, Plasma activation of carbon nanotubes for chemical modification, *J. Phys. Chem. B*, **105**, 618–622 (2001).

[125] B. Kahre, P. Wilhite, B. Tran, E. Teixera, K. Fresquez, D. Nna Mvodo, C. Bauschlicher Jr. and M. Meyyapan, Functionalization of carbon nanotubes via nitrogen glow discharge, *J. Phys. Chem. B*, **109**, 23466–23472 (2005).

[126] B. Ni, R. Andrews, D. Jacques, D. Qian, M.B.J. Wijesundra, Y. Choi, L. Hanley and S.B.A. Sinnott, A combined computational and experimental study of ion-beam modification of carbon nanotube bundles, *J. Phys. Chem. B*, **105**, 12719–12725 (2001).

[127] B. Ni and S.B.A. Sinnott, Chemical functionalization of carbon nanotubes through energetic radical collisions, *Phys. Rev. B*, **61**, R16343–R16346 (2000).

[128] S.S. Wong, E. Joselevich, A.T. Wooley, C.L. Cheung and C.M. Lieber, Covalently functionalized nanotubes as nanometre-sized probes in chemistry and biology, *Nature*, **394**, 52–55 (1998).

[129] S.S. Wong, A.T. Wooley, E. Joselevich, C.L. Cheung and C.M. Lieber, Covalently-functionalized single-walled carbon nanotube probe tips for chemical force microscopy, *J. Am. Chem. Soc.*, **120**, 8557–8558 (1998).

[130] Y. Yang, J. Zhang, X. Nan and Z. Liu, Toward the chemistry of carboxylic single-walled carbon nanotubes by chemical force microscopy, *J. Phys. Chem. B*, **106**, 4139–4144 (2002).

[131] S.S. Wong, A.T. Wooley, E. Joselevich and C.M. Lieber, Functionalization of carbon nanotube AFM probes using tip-activated gases, *Chem. Phys. Lett.*, **306**, 219–225 (1999).

[132] A. Javey, J. Guo, Q. Wang, M. Lundstrom and H. Dai, Ballistic carbon nanotube field-effect transistors, *Nature*, **424**, 654–657 (2003).

[133] S. Latil, S. Roche and J.-C. Charlier, Electronic transport in carbon nanotubes with random coverage of physisorbed molecules, *Nano Lett.*, **5**, 2216–2219 (2005).

[134] J. Liu, A.G. Rinzler, H. Dai, J.H. Hafner, R.K. Bradley, P.J. Boul, A. Lu, T. Iverson, K. Shelimov, C.B. Huffman, F. Rodriguez-Marcias, Y.-S. Shon, T.R. Lee, D.T. Colbert and R.E. Smalley, Fullerene pipes, *Science*, **280**, 1253–1256 (1998).

[135] K.J. Ziegler, Z. Gu, H. Peng, E.L. Flor, R.H. Hauge and R.E. Smalley, Controlled oxidative cutting of single-walled carbon nanotubes, *J. Am. Chem. Soc.*, **127**, 1541–1547 (2005).

[136] H. Li, R.B. Martin, B.A. Harruff, R.A. Carino, L.F. Allard and Y.-P. Sun, Single-walled carbon nanotubes tethered with porphyrins: Synthesis and photophysical properties, *Adv. Mater.*, **16**, 896–900 (2004).

[137] T. Umeyama, M. Fujita, N. Tezuka, N. Kadota, Y. Matano, K. Yoshida and H. Imahori, Electrophoretic deposition of single-walled carbon nanotubes covalently modified with bulky porphyrins on nanostructured SnO_2 electrodes for photoelectrochemical devices, *J. Phys. Chem. C*, **111**, 11484–11493 (2007).

[138] L. Sheeney-Haj-Ichia, B. Basnar and I. Willner, Efficient generation of photocurrents by using CdS/carbon nanotube assemblies on electrodes, *Angew. Chem., Int. Ed.*, **44**, 78–83 (2005).

[139] D.M. Guldi, M. Marcaccio, D. Paolucci, F. Paolucci, N. Tagmatarchis, D. Tasis, E. Vázquez and M. Prato, Single-wall carbon nanotube–ferrocene nanohybrids: Observing intramolecular electron transfer in functionalized SWCNTs, *Angew. Chem., Int. Ed.*, **42**, 4206–4209 (2003).

[140] C. Ehli, S. Campidelli, F.G. Brunetti, M. Prato and D.M. Guldi, Single-wall carbon nanotube porphyrin nanoconjugates, *J. Porphyr. Phthalocyanines*, **11**, 442–447 (2007).

[141] J. Bahr and J.M. Tour, Covalent chemistry of single-wall carbon nanotubes, *J. Mater. Chem.*, **12**, 1952–1958 (2002).

[142] Vögtle, F. *Dendrimers II Architecture, Nanostructure and Supramolecular Chemistry*; Springer, Berlin, Germany (2000).

[143] J.-P. Majoral and A.-M. Caminade, Dendrimers containing heteroatoms (Si, P, B, Ge, or Bi), *Chem. Rev.*, **99**, 845–880 (1999).

[144] A.W. Bosman, H.M. Janssen and E.W. Meijer, About dendrimers: Structure, physical properties, and applications, *Chem. Rev.*, **99**, 1665–1688 (1999).

[145] G.R. Newkome, E. He and C.N. Moorefield, Suprasupermolecules with novel properties: Metallodendrimers, *Chem. Rev.*, **99**, 1689–1746 (1999).

[146] S. M. Grayson and J.M. J. Fréchet, Convergent dendrons and dendrimers: from synthesis to applications, *Chem. Rev.*, **101**, 3819–3867 (2001).

[147] D. A. Tomalia, H. Baker, J.R. Dewald, M. Hall, G. Kallos, S. Martin, J. Roeck, J. Ryder and P. Smith, A new class of polymers: Starburst-dendritic macromolecules, *Polym. J.*, **17**, 117–132 (1985).

[148] H.C. Kolb, M.G. Finn and K.B. Sharpless, Click chemistry: Diverse chemical function from a few good reactions, *Angew. Chem., Int. Ed.*, **40**, 2004–2021 (2001).

[149] C.W. Tornøe, C. Christensen and M. Meldal, Peptidotriazoles on solid phase: [1,2,3]-triazoles by regiospecific copper(I)-catalyzed 1,3-dipolar cycloadditions of terminal alkynes to azides, *J. Org. Chem.*, **67**, 3057–3064 (2002).

[150] V.V. Rostovtsev, L.G. Green, V.V. Fokin and K.B. Sharpless, A stepwise Huisgen cycloaddition process: Copper(I)-catalyzed regioselective "ligation" of azides and terminal alkynes, *Angew. Chem., Int. Ed.*, **41**, 2596–2599 (2002).

[151] R. Huisgen, Cycloadditions - Definition, classification, and characterization, *Angew. Chem., Int. Ed. Engl.*, **7**, 321–328 (1968).

[152] R. Huisgen, In *1,3–dipolar Cycloaddition Chemistry*; Padwa, A. (ed.) John Wiley & Sons, Inc., New York, 1 (1984).

[153] S. Campidelli, Click chemistry for carbon nanotubes functionalization, *Curr. Org. Chem.*, in press (2011).

[154] S. Campidelli, B. Ballesteros, A. Filoramo, D. Díaz-Díaz, G. de la Torre, T. Torres, G.M.A. Rahman, C. Ehli, D. Kiessling, F. Werner, V. Sgobba, D.M. Guldi, C. Cioffi, M. Prato and J.-P. Bourgoin, Facile decoration of functionalized single-wall carbon nanotubes with phthalocyanines via "click chemistry", *J. Am. Chem. Soc.*, **130**, 11503–11509 (2008).

[155] T. Palacin, H. Le Khanh, B. Jousselme, P. Jégou, A. Filoramo, C. Ehli, D.M. Guldi and S. Campidelli, Efficient functionalization of carbon nanotubes with porphyrin dendrons via click chemistry, *J. Am. Chem. Soc.*, **131**, 15394–15402 (2009).

[156] L. Tao, G. Chen, G. Mantovani, S. York and D.M. Haddleton, Modification of multi-wall carbon nanotube surfaces with poly(amidoamine) dendrons: Synthesis and metal templating, *Chem. Commun.*, 4949–4951 (2006).

[157] S.-H. Hwang, C.N. Moorefield, P. Wang, K.-U. Jeong, S.Z.D. Cheng, K.K. Kotta and G.R. Newkome, Dendron-tethered and templated CdS quantum dots on single-walled carbon nanotubes, *J. Am. Chem. Soc.*, **128**, 7505–7509 (2006).

[158] M. Alvaro, P. Atienzar, P. de la Cruz, J.L. Delgado, V. Troiani, H. Garcia, F. Langa, A. Palkar and L. Echegoyen, Synthesis, photochemistry, and electrochemistry of single-wall carbon nanotubes with pendent pyridyl groups and of their metal complexes with zinc porphyrin. Comparison with pyridyl-bearing fullerenes, *J. Am. Chem. Soc.*, **128**, 6626–6635 (2006).

[159] S. Qin, D. Qin, W.T. Ford, J.E. Herrera, D.E. Resasco, S.M. Bachilo and R.B. Weisman, Solubilization and purification of single-wall carbon nanotubes in water by in situ radical polymerization of sodium 4-styrenesulfonate, *Macromolecules*, **37**, 3965–3967 (2004).

[160] D.M. Guldi, G.M.A. Rahman, J. Ramey, M. Marcaccio, D. Paolucci, F. Paolucci, S. Qin, W.T. Ford, D. Balbinot, N. Jux, N. Tagmatarchis and M. Prato, Donor–acceptor nanoensembles of soluble carbon nanotubes, *Chem. Commun.*, 2034–2035 (2004).

[161] D.M. Guldi, G.M.A. Rahman, M. Prato, N. Jux, S. Qin and W.T. Ford, Single-wall carbon nanotubes as integrative building blocks for solar-energy conversion, *Angew. Chem., Int. Ed.*, **44**, 2015–2018 (2005).

[162] D.M. Guldi, G.M.A. Rahman, S. Qin, M. Tchoul, W.T. Ford, M. Marcaccio, D. Paolucci, F. Paolucci, S. Campidelli and M. Prato, Versatile coordination chemistry towards multifunctional carbon nanotube nanohybrids, *Chem. Eur. J.*, **12**, 2152–2161 (2006).

[163] M. J. O'Connell, P. Boul, L.M. Ericson, C. Huffman, Y. Wang, E. Haroz, C. Kuper, J. Tour, K.D. Ausman and R.E. Smalley, Reversible water-solubilization of single-walled carbon nanotubes by polymer wrapping, *Chem. Phys. Lett.*, **342**, 265–271 (2001).

[164] A. Star, J.F. Stoddart, D. Steuerman, M. Diehl, A. Boukai, E.W. Wong, X. Yang, S.-W. Chung, H. Choi and J.R. Heath, Preparation and properties of polymer-wrapped single-walled carbon nanotubes, *Angew. Chem., Int. Ed.*, **40**, 1721–1725 (2001).

[165] A. Star, D.W. Steuerman, J.R. Heath and J.F. Stoddart, Starched carbon nanotubes, *Angew. Chem., Int. Ed.*, **41**, 2508–2512 (2002).

[166] N. Nakashima, Y. Tomonari and H. Murakami, Water-soluble single-walled carbon nanotubes via noncovalent sidewall-functionalization with a pyrene-carrying ammonium ion, *Chem. Lett.*, 638–639 (2002).

[167] Y. Tomonari, H. Murakami and N. Nakashima, Solubilization of single-walled carbon nanotubes by using polycyclic aromatic ammonium amphiphiles in water-strategy for the design of high-performance solubilizers, *Chem. Eur. J.*, **12**, 4027–4034 (2006).

[168] D.M. Guldi, H. Taieb, G.M.A. Rahman, N. Tagmatarchis and M. Prato, Novel photoactive single-walled carbon nanotube-porphyrin polymer wraps: Efficient and long-lived intracomplex charge separation, *Adv. Mater.*, **17**, 871–875 (2005).

[169] K. Saito, V. Troiani, H. Qiu, N. Solladié, T. Sakata, H. Mori, M. Ohama and S. Fukuzumi, Nondestructive formation of supramolecular nanohybrids of single-walled carbon nanotubes with flexible porphyrinic polypeptides, *J. Phys. Chem. C*, **111**, 1194–1199 (2007).

[170] E. Kymakis and G.A.J. Amaratunga, Single-wall carbon nanotube/conjugated polymer photo-voltaic devices, *Appl. Phys. Lett.*, **80**, 112–114 (2002).

[171] J. Chen and C.P. Collier, Noncovalent functionalization of single-walled carbon nanotubes with water-soluble porphyrins, *J. Phys. Chem. B*, **109**, 7605–7609 (2005).

[172] G.M.A. Rahman, D.M. Guldi, S. Campidelli and M. Prato, Electronically interacting single wall carbon nanotube–porphyrin nanohybrids, *J. Mater. Chem.*, **16**, 62–65 (2006).

[173] D.R. Kauffman, O. Kuzmych and A. Star, Interactions between single-walled carbon nanotubes and tetraphenyl metalloporphyrins: Correlation between spectroscopic and FET measurements, *J. Phys. Chem. C*, **111**, 3539–3543 (2007).

[174] F. Cheng and A. Adronov, Noncovalent functionalization and solubilization of carbon nanotubes by using a conjugated Zn–porphyrin polymer, *Chem. Eur. J.*, **12**, 5053–5059 (2006).

[175] F. Cheng, S. Zhang, A. Adronov, L. Echegoyen and F. Diederich, Triply fused Zn^II-porphyrin oligomers: Synthesis, properties, and supramolecular interactions with single-walled carbon nanotubes (SWCNTs), *Chem. Eur. J.*, **12**, 6062–6070 (2006).

[176] R. Chitta, A.D. Sandanayaka, A.L. Schumacher, L. D'Souza, Y. Araki, O. Ito and F. D'Souza, Donor-acceptor nanohybrids of zinc naphthalocyanine or zinc porphyrin noncovalently linked to single-wall carbon nanotubes for photoinduced electron transfer, *J. Phys. Chem. C*, **111**, 6947–6955 (2007).

[177] F. D'Souza, R. Chitta, A.S.D. Sandanayaka, N.K. Subbaiyan, L. D'Souza, Y. Araki and O. Ito, Self-assembled single-walled carbon nanotubes: Zinc–porphyrin hybrids through ammonium ion–Crown ether interaction: Construction and electron transfer, *Chem. Eur. J.*, **13**, 8277–8284 (2007).

[178] M.A. Herranz, C. Ehli, S. Campidelli, M. Gutiérrez, G.L. Hug, K. Ohkubo, S. Fukuzumi, M. Prato, N. Martín and D.M. Guldi, Spectroscopic characterization of photolytically generated radical ion pairs in single-wall carbon nanotubes bearing surface-immobilized tetrathiafulvalenes, *J. Am. Chem. Soc.*, **130**, 66–73 (2008).

[179] C. Ehli, D.M. Guldi, M.A. Herranz, N. Martín, S. Campidelli and M. Prato, Pyrene-tetrathiafulvalene supramolecular assembly with different types of carbon nanotubes, *J. Mater. Chem.*, **18**, 1498–1503 (2008).

[180] D.M. Guldi, G.M.A. Rahman, N. Jux, N. Tagmatarchis and M. Prato, Integrating single-wall carbon nanotubes into donor-acceptor nanohybrids, *Angew. Chem., Int. Ed.*, **43**, 5526–5530 (2004).

[181] C. Ehli, G.M.A. Rahman, N. Jux, D. Balbinot, D.M. Guldi, F. Paolucci, M. Marcaccio, D. Paolucci, M. Melle-Franco, F. Zerbetto, S. Campidelli and M. Prato, Interactions in single wall carbon nanotubes/pyrene/porphyrin nanohybrids, *J. Am. Chem. Soc.*, **128**, 11222–11231 (2006).

[182] V. Sgobba, G.M.A. Rahman, D.M. Guldi, N. Jux, S. Campidelli and M. Prato, Supramolecular assemblies of different carbon nanotubes for photoconversion processes, *Adv. Mater.*, **18**, 2264–2269 (2006).

[183] G.M.A. Rahman, D.M. Guldi, R. Cagnoli, A. Mucci, L. Schenetti, L. Vaccari and M. Prato, Combining single wall carbon nanotubes and photoactive polymers for photoconversion, *J. Am. Chem. Soc.*, **127**, 10051–10057 (2005).

[184] W. Lee, J. Lee, S.-H. Lee, J. Chang, W. Yi and S.-H. Han, Improved photocurrent in Ru(2,2'-bipyridine-4,4'-dicarboxylic acid)2(NCS)2/Di(3-aminopropyl)viologen/single-walled carbon nanotubes/indium tin oxide system: Suppression of recombination reaction by use of single-walled carbon nanotubes, *J. Phys. Chem. C*, **111**, 9110–9115 (2007).

[185] R. Saito, M. Fujita, G. Dresselhaus and M.S. Dresselhaus, Electronic structure of chiral graphene tubules, *Appl. Phys. Lett.*, **60**, 2204–2206 (1992).

[186] Freitag, M. carbon nanotube electronics and devices; In *Carbon Nanotubes: Properties and Applicatons*; O'Connell, M. J. (ed.), CRC Press, Florida, 83–117 (2006).

[187] R. Martel, T. Schmidt, H.R. Shea, T. Hertel and P. Avouris, Single- and multi-wall carbon nanotube field-effect transistors, *Appl. Phys. Lett.*, **73**, 2447–2449 (1998).

[188] S.J. Tans, A.R.M. Verschueren and C. Dekker, Room-temperature transistor based on a single carbon nanotube, *Nature*, **393**, 49–52 (1998).

[189] M. Kanungo, H. Hu, G.G. Malliaras and G.B. Blanchet, Suppression of metallic conductivity of single-walled carbon nanotubes by cycloaddition reactions, *Science*, **323**, 234–237 (2009).

[190] M. S. Strano, C.A. Dyke, M.L. Ursey, P.W. Barone, M.J. Allen, H.W. Shan, C. Kittrell, R.H. Hauge, J.M. Tour and R.E. Smalley, Electronic structure control of single-walled carbon nanotube functionalization, *Science*, **301**, 1519–1522 (2003).

[191] J. Cabana and R. Martel, Probing the reversibility of sidewall functionalization using carbon nanotube transistors, *J. Am. Chem. Soc.*, **129**, 2244–2245 (2007).

[192] J. Kong, N.R. Franklin, C. Zhou, M.G. Chapline, S. Peng, K.J. Cho and H. Dai, Nanotube molecular wires as chemical sensors, *Science*, **287**, 622–625 (2000).

[193] P.G. Collins, K. Bradley, M. Ishigami and A. Zettl, Extreme oxygen sensitivity of electronic properties of carbon nanotubes, *Science*, **287**, 1801–1804 (2000).

[194] P. Qi, O. Vermesh, M. Grecu, Q. Wang, H. Dai, S. Peng and K.J. Cho, Toward large arrays of multiplex functionalized carbon nanotube sensors for highly sensitive and selective molecular detection, *Nano Lett.*, **3**, 347–351 (2003).

[195] E.S. Forzani, X. Li, P. Zhang, N. Tao and R. Zhang, Tuning the chemical selectivity of SWCNT-FETs for detection of heavy-metal ions, *Small*, **2**, 1283–1291 (2006).

[196] E. Bekyarova, M. Davis, T. Burch, M.E. Itkis, B. Zhao, S. Sunshine and R.C. Haddon, Chemically functionalized single-walled carbon nanotubes as ammonia sensors, *J. Phys. Chem. B*, **108**, 19717–19720 (2004).

[197] E. Bekyarova, I. Kalinina, M.E. Itkis, L. Beer, N. Cabrera and R.C. Haddon, Mechanism of ammonia detection by chemically functionalized single-walled carbon nanotubes: *In situ* electrical and optical study of gas analyte detection, *J. Am. Chem. Soc.*, **129**, 10700–10706 (2007).

[198] A. Star, J.C.P. Gabriel, K. Bradley and G. Grüner, Electronic detection of specific protein binding using nanotube FET devices, *Nano Lett.*, **3**, 459–463 (2003).

[199] A. Star, T.-R. Han, V. Joshi, D. Steuerman and J.R. Stetter, Sensing with Nafion coated carbon nanotube field-effect transistors, *Electroanalysis*, **16**, 108–112 (2004).

[200] A. Star, T.-R. Han, V. Joshi, J.C.P. Gabriel and G. Grüner, Nanoelectronic carbon dioxide sensors, *Adv. Mater.*, **16**, 2049–2052 (2004).

[201] A. Star, E. Tu, J. Niemann, J.C.P. Gabriel, C.S. Joiner and C. Valcke, Label-free detection of DNA hybridization using carbon nanotube network field-effect transistors, *Proc. Nat. Acad. Sci. USA*, **103**, 921–926 (2006).

[202] A. Star, V. Joshi, S. Skarupo, D. Thomas and J.C.P. Gabriel, Gas sensor array based on metal-decorated carbon nanotubes, *J. Phys. Chem. B*, **110**, 21014–21020 (2006).

[203] D.S. Hecht, R.J.A. Ramirez, M. Briman, E. Artukovic, K.S. Chichak, J.F. Stoddart and G. Grüner, Bioinspired detection of light using a porphyrin-sensitized single-wall nanotube field effect transistor, *Nano Lett.*, **6**, 2031–2036 (2006).

[204] X. Zhou, T. Zifer, B.M. Wong, K.L. Krafcik, F. Léonard and A.L. Vance, Color detection using chromophore-nanotube hybrid devices, *Nano Lett.*, **9**, 1028–1033 (2009).

[205] G. Berkovic, V. Krongauz and V. Weiss, Spiropyrans and spirooxazines for memories and switches, *Chem. Rev.*, **100**, 1741–1753 (2000).

[206] X. Guo, L. Huang, S. O'Brien, P. Kim and C. Nuckolls, Directing and sensing changes in molecular conformation on individual carbon nanotube field effect transistors, *J. Am. Chem. Soc.*, **127**, 15045–15047 (2005).

[207] J.M. Simmons, I. In, V.E. Campbell, T.J. Mark, F. Léonard, P. Gopalan and M.A. Eriksson, Optically modulated conduction in chromophore-functionalized single-wall carbon nanotubes, *Phys. Rev. Lett.*, **98**, 086802-1-4 (2007).

[208] A. Star, Y. Lu, K. Bradley and G. Grüner, Nanotube optoelectronic memory devices, *Nano Lett.*, **4**, 1587–1591 (2004).

[209] J. Borghetti, V. Derycke, S. Lenfant, P. Chenevier, A. Filoramo, M. Goffman, D. Vuillaume and J.-P. Bourgoin, Optoelectronic switch and memory devices based on polymer-functionalized carbon nanotube transistors, *Adv. Mater.*, **18**, 2535–2540 (2006).

[210] X. Guo, J.P. Small, J.E. Klare, Y. Wang, M.S. Purewal, I.W. Tam, B.H. Hong, R. Caldwell, L. Huang, S. O'Brien, J. Yan, R. Breslow, S.J. Wind, J. Hone, P. Kim and C. Nuckolls, Covalently bridging gaps in single-walled carbon nanotubes with conducting molecules, *Science*, **311**, 356–359 (2006).

[211] A.C. Whalley, M.L. Steigerwald, X. Guo and C. Nuckolls, Reversible switching in molecular electronic devices, *J. Am. Chem. Soc.*, **129**, 12590–12591 (2007).

[212] M. Irie, Diarylethenes for memories and switches, *Chem. Rev.*, **100**, 1685–1716 (2000).

[213] X. Guo, A.C. Whalley, J.E. Klare, L. Huang, S. O'Brien, M.L. Steigerwald and C. Nuckolls, Single-molecule devices as scaffolding for multicomponent nanostructure assembly, *Nano Lett.*, **7**, 1119–1122 (2007).

[214] X. Guo, A.A. Gorodetsky, J. Hone, J.K. Barton and C. Nuckolls, Conductivity of a single DNA duplex bridging a carbon nanotube gap, *Nature Nanotechnol.*, **3**, 163–167 (2008).

[215] E. Katz and I. Willner, Biomolecule-functionalized carbon nanotubes: Applications in nanobioelectronics, *Chem. Phys. Chem.*, **5**, 1084–1104 (2004).

[216] G. Grüner, Carbon nanotube transistors for biosensing applications, *Anal. Bioanal. Chem.*, **384**, 322–335 (2006).

[217] K. Balasubramanian and M. Burghard, Biosensors based on carbon nanotubes, *Anal. Bioanal. Chem.*, **385**, 452–468 (2006).

[218] B.L. Allen, P.D. Kichambare and A. Star, carbon nanotube field-effect-transistor-based biosensors, *Adv. Mater.*, **19**, 1439–1451 (2007).

[219] M. Shim, N. Wong Shi Kam, R.J. Chen, Y. Li and H. Dai, Functionalization of carbon nanotubes for biocompatibility and biomolecular recognition, *Nano Lett.*, **2**, 285–288 (2002).

[220] R.J. Chen, S. Bangsaruntip, K.A. Drouvalakis, N. Wong Shi Kam, M. Shim, Y. Li, W. Kim, P.J. Utz and H. Dai, Noncovalent functionalization of carbon nanotubes for highly specific electronic biosensors, *Proc. Nat. Acad. Sci. USA*, **100**, 4984–4989 (2003).

[221] H.-M. So, K. Won, Y.H. Kim, B.-K. Kim, B.H. Ryu, P.S. Na, H. Kim and J.-O. Lee, Single-walled carbon nanotube biosensors using aptamers as molecular recognition elements, *J. Am. Chem. Soc.*, **127**, 11906–11907 (2005).

[222] K. Besteman, J.-O. Lee, F.G.M. Wiertz, H.A. Heering and C. Dekker, Enzyme-coated carbon nanotubes as single-molecule biosensors, *Nano Lett.*, **3**, 727–730 (2003).

[223] C. Li, M. Curreli, H. Lin, B. Lei, F.N. Ishikawa, R. Datar, R.J. Cote, M.E. Thompson and C. Zhou, Complementary detection of prostate-specific antigen using In_2O_3 nanowires and carbon nanotubes, *J. Am. Chem. Soc.*, **127**, 12484–12485 (2005).

[224] J. Wang, Carbon-nanotube based electrochemical biosensors: A review, *Electroanalysis*, **17**, 7–17 (2005).

[225] G.A. Rivas, M.D. Rubianes, M.C. Rodríguez, N.F. Ferreyra, G.L. Luque, M.L. Pedano, S.A. Miscoria and C. Parrado, Carbon nanotubes for electrochemical biosensing, *Talanta*, **74**, 291–307 (2007).

[226] J. Li, H.T. Ng, A. Cassell, W. Fan, H. Chen, Q. Ye, J. Koehne, J. Han and M. Meyyappan, Carbon nanotube nanoelectrode array for ultrasensitive DNA detection, *Nano Lett.*, **3**, 597–602 (2003).

[227] H. Cai, X. Cao, Y. Jiang, P. He and Y. Fang, Carbon nanotube-enhanced electrochemical DNA biosensor for DNA hybridization detection, *Anal. Bioanal. Chem.*, **375**, 287–293 (2004).

[228] S.G. Wang, R. Wang, P.J. Sellin and Q. Zhang, DNA biosensors based on self-assembled carbon nanotubes, *Biochem. Biophys. Res. Commun.*, **325**, 1433–1437 (2004).

[229] F. Patolsky, Y. Weizmann and I. Willner, Long-range electrical contacting of redox enzymes by SWCNT connectors, *Angew. Chem., Int. Ed.*, **43**, 2113–2117 (2004).

[230] A. Callegari, S. Cosnier, M. Marcaccio, D. Paolucci, F. Paolucci, V. Georgakilas, N. Tagmatarchis, E. Vázquez and M. Prato, Functionalised single wall carbon nanotubes/ polypyrrole composites for the preparation of amperometric glucose biosensors, *J. Mater. Chem.*, **14**, 807–810 (2003).

4

Decorated (Coated) Carbon Nanotubes: (X/CNTs)

Revathi R. Bacsa and Philippe Serp
LCC – ENSIACET, CNRS, University of Toulouse, France

4.1 Introduction

A carbon nanotube (CNT) is an ideal ultra strong 1D nanoscale material with chemical stability, excellent mechanical properties, high surface area and thermal conductivity [1–3]. Due to the high energy of the carbon-carbon double bond, carbon nanotubes (CNTs) resist quite well to acid treatment and are stable in all reductive atmospheres. In addition to the unique electronic and thermal properties of nanotubes, their nanometric dimension and hollow structure result in increased surface/volume ratio when suitably dispersed. Typically, the total surface area of as-grown single-walled carbon nanotubes (SWCNTs) ranges between 400 and 900 m^2 g^{-1} (micropore volume, 0.15–0.3 cm^3 g^{-1}) [4], whereas for as-produced multi-walled carbon nanotubes (MWCNTs) values ranging from 50 to 400 m^2 g^{-1} (mesopore volume, 0.5 and 2 cm^3 g^{-1}) [5,6] have been measured using gas adsorption isotherms. Detailed study of these isotherms have shown that nanotubes have many potential adsorption sites namely, the outer surfaces of tubes, tube ends, intertube sites in a SWCNT bundle and the hollow cavity of open tubes. These adsorption sites can be decorated with a wide variety of atoms and molecules giving rise to a new class of materials that are here collectively designated as 'decorated nanotubes'. The nanostructure of the surface serves as a template to anchor or organize molecules and particles in a highly dispersed state. Due to their interaction with the surface of the nanotube substrate, the deposited

Carbon Meta-Nanotubes: Synthesis, Properties and Applications, First Edition. Edited by Marc Monthioux.
© 2012 John Wiley & Sons, Ltd. Published 2012 by John Wiley & Sons, Ltd.

metals, semiconductors, oxides, organic molecules and biomolecules often present unique structure-property relations. Thus, it has been observed that the curved surface of the SWCNTs modifies the interaction of the substrate with the coated material thereby giving rise to new properties [7]. As a result, these composite materials find a number of applications in catalysis, electronics, environmental pollution control, energy and medicine. The nanotube surface can also be suitably modified to attach specific functional groups thanks to the presence of inherent structural defects. For applications in catalysis where carbon nanotube supports coated with transition metals have been investigated, modifications can be effected on specific sites so as to obtain high catalyst dispersion resulting in an efficient utilization of the entire catalyst loading [8–11]. Also, the possibility of macroscopic shaping makes CNTs attractive as catalyst supports in large-scale applications. Indeed, resistance to abrasion, thermal and dimensional stabilities, and specific adsorption properties are important factors that influence the activity and reproducibility of the catalytic system. In particular, CNTs and carbon nanofibers (henceforth termed CNFs) could replace activated carbon supports in liquid-phase reactions since the surface properties of the latter cannot be easily controlled, and increasing their microporosity to increase their specific surface area has often been detrimental to mass transport [12]. Apart from transition metals, including noble metals, a number of inorganic materials have also been coated onto the nanotube surface targeting applications such as sensors, solar cells or for magnetic applications. Carbon nanotubes interacting with metals and oxides are increasingly gaining interest in the fabrication of ultra sensitive sensors for gases, liquids such as liquid petroleum gas (LPG) and biomolecules such as glucose [13,14]. Nanotubes coated with macromolecules find uses in areas where the specific biomolecular recognition properties are targeted [15]. Understandably, the volume of scientific research on these materials has been increasing each year and review articles are now available that describe the ongoing research in this field [16,17].

In this chapter, our aim is to present a concise overall view of the synthesis, properties and applications of decorated nanotubes illustrated by specific examples. It is important here to distinguish decorated nanotubes from functionalized nanotubes that have been dealt with in detail in Chapter 3. In contrast to functionalized nanotubes where the nanotube walls participate in chemical reactions – most of the time with organic molecules resulting in a new complex structure with specific reactivity, decorated nanotubes consist of principally solids dispersed on the CNT surface, usually covering a large percentage of the surface. The physical and chemical properties of these solids might be modified due to their confinement and interaction with the nanotube structure.

In the following, experimental and theoretical findings on the coating or decoration of the carbon nanotube surfaces with various nanoscale materials will be detailed. As far as the substrate is concerned, mainly single-, double-, and multi-walled tubes have been considered but experimental observations on carbon nanofibers have also been covered wherever found significant. Coating by transition metals (Cu, Ni, Ag, Au, Ru, Rh, Pd, Pt, Ir, and Ti), semimetals (Si, Ge), oxides (ZnO, TiO_2, SnO_2, and SiO_2), nitrides and chalcogenides (ZnS, CdS, CdSe, and ZnSe) have been described wherever these have been used in specific applications. Although not exhaustive, Table 4.1 lists the elements and compounds that have been deposited on CNTs or CNFs along with the commonly used deposition process.

Table 4.1 *Examples of Elements and compounds that have been deposited on CNTs or CNFs along with the related deposition processes.*

Element	Coating method	Particle size (nm)	Support	References
Metals	Wet-impregnation	1–1.5	MWCNT	44, 55
	Polyol process	2–3	SWCNT	60
Pt	MOCVD	2–3	MWCNT	127
	Electrochemical	12	MWCNT, SWCNT	109, 104
Pd, Ru, Rh	Wet-impregnation	1.5–2.5	MWCNT	59, 136, 133,
	Self-assembly	2–10	MWCNT	95–96
	Electrochemical, CVD	30–70, 4–6	MWCNT, SWCNT	108, 110, 116, 126
Pt-Pd, Pt-Ru, Pt-Ni	Wet-impregnation electrochemical	1–5 60–80, 10	AC, MWCNT MWCNT film	56–58 105, 109
Ag, Au, Cu	Wet-impregnation, Electroless,	5–20 3–4	MWCNT MWCNT, SWCNT	44, 43, 62 67, 68, 114
	Self-assembly	2–10	MWCNT, SWCNT	72–74, 76
	Elec. Beam. Evap.	Film	SWCNT	92
Cu	CVD	3–50	MWCNT	128
Ni	Wet-impregnation	4–15	CNF, MWCNT	45, 46
	Electroless plating	40	MWCNT	65
Ti, Cr, Ni	Elec. Beam. Evap.	Film	SWCNT	118–119
Zn	Electrodeposition	Film	MWCNT	112
Oxides	Sol-gel synthesis	Film	MWCNT	50–51
SiO_2, TiO_2, Al_2O_3 Fe_3O_4	Self-assembly	3–8	MWCNT	78, 82
	ALD process	Film	MWCNT	122
SnO_2, TiO_2, Cu_2O	CVD, Self-assembly	10–70	MWCNT	130, 131, 84, 86, 94, 53
ZnO, MgO	Wet-imprg. Polyol process	10–60	MWCNT	93, 82, 54
ZrO_2, Eu_2O_3	Homogeneous deposition precipitation	4–20	MWCNT	51, 52
Semiconductors	Homogeneous Deposition precipitation	20–30	MWCNT	48
InP, CdS,CdSe,	CVD			164
Ag_2S, HgS, WS_2	Self-assembly of particles quantum dots	4–10	SWCNT	79–81, 83
Nitrides, carbides BN, SiCN	CVD	Film	MWCNT	132, 165
Halides TbF_3, EuF_3	Precipitation on the nanotube surface using NaF.	8–12	MWCNT	47

The outline of the chapter is as follows: a brief overview of the theoretical studies of the carbon-metal system is given that models the effect of the carbon nanotube defect structure and functionalization on the density and size distribution of the coating particles. The various experimental procedures used for coating nanotubes are detailed, comparing the merits and drawbacks of each method where applicable. Specific examples are presented for each coating method. The commonly used characterization tools for coated nanotubes are described in Section 4.5 followed by examples of application of coated nanotubes in various domains such as catalysis, sensors, hydrogen storage and fuel cells. Finally, recent developments in the applications of coated nanotubes in biology and medicine are presented.

4.2 Metal-Nanotube Interactions – Theoretical Aspects

Theoretical modeling of metal-coated nanotubes has been carried out with the aim to better control the size and dispersion of the metal particles by optimizing their interaction with the carbon support. Understandably, a detailed knowledge of the structure and stability of the interface between the metal/alloy and the graphene substrate can pave the way for the tailoring of more efficient catalysts. The models that have been developed so far for transition metal-graphene interactions are based on certain experimental observations, namely:

1) Some metals such as Ti and Ni form continuous films on carbon supports whereas others such as Pt or Ru form clusters. Electrical property measurements on nanotube-metal junctions have also shown significant differences concerning the nature of the contact for different metals [18]. In fact, Pd seems to be unique in that it consistently yields ohmic contacts to metallic nanotubes (and zero or even negative Schottky barriers at junctions with semiconducting SWCNTs for FET applications).

2) High resolution TEM studies on coated nanotubes have shown that the metal atom is anchored preferentially on certain sites on the nanotube. It is therefore important to calculate the relative binding energies for the different potential adsorption sites present in carbon nanotubes and nanofibers so as to predict the ease of, say, decorating the inner wall of a nanotube when compared to the outer surface or intertube spaces in a bundle.

3) Functionalizing the nanotubes using acid treatment or polymer wrapping greatly increases the density of coating and particle size control irrespective of the coated material. This last observation implies that introducing defects/functional groups in the graphene plane should necessarily change the binding energy and must be included in order to arrive at a credible result.

In view of the above mentioned three experimental observations, we will briefly describe the results of metal-carbon interaction calculations, firstly, on the different sites of a graphene sheet. Subsequently, a brief discussion on the influence of curvature on the metal-nanotube interaction will be presented and lastly, the effect of introduction of defects and functional groups on the binding energy will be described. It is important to note that only representative examples have been given and readers are referred to individual papers and references sited therein for further information. Also, very little theoretical information is available for coating by nonmetals or oxides and therefore, these materials could not be addressed.

(a)

(b)

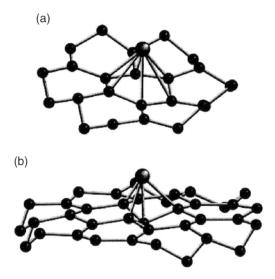

Figure 4.1 *(a) Stable hole and (b) atop positions for a Ni atom interacting with a graphene surface, the latter being simulated by a planar cluster of 22 and 37 atoms respectively. Reprinted with permission from [19] Copyright (1999) Elsevier Ltd.*

The simplest model involves essentially the calculation of the binding energy of an isolated metal atom on potential adsorption sites of the support. For large diameter multi-walled nanotubes and nanofibers, the graphene approximation is found to be adequate. There are large variations in the calculated energies of interaction for the non-noble transition metals (Ni, Cu) when compared to noble metals (Pt, Ru, Pd, Au and PtRu alloys). Among the noble metals, Pt has been the most studied element; among the non noble transition elements, Ni has been the most studied [19]. For example, the interactions between nickel atoms and a planar graphite sheet have been modeled using the tight binding molecular dynamics method. Several different types of metal-graphene configurations were considered. These consisted of a Ni atom (i) above the center of a hexagon (hole), (ii) directly above a C atom (atop) and lastly, over a C–C bond. The size of the corresponding planar graphitic cluster was taken to consist of 24, 37 and 42 atoms, the relaxed structure for (i) and (ii) are shown in Figure 4.1. For each of these cases, the magnetic moment and the Ni–C distances were calculated for the relaxed state. The strong Ni–C interaction caused considerable distortions in the planar structure beneath the Ni atom changing the Ni–C separation distances and resulted in a 70% reduction of the magnetic moment relative to its value in bulk Ni.

More recently [20,21], the density functional theory (DFT) was used to calculate the binding energies of platinum supported on planar graphite. In theoretical modeling, the DFT is used in the calculation of the energy and electronic structure of many body systems applying quantum mechanical methods. The main advantage of this theory lies in the replacement of a complicated wave function using 3N variables (where N is the number of electrons in the system) by a single quantity that is defined as the electron density function. A system of N interacting electrons is reduced to one of non interacting electrons moving

Table 4.2 *Binding energies for Au, Pd and Pt on a 2×2 graphene sheet [24]. Eb_1 is the binding energy (normalized per metal atom) with respect to the isolated metal atom and Eb_2 is the energy computed (per substrate-adjacent metal atom) with respect to the isolated metal film. E_{coh} is the computed cohesive energy. All energy values are in eV per metal atom for the bulk crystal.*

Type of coating	Au		Pd		Pt	
	Eb_1	Eb_2	Eb_1	Eb_2	Eb_1	Eb_2
Isolated atom	0.30	–	0.94	–	1.65	–
Monolayer	1.51	0.014	2.00	0.045	4.15	0.070
Bilayer	1.56	0.011	2.38	0.038	4.26	0.038
Trilayer	1.57	0.009	2.54	0.039	4.15	0.029
E_{coh}	3.13		3.64		5.88	

in an effective potential. It is not the purpose of this chapter to go into the details of how this potential is calculated, for which readers are referred to excellent text books and reviews [22,23]. It suffices here to note that DFT with the local-density approximation yields results that compare well with experimental data for solid state calculations. These theoretical calculations have mainly shown that the binding energies vary from one transition metal to the other due to differences in the energy of the hybrid orbitals arising from the overlap of the carbon p_z and the d orbital of the metal atom. Often the metal-carbon interaction, while very strong for the first monolayer of metal, gets weaker when the second layer is deposited. For platinum, it turns out that the calculated binding energy for a multilayer film is lower than the cohesive energy of Pt-Pt clustering, resulting in the formation of clusters. Results from binding energy calculations have been used to predict the ease of formation of a continuous film of the transition elements on the nanotube surface [24]. The values of calculated binding energies for Au, Pd and Pt on a 2x2 graphene sheet are given in Table 4.2.

It is seen that the binding energy increases from Au to Pt from 0.3 to 1.65 eV when isolated atoms are considered. In all the cases, the binding energy for metal layers on graphene is higher than for isolated atoms. It is interesting to note that for Pt, the binding energy decreases rapidly with film thickness. This decrease has been attributed to the high cohesive energy for Pt metal when compared to Pd and Au indicating that Pt films beyond a certain thickness are unstable and tend to form clusters.

4.2.1 Curvature-Induced Effects

The binding energies for SWCNTs of small diameters on flat metal surfaces of Au, Pt and Pd have been calculated in order to investigate the influence of curvature on the metal-carbon interactions. For small diameter SWCNTs with n and m values of (5,0) and (8,0) (corresponding to diameters of 0.39 nm and 0.62 nm respectively) on Au, the calculated metal-carbon distances are much lower than for graphite, but for Pt and Pd the calculated binding energies did not differ very much from that obtained for planar graphite. For small diameter tubes, due to the increased curvature, a local sp^2-sp^3 transition of the hybridization of the carbon atoms in contact with the transition element forming near covalent bonds has been proposed. Similar calculations for the interaction of a Ni cluster on a surface of armchair (5,5) and (10,10) tubes have shown that while the Ni-Ni bond is greatly weakened upon chemisorption on a planar graphite sheet, this weakening is suppressed by the curvature of the nanotube indicating that charge transfer between the nanotube and transition metal is increased when compared to graphite [25]. Curvature-induced stabilization has also been predicted for Pt coated on subnanometer (5,5) or (8,0) SWCNTS where a metal-carbon binding energy difference of 1.5 eV has been observed when compared to graphite. This difference is only 0.6 eV for larger SWCNTs such as (10,10) [26].

4.2.2 Effect of Defects and Vacancies on the Metal-Graphite Interactions

Experimental observations have clearly shown that defects present inherently or induced during purification play an important role in the adsorption of transition metal atoms on carbon nanotubes in contrast to pristine nanotube surfaces that are impervious to any sort of coating. Hence, modeling of the metal-nanotube system also includes the estimation of the change in binding energy of a metal adsorbant due to the introduction of a defect in the form of a Stone-Wales defect or a vacancy in the lattice. The interface between graphene and a Pt_{13} or Au_{13} cluster has been investigated using density functional calculations. The results revealed that the introduction of a carbon vacancy into the graphene sheet (the presence of a five- or seven-membered ring) increased the adsorption energy of the clusters. Similar calculations have been carried out for the interaction of a nickel atom with (5,5) and (10,10) armchair SWCNTs containing three intrinsic defects, namely, single vacancy, double vacancy, and Stone-Wales defect [27]. While the bonding of Pt multilayer on nanotubes has been shown to be weak, for Pt/Ru multilayers, it was found that the ruthenium atoms stabilized the carbon-metal interface and that vacancies on the graphene enhance the interaction between the graphene and multilayer metal surfaces. The energy barrier for diffusion of Pt through the side wall of the nanotube has been estimated to be as high as 6 eV showing that Pt diffusion through the side wall into the tube is negligible.

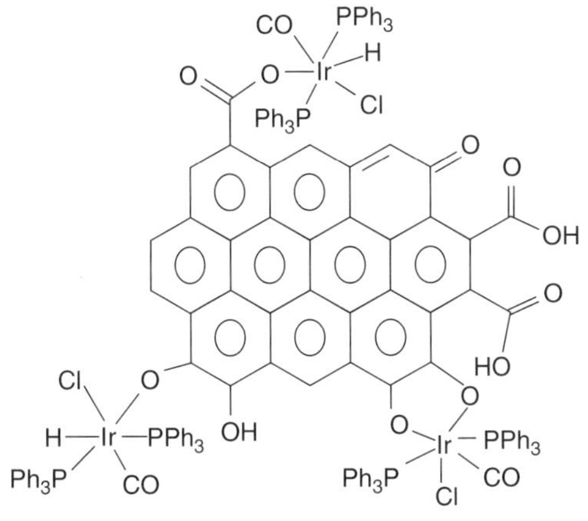

Figure 4.2 *Complexation of functionalized nanotubes with Vaska's complex. Reprinted with permission from [30] Copyright (2002) American Chemical Society.*

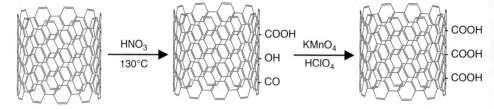

Figure 4.3 *Nanotube functionalization scheme. Adapted with permission from [78] Copyright (2004) American Chemical Society.*

In addition to vacancies, functional groups present on the nanotubes are also shown to bind effectively to adduct. The binding of an electron rich transition metal complex such as the Vaska's complex *trans*-[Ir(CO)Cl(PPh$_3$)$_2$)] onto a (9,0) SWCNT with finite length has been modeled using DFT [28,29]. It was found that a stable bound adduct was not obtained when a perfect hexagonal network was considered showing that the nanotube side wall was relatively unreactive to the ligand. Only a weakly bound complex with π–π interaction between the phenyl rings and the CNT side was obtained. However, nanotube end caps or defective sites on the side wall showed higher interaction energy and a stable complex is more likely to be formed if at least one of the coordinating carbon atoms belongs to a pentagonal ring formed by the presence of a defect in the graphite lattice (Figure 4.2). The formation of this complex on functionalized carbon nanotubes has been experimentally shown [30].

4.3 Carbon Nanotube Surface Activation

Both experimental and theoretical results hitherto obtained conclude that a surface activation in the form of functionalization or attachment of active groups on the surface of carbon nanotubes is essential to achieve a high coating density and for improved catalyst dispersion. While most of the details can be found in Chapter 3, a few important results are summarized below. The most common method of introducing surface carbonyl and carboxyl group is by oxidative acid reflux that purifies and functionalizes the nanotubes in a single step. The procedure was initially developed to remove the metal catalyst impurity present in the raw nanotube samples. Nanotubes were sonicated/refluxed either in concentrated nitric acid or in a mixture of sulfuric and nitric acids. Any amorphous carbon present is oxidized and unreacted and accessible transition metal catalysts such as Fe, Co and Ni can be removed. Infrared spectra of acid treated samples show the presence of carboxyl and carbonyl groups [31–33]. A prolonged reflux induces the opening of MWCNT tips, damages the walls and slightly increases the BET surface area [34]. Diluted nitric acid (6 M) is often used for functionalizing SWCNTs with reduced reflux times to prevent permanent damage to the wall structure. Other common oxidizing agents such as potassium permanganate, osmium tetroxide, hydrogen peroxide and ozone have also been used [35] (Figure 4.3).

Alternatively, gas phase oxidation in air or oxygen can be performed but this method is found to generate phenolic or carbonyl groups instead of carboxylic groups [36].

Hydrothermal oxidation on carbon nanotubes has also been reported [37]. Functional groups can also be attached to the nanotube surface by van der Vaals type forces by simple mixing or grinding with the appropriate molecules. Thus, polymers such as polyvinyl alcohol (PVA), polyacrylic acid (PAA), polyvinylpyrollidone (PVP) or polyethylene glycol (PEG) have been wrapped around nanotubes by simple mixing after which the excess polymer is removed by centrifugation [38]. Microwave radiation has also been used for the quick functionalization of carbon nanotubes [39–40]. Microwave energy decreases the kinetic barrier for reactions by typically altering the vibrational energies of specific reactants. Functional groups can be formed on MWCNTs dispersed in water exposed to sequential microwave radiation for 4500 s by runs of 20 s separated by a 10 s off-time. Infrared spectra of the treated sample showed the presence of allyl, carbonyl, carboxyl and hydroxyl groups. Other sources of excitation have been used including sonication or UV radiation [41]. Ball milling under reactive atmosphere has also been found to produce shortened functionalized nanotubes. A quantitative estimate of the functionalization along with the spectroscopic determination of the different functional groups has been detailed by Solhy et al. [42].

4.4 Methods for Carbon Nanotube Coating

The choice of the coating method depends largely on the application of the coated nanotubes and the scale of operation. Electrochemical coating methods are obvious choices in eletrocatalytic applications whereas gas phase methods seem to be more compatible with electronic applications. The wet-impregnation method, the most widely used technique, is a multistep process that consists of the deposition of a precursor of the coated material followed by calcination and reduction. Self-assembly techniques, though efficient and precise, are difficult to upscale due to the high cost of the reagents. In this section, the different experimental methods developed for coating elements such as metals, semimetals or compounds have been described in some detail. Each method is illustrated by representative examples and the list of references is not exhaustive. In all cases it has been found that an efficient coating is possible only when the nanotube surfaces are suitably functionalized or modified.

4.4.1 Deposition from Solution

4.4.1.1 Wet-Impregnation
The wet-impregnation method involves the impregnation of the nanotube surface with a solution of the metal precursor salt/organometallic complex, followed by drying, calcination, and reduction to result in metal particles anchored to the carbon nanotube surface. The reduction is carried out either by using H_2 gas or by common reducing agents such as sodium borohydride. This method is the most commonly used, principally due to its simplicity, cost and ease of upscaling. A first example of such a coating was reported as early as 1996 [43] where carbon nanotubes were coated by uranyl acetate and silver particles. Uranyl acetate was used due to its relative ease of observation by electron microscopy even at low concentrations. Arc produced multi-walled carbon nanotubes were treated with concentrated nitric acid that not only dispersed the nanotubes in solution but

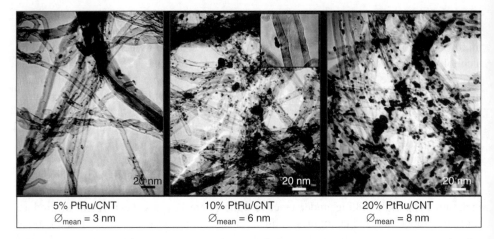

| 5% PtRu/CNT | 10% PtRu/CNT | 20% PtRu/CNT |
| $\varnothing_{mean}$ = 3 nm | $\varnothing_{mean}$ = 6 nm | $\varnothing_{mean}$ = 8 nm |

Figure 4.4 *TEM images of PtRu nanoparticles coating acid-treated MWCNTs by wet-impregnation from metal salt solutions.*

also created active carboxyl groups that were replaced by uranyl groups by simple ion exchange. The raw nanotubes on the other hand were not coated by uranyl acetate thus leading to the conclusion that oxidizing the nanotubes is an essential preparative step before coating. A similar pretreatment was used in silver electroless plating reaction to coat carbon nanotubes with 10 nm silver particles. Following this work, acid treated carbon nanotubes were decorated with nanoscale Pt, Au, Ag clusters by reaction in solution of their corresponding oxo-acids followed by hydrogen reduction. Once again, the importance of acid functional groups on the nanotube surface for effective coating of the nanoparticles was demonstrated [44]. Various experimental procedures have been used for the anchoring of the precursor on the functionalized nanotubes and these will be described below. Examples of the deposition of PtRu alloy particles on acid-treated MWCNTs are shown in Figure 4.4.

4.4.1.2 *Homogenous Deposition Precipitation Method (HDP)*
While coating was efficient only on activated carbon nanotube surfaces, the nature of the precursor was also shown to play an important role in the control of particle size. For example, two different methods for the coating of surface-activated carbon nanofibres by Ni using the wet-impregnation of nickel nitrate have been compared for the density and size distribution of Ni particles [45]. In the first one, nickel nitrate solutions were stirred with acid functionalized carbon nanofibers and the composite was dried and reduced in H_2+Ar atmosphere at 773 K to generate nickel particles on the surface of the CNF. The second method consisted of reacting nickel nitrate and urea in the presence of carbon nanofibers so as to precipitate a layer of nickel oxide on the nanofibers that was subsequently reduced in H_2+Ar atmosphere as described above. The latter method was termed by the authors as homogenous deposition precipitation method (HDP). It was reported that the HDP process yielded a better control of particle size when compared to simple wet-impregnation. For low Ni loadings, the HDP method gave a particle size distribution of 5–7 nm when

compared to 9–11 nm for the wet-impregnation. For higher Ni loadings (30%) the HDP procedure allowed a much better control of particle size (8–10 nm) compared to the wet-impregnation (8–60 nm). Following this, there have been efforts to coat Ni on nanotubes without going through the process of surface activation. $Ni(NO_3)_2$ was heated to dissociation in the presence of the nanotubes at 773 K followed by reduction at 923 K to obtain a coating by Ni particles of size in the range 4–15 nm. The coating process owed its success to the formation of a thin oxide layer between the nanotubes and metal particles that increased the adhesion of the latter to the nanotube surface [46]. By the same principle, the precipitation of fluorides of rare earth elements such as Eu and Tb on the surface of the nanotube stabilized with a surfactant (sodium dodecyl sulfate) gave rise to nanosized particles of Eu and Tb with controlled size distribution in the range of 8–12 nm [47]. A similar method using sodium sulfide as the sulfur source was used to precipitate CdS, Ag_2S and HgS nanoparticles [48]. Among the oxides, silica has been the most studied. The most common method used for coating nanotubes with silica is the sol gel synthesis technique wherein polymer-wrapped MWCNTs are stirred along with tetra-ethyl or tetra-isopropyl orthosilicate in the presence of ammonia catalyst [49]. Sol gel techniques have also been used for successfully coating multiwall carbon nanotubes with titanium oxide [50]. Other ceramic oxides such as cerium and zirconium oxides have been coated on MWCNTs. ZrO_2-CNT composites were prepared by the decomposition of $Zr(NO_3)_4.5H_2O$ in supercritical carbon dioxide–ethanol solution or by hydrothermal heating of $ZrOCl_2.8H_2O$+MWCNT mixture at 523 K [51,52]. Several authors have reported on the coating of carbon nanotubes by Cu_2O, that cite the use of nanotubes as a template to achieve a better size control of nanoscale cuprous oxide which is a direct band gap semiconductor having numerous applications in photocatalysis and hydrogen production. The Cu_2O coating has been carried out by wet-impregnation, polyol process (see Section 4.4.1.4) and by electrochemical deposition of Cu. Under different experimental conditions, Cu_2O on the surface of MWCNTs changed from monodisperse nanoparticles, leaf-like particles, to larger spheres [53]. Recently, a hydrothermal synthesis method has been reported for the deposition of ZnO on carbon nanotubes [54].

4.4.1.3 Organic Functionalization

Organic functional groups have been introduced to anchor an appropriate precursor to the nanotube surface, which upon reduction could give rise to a high density coating. Thus, platinum particles in the size range 1–1.5 nm with a narrow size distribution were coated on a MWCNT support modified by surface thiolation [54]. Acid functionalized nanotubes were reacted with thionyl chloride at 343 K, and after evaporation of excess thionyl chloride, thiolation was achieved by reacting with $NH_2C_6H_4SH$. A sketch of a thiol-functionalized nanotube surface is shown in Figure 4.5. A very high Pt loading (40 wt.%) could be achieved using this method.

The thiolated nanotubes were impregnated with a solution of H_2PtCl_6 under sonication followed by reduction with sodium borohydride. The presence of sulfur in the nanotubes was confirmed by XPS and EXAFS measurements. XPS spectra showed a significant increase in intensity in the Pt4f peak accompanied by a shift in binding energy for the thiolated nanotubes. This shift has been explained to be due to two combined effects: the decrease in particle size and the interaction of Pt particles with the sulfur of the thiol group. A further confirmation of this interaction comes from the shift of the S2p peak in the

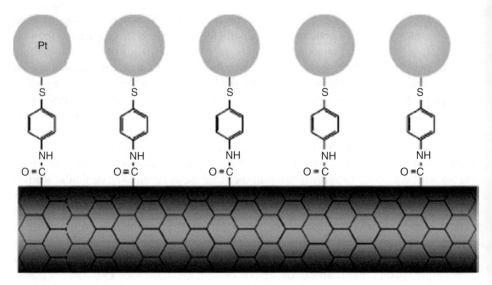

Figure 4.5 *A schematic diagram of a nanotube sidewall functionalized by thiol groups coated with platinum. Reprinted with permission from [55] Copyright (2006) Elsevier Ltd.*

thiolated nanotubes towards low binding energy after Pt deposition. In addition, the radial distribution function (RDF) of the Fourier transformed EXAFS showed the presence of a Pt-S bond corresponding to a distance of 1.9 Å compared to the Pt-Pt bond distance of 2.7 Å predominant in non thiolated nanotubes. Thus the thiol groups firmly anchor the Pt particles to the nanotube surface preventing their agglomeration.

4.4.1.4 Polyol Process

In this process metal salt solutions are reduced by heating them with a liquid polyol such as ethylene glycol [56–58]. The size distribution of the metal particles depends largely on the experimental conditions – mainly the polyol/water concentration ratio. A microwave-assisted polyol process was employed to coat Vulcan XC-72 carbon black and carbon nanotubes with Pt-Ru alloy nanoparticles of uniform size. In this process, the polyol solution containing the metal salt is refluxed at 393–413 K to decompose ethylene glycol that acts as a reducing agent leading to the formation of ultra fine nanoparticles of Pt-Ru alloy. Transmission electron microscopy images have shown a high density of uniformly sized Pt-Ru particles (Figure 4.6).

It appears that the size of the metal nanoparticles is determined by the rate of reduction of the metal precursor and hence the use of microwave techniques for the decomposition reaction is significant. The dielectric constant (41.4 at 298 K) and the dielectric loss of ethylene glycol are high, and hence rapid heating occurs easily under microwave irradiation. The fast heating by microwave accelerates the reduction of the precursor and results in the instantaneous nucleation of the metal clusters. Additionally, the homogeneous microwave heating of liquid samples reduces the temperature and concentration gradients in the reaction. Decomposition of ruthenium chloride by ethylene glycol on MWCNTs stabilized by SDS shows that the sulfonic groups of the SDS molecules play a key role in the deposition

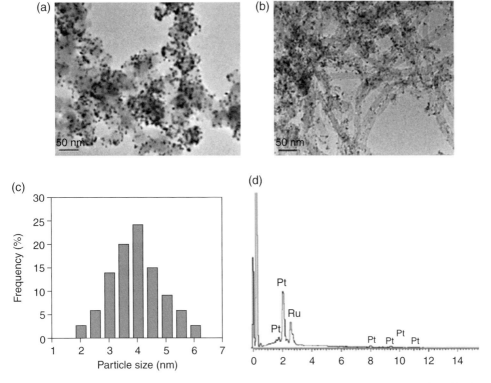

Figure 4.6 *TEM images of PtRu nanoparticles supported on (a) Vulcan XC-72 carbon black and (b) MWCNTs using the polyol process; (c) and (d) show the particle size distribution and the EDX spectra of PtRu nanoparticles on Vulcan XC-72. Reprinted with permission from [56] Copyright (2004) American Chemical Society.*

of ruthenium nanoparticles. A very high density dispersion of monodisperse ruthenium nanoparticles coating was achieved by this method [59].

For SWCNTs, it is necessary to exfoliate bundles into individual tubes by using a suitable surfactant or a stabilizer molecule to achieve uniform coating. A modified polyol process for the coating of Pt particles on SWCNTs has been reported where wrapping a polymer around the nanotubes has been used for the effective dispersion of SWCNTs and controlling of the deposition of Pt catalysts [60]. SWCNT bundles were dispersed into individual tubes by dispersing them with polystyrene sulfonate on which Pt was deposited by wet-impregnation of H_2PtCl_6, and subsequent reduction to Pt was achieved using ethylene glycol at pH 12.5. The Pt loading could be increased up to 30wt.% and 2–3 nm size Pt particles were deposited in high density.

4.4.1.5 Organometallic Grafting

Organometallic complexes have been used for the deposition of metal particles onto functionalized carbon nanotubes. By anchoring these complexes, a better control of the metal particle size was expected. One of the first species to be grafted was the Vaska's

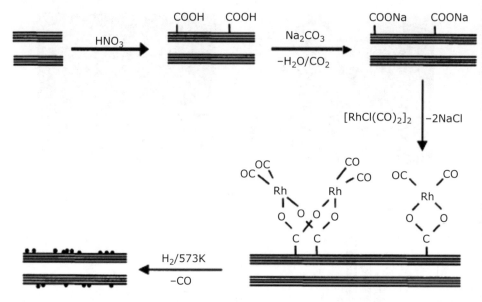

Figure 4.7 *Surface mediated organometallic reaction between [Rh₂(μ-Cl)₂(CO)₄] and MWCNT-COONa. Reprinted with permission from [61] Copyright (2003) John Wiley and Sons.*

complex [IrCl(CO)(PPh₃)₂] that was added to nitric acid-functionalized carbon nanotubes [30]. The attachment of the complex takes place by the coordination of the nanotube surface to the iridium center through the oxygen atoms. The organometallic reaction of the rhodium complex [Rh₂Cl₂(CO)₄] with a modified MWCNT surface bearing sodium carboxylate groups was found to result in highly dispersed Rh nanoparticles after a reduction step [61]. The acid-treated nanotube surface was further modified with sodium carbonate base to produce the sodium salt of the acidified nanotubes. After impregnation of the complex, and filtration, the resulting solution was always colorless, pointing to a complete adsorption of the rhodium complex. Small highly dispersed (1.5–2.5 nm for a 1% w/w loading) rhodium nanoparticles were produced that retained the size distribution even for rhodium loadings up to 9.5% w/w. Figure 4.7 shows a schematic diagram of the preparation procedure. The decomposition of organometallic gold complex precursors and their deposition on functionalized multi-walled carbon nanotubes has been reported recently [62].

4.4.1.6 Electroless Deposition

Electroless plating has been described as the process involving the spontaneous reduction of metal ions to metallic particles and films in the absence of electric current from an external source [63]. Using this deposition method, even insulating surfaces can be coated with metal films and hence this method has found a wide application in the metallization of plastics used extensively in semiconductor industry. There are three types of electroless processes namely, autocatalytic, substrate catalyzed and galvanic displacement processes [64]. Autocatalytic and substrate catalyses processes use an activator in the form of a reducing agent and a metal salt solution usually tin chloride and palladium chloride to achieve the surface activation of the substrate. Once the surface is activated, spontaneous

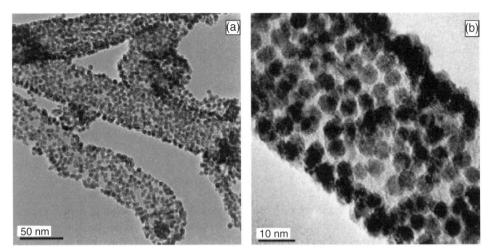

Figure 4.8 *Two TEM images at different magnification of gold nanoparticles deposited on MWCNTs by the electroless process. Reprinted with permission from [67] Copyright (2006) Elsevier Ltd.*

coating is achieved; often metal films with thicknesses of the order of a few microns are produced. This is the process used for metallization of ceramics and polymer bodies in electronics industry. This type of electroless plating has been used to obtain a uniform coating of Ni on the surface of MWCNTs [65]. For nickel coating, the plating bath used was composed of a solution of nickel sulfate ($0.1\,M$ $NiSO_4.6H_2O$), a $0.2\,M$ solution of sodium dihydrogen phosphate, $0.5\,M$ $C_6H_5Na_3O_7$, and $0.5\,M$ $(NH_4)_2SO_4$. Acid purified MWCNTs were used as substrates. Prior to coating, the nanotubes were sensitized by an aqueous solution containing $8.9\times10^{-3}\,M$ $SnCl_2$ and $2.4\times10^{-3}\,M$ HCl for 10 min that caused the adsorption of Sn^{2+} ions on the surface of the nanotubes. A solution of palladium chloride was added to the mixture upon which Pd^{2+} ions react with the Sn^{2+} to form metallic palladium particles on the surface of the nanotubes. The nanotubes were subsequently immersed in the nickel bath at 293 K to provide a uniform layer of Ni on the nanotubes. The chemical composition of the coating is found to be a mixture of Ni metal and Ni_3P, an intermetallic compound of nickel and phosphorous. It was possible, as the authors report in a later publication [66], to prepare discrete spherical Ni-P particles of diameter around 40 nm by rapidly cooling the bath after 1 min of coating. This cooling stops the coating process resulting in discrete particles instead of a film. Similarly, sensitized and activated MWCNT surfaces (previously functionalized by $HNO_3+H_2SO_4$ treatment) have been coated with gold particles using a gold cyanide electroless plating bath with sodium borohydride as a reducing agent. Uniformly sized spherical nanoparticles of gold with diameters in the range of 3–4 nm were obtained by this method (Figure 4.8). The remarkable uniformity in particle size and density has been attributed to the presence of numerous functional groups on the surface of the carbon nanotubes formed during the functionalization due to the reaction of the oxidants with reactive pentagons present at these defect sites [68].

Another electroless process is the galvanic displacement process. There is no added reducing agent, and the reduction of the metal ion to be coated takes place by electron transfer from the substrate. Deposition can go on until such an electron transfer becomes impossible, presumably due to the formation of a layer of oxidized substrate on the surface. An elegant demonstration of this deposition process on single wall carbon nanotubes was done by the group of Dai [68]. The nanotube sample used was a thin film of SWCNTs grown on a SiO_2 substrate. Various salt solutions such as $HAuCl_4$, Na_2PtCl_4 in a 1:1 solution of water to ethanol were used and the nanotube film was immersed into the salt solution at room temperature. Nanoparticles of Au or Pt were formed selectively on the nanotube surface and were absent on the SiO_2 substrate showing that the reduction takes place on the surface of the SWCNT. The maximum particle density was attained at 30 s after immersion. The authors have hence determined that the Fermi level of a SWCNT is +0.5 V above the potential of SHE (Standard Hydrogen Electrode) and is therefore well above the reduction potentials of $AuCl_4^-$ and $PtCl_4^{2-}$ (1.002 V and +0.775 V with respect to SHE respectively) and hence a spontaneous reduction of these solutions to the corresponding metals should take place in contact with a nanotube surface. It is of interest to note that the nanotube surfaces were as-prepared and untreated. If a bias of 10 mV was applied across the nanotube, it was seen that the current through the nanotube increased by 600 nA within the first 2 s indicative of the process of hole injection into the SWCNTs. Heating the film to 873 K in air caused the oxidation of the nanotubes leaving behind 2 nm diameter wires of Pt or Au proving that the nanotubes can act as efficient templates for nanowire synthesis.

4.4.2 Self-Assembly Methods

Recent years have seen an unprecedented development in the self-assembly of nanoparticles to form ordered arrays of functional surfaces [69–71]. These methods take advantage of the electrostatic interaction between the surface of the nanoparticle and the chemically functionalized sites present on the substrate, typically the walls, the tips, and the cavity, in the case of nanotubes. Both metal particles and semiconductor quantum dots have been organized on the nanotube surface using suitable capping agents to provide electrostatic bonding to specified sites on the nanotube surface. The electrostatic attraction causes the nanoparticles to anchor themselves on the surface of the nanotubes and organize themselves to form interesting ordered arrangements. Needless to say, the nanoscale nature of the tube provides an excellent template for a bottom-up organization of a nanocomposite.

4.4.2.1 Self-Assembly of Prefabricated Metal Nanoparticles
The self-assembly technique has been particularly successful for gold coatings due to the availability of prefabricated stable gold colloids with controlled particle sizes and the ease of anchoring these particles to thiol functional groups; a few selected examples of self-assembly of nanocomposites will be described below. The sidewalls of acid treated MWCNTs, coated with a cationic polyelectrolyte such as PDADMAC [poly (diallyldi-methylammonium chloride)], were used as substrate to anchor 10 nm gold particles by dipping in a negatively charged gold colloid [72]. Well-dispersed gold particles decorated the walls of the nanotubes quite uniformly. The interaction between the gold particles and the nanotubes was found to be quite strong and was not washed away by simple rinsing. In some cases, linear polymers such as polyethyleneimine have been used both for surface activation of the nanotube and also as a reducing agent for gold salts [73]. A simple modification of the experimental conditions to produce either gold clusters or uniformly coated gold films has also been achieved. Gold nanowires were formed by the templated

self-assembly of gold nanoparticles surface coated with tetraoctylammonium bromide. On heating the composite, the gold nanoparticles could be sintered to produce wires. The strong nanotube-gold interaction is attributed to electron transfer between the gold nanoparticles to the continuum of π^* states in the nanotubes [74]. Alternatively, linker molecules such as MUA (1-mercaptoundecanoic acid) or NDT (nonane dithiol) can be used to effectively link decane thiolate stabilized nanometric size gold particles onto carbon nanotubes in a single step. It seems that the size distribution of the particles is changed by changing the concentration of the metal colloid with respect to the concentration of CNTs. As expected, the assembled particles were separated by a distance equal to the length of the linker. Particle densities up to 1.6–1.9×10^{12} particles per cm^2 could be obtained [75]. In the case where gold particles are deposited by self assembly onto thiol-functionalized MWCNTs, the nature of the thiol determined the particle size distribution. Small particles with narrow size distributions (2–5 nm) were obtained for 1,6-hexanedithiol whereas aminothiol functionalization increased the particle sizes [76]. An interesting example of self-assembly is the tip selective coating of gold particles on vertically aligned MWCNTs where the side walls were protected by impregnation with polystyrene [77].

4.4.2.2 Self-Assembly of Compounds

Self-organization of oxides, sulfides and selenides can be also be effectively carried out by surface modification followed by deposition of nanoparticles. Surface modification by AEPA (2-aminoethylphosphonic acid) and APTEOS (3-aminopropyltriethoxysilane) was used for the coating of carbon nanotubes with TiO_2 and SiO_2 particles [78]. The first step involves the formation of amide groups on acid functionalized CNTs by reaction with EDAC (ethyldimethylaminopropylcarbodiimide). The amide was treated with AEPA or APTEOS in a second step. Partially hydrolyzed titanium or silicon oxide colloids in organic medium were deposited by simply mixing them with the functionalized nanotube suspension. In the present case, the chemisorption of the alkoxide groups on the surface of the modified MWCNTs is the driving force for self assembly.

For quantum dots of sulfides and selenides, it turns out that the capping agent used for the stabilization of quantum dots plays an important role in the decoration of nanotubes with these nanoparticles. For example, CdSe nanocrystals stabilized by triphenyl phosphinc oxide (TPO) did not coat on nanotubes, whereas changing the cap to pyridine caused their self-assembly on the nanotube surface. The strong interaction between the pyridine molecules and the SWCNT surface resulting from the π–π stacking interaction with the electron-rich SWCNTs is the primary driving force for the self assembly [79]. Similar results were observed for InP and CdSe quantum dots stabilized by other organic ligands such as mercaptoacetic acid [80,81] (Figure 4.9).

Magnetic nanoparticles of Fe_3O_4 were coated with increased density on MWCNTs functionalized with carboxylate terminated alkyl chains using poly(2-vinylpyridine) chains [82]. Noncovalent functionalization of carbon nanotubes using polyelectrolytes such as polyallylamine hydrochloride allowed the formation of a stable CNT suspension that could be transferred to an organic or aqueous solvent. It was possible to coat CNTs with a variety of nanoparticles of sulfides, selenides and oxides by simply adding a stable suspension of the particles to the dispersed nanotubes (Figure 4.10) [83].

Quantum dots of ZnO or CdSe have also been deposited on silica coated MWCNTs and their photoluminescence spectra have been recorded. Carbon nanotubes act as electron

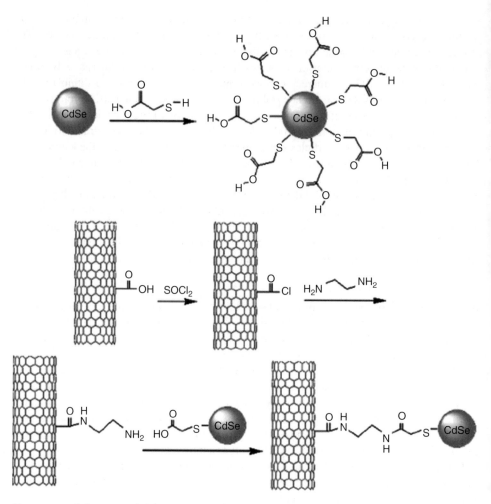

Figure 4.9 *Schematic of CdSe quantum dots coating MWCNTs. Reprinted with permission from [80] Copyright (2006) American Chemical Society.*

acceptors in their photoexcited state and electron transfer between the nanotubes and the coating can occur, resulting in quenching the luminescence process [84–86]. The work-function of multi-wall nanotubes has been estimated to be in the range 4.4–5.1 eV and the Fermi level has been calculated to lie within the band gap of semiconductor oxide materials such as CdSe [87,88]. The introduction of a silica layer between the nanotubes and the coating restores the luminescence properties while maintaining particle size control. The large band gap and thickness of the silica layer prevent electron transfer between nanotubes and the conduction band of the semiconductor [89].

While most of the coating work described above has been done on MWCNTs, the metal coating on SWCNTs is complicated due to their increased tendency to bundle, and dispersing agents must be used to access individual tube walls. Polymer wrapping is generally used to separate the SWCNT bundles before coating them with capped metal

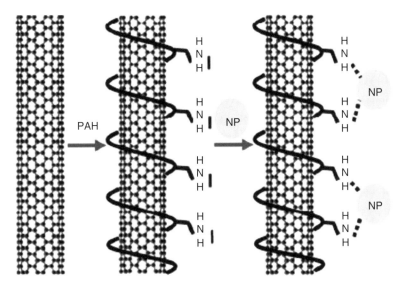

Figure 4.10 *Coupling of nanoparticles to MWCNTs wrapped with PAH. Reprinted with permission from [83] Copyright (2007) Elsevier Ltd.*

particles [90]. Alternatively, very dilute SWCNT solutions can be produced by the ultracentrifugation of a suspension wherein only the individual tubes stabilized by a surfactant are retained [91]. In an interesting experiment, Lee et al. showed that gold nanoparticles in an aqueous medium were transferred to SWCNTs suspended in an organic solvent by forming a stable nanocomposite film at the interphase. The aqueous medium was rendered colorless, showing that all the gold particles were efficiently transferred to the interphase. The metallic luster of the film showed the electronic coupling of the gold nanoparticles thus forming a stable network [92]. It is interesting to note that the ordering of the SWCNT bundles in an array influences the nature of the gold coating showing that nanotube bundles are efficient one dimensional templates. On highly ordered CNT bundles prepared by laser ablation, self-organization is observed for Au, Ag and Cu ultra fine nanoparticles and Ti, Mo or Zr nanowires. For disordered arrays, no particle arrays were formed. Predictably, a high ordering of the bundles was needed to produce good results. The cluster sizes in an array were uniform but varied between bundles or between different tubes within the same bundle [93] (Figure 4.11).

4.4.2.3 Self-Assembly using Microemulsion Templated Deposition

MgO and ZnO nanoparticles have been anchored onto nanotubes using a novel technique where a water-cyclohexane system was used to create a microemulsion. In this microemulsion, fine microdroplets of an aqueous phase are trapped within assemblies of surfactant molecules dispersed in continuous organic oil phase (Triton X-114 in cyclohexane) [94]. When the aqueous phase contains a divalent metal cation such as Co(II), Mn(II) or Cu(II), it is extracted by the surfactant into the water-oil interface. On the addition of a reducing agent, uniform nanoparticles of the metal or the hydroxide are formed. Calcination of the composite yields 5 nm diameter spheres of ZnO or 30–40 nm particles

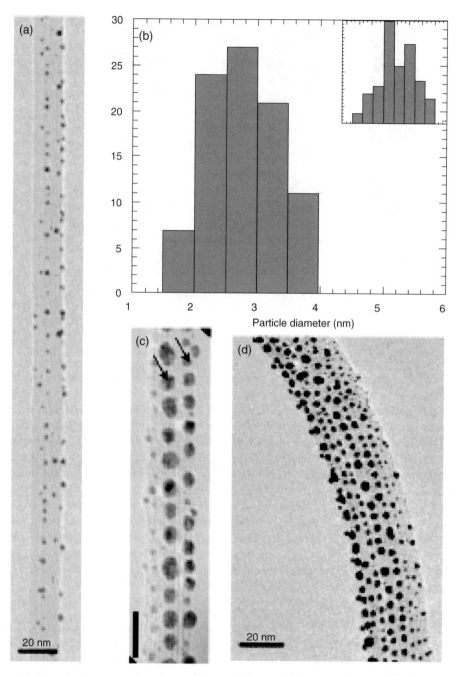

Figure 4.11 *TEM images of ordered arrays of gold clusters in the intertube groove spaces of a SWCNT bundle along with the particle size distribution. For comparison, the inset shows the size distribution of Au clusters formed on an amorphous carbon film in the same sample using the same x- and y-axis setting. Reprinted with permission from [93] Copyright (2006) John Wiley and Sons.*

of MgO decorating the surface of the nanotubes [94]. The microemulsion-templated deposition has also been used for the deposition of 2–10 nm particles of Pd and Rh on CNTs. The metal nanoparticles were formed in a water-hexane microemulsion with AOT (Sodium bis-2ethyl hexyl-sulfosuccinate) surfactant and were transferred to the CNT surface by stirring [95,96]. Silica nanobeads were formed by controlled functionalization of the nanotubes using APTEOS (aminopropyltriethoxysilane) by the water-oil emulsion technique to control the size of the nanobeads. Active groups such as magnetic, fluorescent, or biomolecules with potential applications in nanotechnology and drug delivery can be successfully grafted onto the silica layer [97,98].

Self-assembly techniques, though useful for controlling particle sizes have several drawbacks. The principal disadvantage is the presence of an organic capping agent added to stabilize the nanoparticles. The presence of these stabilizing agents may decrease the efficiency of the coating for applications. While the effect of these complex molecules on the catalytic activity is not well studied, it is clear that they will have a negative influence on the electrical conductivity and the sensing capability of the coating. Another disadvantage is the high cost of synthesis of the complex organic capping agent for large scale applications.

4.4.3 Electro- and Electrophoretic Deposition

Electrodeposition is the process whereby metal particles are produced during the reduction of a metal salt solution at the electrode (cathode). In conventional electroplating, the anode is made of the coating material that dissolves in the electrolyte. The polarization of the cathode results in the metal salts losing their charge and plating out on the surface of the cathode. However, the reduction or oxidation can also occur in solution, the anode or cathode merely serving to transport the charges. Another common electrochemical method is electrophoretic deposition wherein colloidal particles of metal suspended in a liquid medium migrate under the influence of an electric field and are deposited onto an electrode that is the substrate to be coated. In contrast to electrodeposition predominantly used for metal coating, all colloidal particles that can be used to form stable suspensions and that can carry a charge can be used in electrophoretic deposition.

Electrochemical methods would have been the obvious choice for plating with metals, especially noble metals since they do not involve a gas phase reduction step and correspond to a one-step process. Chemical methods involving impregnation, filtering, then reduction can be tedious. In addition, electrochemical methods offer a great potential for particle size and density control by a simple variation of deposition parameters. However, the difficulty of its use on a large scale and the difficulties involved in making good electrical contacts have made that only very few experiments have been conducted on nanotubes, and mostly with noble metals such as Au, Pt, Pd and bimetallic Pd-Ru. Electrodeposition of noble metals on activated carbon and graphite has been extensively studied [99,100]. Both electrochemical and electroless deposition have been described. It has been envisaged that, by changing the parameters of deposition, the rate of nucleation and hence the control of particle size can be achieved. It appears that the most important step in the electrochemical coating process is the preparation of the nanotube working electrode. Hence, the different methods that have been used to prepare this electrode with both SWCNTs and MWCNTs are described below.

4.4.3.1 Fabrication of the Nanotube-Based Working Electrode

Nanotube electrodes have been formed by using methods similar to those for graphite powders. First experiments have been conducted with buckypapers prepared by filtering a suspension of nanotubes under low pressure in order to obtain a filter cake that is a few tens of microns thick and contains a mesh of entangled carbon nanotubes. The porosity of the electrode can be lowered to around 20% but remains a problem for achieving uniform current density [101,102]. In later applications of nanotubes such as for fuelcells and electrocatalysis, two kinds of CNT electrodes have been developed: binder-free and binder-enriched. The binder-free electrode is very similar to the buckypaper where carbon nanotubes are pressed together to contact each other. The main drawback in this type of electrode is the insufficient electrical contact between the nanotubes due the presence of pores. A simple form of nanotube electrode using a binder consists of a paste of single-wall carbon nanotubes with mineral oil in the ratio 60:40 by weight. The paste is filled into a 3 mm Teflon tube. A copper wire dipping into the paste provides electrical contact. While this electrode ensures improved contact between the nanotubes, the electrochemical performances may be compromised due to the presence of oil remnants as organic impurities. For small scale applications, aligned MWCNTs [103] or SWCNTs directly grown on conducting substrates have been used [104]. Nanotube films coated on graphite substrates have also been investigated [105], the graphite providing adherence to the nanotube film. Nanotubes grown directly on a patterned substrate have the advantage of a substrate patterned by using a photo resist that makes it possible to expose a precise area of the electrode to the electrolyte and thereby offers a better control of the deposition parameters. Thus, CNTs grown on SiO_2 were used as electrodes connected by a Ti wire to a potentiostat, the nanotubes themselves acting as nanowires to produce the connectivity for the nanoparticles formed in the solution during the reduction of the metal salt [106,107]. Deposited chromium and gold electrodes have also been found to provide good ohmic contacts to the nanotubes. An important necessity in both these cases is the presence of a good electrical connectivity over the entire surface of the electrode. Usually, this is tested by measuring the potential drop across the nanotube sample; a linear potential drop with distance is desirable. A bottom-up approach to the formation of a nanoelectrode formed in situ by the electrodeposition of a noble-metal nanoparticle at the exposed end of a SWCNT is shown in the Figure 4.12. The nanotube end serves as template for nanoparticle electrodeposition, while the tube itself serves as a nanometer wire to interface the deposited nanoparticle with macroscopic leads [107].

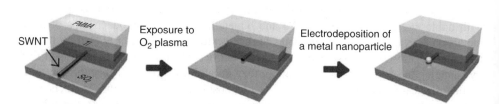

Figure 4.12 *Sketch of the electrodeposition of a metal particle at the end of a SWCNT. Reprinted with permission from [107] Copyright (2006) John Wiley and Sons.*

4.4.3.2 Electrochemical Coating of Nanotube Surfaces: Some Examples

No surface modification was needed for the direct formation of Pt or Ru particles by electrodeposition on carbon nanotubes. Pt was electrodeposited on MWCNT arrays by the potential-step deposition method from N_2 saturated 7.7 mM $H_2PtCl_6 + 0.5$ M HCl aqueous solution and it was observed that while a similar deposition condition gave rise to 100–150 nm particles on graphite surface, particles in the size range 30–70 nm were observed on MWCNTs [108]. It was also found that the adhesion of the particles was much higher in the case of the CNT surface whereas on graphite they had a tendency to peel off. In addition, the Pt/CNT system gave a higher catalytic activity for the oxygen reduction reaction when compared to the Pt/graphite system by a factor of two. Similarly, Pt/Ru alloy particles were deposited on a nanotube film grown on a graphite surface [105], where a mixture of 1.3 mM chloroplatinic acid and 1.3 mM ruthenium chloride in 0.5 M sulfuric acid was used as the electrolyte and the deposition charge was 1.17×10^{-3} C per cm^2. Particles in the 60–80 nm size range were obtained and EDS analysis showed the codeposition of Pt and Ru. In both these cases, the graphite electrode gave a much larger particle size distribution when compared to CNTs, showing that the presence of mesopores present in CNTs played a major role in controlling the particle size of the deposit. High density of less than 10 nm size particles of PtNi alloy were electrodeposited on MWCNTs and were found to have a high electrocatalytic activity for methanol oxidation [109].

As in the cases of wet-impregnation and self-assembly, It has been observed that the particle size can be better controlled if the reduction were to be carried out on the surface of the nanotubes by exchange of electrons through a specific molecule attached to selected CNT sites. In this way, particles with sizes less than 10 nm have been obtained. One example is shown in Figure 4.13, as an efficient Pt/Pd coating method involving three steps:

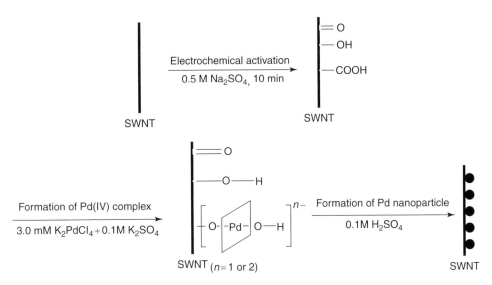

Figure 4.13 *Sketch illustrating the three-step electrochemical synthesis and deposition of Pd nanoparticles on SWCNTs. Reprinted with permission from [110] Copyright (2005) Elsevier Ltd.*

(i) electrochemical surface modification of a CNT electrode, (ii) formation of Pd(IV) complex on the surface of the nanotube followed by (iii) the reduction of the complex to Pd nanoparticles on the SWCNT surface.

Well-dispersed Pd nanoparticles in the size range of 4–6 nm were obtained by this method [110]. It was seen that the nanoparticles were preferentially distributed on the ends, kinks and connecting regions of the CNT due to the higher degree of functionalization on these sites when compared to the bulk of the electrode. In contrast experiments on the deposition of Pd and Pt on pristine SWCNT electrodes [104] have shown that the nanoparticles are deposited uniformly on the tips and on the side walls of the nanotubes. Hence, it appears that the deposition mechanism is dependent on both the deposition parameters and the chemical functional groups – if present – on the surface of the nanotube electrode. The geometry of the electrode also plays a role. Whereas nanoparticles of Pt deposited uniformly on regular CNT electrodes, they tended to cluster on the tips of the MWCNTs when the electrode is made of vertically aligned CNTs grown on a substrate [111]. In some cases, particle sizes are lowered by allowing the reduction of the metal salt to take place on the nanotubes suspended in the electrolyte using a surfactant. A composite coating of Zn-MWCNTs coated on steel was obtained by electroplating Zn in the presence of CNTs dispersed in the electrolyte, using a surfactant such as cetyltrimethylammoniumbromide. The composite coating was found to improve the corrosion properties of the steel cathode [112]. A two-step electrochemical process was used for the synthesis of polyaniline/carbon nanotube/gold composite. In the first step, aniline and MWCNTs were refluxed, after which the composite was coated electrochemically. Following this, the composite film was dipped into a $HAuCl_4$ solution and the deposition of the gold particles was studied by pulse voltammetry [113].

4.4.3.3   Growth Mechanisms in Electrochemical Deposition

In order to achieve uniform size and a controlled dispersion of the nanoparticles on the surface of CNTs, their formation mechanism and kinetics have to be understood. The nucleation and growth mechanism of Ag and Pt on dense networks of SWCNTs has been described by Day et al. in a detailed article on electrochemical templating of metal nanoparticles by SWCNTs [114]. In order to precisely control the electrode area, a microelectrochemical cell with a microcapillary was used, firstly to define the area of deposition precisely and secondly, to minimize capacitance and network resistance effects. Ultra small Ag and Pt particles were generated by high current densities for a short time and a parametric study on the influence of time and current density on the size and density of the particles deposited on a SWCNT network was carried out. The growth of Ag nanoparticles was found to be rapid and progressive with an increasing nanoparticle density with time, whereas Pt deposition was characterized by lower nucleation densities and slower growth rates with a tendency for larger particles to be produced over long deposition times. Thus, the Ag particles were more monodisperse than the Pt particles. In the vicinity of the electrodes, a high density of nanoparticles was deposited tending to form nanowires whereas at farer distances the nanoparticle density decreased. In other experiments where electrodes made of a low density of SWCNTs were used instead of a dense network of interconnected SWCNTs, the electrochemical behavior of metallic and semiconducting SWCNTs could be compared [115].

4.4.3.4 Tip Selective Electrochemical Deposition

When a substrate is placed between two electrodes in an electrolyte, and the conductivity of the electrolyte is less than that of the substrate, the substrate is polarized resulting in a potential difference between the opposite poles of the substrate in line with the field. If this potential difference is high enough to drive a redox couple present in the solution, electrochemical reactions can occur at the opposite poles of the substrate. The advantage of bipolar electrochemical deposition is that it requires no direct contact between the nanotubes and the power supply, and offers an alternative method to control the site of deposition. This principle was used [116] for the deposition of Pd particles onto the tips of CNFs and CNTs prepared by CVD. Different morphologies were observed varying from small particles that just covered the tip of the tube up to large deposits of faceted particles. Carbon nanostructures immobilized on glass or track-etched membranes were placed between two platinum electrodes separated by a distance of 5 mm. Palladium chloride dissolved in dry acetonitrile and toluene (30:70 v/v) was used as the electrolyte. The deposition of palladium on the CNFs were carried out in DC mode and a unipolar pulsed field was used for Pd deposition on the CVD-grown CNTs. Best results were obtained with low depositing times where the potential was at its maximum at the tip of the nanotube, giving rise to 1–3 nm large particles.

4.4.4 Deposition from Gas Phase

Though not as prevalent as wet-impregnation and electrochemical coating, gas phase coating technologies such as electron beam evaporation, sputtering, Atomic Layer Deposition (ALD) and Chemical Vapor Deposition (CVD) have been employed for the coating of metals and oxides on nanotube surfaces. These methods have the advantage of being a single step process with high coating speeds and can be particularly adapted to the large scale production of coated samples provided a good gas-solid contact can be achieved. In principle, by tuning the evaporation conditions, highly reproducible materials can be obtained if suitable precursors are available that are sufficiently volatile and do not decompose in the homogeneous phase under the experimental conditions. However, the substrate size is often limited by stringent experimental conditions such as high vacuum in the case of electron beam evaporation and the use of expensive and often toxic precursors in organometallic CVD reactions. Nevertheless, for selected applications, gas phase techniques have shown a definite advantage. For instance, a relatively precise tip-selective coating can be achieved by electron beam evaporation and has been thus employed for the metal coating of nanotube tips for magnetic force microscopy [117]. Gas phase techniques at times yield purer products without the formation of insoluble residues that come with wet-impregnation, electroless or electrochemical plating.

4.4.4.1 Electron Beam Evaporation

Electron beam techniques have been used for the coating of electrodes on bundles of SWCNT and MWCNTs for electrical property measurement. Among the metals used, electron beam evaporated Ti was found to provide the lowest contact resistance of 12 K-Ω [118,119]. Ohmic contacts were obtained for both Ni and Ti but those with Al and Au were non-ohmic, showing high contact resistance. A detailed study revealed that while Ti formed a continuous wire on SWCNTs, Au, Al and Fe formed disconnected crystalline particles and some of the gold particles were larger than the film thicknesses showing that larger particles are formed by the migration and coalescence of smaller ones. The type of coating depended on the metal used demonstrating the importance of the metal-nanotube interaction in controlling the morphologies of the coating. The fact that Ti and Ni can coat SWCNTs uniformly over the entire surface shows a strong, near covalent bonding between carbon and titanium or nickel. This result is not entirely surprising since it is well known that titanium has a tendency to form carbides with a high heat of formation and is the preferred metal for forming metal contacts on diamond [120]. In the case of small SWCNTs, curvature-induced effects can further strengthen these interactions.

4.4.4.2 Atomic Layer Deposition

In this process, a spray of an appropriate precursor or a reactive gas is alternately pulsed onto a substrate and the chamber is purged with inert gas in between the cycles. All the processes take place at relatively low temperatures making the process compatible with semiconductor technology. The precursor coats the active sites on the substrate as a monolayer making it a self-terminating reaction. The precursor in excess is removed by the purging gas, which helps in controlling the film thickness and also results in the uniform coverage of substrate inhomogeneities such as trenches. The reactant gas is sent in and the precursor molecules react to produce a thin film. The number of pulses determines the thickness of the film. Today, this technique is mainly used for coating high k dielectric ultra thin films on semiconductors and its use is mainly limited by its low deposition rate. Even so, this process has been used to coat ruthenium oxide onto CVD-grown nanotubes. The ruthenium precursor used was $Ru(od)_3$n-butyl-acetate solution (od = octane-2,4 –dionate) and the reactant gas was oxygen. A 0.1 M solution kept in a canister pressurized with argon fitted to an injector was employed to pulverize the ruthenium source. Typically, 0.01 ml of this solution was injected and carried by argon and alternative pulses of 2 s duration of argon and oxygen were used; a total pressure of 1 Torr was maintained in the chamber. High resolution electron microscopy images showed that ruthenium films were observed on the inner and outer walls of the nanotubes [121]. The low temperature of the reaction prevented the oxidation of the nanotube template. The ALD technique has also been used for (i) coating MWCNTs with a dielectric layer of aluminum oxide; (ii) depositing multiple insulating and conducting layers, (iii) coating nanotubes with a seed layer of Al_2O_3 to increase the concentration of surface hydroxyl groups for the attachment of functional molecules [122–124]. Carbon nanotubes were first exposed to trimethylaluminum $Al(CH_3)_3$ and then to water vapor. This pattern was repeated to obtain the desired film thickness. At 550 K, the growth rate for Al_2O_3 by ALD was 0.12 nm per cycle. Figure 4.14 shows a MWCNT coated with a 20 nm thick aluminum oxide layer by the ALD process.

4.4.4.3 CVD Coating Methods

To date, CVD methods are limited in applications mainly due to experimental difficulties encountered in obtaining a uniform coating on carbon nanotubes. Generally, organometallic complexes are needed to ensure high volatility of the precursor [125]. Luckily, a variety of organometallic precursors can be efficiently synthesized and are available readily for noble metals and some representative examples are presented. Pd/C catalysts were obtained by the chemical vapor deposition of [Pd(allyl)Cp] in helium atmosphere at 323 K on a CNT-VGCF (VGCF = vapor

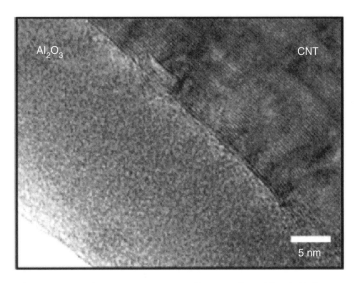

Figure 4.14 *Aluminium oxide coating on a MWCNT produced by ALD. Reprinted with permission from [122] Copyright (2003) Elsevier Ltd.*

grown carbon fiber) support system in a fixed bed reactor [126]. Controlled-growth of platinum nanoparticles on graphene-based substrates such as carbon nanospheres or MWCNTs was achieved by the CVD of [PtMe$_2$(COD)] in a fluidized bed reactor. Under the operating conditions, the platinum germination is controlled by surface chemical reactions, and it was necessary to create oxygen containing groups to deposit Pt. Highly dispersed metallic Pt nanoparticles were deposited on MWCNT and carbon nanospheres (2–3 and 5–6 nm mean particle size respectively) [127]. Recently, Pt particles were deposited on carbon paper/carbon nanotubes composite by the reduction of Pt(acac)$_2$ precursor with acetic acid. Transmission electron microscopy (TEM) images show that the Pt nanoparticles with high density and relatively small sizes (2–4 nm) were homogeneously dispersed on the surface of CNTs. In CVD coating, the vapor deposition may be either carried out directly in the presence of a reducing agent resulting in the immediate reduction of the adsorbate or as a separate step. For Pd, the latter tended to result in smaller Pd particles and led to higher catalytic activity. As in the case of wet-impregnation, oxygen-containing functional groups were essential for successful deposition of Pd on the nanotube substrate. Online mass spectroscopy during the deposition indicated that carboxylic groups present on functionalized nanotubes played a major role in the dissociative adsorption of the precursor resulting in replacing the strongly bound allyl and cyclopentadienyl (Cp) ligands from the Pd. For the deposition of Cu nanoparticles on MWCNTs, sublimation of [Cu(hfac)(TMVS)] [(Trimethylvinylsilyl)-hexaflouroacetylacetonate-Cu] at 393–573 K followed by hydrogen reduction has been used. The influence of different pretreatment steps such as thermal and plasma activation has been studied in detail. In all cases no copper wires were obtained and the size of the particles depended on the substrate temperature, deposition time, and the number and type of functional groups on the CNTs [128]. Other commonly used precursors are ruthenium acetyl acetonate [129] and titanium isopropoxide [130].

There are few examples of deposition from inorganic precursors: volatile compounds such as hydrides have been used leading to the coating of inorganic oxides. Due to the excellent electronic properties of tin oxide as sensor, SnO_2 coated carbon nanotube sensors have been investigated. A gas phase precursor such as SnH_4 has been used to coat SnO_2 nanoparticles on the nanotube surface by CVD at 823 K. The particle sizes were found to be dependent on the deposition time and varied between 10–70 nm. At high deposition temperatures, organized chain-like structures of SnO_2 were observed that could be separated by burning the MWCNT template [131]. A plasma assisted CVD method has been recently developed for the coating of boron nitride on vertically aligned MWCNTs [132].

4.4.5 Nanoparticles Decorating Inner Surfaces of Carbon Nanotubes

Increased catalytic activity of catalyst particles decorated on the walls of the inner cavity of nanotubes has been reported recently [133]. Narrow inner diameter (4–8 nm) MWCNTs were filled with nanoparticles of a mixture of Rh, Mn, Li and Fe in the ratio 1:1:0.075:0.05 by sonicating functionalized nanotubes with a solution of the metal salts wherein a significant increase in the rate of conversion of CO and H_2 to ethanol was observed. Palladium nanoparticles (4–6 nm) were deposited inside MWCNTs (inner diameter ~50 nm) by a wet-impregnation technique using an aqueous solution of palladium nitrate, followed by calcination to form the oxide followed by reduction under hydrogen at 673 K [134]. Electron microscopy revealed the presence of a significant number of Pd particles in the cavity of the nanotubes. The relatively large inner cavity of the nanotube was reported to be responsible for the high (30%) rate of filling. The confinement in the cavity also induced a better control of particle size when compared with palladium particles dispersed on activated carbon supports. The catalyst showed a higher selectivity in the hydrogenation of cinnamaldehyde, an effect that has been attributed to the absence of functional oxygen groups that are generally present on the outer surface of the nanotubes. In some works, the density of particles coating the inner surface of the nanotube cavity has been increased by using supercritical solvents to decrease the viscosity, diffusivity and surface tension so as to favor the insertion of nanoparticles in the nanotube cavity. Thus, CNTs were decorated with Pd, Ru and Rh by the hydrogen reduction of organometallic precursors in supercritical carbon dioxide [135]. Recently, a simple and highly efficient method for selectively depositing a high proportion (>80%) of preformed PtRu nanoparticles has used surface chemistry to favor the coating of the nanoparticles selectively on the inner surface of an open tube by attraction while simultaneously using functional groups to repel the nanoparticles from the outer surface. Three-dimensional TEM analysis was used to characterize the samples [9,136]. Tessonnier et al. have described a general method for the deposition of metal nanoparticles selectively either inside or outside the carbon nanotubes (CNTs) by making use of the difference in interfacial energies of organic and aqueous with the carbon nanotubes surface. It was observed that the organic solvent wetted the nanotubes surface and penetrated into the inner volume whereas nanoparticles suspended in the aqueous medium remained on the outer surface. The decorated nanotubes were characterized by state of the art electron microscopy techniques and showed that a selectivity of up to 75% could be achieved by this method [137].

4.5   Characterization of Decorated Nanotubes

Simple chemical methods such as acid-base titrations have been used for determining the number of acid functional groups on the nanotube surface and it is found that a strong correlation exists between the number of acid groups present to the quantity of metal coated. A fairly precise quantitative estimation of the latter is easily obtained by treating the coated nanotubes with strong acids to dissolve the metal particles which subsequently are estimated by Inductively Coupled Plasma Atomic Emission/Optical Spectroscopy (ICP-AES/OES) analysis assuming that any metal particles coated on the inside walls of the nanotubes would be leached-out during the acid treatment [35]. Alternatively, a burn-off of the carbon can be attempted, after which elements making the residue are estimated by ICP analysis. Systematic quantitative estimation has led to certain general observations after which the quantity of coating metal was directly related to the stability constants of the corresponding complexes with acid containing ligands such as COO- or HO_2C-COO- [138,139]. A chemical titration of the defect concentration can also be achieved by treating the functionalized carbon nanotubes with hydrogen selenide where upon Se nanoparticles are formed preferentially on the oxygen functionalities attached to the nanotube followed by AFM mapping of these groups [140]. A similar study has also been reported for gold and titanium dioxide nanoparticles decorated carbon nanotubes [141,142].

4.5.1   Electron Microscopy and X-ray Diffraction

Apart from chemical methods, standard materials characterization techniques such as electron microscopy and spectroscopy have been extensively used for the characterization of coated nanotubes. High resolution transmission electron microscopy has been the main tool for the determination of the coating density and particle size dispersion. Most of the metal particles coat in the form of single crystals thereby giving detailed information on preferred crystallographic planes if any. Figure 4.15a shows Pt nanoparticles on multi-wall carbon nanotubes. Figure 4.15b shows a high resolution TEM image of the Pt nanocrystals. The TEM images show clearly that the contact area between the particles and the nanotube surface was found to be large indicating the high interaction of the particles with the nanotube surface.

As far as coating nanotubes is concerned, determining whether the coating material is actually attached to the outer or the inner nanotube surface is important. It has been demonstrated that a gradual tilting of the sample with respect to the electron beam can help in distinguishing metal particles coated on the inner cavity of the nanotubes when compared to those present on the surface [143]. When a metal particle is located on the outer surface, during tilting the sample holder, the position of the particle is largely determined by the tilt angle. In contrast, little or no translation occurs when the particle is located on the inner surface of the nanotube. Recently, Pt/ and Pd/MWCNTs have been imaged using the 3D TEM technique wherein a high tilt sample holder was used to image the nanotubes at different tilt angles followed by image processing and reconstruction to obtain 3D

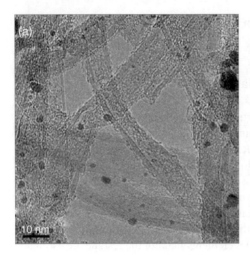

Figure 4.15 *(a) TEM image of Pt/MWCNTs; (b) High resolution TEM image of the Pt/MWCNT surface contact. MWCNTs are pyrograph grade (Applied Sciences Inc.).*

information on the particles' size and location [144–146]. The investigation concluded that the introduction of nanoparticles inside the nanotube cavity (for subsequent grafting onto the inner surface) was strongly dependent on the nanotube diameter, and filling is nearly impossible under mild conditions for diameters less than 30 nm. It was also observed that the functional groups introduced during the functionalization are distributed equally on the inner and the outer surface of the nanotube. A qualitative estimate of the interaction between the metal and support can also be made by measuring the change in interplanar distances at the edges when compared to the center of the particle [147]. In some cases the fringes of the crystals were aligned in the same direction as the nanofibers whereas certain inter-planar distances could not be indexed by standard patterns, showing the change in lattice spacing of the particles on the substrate.

For polycrystalline particles, crystallite sizes calculated from X-ray powder diffracto-grams have helped in getting an overall view of the crystalline phase and ordering in the coating, especially in the case of oxide and sulfide coatings [48,51]. XRD has also been used to study the relative concentration of the corresponding metals in bimetallic systems such as Pt-Ru that have been used as efficient hydrogenation catalysts [148]. Crystallite sizes can be calculated from the FWHM of the peaks in the diffractograms using the Scherrer equation and compared with those obtained from electron microscopy data. In certain cases, preferred orientation can also be identified using XRD data.

4.5.2 Spectroscopic Methods

Though an estimate of the number of attached functional groups can be carried out by simple acid-base titration, the nature of the groups and their bonding to the CNTs is better studied using FTIR, XPS and TPD/MS. IR spectroscopy is routinely used to detect and follow the formation and reaction of surface functional groups that are essential for the successful

Table 4.3   *Some representative IR data for as-received, purified and oxidized SWCNTs showing peak positions [150].*

Mode Assignment	As-received SWCNTs (transmission IR) per cm^{-1}	Purified SWCNTs (transmission IR) per cm^{-1}	Oxidized SWCNTs (transmission IR) per cm^{-1}
vCH	3429	3439	3438
vaCH$_3$	2915	2921	2923
vsCH$_3$	2847	2848	2853
vC=O	–	1732	1735
vC=O	1635	1628	1633
vC=O	–	–	1383

Table 4.4   *Representative XPS data obtained from pure and acid-treated MWCNT samples; (a) XPS Intensity for the given binding energies (% of total amount), (b) Atomic percentage of the given elements [61].*

A Materials	XPS intensity (%) with respect to activated carbon treated with nitric acid for the following binding energy (eV) values			
	530.9 eV	532.1 eV	533.2 eV	534.2 eV
MWCNT	55	45	–	–
MWCNT-COOH	23	44	24	9
MWCNT-COONa	25	49	–	14

B Materials	S	O	C	Na
MWCNT	0.14	0.81	99.05	–
MWCNT-COOH	–	3.23	96.77	–
MWCNT-COONa	–	5.29	93.04	1.67

anchoring of the coating material, the most important being the carbonyl and the carboxyl groups. Table 4.3 shows the IR frequencies for acid-purified and oxidized SWCNTs.

IR spectra have also been used to study the interaction between nanoparticles of Pt and the functional groups present in the nanotubes. For example, it was found that for Pt-coated nanotubes, the carbonyl C=O stretching at 1700 cm^{-1} is red-shifted to 1550 cm^{-1} following Pt cluster deposition. In addition, changes in the C-O structural features at ~1030 and 1150 cm^{-1} were observed, indicative of Pt cluster binding with the oxygen atoms on the carboxylate or ester groups on the nanotubes [149].

To determine the quantity and oxidation state of elements present on the surface, X-ray photon spectroscopy (XPS) is the most widely used technique. The concentration profiles obtained by XPS are often higher than that obtained by acid-base titrations showing the possible presence of groups/molecules in the subsurface region of the sample [62,150]. Table 4.4 shows representative data obtained from functionalized MWCNT samples:

individual samples of nanotubes may show variations depending on the sample preparation and functionalization conditions.

Since XPS is specific to the surface of the sample, its importance in understanding the metal-support interactions in catalysis is easily appreciated. Thus, XPS is routinely used to determine the identity and quantity of the elements present within ~10 nm of the sample surface, their oxidation states, and the binding energies of one or more electronic states of these elements. In addition, the technique allows calculating the density of electronic states of the required elements and the thickness of the coating. Combined with EXAFS, it is widely used to get insight into metal-support distances and their electronic interactions. The reader is referred to excellent monographs on these applications and only a few examples are cited in the following to illustrate the importance of these techniques in understanding the mechanism of heterogeneous catalysis [151–154]. XPS spectra of thiol-functionalized CNTs have yielded valuable information on the concentration and valence states of sulfur that could not be obtained by IR spectroscopy. The formation of S-Pt covalent bonds in Pt deposited on thiol-functionalized CNTs was evidenced by XPS. In addition, when aromatic thiols are used, the phenyl group participates in a π–π interaction with the walls of the CNT and strongly interacts with the Pt nanoparticles possibly through the formation of a Pt-S bond resulting in high Pt nanoparticle loading and narrow size distribution as was observed by electron microscopy. The Pt-S bond formation was detected due to the change in the width of the C_{1s} spectrum and the change in binding energy of the S_{2p} spectra of the functionalized CNTs after the coating by Pt nanoparticles. Both these observations have indicated the formation of a covalent S-Pt bond with charge transfer [55,155,156].

Extended X-ray absorption fine structure spectroscopy (EXAFS) has often been used to study the metal-support interactions in supported catalyst systems such as Rh or Pt/Al_2O_3, Rh/TiO_2 and Pt/zeolite. One of the principal advantages of EXAFS analysis is its ability to determine near-neighbor structures: this makes it an excellent tool for studying the short range ordering in amorphous materials and composite structures. By tuning the X-rays to the absorption edges of the different component species in the sample, it is possible to determine coordination numbers, interatomic distances, and the degree of disorder for each component. The major experimental drawback that has prevented the routine use of this technique is the need for a suitable X-ray source with a high intensity beam for a good signal/noise ratio which implies the use of a synchrotron radiation source. In addition, EXAFS is a theoretically complex process, involving the constructive and destructive interferences of the waveforms that originate from the photoelectrons produced, which change the absorption process. In catalysis, the main application of EXAFS is in the determination of the long distance (2.5–2.8 Å) metal-support interactions and enables the study of the extended environment of the metal catalyst particle with respect to the support, thus enabling the determination of the extent of coordination of the metal with the support. This technique was used especially in the study of Pt supported catalysts used in fuel cell applications. Using EXAFS spectra, it was found that the metal-support distances increased during cell operation. The elongation was explained as due to the presence of adsorbed hydrogen on the interface. Heating the catalyst in vacuum regenerated the original metal-support distances for both CNF and activated carbon supports [157,158]. Metal-nanotube interactions have also been studied using X-ray absorption near edge structures (XANES). XANES measures the changes in the absorption coefficient at a particular core level of an atom in a chemical environment and has been used as a tool to study chemical bonding and surface chemistry [159]. In contrast to EXAFS, that deals with single photoelectron backscattering events, in the XANES region, starting about 5 eV beyond the absorption threshold, because of the low kinetic energy range (5–150 eV), multiple scattering of the photoelectron by its neighbors becomes important. The carbon K-edge and Pt M_3-edge have been applied to elucidate the interactions between Pt nanoparticles and carbon nanotubes and compared with values obtained on HOPG and silica substrates. In the case of the Pt M_3-edge spectra, a very small reduction in intensity was observed for the Pt/CNT relative to the Pt metal in contrast to silica supports where this difference in intensity was much larger [160]. This discrepancy between Pt on porous Si and Pt on CNT was explained to be due to the difference in the nature of the Pt/support interactions in the two cases. It was shown that crystalline Pt nanoparticles interact with CNTs through synergic bonding involving charge redistribution between C 2p-derived states and Pt 5d bands. The XANES features on Pt coated CNTs did not vary significantly from Pt deposited on HOPG showing that the graphitic features in the CNTs remain intact after Pt coating [161].

Electron energy loss spectroscopy (EELS) has been used for the identification of elements and the determination of their binding state to their environment. Even with pristine nanotubes on which EELS has been used to detect the presence of amorphous carbon or impurities at the atomic scale [162]. The advantage of EELS over EDX is in its sensitivity to lighter elements such as boron and nitrogen. Spatially resolved EELS is an exceptionally sensitive technique for the study of decorated nanotubes wherein a precise mapping of the various elements provides an atomic scale resolution of the coating and gives valuable information on the uniformity and crystal structure of the coating. For instance, the interface between the carbon and the SiO_x in silica-coated nanotubes has been mapped by EELS [163]. Other widely studied coating materials include BN and WS_2, both of which are layered materials as graphite. In BN and WS_2, the variation of the plasmon energy (corresponding to the collective excitation of valence electrons) was studied as a function of the coating thickness using EELS. It was found that, for thicknesses over five monolayers, bulk dielectric properties were achieved [164]. More recently, silicon carbonitride (SiCN) -coated multi-walled carbon nanotubes in a core-shell structure were characterized by the combination of high resolution TEM, SEM and EELS. The local composition of the composite coating was elucidated from an EELS map obtained using energy filtering TEM. The SiCN coating on the nanotube could be clearly observed using the Si-energy window (or slit) in the EELS mapping images [165].

In cases where the coating contains active chromophores, optical spectroscopy methods such as Raman, fluorescence, and absorption can be used for quick routine characterization on a large scale. Certain π chromophores that are strongly fluorescent have been used as tags to determine the chemical nature of the nanotube surface and the number of active sites. The electronic state of a chromophore such as naphthalimide as determined by luminescence spectra changes drastically in the presence of acid treated carbon nanotubes, causing a large red shift for the luminescence maximum showing extensive electronic interaction. A preliminary semi-empirical calculation for a simplified model system predicted a face-to-face arrangement of the two π electron systems. It was indicated that the lone pair of electrons on the nitrogen atom also contributed to the π-delocalization stabilizing the composite [166].

Similarly, UV visible absorption spectroscopy has proved useful in the characterization of Ru- and Os- based polymer-coated nanotubes, where the high intensity charge transfer spectra are modified due to their interaction with the nanotube surface.

Since the discovery of SWCNTs, Raman spectroscopy has been a powerful tool for the study of the C-C bond structure. In addition to providing valuable information of the nature of the carbon-carbon bonding and sample purity, Raman spectra have been used to study the intertube interaction in SWCNT bundles and also determine the extent of noncovalent side-wall functionalization that produced significant detectable changes in the vibrational frequencies of the breathing modes [167]. The ratio of the intensities of the D band with respect to the G band has been used as a parameter to measure the purity of ball milled and functionalized nanotubes.

4.5.3 Porosity and Surface Area

Nitrogen adsorption isotherms have been used in catalysis to measure the specific surface area and pore volume that are intricately related to the distribution of active sites for the adsorption of reactants. Due to their low density, nanodimensions and hollow structure, carbon nanotubes have a high specific surface area. However, the surface area is a function of the diameter, number of walls, nature of dispersion of the tubes (whether individual or in bundles) and the presence of functional groups. The porosity measurements thus give important information on the state of agglomeration and side-wall functionalization of the nanotubes and therefore are routinely used as characterization methods. In addition, open MWCNTs often show an increase in S_{BET} and changes in pore size distributions whereas, open or cut SWCNTs show a decrease in S_{BET} when compared to pristine tubes due to the blocking of active sites by the functional groups or by the introduction of impurities during cutting. Hydrogen adsorption studies have been performed with the aim of using nanotubes for hydrogen storage [168]. N_2O adsorption isotherms have been measured to determine the adsorption due to Cu particles deposited on different types of carbonaceous substrates such as CNFs, CNTs and activated carbon. The adsorptive decomposition of N_2O can take place on the metallic surface of Cu to produce N_2: $N_2O + 2Cu = N_2(g) + Cu-O-Cu$.

The oxidation of Cu is found to proceed only on the surface and the number of moles of N_2 evolved is twice that of the number of surface Cu atoms thus allowing a precise determination of the surface deposited copper [169]. Other chemisorptions such as CO or H_2 can also be used to determine the surface metal concentrations.

4.6 Applications of Decorated Nanotubes

4.6.1 Sensors

In addition to their unique electronic, thermal and mechanical properties, carbon nanotubes have a high surface-volume ratio that makes them particularly attractive sensor materials capable of detecting small concentrations of molecules with high sensitivity under ambient conditions. Adsorption of both electron withdrawing molecules such as NO_2 or O_2 and electron-donating molecules such as NH_3 can cause sharp changes in conductance due to charge transfer between the nanotube and the adsorbed molecule [1–3,170]. There are numerous reports on the use of carbon nanotubes for gas detection, strain monitoring, piezoelectric sensors and biosensors. The particular case of biosensors will be described in Section 4.8. As far as gas detection is concerned, a variety of gases have been tested including ammonia, carbon dioxide, oxygen, hydrogen, nitrogen and vapors of compounds such as ethanol, chloromethane and propionic acid. For more detailed information, readers are referred to reviews by Sinha et al. and Terrones [171,172]. Early experiments proposed that pristine nanotube surfaces are not sensitive to specific interactions with particular gases and fail to discriminate them. On the contrary, nanotube sensors with molecular specificity can be obtained through appropriate coating or composite formation with desired molecular groups or clusters of compounds [173]. Compared to pristine nanotubes, nanotube-based composites including active sensor materials such as oxides or sulfides as a coating perform much better in terms of stability and reproducibility due to their low defect content and better electronic responses [174–176]. Theoretical modeling of a metal cluster on a nanotube exposed to a molecule of ammonia showed that CNT-metal assemblies could be, in principle, tailored to adsorb and recognize chemical agents with high selectivity and specificity [177]. Thus CNTs serve as nanosized filamentous substrates, mold or template for the stabilization of nanoparticles of active materials and a great variety of electronic material coatings such as ultra thin films, quantum dots and nanowires of oxides, sulfides and complex biomolecules have been tested for sensor applications. The nanotubes serve to anchor the sensor particles preventing their size to increase or agglomeration. Since gas sensing properties are dependent on the adsorption of the gas on the sensor surface, the nanostructure of the sensing material is expected to influence the response time and sensitivity of the device.

The key principle behind the application is to use relatively small clusters that donate or accept a significant amount of charge on adsorption of a target molecule that in turn modifies the electron transport through the nanotube and causes a change in its conductance. For example, Pd metal nanoparticles attached to nanotube surfaces show sensitivity for hydrogen with a threshold of 400 ppm. When exposed to air containing 400 ppm of hydrogen, the conductance of the Pd/SWCNTs decreased rapidly (5–10 s) to half the initial value and reversed slowly (400 s) when the hydrogen flow was turned off. At room temperature, hydrogen molecules dissociate on the surface of Pd and the nascent hydrogen dissolves into the Pd crystals. This lowers the work function of Pd, causing electron transfer from the Pd into the SWCNTs and changing the conductance of the latter [178]. The recovery is probably due to the oxygen in the air that removes the hydrogen from the Pd and restores its work function to its initial value. Similarly, CNTs coated with Pt particles showed improved sensitivity to hydrogen peroxide when compared to CNT electrodes prepared without Pt coating [179].

SnO_2, a n-type semiconductor with a wide band gap energy, has been extensively used as a gas sensing material to detect combustible, toxic and pollutant gases. SnO_2 nanoparticles coating MWCNTs have shown high sensitivity to ethanol (used in breath analyzers) and LPG (liquefied petroleum gas). The resistivity of the sensor was reduced by 50% in the presence of 100 ppm of LPG [180]. Large molecules have also been coated on nanotubes for use in sensor applications [181]. A DNA-decorated carbon nanotube has also been proposed for sensor applications where single-stranded DNA (ss-DNA) is used as the chemical recognition site and single-walled carbon nanotube field effect transistors (SWCNT-FETs) act as the electronic read-out component. SWCNT-FETs with a nanoscale coating of ss-DNA show significant and reproducible responses to odors, that is, specific vapors.

4.6.2　Catalysis

Carbon based materials such as graphite, activated carbon, Vulcan XC-72 R carbon black have been traditionally used in heterogeneous catalysis, to disperse and stabilize metallic particles [7]. Catalytic properties of these solids are known to be dependent upon the interaction between the carbon support and the metal particles and hence the development of high performance supports is a high priority activity for the chemical industry. As mentioned in Section 4.1, among the different types of carbonaceous supports, CNTs have high mechanical strength, thermal conductivity, and show mesoporosity that can be suitably tailored. Not surprisingly, CNTs and CNFs have been used as catalyst supports for a number of well-known organic reactions. and it has been universally observed that both the yield and selectivity depend extensively on the catalyst vs support interactions. Carbon nanotubes appear as unique nanostructured carbon material with tunable porosity and surface structure. The literature on this subject is vast and only selected representative examples illustrating the advantages and disadvantages of CNTs versus other supports such as alumina, silica and graphite will be briefly discussed with respect to a few well known organic reactions. In particular, much attention has been dedicated to liquid-phase reactions with MWCNT and CNF supported catalysts where their high external surface and their mesoporosity would result in a significant decrease in mass-transfer limitations when compared to activated carbon. Relatively few studies on SWCNT-supported catalytic systems have been reported, due either to their microporosity or to the fact that it is still very difficult to obtain large amounts of pure material required for systematic studies. Existing results on MWCNTs and CNFs show that, in general, CNT-based supports show increased loading and better dispersion of the catalyst leading to improved performance when compared to other supports such as alumina and silica. A range of catalysts have been so far deposited on the surface of carbon nanotubes. Most commonly reported are metals such as palladium, platinum, silver, gold, iron, lead, nickel and ruthenium. MWCNTs have also been used as supports for bimetallic nanoparticles, such as Pt–Ru, Pt–Fe and Pt–Sn systems. Besides these elements, CNT supports for anatase TiO_2, magnesia, Vaska's complex $[IrCl(CO)(PPh_3)_2]$, rhodium complex $[Rh_2Cl_2(CO)_4]$, or Wilkinson's complex $[RhCl(PPh_3)_3]$ have also been described. These catalysts have been tested in important organic reactions such as hydrogenation, hydroformylation, redox and decomposition reactions that are extremely important for the efficient synthesis of a wide variety of products in daily use and for protection of the environment. Some representative reaction types are treated below.

4.6.2.1　Hydrogenation Reactions

Hydrogenation is used to form a wide variety of commercial products, including olegochemicals (edible fats and oils, fatty acids, fatty amines, etc.), specialty chemicals (pharmaceuticals, flavors and fragrances, herbicides pesticides, etc.) and petrochemicals, where selective hydrogenation catalysts are important for the viability of these processes [134]. Several studies report that, for liquid-phase hydrogenation, the use of CNT or CNF supports with high mesoporosity leads to better catalytic activity than microporous activated carbon for which mass transfer limitations have been observed. Early studies on the hydrogenation of but-1-ene and buta-1,3-diene on nickel catalysts supported on different types of CNFs, α-alumina and activated carbon have shown that the catalyst supported on CNFs gives higher conversion when compared to α-alumina and activated carbon, even though the average metal catalyst particle size is larger (6.4–8.1 nm on CNFs, 5.5 nm on activated carbon, and 1.4 nm on Al_2O_3) [182–184]. These results indicate that the catalytic hydrogenation might be structure-sensitive. It was further observed from HRTEM studies that the metal particles on CNFs were thin hexagonal plate-like structures typically generated by strong support-metal interactions whereas discrete spherical shapes characteristic of weak support-metal interactions were observed on alumina.

Ultra-small rhodium nanoparticles (1.1–2.2 nm) supported on oxidized CNFs were used in the hydrogenation of cyclohexene [185]. These catalysts turned-out to be extremely active even under low hydrogen pressures at low metal loadings (1% w/w) and low cyclohexene concentrations (1% v/v). The selective hydrogenation of α–β -unsaturated aldehydes such as cinnamaldehyde to unsaturated alcohols is a key reaction in the preparation of various fine chemicals and is often used in the literature as a sensitive test reaction. Cinnamaldehyde contains both a C=C bond and a C=O bond in a α–β unsaturated arrangement. The possible reaction pathways may produce the unsaturated alcohol which is, in general, the targeted molecule, the saturated aldehyde and the saturated alcohol.

The selective hydrogenation of cinnamaldehyde has been studied on different supported catalysts including monometallic Pd [186–189], Pt [190–192], Ru [10,193] and bimetallic PtCo [194], PtNi [195], PtRu [196], and PdRu [197] systems. Pt and Pd deposited on CNFs and MWCNTs turned out to be only slightly more active than similar catalysts deposited on activated carbon but the Pd/MWCNTs showed a significantly increased selectivity (80%) to hydrocinnamaldehyde. A higher selectivity to cinnamyl alcohol for the Pt/MWCNTs can be realized if the Pt nanoparticles are confined to the inner cavities of large diameter tubes (60–100 nm) when compared to systems where the Pt particles are coated on the surface [190]. Finally, the use of PtRu/MWCNT (175 m² per g, average particle size ~2 nm) leads to activities higher than with PtRu-decorated activated carbon, and a selectivity to cinnamyl alcohol > 90% provided the catalysts are heated to remove oxygen containing functional groups [196] demonstrating the importance of surface chemistry on the reaction mechanism.

4.6.2.2 Ammonia Decomposition

The ammonia decomposition reaction has been studied for the CO-free on-site generation of hydrogen thereby reducing the emission of greenhouse gases and the unconverted ammonia can be reduced to less than 200 ppb level by using suitable absorbers. The main drawback has been the high temperature of the reaction (> 873K) that makes difficult the activity of the catalyst particles to be maintained and their sintering and coalescence to be avoided [198–200]. However, recently, a finely divided (~2 nm) Ru catalyst promoted by

KOH and supported on high surface area MWCNTs ($224 \, m^2$ per g) has shown a high activity for the reaction [201]. The catalytic activities rank as: Ru/MWCNT > Ru/MgO > Ru/TiO$_2$ > Ru/Al$_2$O$_3$ > Ru/ZrO$_2$ > Ru/activated carbon, indicating the need for a conducting support that permits electron transfer from promoter and/or support to Ru that aids the desorption of N atoms and their recombination. More results on this reaction have been summarized in a short review [202].

4.6.2.3   Reactions Involving CO/H$_2$

The important reactions in this category where CNTs have been employed as supports are the Fischer-Tropsch synthesis, methanol or higher alcohol synthesis, and hydroformylation reactions. The Fischer-Tropsch process is a chemical reaction whereby carbon monoxide and hydrogen are combined in the presence of a cobalt or iron catalyst into liquid hydrocarbons containing a mixture of alkanes and olefins. The main goal of this reaction is to produce a synthetic petroleum substitute for use as a fuel or a lubricant. Copper-promoted Fe/MWCNT catalysts prepared by wet-impregnation are active in Fischer-Tropsch syntheses at 493 K, and an olefin (C$_2$–C$_{10}$) content of 40 to 60 mol.% in the hydrocarbon fraction was obtained [203]. Compared to Fe/activated carbon catalysts, Fe/MWCNT systems (S$_{BET}$ = 20 m^2 per g) lead to a lower selectivity to methane, and to a high selectivity to olefins [204]. A novel structured catalyst, based on aligned MWCNT arrays in a microchannel reactor, has been used for Fischer-Tropsch reactions [205]. The catalyst, CoRe/Al$_2$O$_3$, was deposited on the MWCNT material by dip-coating. When integrated into a microchannel reactor, this microstructured catalyst exhibited a four-fold enhancement in Fischer-Tropsch reaction activity compared to the values observed for similar catalytic system without CNT arrays. Such an enhancement is correlated to the (i) higher specific surface area, or (ii) the efficiency of MWCNTs to act as hydrogen reservoir, that favors CO/CO$_2$ hydrogenation reactions. Rh-ZnO/ MWCNT systems are active and selective for methanol synthesis under 100 bar pressure and at 523 K [206]. Rhodium-supported catalysts on fishbone-CNFs, platelet-CNFs, or ribbon-like CNFs have been used for the hydroformylation of ethylene and compared to Rh/SiO$_2$ system [207]. Although silica has a higher specific surface area and Rh particles are smaller, CNF-supported catalysts exhibit higher selectivity while maintaining the same catalytic activity. The enhanced selectivity has been attributed to the unique hexagonal shape of the rhodium crystals when deposited on the carbon nanofiber surface. The complex [HRh(CO)(PPh$_3$)$_3$] has been grafted onto MCWNTs and fishbone-CNFs, and the catalytic activity in the hydroformylation of propene has been compared to that of similar systems prepared on activated carbon, carbon molecular sieves and silica [208]. Higher conversions and regioselectivity towards n-butylaldehyde have been reported for CNT- and CNF-supported systems. The authors propose that the rhodium complex can be accommodated into the oxidized channels because of their size and that the extensive capillary condensation of n-butyraldehyde is prevented compared to other supports.

4.6.2.4   Polymerization Reactions

Polymer-nanotube composites have potential applications in industry as light materials with high conductivity, high tensile strength, and high Young's modulus and hence various synthesis methods have been developed for the large scale production of these composites

[1–3]. To prevent the aggregation of nanotubes during the composite formation, nanotubes are often coated with the monomer and the polymerization reaction is implemented in situ. For example, in situ polymerization of ethylene catalyzed by a highly active metallocene complex physicochemically anchored onto the nanotube surface results in the formation of CNTs uniformly coated with in situ-grown polyethylene that finally breaks up the bundles into individual tubes. This method has been previously applied to graphite [209–210]. For nanotubes, the technique consists of anchoring methylaluminoxane (MAO) activator followed by the addition of the metallocene catalyst to the surface-activated carbon nanotubes suspended in n-heptane. In this study, $[Cp_2^*ZrCl_2]$ (bis(pentamethyl-η^5-cyclopentadienyl)zirconium(IV)dichloride) was used as a typical polymerization catalyst. Upon ethylene addition, polyethylene (PE) is exclusively formed near the carbon nanotube surface and, with increasing molecular mass, precipitates onto the nanotubes to coat them and ultimately separate them.

During the polymerization of ethylene on a MWCNT surface catalyzed by a highly active metallocene complex physicochemically anchored onto the nanotube surface, it was observed that after one hour of polymerization, ethylene consumption for the catalyst supported on MWCNTs was 28% more than for ethylene polymerization in the absence of any filler, showing that supporting the catalytic system onto MWCNTs significantly increased the ethylene polymerization rate [211]. Efficient and homogeneous polyethylene coating has been observed also for SWCNTs and DWCNTs. Again, higher catalyst activities were detected for the surface-treated nanotubes, with large increases (by 36% for SWCNTs and by 94% for DWCNTs) when compared to homogeneous ethylene polymerization carried out under the same experimental conditions.

4.6.2.5 *Environmental Catalysis and Oxidation Reactions*

The high surface area and the electrical conductivity of MWCNTs could be of interest to improve the activity of TiO_2 photocatalysts that are routinely used in environmental pollution control for the decomposition of organic pollutants. Nanotube-titania composites have been tested for the photodegradation of phenol [212–214], acetone [215], or azo dyes [216]. These studies showed that the presence of MWCNTs in the composite permits:

1) an increase in the effective surface area of the photocatalyst,
2) the stabilization of the nanoparticles of TiO_2 in the active state,
3) a strong interface interaction that favors adsorption [212] and electron transfer.

In these reactions, MWCNTs could be considered as sensitizers that promote the transport of electron/hole pairs along the tubes, thus decreasing their recombination rate and improving the photocatalytic activity of TiO_2. It has also been shown that an optimal amount of CNTs to incorporate in the composite, generally between 10–20% [212] is needed to obtain a significant improvement in the catalyst efficiency. Above this critical amount, the photocatalytic activity decreases due to the increased absorbing and scattering of photons by the carbon support. MWCNTs coated with Ni nanoparticles grown on titanium dioxide substrates have also been used as sensitizers for the transfer of electrons from the nanotubes to the conduction band of TiO_2, improving its photocatalytic activity in

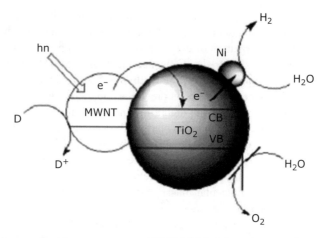

Figure 4.16   *Process for H$_2$ generation on MWCNT-TiO$_2$-Ni composite photocatalyst.
'D' stands for methanol molecules. Reprinted with permission from [217] Copyright (2006)
Elsevier Ltd.*

the decomposition of methanol to give hydrogen (Figure 4.16) [217]. Methanol molecules
are oxidized by the transfer of electrons to the MWCNTs. There is no hole formed in the
valence band of the titanium dioxide and hence there is no generation of oxygen.

4.6.3   Fuel Cells

Proton Exchange Membrane (PEM) fuel cells (Figure 4.17) are power generators based on
two electrochemical reactions, namely, the oxidation of hydrogen at the anode and the
reduction of oxygen at the cathode. They have important advantages over chemical batteries
in terms of low emission and high energy density [218]. A polymer membrane such as
Nafion is used as the electrolyte. Due to the high acidity of the electrolyte, Pt or Pt alloys
have been the only catalyst that has been used up to now. While Pt works as an efficient
catalyst for hydrogen oxidation, it has a low activity for oxygen reduction and hence high
Pt loadings are needed at the cathode for the cell to operate. Nanosized platinum or platinum
alloy particles supported on carbon blacks have been the most frequently used electrocata-
lyst for this purpose, resulting in prohibitive manufacture costs that have been the main
drawback for the large scale utilization of these cells in automotive applications [219].
Consequently, research in this area is focused on the reduction of the Pt loading at the cathode either by alloying with other metals (for which the choice is limited) or by using
different supports for optimizing the catalyst performance. A variety of catalyst supports
such as CNFs, MWCNTs, SWCNTs, and DWCNTs are currently being tested [220–223].
Due to their higher electronic conductivity, tensile strength, resistance to electrochemical
corrosion and higher mesoporosity compared to Vulcan XC-72R carbon black that is
mainly microporous, CNTs and CNFs could provide better supports for high platinum
loading and facilitate electron transfer. In addition, the microporosity of the carbon blacks
could result in trapping a fraction of the Pt nanoparticles thus rendering them inactive.
Also, oriented nanotubes are expected to have higher gas permeability. While results
greatly vary with the source of the carbon nanotube and the preparation of the electrode,

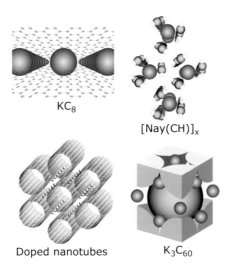

KC₈

[Naᵧ(CH)]ₓ

Doped nanotubes

K₃C₆₀

Plate 1: Figure 2.2 *Schematic structural models for alkali-intercalated carbonaceous host lattices. Clockwise from upper right: stage 1 potassium-intercalated graphite KC_8; saturated phase of Na-intercalated polyacetylene in which close-packed Na chains occupy all available interstitial channels; potassium fulleride with both tetrahedral and the single octahedral site in the fcc host are singly occupied; hypothetical saturation-intercalated SWCNT triangular lattice corresponding approximately to KC_{13}. All but the last one have been identified crystallographically. Adapted from [14] Copyright (2000) American Chemical Society.*

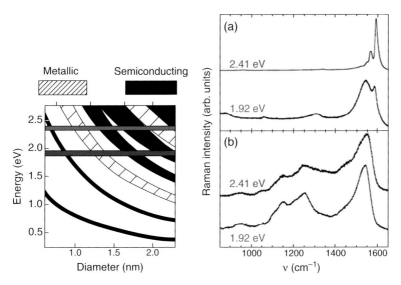

Plate 2: Figure 2.26 **(Left)** *Allowed optical transitions as a function of the diameter of SWCNTs.* **(Right)** *Raman spectra in the TM range of (a) pristine SWCNT and (b) saturated Rb:SWCNT bundles at laser excitation energies 2.41 eV (top) and 1.92 eV (bottom). Adapted with permission from [65] Copyright (2003) Elsevier Ltd.*

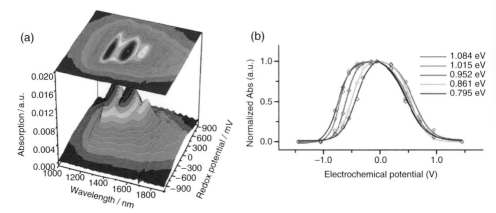

Plate 3: Figure 2.30 *(a) Smoothed surface plot of optical absorption intensity as a function of wavelength and electrode potential in the S11 region for a solution of a potassium salt of HiPco SWCNTs in DMSO. (b) Intensity of each of the five bands selected in the spectra above after normalization as a function of potential. The fitting curves (full lines) were calculated by assuming that only the neutral tubes absorb, and by using a simple Beer-Lambert law wherein the concentrations of reduced/neutral and neutral/oxidized tubes where expressed as a function of electrode potential, using Nernst equation [114]. This way, oxidation and reduction potentials could be determined (corresponding to the inflection points in each sigmoidal curve). In all plots, raw electrochemical data, that is, uncorrected for ohmic drop, are referenced to standard Calomel electrode. Reprinted with permission from [6] Copyright (2008) Royal Society of Chemistry.*

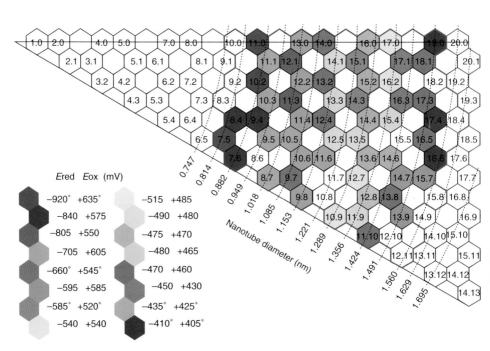

Plate 4: Figure 2.31 *Chirality map displaying the average standard potentials associated to each of the SWCNT structures identified in this work. HiPco SWCNTs are located inside the red line, while arc-discharge SWCNTs are inside the blue line. Reprinted with permission from [114] Copyright (2008) American Chemical Society.*

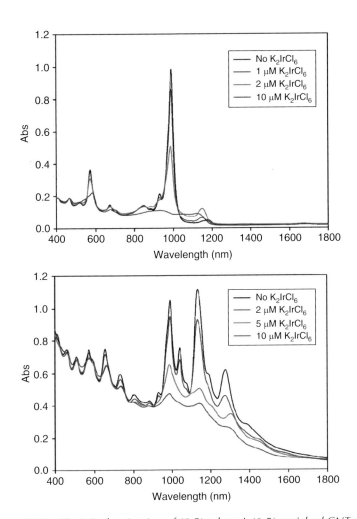

Plate 5: Figure 2.51 **(Top)** *Redox titration of (6,5) tubes. A (6,5)-enriched CNT solution was centrifugated using a Microcon spin filter YM100 (Millipore) and was resuspended in D$_2$O. Freshly made K$_2$IrCl$_6$ solution in D$_2$O was added to 100 ml of the CNT solution to the indicated final concentration. The spectrum was recorded after 10 min of incubation.* **(Bottom)** *Redox titration of unfractionated HiPco tubes under similar experimental conditions. Reprinted with permission from [201] Copyright (2004) American Chemical Society.*

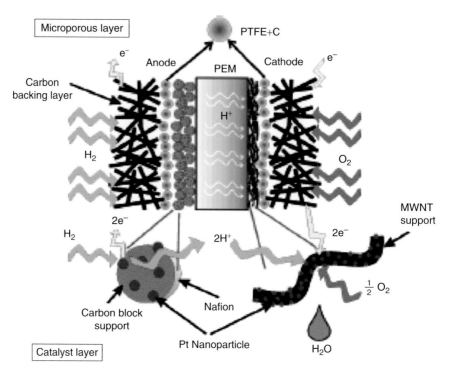

Plate 6: Figure 4.17 *Sketch of the fuel cell architecture for the ultra-low Pt loading thin film Pt/MWCNT PEMFC. Reprinted with permission from [230] Copyright (2007) American Chemical Society.*

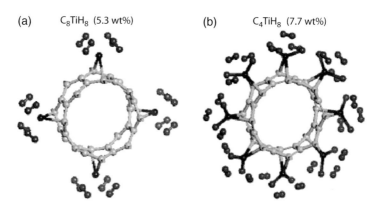

Plate 7: Figure 4.18 *A model of the high density coverage of hydrogen on Ti/SWCNT. Reprinted with permission from [236] Copyright (2005) American Physical Society.*

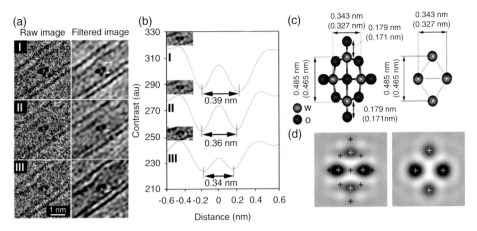

Plate 8: Figure 5a.10   (a) A sequence of HRTEM images (left) obtained from a single $[W_6O_{19}]^{2-}$ anion locked into position within the capillary of a DWCNT shown with Fourier-filtered images (right). (b) Line profiles produced through equatorial spots corresponding to pairs of W atoms. (c) Structure models of the $[W_6O_{19}]^{2-}$ anion with oxygen included (left) and excluded (right). Selected bond distances are included for the unrelaxed anion (in brackets) and the relaxed anion (from MD). (d) HRTEM image simulations produced at ideal defocus from the structure models in (c) showing the effect of including oxygen in the image simulation calculation. When the oxygen atoms are included (left simulation), the centre of the spot contrast in the equatorial W atom pairs are distorted outwards towards the three terminal oxygen atoms on each side of the anion. When the oxygen atoms are not included (right simulation), there is relatively little distortion of the W positions (red crosses = O; green crosses = W). Reprinted with permission from [118] Copyright (2008) American Chemical Society.

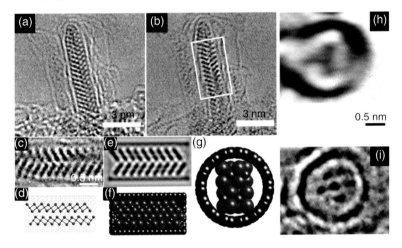

Plate 9: Figure 5a.13   (a) and (b) High-resolution TEM images of a tip of a SWCNT filled with PbI_2 for two different focus values of the objective lens. (c) Enlarged image of the area framed in (b). (e) Corresponding calculated image, based on the model in (d). Although bulk PbI_2 structure is a two-dimensional-layered type, encapsulation in SWCNTs enforces it to become a one-dimensional polyhedral chain. (f) and (g) Ball and stick models of the structure modelled in (d), as a side and a cross-section views respectively. (h) High-resolution TEM image of a cross-section of a SWCNT filled with CrO_x. The filling material structure exhibits a three-fold symmetry in which each lobe is assumed to be the projection of a chain of CrO_4 tetrahedra. (reprinted with permission from [18] Copyright (2001) Elsevier Ltd) (i) High-resolution TEM image of a cross-section of a SWCNT filled with $HoCl_3$. The filling material structure exhibits a somewhat five-fold symmetry, yet is deformed. Based on the observed structural similarities with $NdCl_3$ [24], each dark spot is assumed to a projection of a single 1D chain of $HoCl_x$ polyhedra.

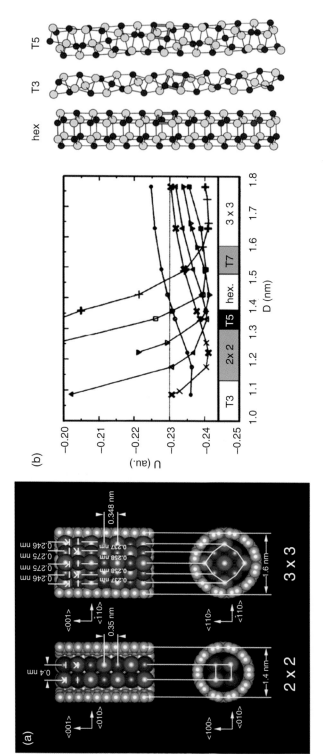

Plate 10: *Figure 5a.14* (a) *Structure models of 2 × 2 and 3 × 3 layer KI@SWCNT hybrid nanotubes determined experimentally.* (b) *Minimum energy (U) phase diagram depicted for structure types of KI predicted to fill SWCNTs of varying diameter (D). T3, T5 and T7 are helical KI 1D crystals with three-, five- and seven-membered rings in cross-section, respectively. Reprinted with permission from [163] Copyright (2002) Elsevier Ltd.*

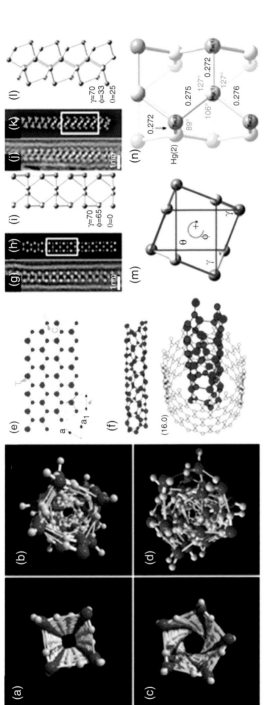

Plate 11: Figure 5a.15 (a) and (b) Square and hexagonal ice tunnels formed within (14,14) and (15,15) SWCNTs respectively [34]. The containing SWCNTs are not shown for clarity. (c) and (d) show the corresponding liquid phases, respectively (reprinted with permission from [34] Copyright (2001) Macmillan Publishers Ltd). (e) to (f): The construction of an inorganic nanotube from a corresponding hexagonal planar structure. The red and blue circles represent the two ionic species. C_h is the chiral vector along which the structure must be folded in order to form a (3,2) hexagonal nanotube. The bottom sketch in (f) shows a perspective view of a (3,2) inorganic nanotube within a (16,0) SWCNT. This nanostructure turned out to be nearly identical to a similar nanostructure later reported within a SWCNT by HgTe [27]. (g), (h) and (i) show a reconstructed phase image, corresponding image simulation, and structure model respectively, of one projection of a twisted, chiral HgTe crystal in a SWCNT. (j), (k) and (l) show similarly a phase image, simulation, and model from a second twisted HgTe crystal in a separate carbon nanotube. (m) and (n) show the end on views of the essential structure motif and corresponding DFT-optimized model corresponding to the materials shown above them respectively. There is a close correspondence between this model and the model shown in (f) (reprinted with permission from [164] Copyright (2004) Elsevier Ltd).

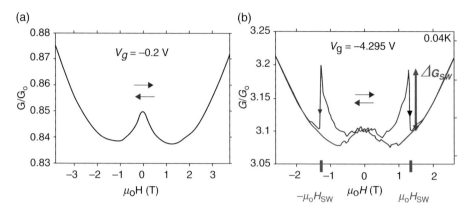

Plate 12: Figure 5a.16 *Variation of conductance versus cycling magnetic field measured on individual Co@SWCNT bundles contacted (with Pd) following the transistor-type device, for an angle of 25° between the direction of the outer magnetic field and the elongation axis of the contacted bundle and at a temperature of 40 mK. (a) 90% of the devices do not show a hysteretic behaviour, as illustrated, whatever the gate voltage. (b) 10% of the devices show a hysteretic behaviour, as illustrated. Conductance curves indeed exhibit a hysteresis with two conductance jumps. They correspond to the magnetic field value at which the magnetization direction of a nanomagnet (presumably one, or several but aligned, encapsulated cobalt wire segment(s)) reverses (reprinted with permission from [179] Copyright (2011) American Chemical Society).*

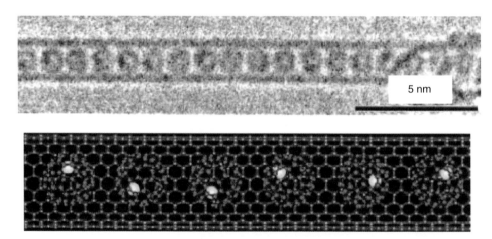

Plate 13: Figure 5b.4 *HR-TEM micrograph and a schematic representation (not to scale) of $Gd@C_{82}$ metallofullerene peapods. Dark spots seen on the fullerene cages correspond to the heavy Gd atoms. Reprinted with permission from [26] Copyright (2000) American Physical Society.*

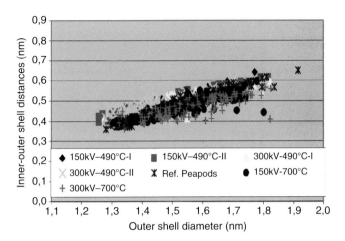

Plate 14: Figure 5b.14 *Plot of the inner-outer shell distance with respect to the outer diameter for various temperature/electron energy (1V = 1eV) condition couples. The trend for 490°C experiments, regardless of the voltage value (specifically for the set of experiments II, that is, a pink square for 150kV, and a light blue cross for 300kV) correspond to larger inner-outer shell distances, as opposed to the trend for 700°C experiments (brown dots for 150kV, and dark 'plus' sign for 300kV). Reprinted with permission from [65] Copyright (2005) American Institute of Physics.*

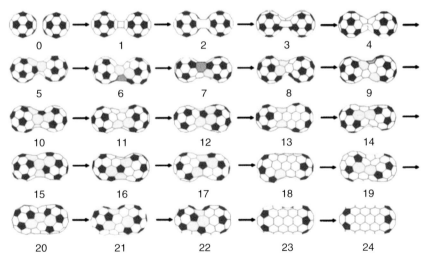

Plate 15: Figure 5b.15 *Schematics of the pathway for the $2C_{60} \rightarrow C_{120}$ fusion. The minimum possible Stone-Wales transformations were found using a geometrical search. Note that the starting point of the fusion is the formation of a (2 + 2) cycloadditional bond between the adjacent C_{60}s. Reprinted with permission from [67] Copyright (2004) American Physical Society.*

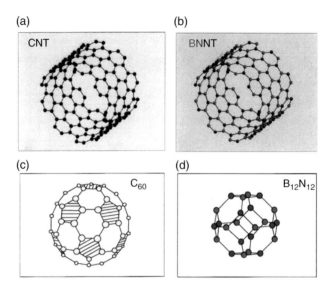

Plate 16: Figure 6.1 *Structural models of (a) carbon and (b) boron nitride nanotubes. (c) corresponding C and (d) BN closed atomic cages optimally sealed. The latter reveals a striking difference in the defective atomic ring nature, that is, a pentagon (odd-member ring defect) for the C system and a square (even-number ring defect) for the BN system.*

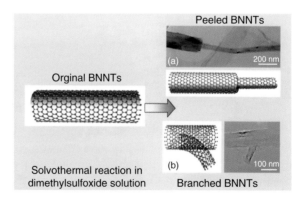

Plate 17: Figure 6.20 *A strategy towards (a) peeling and (b) branching multi-walled BNNTs using a solvothermal reaction in dimethylsulfoxide and the related experimental TEM images.*

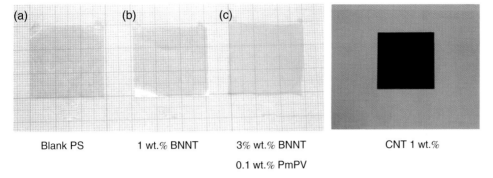

(a)   (b)   (c)

Blank PS   1 wt.% BNNT   3% wt.% BNNT   CNT 1 wt.%

0.1 wt.% PmPV

Plate 18: Figure 6.23 *(a) to (c) fully transparent polystyrene films made with the use of various loading fractions of multi-walled BNNTs; far right side: a black non-transparent film with a similar CNT loading fraction (1 wt.%) is provided for comparison.*

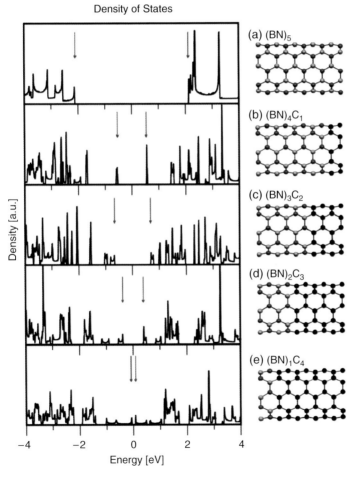

Density of States

(a) $(BN)_5$

(b) $(BN)_4C_1$

(c) $(BN)_3C_2$

(d) $(BN)_2C_3$

(e) $(BN)_1C_4$

Density [a.u.]

−4   −2   0   2   4

Energy [eV]

Plate 19: Figure 6.28 *Densities of states for C*BNNT heterojunctions with different stoichiometries, from pure BN to $(BN)C_4$. The top valence and bottom conduction states are indicated with the pairs of arrows. The energy gap in these NTs varies from 0.15–4.2 eV as the BN content is increased, with the exception of $(BN)_4C$, where two defect-like non-dispersive energy states reduce the energy gap.*

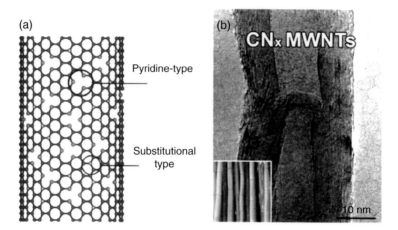

Plate 20: Figure 6.33 *(a) Molecular model of a N*CNT, exhibiting two types of N sites: 1) pyridine-type in which each N atom is bonded to two C atoms, responsible of creating cavities and corrugation in the nanotube structure (atoms in pink), and 2) substitutional N atoms which are bonded to three C atoms (green spheres); (b) HRTEM image of a N*MWCNT exhibiting the bamboo-type morphology, and (inset) image of natural bamboo trees whose appearance is similar to N*MWCNTs.*

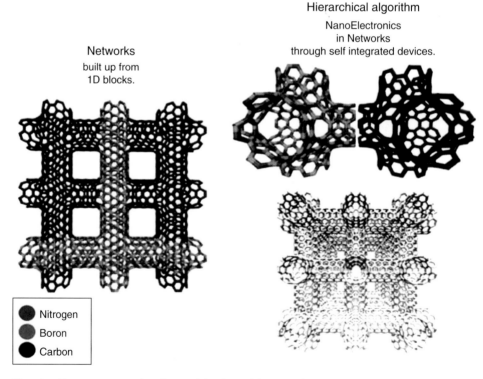

Plate 21: Figure 6.51 *Molecular models of possible 2D and 3D networks containing nanotubes made of C and BN fragments. These devices will certainly exhibit novel electronic and mechanical properties.*

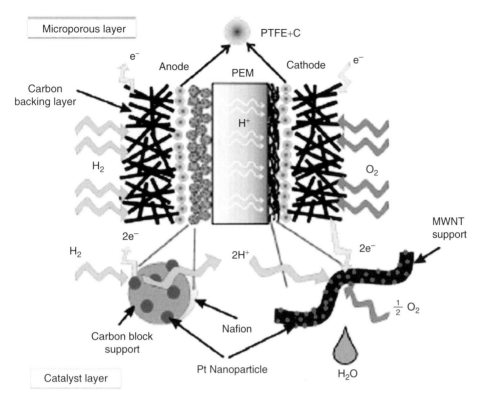

Figure 4.17 *Sketch of the fuel cell architecture for the ultra-low Pt loading thin film Pt/MWCNT PEMFC. For a better understanding of the figure, please refer to the colour plate section. Reprinted with permission from [230] Copyright (2007) American Chemical Society.*

the general tendency observed is that the catalysts prepared on CNTs or CNFs are more active and in some cases present a better resistance to poisoning than those prepared on conventional carbon supports such as Vulcan XC-72 [224]. In some cases it is also possible to obtain similar or better performances with significant reduction of Pt loadings [225]. Both the fabrication of the nanotube-based electrode and the coating method for Pt are also found to influence the performance of the cell, as reviewed by Lee et al. [226]. The Pt particles can be anchored onto the carbon nanotubes by the polyol process (cf. Section 4.5.1.4) whereby high loadings (30–40%) of 3–5 nm sized Pt particles are obtained on the surface of the CNTs. Pressed films of Pt/CNTs are transferred onto a Nafion film by hot pressing in the presence of Nafion solution. Alternatively, Pt could be deposited onto CNTs grown on carbon paper by a CVD process [227]. These electrodes have been tested in hydrogen and methanol fuel cells [127,228,229]. Fuel cell membrane electrode assemblies with Pt loading of 0.2 mg Pt per cm^2 at the anode and ultra low Pt loadings of 6 μg Pt per cm^2 and 12 μg Pt per cm^2 at the cathode fabricated by using thin films of Pt/MWCNTs using Nafion NRE-212 as the electrolyte have been reported [230] (Figure 4.17). Such a cell could achieve power densities that were only slightly lower than cells that contained nearly 10 times higher platinum loading.

(a) C_8TiH_8 (5.3 wt%)   (b) C_4TiH_8 (7.7 wt%)

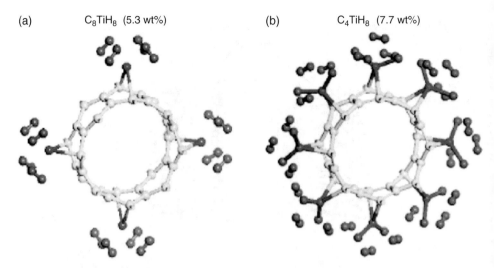

Figure 4.18 *A model of the high density coverage of hydrogen on Ti/SWCNT. For a better understanding of the figure, please refer to the colour plate section. Reprinted with permission from [236] Copyright (2005) American Physical Society.*

4.6.4 Hydrogen Storage

It has long been acknowledged that the efficient use of hydrogen as a fuel will require cost-effective and safe storage means. Intense activity followed the initial reports on high hydrogen storage by carbon nanotubes [231,232]. Unfortunately, later detailed investigations have confirmed that pure carbon nanotubes stored less than 1 wt.% of hydrogen [168,233–235] and the initial results obtained were shown to be due to either inaccuracies in measurement or to impure samples. However, later publications reporting theoretical modeling studies have shown that nanotube-metal composites could be better hydrogen storage materials when compared to pristine carbon nanotubes [236]. In this case, the adsorbed hydrogen is molecular, with a H-H bond length of 0.756Å. A theoretical model based on a first principles study showed that a single titanium atom adsorbed on a SWCNT can strongly bind up to four hydrogen molecules. The strong bonding is reported to be due to a concerted interaction between the H, Ti and SWCNT arising out of a unique hybridization between Ti-d, hydrogen π^* antibonding and SWCNT C-p orbitals. Figure 4.18 shows high density hydrogen coverage on Ti/SWCNTs.

It was shown that at large Ti coverage, a (8,0) SWCNT can store hydrogen molecules up to 8 wt.%. Quantum molecular dynamics simulations indicated that this stored hydrogen can be released upon heating. It was also predicted that a single Ni atom located on a carbon nanotube surface can retain up to five hydrogen molecules [237]. These predictions have given a new impetus to hydrogen storage experiments on metal decorated nanotubes and it was observed that 1 wt.% hydrogen is stored in CNTs coated with Pd particles at 1 atm. and 573 K. Similarly, Ni nanoparticle-decorated MWCNTs result in an enhancement in hydrogen uptake capacity. Measuring hydrogen adsorption isotherms at high pressure

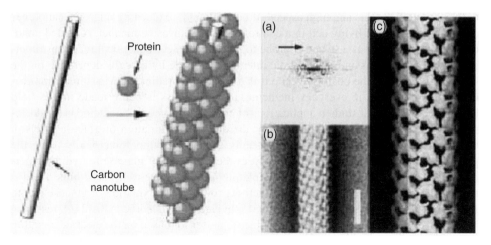

Figure 4.19 *Nanotubes coated with streptavidin. Reprinted with permission from [242] Copyright (1999) John Wiley and Sons.*

showed that Pt/MWCNTs have a hydrogen uptake of up to 3.1 wt.% at 298 K and at 90 bar. The increased adsorption has been attributed to the catalytic activity of the Pt particles in the dissociation of molecular hydrogen into hydrogen atoms increasing chemisorption at the defect sites present on the nanotubes [238–239]. Alternatively, SWCNTs have been added to other materials known for their high hydrogen storage potential such as metal hydrides. With MgH_2, it was found that the SWCNTs were able to alter the hydrogen-sorption kinetics. However, it was far from clear as to how the presence of SWCNTs affects the hydrogen adsorption in MgH_2 [240]. In this context, a first-principle study on the hydrogen storage in aluminum hydride (AlH_3)-coated (5,5) SWCNTs showed that each Alane (AlH_3) molecule can bind up to four hydrogen molecules [241]. The hydrogens could be extracted from the system without breaking the C-Al link. At half coverage of AlH_3, the hydrogen storage capacity of the SWCNT was found to be 8.3 wt.%. The adsorption of H_2 was shown to be molecular, with a H–H bond length of 0.756 Å. It appears that at full coverage, even though the hydrogen storage capacity increases, the binding of H_2 is weak. However, these theoretical calculations have not been substantiated by experimental results. In summary, the results are inconclusive with respect to the actual hydrogen storage capacity of coated nanotubes, yet the recent developments seem encouraging.

4.7 Decorated Nanotubes in Biology and Medicine

Not long after the discovery of nanotubes, it was observed that biomolecules could be made to adopt specific conformations when attached to nanotubes. The helical configuration of streptavidin coated on nanotube is shown in Figure 4.19 [242].

The electronic properties of nanotubes coupled with the specific recognition properties of the immobilized biosystems would indeed make an ideal miniaturized sensor. Thanks to their small diameters and high aspect ratio with lengths exceeding microns, nanotubes can be inserted into a living cell in a controlled but noninvasive manner. A coated nanotube can thus operate as a selective probe to investigate electrochemical reactions, intermolecular interactions and electrostatic interactions inside living cells depending on the functional nature of the coating. While it is too early to predict the widespread application of these sensors in everyday medicine, rapid progress is being made in this field [243]. One of the most studied applications of coated nanotubes in medicine is the detection and analysis of dissolved glucose for the quick determination of glucose levels in diabetic patients. Glucose detection is usually carried out amperometrically using the glucose oxidase (GOx) enzyme that catalyzes the oxidation of glucose to hydrogen peroxide [244–245]. However, conventional platinum or gold electrodes used hitherto have the shortcoming of low electro catalytic activity towards the oxidation of glucose due to the poisoning effect of reaction intermediates, in particular, adsorbed CO [246], resulting in low sensibility and poor selectivity. A glucose biosensor based on the electrodeposition of platinum nanoparticles and GOx onto Nafion-solubilized carbon has been reported [247]. The deposition of GOx onto the electrode surface was carried out by the adsorption of GOx on gold nanoparticles and the electrochemical behavior of the system was studied. Different electrode configurations such as gold, CNT/gold, Pt/gold and Pt/CNT/gold electrodes have been investigated by the amperometric method. A good biosensing capability was reported for the Pt/CNT/gold electrode that was unaffected by the presence of uric acid or ascorbic acid. Glucose detection has also been carried out by the attachment of the enzyme glucose oxidase to the walls of a functionalized nanotube through a linking molecule (1-pyrenebutanoic acid succinimidyl ester) [248]. The enzyme-coated tube is found to act as a pH sensor with large and reversible changes in conductance upon the addition of glucose. A nonenzymatic glucose sensor has been reported where Pt nanoparticle-coated MWCNTs were used as electrodes in the direct electrocatalytic oxidation of glucose in alkaline solution [249]. The Pt/MWCNT electrode covered with 1.5% Nafion detected glucose at zero potential, as measured against standard calomel electrode at 298 K. A linear amperometric response was observed for glucose concentrations in the range of 1–26.5 mM. More recently, a tyrosinase (Tyr) biosensor has been reported based on Fe_3O_4 magnetic nanoparticle-coated CNTs that have been applied to detect the concentration of coliforms with a flow injection assay system. Negatively charged magnetic nanoparticles were absorbed onto the surface of the CNTs which were wrapped with cationic polyelectrolyte poly(dimethyldiallylammonium chloride) (PDDA). This Fe_3O_4/CNT nanocomposite was modified on the surface of the glassy carbon electrode. The biosensor was tested for the detection of coliforms represented by *E. Coli*. The principle of the sensor was the detection of phenol produced during the enzymatic reaction of *E. Coli* in water. This sensor has a detection limit for bacteria of 10 cfu [ml-1] and could find application in water quality analysis [250].

Another application of decorated nanotubes in medicine is drug delivery. Experiments have shown that carbon nanotubes can be used to deliver therapeutic molecules to targeted cells and organs and can reduce induced immune responses and toxicity [251]. It has been demonstrated that coating MWCNTs with an antibiotic such as amphotericin B enabled its delivery across the cytoplasmic membrane with lower damage to cells when compared to traditional methods for drug encapsulation [252]. Analogously, the use of DNA/SWCNTs for the delivery of plasmid DNA into an A549 cell led to gene expression levels ten times higher for the CNT-DNA complex than for DNA alone [253].

SWCNTs show intense fluorescence in the infrared region and hence can be used as markers to study biological reactions since near infrared wavelengths are not blocked by human tissue or biological fluids [254–255]. It is well known that certain DNA oligonucleotides transform from the native right-handed B form to the left-handed Z form in the presence of certain cations [256–257]. When the former coat SWCNTs, the transformation into the latter modulates the dielectric environment of the SWCNTs and decreases their near-IR emission energy up to 15 meV. Lysophospholipids have been coated on SWCNTs to increase the biocompatibility of a device wherein electronic properties of the SWCNTs can be combined with the activity of the lipids for biosensing since the headgroups of lysophospholipids can be functionalized with antioxidants and monoclonal antibodies. SWCNTs have been found to bond very well with lipids and actually disperse much better when coated with lipids compared to common surfactants such as SDS, and assuming the size of the lyso-phosphatidylcholine (LPC) head group to be 0.6 nm and assuming the wrapping of half the CNT cylinder surface, an average of 21 000 lipids can coat each tube [258]. A theoretical model for the lipid-nanotube interaction in water and the self-assembly of the lipid-coated nanotubes has been proposed [259].

While dealing with the application of decorated nanotubes in biology and medicine, it is important to take into consideration the toxicity of these composite materials. There have been a number studies on the toxicity of pristine (or believed so) carbon nanotubes, and it is to be expected that the coated nanotubes would have similar if not higher toxicity. A summary of the results on toxicity studies on nanotubes is found in an early review [260]. Several studies on the ingestion of MWCNT dusts by rodents have been performed accompanied by in vitro studies to assess the toxicity of CNT to skin cell cultures. A recent example is a comparative study of carbon nanotubes with asbestos [261]. However, the results of all the studies published so far have still failed demonstrating unambiguously that pristine CNTs and/or modified CNTs are actually toxic under real-life exposure conditions. Establishing standard testing conditions is still needed, as well as a close collaboration between carbon scientists and biologists. However, it is likely that CNTs and materials derived therefrom could have adverse effects on human health.

4.8 Conclusions and Perspectives

We have shown that several successful experimental methods have been developed to decorate CNT surfaces with metals, semimetals, inorganic and biomolecules on carbon nanotube surfaces. The choice of the coating method is generally determined by the nature of the subsequent application, the scale of operation, and cost considerations. It is clearly demonstrated that a prior functionalization or activation of the CNT surface is necessary to achieve uniform coatings. Wet-impregnation methods, though multistep processes, are most commonly used due to their simplicity and suitability for large scale coating. Gas phase coating methods on the other hand have been adopted mainly for the in situ fabrication of electronic devices. Electrochemical methods are becoming increasingly attractive for application in electrocatalytic reactions and fuel cells. Even though oxidative surface treatments on the supports seemingly improve the dispersion of the metallic phase, it is important to stress that a better knowledge of the concentration and distribution of the attached functional moieties will help to understand the dependence of CNT and CNF reactivity on their structure. Coated nanotubes have not been prepared until now on an industrial scale and the progress in the upscaling of these methods would need large quantities of well characterized carbon nanotubes with controlled surface and electronic properties. Also, reliable methods need to be developed to control precisely the size, shape and agglomeration of the deposited materials.

The main applications of these coated nanotubes have been in the field of catalysis and for sensor applications. Concerning the preparation of supported catalysts, several methods have been successfully used and, as in the case of activated carbon, the role of surface pretreatments on the final dispersion of the metal nanoparticles has been clearly demonstrated. The possibility of shaping these nanomaterials also offers some interesting perspectives, including for the design of structured microreactors [262]. Experimental as well as theoretical works would be necessary in order to have a better understanding of charge transfer phenomena and strength of metal support interaction in catalytic systems involving CNTs and CNFs. The catalytic studies conducted on CNT- or CNF-based systems have shown encouraging results in terms of activity and selectivity. In particular, high selectivity has been obtained on catalytic systems displaying a different metal-support interaction and/or charge transfer phenomena than those observed on other supports such as activated carbon or alumina. For electrocatalysis, a combination of specific support morphology and electrical conductivity is the key to achieving high activities. A recent detailed report published by Global Industry Incorporated estimates the global market for nanocatalysts to reach US\$6 billion in 2015. Though at present nanocatalysts supported on CNTs form a very small percentage (around 2%) of this market, their market share has been projected to triple in the coming years. In the field of sensors, coated nanotubes have been shown to be more active and sensitive when compared to pristine tubes and are likely to make a great impact. Other fields such as medicine and energy storage are still in the research stage and their impact on technology is to be anticipated.

References

[1] D. Tomànek and R.J. Enbody (eds.), *Science and Application of Nanotubes,* Kluwer Academic/Plenum Publishers, New York (2000).

[2] J.P. Issi and J.C. Charlier, in *The Science and Technology of Carbon Nanotubes*, K. Tanaka, T. Yamabe and K. Fukui (eds.), Elsevier, Amsterdam (1999).

[3] M. Monthioux, P. Serp, E. Flahaut, C. Laurent, A. Peigney, M. Razafinimanana, W. Bacsa and J.-M. Broto, Introduction to carbon nanotubes, In *Springer Handbook of Nanotechnology*, Second revised and extended edition, B. Bhushan (ed.), Springer-Verlag, Heidelberg, 43–112 (2007).

[4] R.R. Bacsa, Ch. Laurent, A. Peigney, W. Bacsa, T. Vaugien and A. Rousset, High specific surface area carbon nanotubes from catalytic chemical vapour deposition process, *Chem. Phys. Lett.*, **323**, 526–530 (2000).

[5] Q.-H. Yang, P.X. Hou, S. Bai, Z. Wang and H.M. Cheng, Adsorption and capillarity of nitrogen in aggregated multiwalled carbon nanotubes, *Chem. Phys. Lett.*, **345**, 18–24 (2001).

[6] E. Raymundo-Piñero, P. Azaïs, T. Cacciaguerra, D. Cazorla-Amorós, A. Linares-Solano and F. Béguin, KOH and NaOH activation mechanisms of multiwalled carbon nanotubes with different structural organisation, *Carbon*, **43**, 786–795 (2005).

[7] P. Serp and J.L. Figueredo, *Carbon Materials for Catalysis*, John Wiley & Sons, Inc., New York (2008).

[8] M. Terrones, Carbon nanotubes: synthesis and properties, electronic devices and other emerging applications, *Int. Mater. Rev.*, **49**, 325–377 (2004).

[9] E. Castillejos and P. Serp, in *Carbon Nanotubes and Related Structures*, D.M. Guldi and N. Martin (eds.), John Wiley & Sons, Inc., New York, 321–347 (2010).

[10] P. Serp, M. Corrias and P. Kalck, Carbon nanotubes and nanofibers in catalysis, *Appl. Catal. A: Gen.*, **253**, 337–358 (2003).

[11] C.H. Liang, Z.L. Li, J.S. Qiu and C. Li, Graphitic nanofilaments as novel support of Ru–Ba catalysts for ammonia synthesis, *J. Catal.*, **211**, 278–282 (2002).

[12] C. Pham-Huu, N. Keller, G. Ehret, L.J. Charbonniere, R. Ziessel and M.J. Ledoux, Carbon nanofiber supported palladium catalyst for liquid-phase reactions: An active and selective catalyst for hydrogenation of cinnamaldehyde into hydrocinnamaldehyde, *J. Mol. Cat. A: Chemical*, **170,** 155–163 (2001).

[13] Y. Wang and J.T.W. Yeow, A review of carbon nanotubes-based gas sensors, *J. Sensors*, 1–25 (2009).

[14] Y. Liu, S. Wu, H. Ju and L. Xu, Amperometric glucose biosensing of gold nanoparticles and carbon nanotube multilayer membranes, *Electroanal.*, **19**, 986–992 (2007).

[15] K. Keren, R.S. Berman, E. Buchstab, U. Sivan and E. Braun, DNA-templated carbon nanotube field-effect transistor, *Science*, **302**,1380–1382 (2003).

[16] G.G. Wildgoose, C.E. Banks and R.G. Compton, Metal nanoparticles and related materials supported on carbon nanotubes: methods and applications, *Small*, **2**, 182–193 (2006).

[17] V. Georgakilas, D. Gournis, V. Tzitzios, L. Pasquato, D.M. Guldi and M. Prato, Decorating carbon nanotubes with metal or semiconductor nanoparticles, *J. Mater. Chem.*, **17**, 2679–2694 (2007).

[18] D. Mann, A. Javey, J. Kong, Q. Wang and H. Dai, Ballistic transport in metallic nanotubes with reliable Pd ohmic contacts, *Nano Lett.*, **3**, 1541–1544 (2003).

[19] A.N. Andriotis, M. Menon, G.E. Froudakis and J.E. Lowther, Tight-binding molecular dynamics study of transition metal carbide clusters, *Chem. Phys. Lett.*, **301**, 503–508 (1999).

[20] Y. Okamoto, Density-functional calculations of icosahedral M_{13} (M = Pt and Au) clusters on graphene sheets and flakes, *Chem. Phys. Lett.*, **420**, 382–386 (2006).

[21] Y. Okamoto, Density-functional calculations of graphene interfaces with Pt (111) and Pt (111)/Ru_{ML} surfaces, *Chem. Phys. Lett.*, **407**, 354–357 (2005).

[22] R. Dreizler and E. Gross, *Density Functional Theory*, Plenum Press, New York (1995).

[23] W. Koch and M.C. Holthausen, *A Chemist's Guide to Density Functional Theory*, Second edition, Wiley-VCH, Weinheim (2002).

[24] A. Maiti and A. Ricca, Metal-nanotube interactions - binding energies and wetting properties, *Chem. Phys. Lett.*, **395**, 7–11 (2004).

[25] M. Menon, A.N. Andriotis and G.E. Froudakis, Curvature dependence of the metal catalyst atom interaction with carbon nanotubes walls, *Chem. Phys. Lett.*, **320**, 425–434 (2000).

[26] G. Chen and Y. Kawazoe, Interaction between a single Pt atom and a carbon nanotube studied by density functional theory, *Phys. Rev. B*, **73**, 125410–125416 (2006).

[27] S.H. Yang, H. Shin, J.W. Lee, S.Y. Kim, S.I. Woo, J.K. Kang, Interaction of a transition metal atom with intrinsic defects in single-walled carbon nanotubes, *J. Phys. Chem. B*, **110**, 13941–13946 (2006).

[28] F. Mercuri and A. Sgamellotti, Functionalization of carbon nanotubes with Vaska's complex: a theoretical approach, *J. Phys. Chem. B*, **110**, 15291–15294 (2006).

[29] E.L. Sceats and J.C. Green, Noncovalent interactions between organometallic metallocene complexes and single-walled carbon nanotubes, *J. Chem. Phys.*, **125**, 154704–154712 (2006).

[30] S. Banerjee and S. Wong, Functionalization of carbon nanotubes with a metal-containing molecular complex, *Nano Lett.*, **2**, 49–53, (2002).

[31] P.M. Ajayan, T.W. Ebbesen, T. Ichihashi, S. Iijima, K. Tanigaki and H. Hiura, Opening carbon nanotubes with oxygen and implications for filling, *Nature*, **362**, 522–524 (1993).

[32] T. Saito, K. Matsushige and K. Tanaka. Chemical treatment and modification of multi-walled carbon nanotubes, *Physica B*, **323**, 280–283 (2002).

[33] B.C. Satishkumar, A. Govindaraj, J. Mofokeng, G.N. Subbanna and C.N.R Rao, Novel experiments with carbon nanotubes: opening, filling, closing and functionalizing nanotubes, *J Phys B: A, Mol. Opt. Phys.*, **29**, 4925–4934 (1996).

[34] I.D. Rosca, F. Watari, M. Uo and T. Akasaka, Oxidation of multiwalled carbon nanotubes by nitric acid, *Carbon*, **43**, 3124–3131 (2005).

[35] L. Zhang, V.U. Kiny, H. Peng, J. Zhu, R.F.M. Lobo, J.L. Margrave and V.N. Khabashesku, Sidewall functionalization of single-walled carbon nanotubes with hydroxyl group-terminated moieties, *Chem. Mater.*, **16**, 2055–2060 (2004).

[36] K. Behler, S. Osswald, H. Ye, S. Dimovski and Y. Gogotsi, Effect of thermal treatment on the structure of multiwall carbon nanotubes, *J. Nanopart. Res.*, **8**, 615–625 (2006).

[37] G.S Duisberg, R. Graupner, P. Downes, A. Minett, L. Ley, S. Roth and N. Nicoloso, Hydrothermal functionalisation of single-walled carbon nanotubes, *Synth. Met.*, **142**, 263–266 (2004).

[38] M.J. O'Connell, P. Boul, L.M. Ericson, C. Huffman, Y.H. Wang, E. Haroz, C. Kuper, J. Tour, K.D. Ausman and R.E. Smalley, Reversible water-solubilization of single-walled carbon nanotubes by polymer wrapping, *Chem. Phys. Lett.*, **342**, 265–271 (2001).

[39] M.S. Raghuveer, S. Agrawal, N. Bishop and G. Ramanath, Microwave-assisted single-step functionalization and in situ derivatization of carbon nanotubes with gold nanoparticles, *Chem. Mater.*, **18**, 1390–1393 (2006).

[40] F.G. Brunetti, M.A. Herrero, J. de M. Muoz, A. Daz-Ortiz, J. Alfonsi, M. Meneghetti, M. Prato and E. Vazquez, Microwave-induced multiple functionalization of carbon nanotubes, *J. Am. Chem. Soc.*, **130**, 8094–8100 (2008).

[41] M. Grujicic, G. Cao, A.M. Rao, T.M. Tritt and S. Nayak, UV-light enhanced oxidation of carbon nanotubes, *Appl. Sur. Sci.*, **214**, 289–303 (2003).

[42] A. Solhy, B.F. Machado, J. Beausoleil, Y. Kihn, F. Goncalves, M.F.R. Pereira, J.J.M. Orfao, J.L. Figueiredo, J.L. Faria and P. Serp, MWCNT activation and its influence on the catalytic performance of Pt/MWCNT catalysts for selective hydrogenation, *Carbon*, **46**, 1194–1207 (2009).

[43] T.W. Ebbesen, H. Hiura, M.E. Bisher, M.M.J. Treacy, J.L. Shreeve-Keyer and R.C. Haushalter, Decoration of carbon nanotubes, *Adv. Mater.*, **8**, 155–157 (1996).

[44] B.C. Satishkumar, E.M. Vogl, A. Govindaraj and C.N.R. Rao, The decoration of carbon nanotubes by metal nanoparticles, *J. Phys. D Appl. Phys.*, **29**, 3173–3176 (1996).

[45] J.H. Bitter, M.K. van der Lee, A.G.T. Slotboom, A.J. van Dillen and K.P. de Jong, Synthesis of highly loaded highly dispersed nickel on carbon nanofibers by homogeneous deposition-precipitation, *Catal. Lett.*, **89**, 139–142 (2003).

[46] P. Ayala, F.L. Freire, Jr., L. Gu, David J. Smith, I.G. Solórzano, D.W. Macedo, J.B. Vander Sande, H. Terrones, J. Rodriguez-Manzo and M. Terrones, Decorating carbon nanotubes with nanostructured nickel particles via chemical methods, *Chem. Phys. Lett.*, **431**, 104–109 (2006).

[47] X.-W. Wei, J. Xu, X.-J. Song, Y.-H. Ni, P. Zhang, C.-J. Xia, G.-C. Zhao and Z.-S. Yang, Multi-walled carbon nanotubes coated with rare earth fluoride EuF_3 and TbF_3 nanoparticles, *Mater. Res. Bull.*, **41**, 92–98 (2006).

[48] X.-W. Wei, J. Xu, X.-J. Song, Y.H. Ni and P. Zhang, Coating multi-walled carbon nanotubes with metal sulfides, *Mater. Chem. Phys.*, **92**, 159–163 (2005).

[49] M. Bottini, L. Tautz, H. Huynh, E. Monosov, N. Bottini, M.I. Dawson, S. Bellucci and T. Mustelin, Covalent decoration of multi-walled carbon nanotubes with silica nano particles, *Chem. Commun.*, 758–760 (2005).

[50] S. Hussein, S. Zein and A.R. Boccaccini, Synthesis and Characterization of TiO_2 coated multiwalled carbon nanotubes using a sol-gel method, *Ind. Eng. Chem. Res.*, **47**, 6598–6606 (2008).

[51] Z. Sun, X. Zhang, N. Na, Z. Liu, B. Han and G. An, Synthesis of ZrO_2-carbon nanotube composites and their application as chemiluminescent sensor material for ethanol, *J. Phys. Chem. B.*, **110**, 13410–13414 (2006).

[52] Y. Shan and L. Gao, Synthesis and characterization of phase controllable ZrO_2–carbon nanotube nanocomposites, *Nanotechnol.*, **16**, 625–630 (2005).

[53] Y.-S. Luo, Q.-F. Ren, J.-F. Li, Z.-J. Jia, Q.-R. Dai, Y. Zhang and B.-H. Yu, Synthesis and optical properties of multiwalled carbon nanotubes beaded with Cu_2O nanospheres, *Nanotechnol.*, **17**, 5836–5840 (2006).

[54] W. Ma and D. Tian, Direct electron transfer and electrocatalysis of hemoglobin in ZnO coated multiwalled carbon nanotubes and Nafion composite matrix, *Bioelectrochem.*, **78**, 106–112 (2010).

[55] Y.-T. Kim and T. Mitani, Surface thiolation of carbon nanotubes as supports: A promising route for the high dispersion of Pt nanoparticles for electrocatalysts, *J. Catal.*, **238**, 394–401 (2006).

[56] Z. Liu, J.Y. Lee, W. Chen, M. Han and L.M. Gan, Physical and electrochemical characterizations of microwave-assisted polyol preparation of carbon-supported PtRu nanoparticles, *Langmuir*, **20**, 181–187 (2004).

[57] W.X. Chen, J.Y. Lee and Z. Liu, Microwave-assisted synthesis of carbon supported Pt nanoparticles for fuel cell applications, *Chem. Commun.*, 2588–2589 (2002).

[58] A. Miyazaki, I. Balint, K.I. Aika and Y. Nakano, Preparation of Ru nanoparticles supported on γ-Al_2O_3 and its novel catalytic activity for ammonia synthesis, *J. Catal.*, **204**, 364–371 (2001).

[59] J. Lu, Effect of surface modifications on the decoration of multi-walled carbon nanotubes with ruthenium nanoparticles, *Carbon*, **45**, 1599–1605 (2007).

[60] A. Kongkanand, K. Vinodgopal, S. Kuwabata and P.V. Kamat, Highly dispersed Pt catalysts on single-walled carbon nanotubes and their role in methanol oxidation, *J. Phys. Chem. B*, **110**, 16185–16191 (2006).

[61] R. Giordano, P. Serp, P. Kalck, Y. Kihn, J. Schreiber, C. Marhic and J.-L Duvail, Preparation of rhodium catalysts supported on carbon nanotubes by a surface mediated organometallic reaction, *Eur. J. Inorg. Chem.*, 610–617 (2003).

[62] E. Castillejos, R. Chico, R. Bacsa, S. Coco, P. Espinet, M. Perez-Cadenas, A. Guerrero-Ruiz, I. Rodríguez-Ramos and P. Serp, Selective deposition of gold nanoparticles on or inside carbon nanotubes and their catalytic activity for preferential oxidation of CO, *Eur. J. Inorg. Chem.*, 5096–5102 (2010).

[63] E.J. O'Sullivan, Fundamental and practical aspects of the electroless deposition reaction, *Adv. Electrochem. Sci. Eng.*, **7**, 225–273 (2001).

[64] L.A. Porter, Jr., H.C. Choi, A.E. Ribbe and J.M. Buriak, Controlled electroless deposition of noble metal nanoparticle films on germanium surfaces, *Nano Lett.*, **2**, 1067–1071 (2002).

[65] F. Wang, S. Arai and M. Endo, The preparation of multi-walled carbon nanotubes with a Ni–P coating by an electroless deposition process, *Carbon*, **43**, 1716–1721 (2005).

[66] F. Wang, S. Arai, K.C. Park, K. Takeuchi, Y.J. Kim and M. Endo, Carbon with high thermal conductivity, prepared from ribbon-shaped mesophase pitch-based fibers, *Carbon*, **44**, 1298–1352 (2006).

[67] X. Ma, X. Li, N. Lun and S. Wen, Synthesis of gold nano-catalysts supported on carbon nanotubes by using electroless plating technique, *Mater. Chem. Phys.*, **97**, 351–356 (2006).

[68] H. C. Choi, M. Shim, S. Bangsaruntip and H. Dai, Spontaneous reduction of metal ions on the sidewalls of carbon nanotubes, *J. Am. Chem. Soc.*, **124**, 9058–9059 (2002).

[69] G. Schmid, *Nanoparticles*, Wiley-VCH, Weinheim (2004).

[70] V.M. Rotello, *Nanoparticles: Building Blocks for Nanotechnology*, Kluwer Academic/Plenum Publishers, New York (2004).

[71] K.J. Klaubunde, *Nanoscale Materials in Chemistry*, John Wiley & Sons, Inc., New York, (2001).

[72] K. Jiang, A. Eitan, L.S. Schadler, P.M. Ajayan, R.W. Siegel, N. Grobert, M. Mayne, M. Reyes-Reyes, H. Terrones and M. Terrones, Selective attachment of gold nanoparticles to nitrogen-doped carbon nanotubes, *Nano Lett.*, **3**, 275–277 (2003).

[73] X. Hu, T. Wang, X. Qu and S. Dong, In situ synthesis and characterization of multiwalled carbon Nanotube/Au nanoparticle composite materials, *J. Phys. Chem. B*, **110**, 853–857 (2006).

[74] S. Fullam, D. Cottell, H. Rensmo and D. Fitzmaurice, Carbon nanotube templated self-assembly and thermal processing of gold nanowires, *Adv. Mater.*, **12**, 1430–1432 (2000).

[75] L. Han, W. Wu, F.L. Kirk, J. Luo, M.M. Maye, N.N. Kariuki, Y. Lin, C. Wang and C.-J. Zhong, A direct route toward assembly of nanoparticle-carbon nanotube composite materials, *Langmuir*, **20**, 6019–6025 (2004).

[76] R. Zanella, E. Basiuk, P. Santiago, V.A. Basiuk, E. Mireles, I. Puente-Lee and J.M. Saniger, Deposition of gold nanoparticles onto thiol-functionalized multiwalled carbon nanotubes, *J. Phys. Chem. B*, **109**, 16290–16295 (2005).

[77] N. Chopra, M. Majumder and B.J. Hinds, Bifunctional carbon nanotubes by sidewall protection, *Adv. Funct. Mater.*, **15**, 858–864 (2005).

[78] T. Sainsbury and D. Fitzmaurice, Templated assembly of semiconductor and insulator nanoparticles at the surface of covalently modified multiwalled carbon nanotubes, *Chem. Mater.*, **16**, 3780–3790 (2004); T. Sainsbury and D. Fitzmaurice, Carbon-nanotube-templated and pseudorotaxane-formation-driven gold nanowire self-assembly, *Chem. Mater.*, **16**, 2174–2179 (2004).

[79] Q. Li, B. Sun, I.A. Kinloch, D. Zhi, H. Sirringhaus and A.H. Windle, Enhanced self-assembly of pyridine-capped CdSe nanocrystals on individual single-walled carbon nanotubes, *Chem. Mater.*, **18**, 164–168 (2006).

[80] C. Engtrakul, Y.-H. Kim, J.M. Nedeljkovic, S.P. Ahrenkiel, K.E.H. Gilbert, J.L. Alleman, S.B. Zhang, O.I. Micic, A.J. Nozik and M.J. Heben, Self-assembly of linear arrays of semicon-ductor nanoparticles on carbon single-walled nanotubes, *J. Phys. Chem. B*, **110**, 25153–25157 (2006).

[81] B. Pan, D. Cui, R. He, F. Gao and Y. Zhang, Covalent attachment of quantum dot on carbon nanotubes, *Chem. Phys. Lett.*, **417**, 419–424 (2006).

[82] X. Lou, C. Detrembleur, C. Pagnoulle, R. Jerome, V. Bocharova, A. Kiriy and M. Stamm, Surface modification of multiwalled carbon nanotubes by poly(2-vinylpyridine): dispersion, selective deposition, and decoration of the nanotubes, *Adv. Mater.*, **16**, 2123–2127 (2004).

[83] M. Olek, M. Hilgendorff and M. Giersig, A simple route for the attachment of colloidal nanocrystals to noncovalently modified multiwalled carbon nanotubes, *Colloids Surf. A*, **292**, 83–85 (2007).

[84] H. Ago, M.S.P. Shaffer, D.S. Ginger, A.H. Windle and R.H. Friend, Electronic interaction between photoexcited poly(*p*-phenylene vinylene) and carbon nanotubes, *Phys. Rev. B*, **61**, 2286–2290 (2000).

[85] L. Cao, H.Z. Chen, M. Wang, J.Z. Sun, X.B. Zhang and F.Z. Kong, Photoconductivity Study of modified carbon nanotube/oxotitanium phthalocyanine composites, *J. Phys. Chem. B*, **106**, 8971–8975 (2002).

[86] L. Cao, H.Z. Chen, H.B. Zhou, L. Zhu, J.Z. Sun, X.B. Zhang, J.M. Xu and M. Wang, Carbon-nanotube-templated assembly of rare-earth phthalocyanine nanowires, *Adv. Mater.*, **15**, 909–913 (2003).

[87] H. Ago, T. Kugler, F. Cacialli, W.R. Salaneck, M.S.P. Shaffer, A.H. Windle and R.H. Friend, Work functions and surface functional groups of multiwall carbon nanotubes, *J. Phys. Chem. B*, **103**, 8116–8121 (1999).

[88] H. Ago, T. Kugler, F. Cacialli, K. Petritsch, R.H. Friend, W.R. Salaneck, Y. Ono, T. Yamabe and K. Tanaka, Workfunction of purified and oxidised carbon nanotubes, *Synth. Met.*, **103**, 2494–2495 (1999).

[89] M. Olek, T. Büsgen, M. Hilgendorff, and M. Giersig, Quantum dot modified multiwall carbon nanotubes, *J. Phys.Chem B*, **110**, 12901–12904 (2006).

[90] V. Tzitzios, V. Georgakilas, E. Oikonomou, M. Karakassides and D. Petridis, Synthesis and characterization of carbon nanotube/metal nanoparticle composites well dispersed in organic media, *Carbon*, **44**, 848–853 (2006).

[91] D.S. Kim, T. Lee and K.E. Geckler, Hole-doped single-walled carbon nanotubes: ornamenting with gold nanoparticles in water, *Angew. Chem. Int. Ed.*, **45**, 104–107 (2006).

[92] K. Lee, M. Kim, J. Hahn, J. Suh, I. Lee, K. Kim and S. Han, Assembly of metal nanoparticle-carbon nanotube composite materials at the liquid/liquid interface, *Langmuir*, **22**, 1817–1821 (2006).

[93] W. Huang, H. Chen and J.-M. Zuo, One-dimensional self-assembly of metallic nanostructures on single-walled carbon-nanotube bundles, *Small*, **2**, 1418–1421 (2006).

[94] J. Sun, L. Gao and M. Iwasa, Noncovalent attachment of oxide nanoparticles onto carbon nanotubes using water-in-oil micro emulsions, *Chem. Commun.*, 832–833 (2004).

[95] B. Yoon and C. Ma, Microemulsion-templated synthesis of carbon nanotube-supported Pd and Rh Nanoparticles for catalytic applications, *J. Am. Chem. Soc.*, **127**, 17174–17175 (2005).

[96] B. Yoon, H. Kim and C.M. Wai, Dispersing palladium nanoparticles using a water-in-oil microemulsion-homogenization of heterogeneous catalysis, *Chem. Commun.*, 1040–1041 (2003).

[97] R.P. Bagwe, C. Yang, L.R. Hilliar and W. Tan, Optimization of dye-doped silica nanoparticles prepared using a reverse microemulsion method, *Langmuir*, **20**, 8336–8342 (2004).

[98] H.H. Yang, S.Q. Zhang, X.L. Chen, Z.X. Zhuang, J.G. Xu and X.R. Wang, Magnetite-containing spherical silica nanoparticles for biocatalysis and bioseparations, *Anal. Chem.*, **76**, 1316–1321 (2004).

[99] S. Gorer, H. Liu, B. Stiger and R.M. Penner, in: *Handbook of Metal Nanoparticles*, D. Feldheim and C. Foss (eds.), Marcel Dekker, New York (2000).

[100] H. Liu, F. Favier, K. Ng, M.P. Zach and R.M. Penner, Size-selective electro-deposition of meso-scale metal particles: a general method, *Electrochim. Acta*, **47**, 671–677 (2001).

[101] R.H. Baughman, C. Cui, A.A Zakhidov, Z. Iqbal, J.N. Barisci, G.M. Spinks, G.G. Wallace, A. Mazzoldi, D. De Rossi, A.G. Rinzler, O. Jaschinski, S. Roth and M. Kertesz, Carbon nanotube actuators, *Science*, **284**, 1340–1344 (1999).

[102] U. Vohrer, I. Kolaric, M.H. Haque, S. Roth and U. Detlaff-Weglikowska, Carbon nanotube sheets for the use as artificial muscles, *Carbon*, **42**, 1159–1164 (2004).

[103] Z.F. Ren, Z.P. Huang, J.W. Xu, J.H. Wang, P. Bush, M.P. Siegal and P.N. Provencio, Synthesis of large arrays of well-aligned carbon nanotubes on glass, *Science*, **282**, 1105–1107 (1998).

[104] T.M. Day, P.R. Unwin and J.V. Macpherson, Factors controlling the electrodeposition of metal nanoparticles on pristine single walled carbon nanotubes, *Nano Lett.*, **7**, 51–57 (2007).

[105] Z. He, J. Chen, D. Liu, H. Zhou and Y. Kuang, Electrodeposition of Pt–Ru nanoparticles on carbon nanotubes and their electrocatalytic properties for methanol electro oxidation, *Diamond Relat. Mater.*, **13**, 1764–1769 (2004).

[106] J. Kong, H.T. Soh, A.M. Cassell, C.F. Quate and H. Dai, Synthesis of individual single-walled carbon nanotubes on patterned silicon wafers, *Nature*, **395**, 878–881 (1998).

[107] B.M. Quinn and S.G. Lemay, Single-walled carbon nanotubes as templates and interconnects for nanoelectrodes, *Adv. Mater.*, **18**, 855–859 (2006).

[108] H. Tang, J.H. Chen, Z.P. Huang, D.Z. Wang, Z.F. Ren, L.H. Nie, Y.F. Kuang and S.Z. Yao, High dispersion and electrocatalytic properties of platinum on well-aligned carbon nanotube arrays, *Carbon*, **42**, 191–197 (2004).

[109] Y. Zhao, Y.E.L. Fan, Y. Qiu and S. Yang, A new route for the electrodeposition of platinum–nickel alloy nanoparticles on multi-walled carbon nanotubes, *Electrochim. Acta*, **52**, 5873–5878 (2007).

[110] D.-J. Guo and H.-L. Li, High dispersion and electrocatalytic properties of Pd nanoparticles on single walled carbon nanotubes, *J. Colloid Interface Sci.*, **286**, 274–279 (2005).

[111] H.-F. Cui, J.-S. Ye, W.-D. Zhang, J. Wang and F.-S. Sheu, Electrocatalytic reduction of oxygen by a platinum nanoparticle/carbon nanotube composite electrode, *J. Electroanal. Chem.*, **577**, 295–302 (2005).

[112] B.M. Praveen, T.V. Venkatesha, Y.A. Naik and K. Prashantha, Corrosion studies of carbon nanotubes–Zn composite coating, *Surf. Coat. Tech.*, **201**, 5836–5842 (2007).

[113] Z. Wang, J. Yuan, M. Li, D. Han, Y. Zhang, Y. Shen, L. Niu and A. Ivaska, Electropolymerization and catalysis of well-dispersed polyaniline/carbon nanotube/gold composite, *J. Electroanal. Chem.*, **599**, 121–126 (2007).

[114] T.M. Day, P.R. Unwin, N.R. Wilson and J.V. Macpherson, Electrochemical templating of metal nanoparticles and nanowires on single-walled carbon nanotube networks, *J. Am. Chem. Soc.*, **127**, 10639–10647 (2005).

[115] T.M. Day, N.R. Wilson and J.V. Macpherson, Electrochemical and conductivity measurements of single-wall carbon nanotube network electrodes, *J. Am. Chem.Soc.*, **126**, 16724–16725 (2004).

214 *Carbon Meta-Nanotubes*

[116] J.-C. Bradley, S. Babu and P. Ndungu, Contactless tip-selective electrodeposition of palladium onto carbon nanotubes and nanofibers, *Fuller. Nanotubes Carbon Nanostruct.*, **13**, 227–237 (2005).

[117] Z. Deng, E. Yenilmez, J. Leu, J.E. Hoffman, E.W.J. Straver, H. Dai and K.A. Moler, Metal-coated carbon nanotube tips for magnetic force microscopy, *Appl. Phys. Lett.*, **85**, 6263–6267 (2004).

[118] C. Zhou, J. Kong and H. Dai, Intrinsic electrical properties of individual single-walled carbon nanotubes with small band gaps, *Phys. Rev. Lett.*, **84**, 5604–5607 (2000).

[119] Y. Zhang, N.W. Franklin, R.J. Chen and H. Dai, Metal coating on suspended carbon nanotubes and its implication to metal–tube interaction, *Chem. Phys. Lett.*, **331**, 35–41 (2000).

[120] T. Tachibana, B. Williams and J. Glass, Correlation of the electrical properties of metal contacts on diamond films with the chemical nature of the metal-diamond interface. II, titanium contacts: A carbide-forming metal, *Phys. Rev. B*, **45**, 11975–11981 (1992).

[121] Y. Min, E.J. Bae, S. Jeong, Y.J. Cho, J. Lee, W.B. Choi and G. Park, Ruthenium oxide nanotube arrays fabricated by atomic layer deposition using a carbon nanotube template, *Adv. Mater.*, **15**, 1019–1022 (2003).

[122] J.S. Lee, B. Min, K. Cho, S. Kim, J. Park, Y.T. Lee, N.S. Kim, M.S. Lee, S.O. Park and J.T. Moon, Al_2O_3 nanotubes and nanorods fabricated by coating and filling of carbon nanotubes with atomic-layer deposition, *J. Cryst. Growth*, **254**, 443–448 (2003).

[123] C.F. Hermann, F.H. Fabreguette, D.S. Finch, R. Geiss and S.M. George, Multilayer and functional coatings on carbon nanotubes using atomic layer deposition, *Appl. Phys. Lett.*, **87**, 123110–123114 (2005).

[124] C.F. Herrmann, F.W. DelRio, V.M. Bright and S.M. George, Conformal hydrophobic coatings prepared using atomic layer deposition of seed layers and non-chlorinated hydrophobic precursors, *J. Micromech. Microeng.*, **15**, 984–992 (2005).

[125] P. Serp, R. Feurer, Y. Kihn, P. Kalck, J.L. Faria and J.L. Figueiredo, Novel carbon supported material: highly dispersed platinum particles on carbon nanospheres, *J. Mater. Chem.*, **11**, 1980–1981 (2001).

[126] W. Xia, O.F.-K. Schlüter, C. Liang, M.W.E. van den Berg, M. Guraya and M. Muhler, The synthesis of structured Pd/C hydrogenation catalysts by the chemical vapor deposition of Pd(allyl)Cp onto functionalized carbon nanotubes anchored to vapor grown carbon microfibers, *Catal. Today*, **102–103**, 34–39 (2005).

[127] M. Saha, R. Li and X. Sun, High loading and monodispersed Pt nanoparticles on multiwalled carbon nanotubes for high performance proton exchange membrane fuel cells, *J. Power Sources*, **177**, 314–322 (2008).

[128] C. Taschner, K. Biedermann, A. Leonhardt, B. Buchner, T. Gemming and K. Wetzig, Decorating multiwalled carbon nanotubes with copper particles using MOCVD, *Electrochem. Soc. Proc.*, 396–401 (2005).

[129] J.D. Kim, B.S. Kang, T.W. Nob, J.-G. Yoon, S.I. Baik and Y.-W. Kim, Controlling the nanostructure of RuO_2/carbon nanotube composite by gas annealing, *J. Electrochem. Soc.*, **152**, D23–D25 (2005).

[130] S. Orlanducci, V. Sessa, M.L. Terranova, G.A. Battiston, S. Battiston and R. Gerbasi, TiO_2 on single walled carbon nanotube arrays: Towards the assembly of organized C/TiO_2 nanosystems, *Carbon*, **44**, 2839–2843 (2006).

[131] Q. Kuang, S.F. Li, Z.X. Xie, S.C. Lin, X.H. Zhang, S.Y. Xie, R.-B. Huang and L.-S. Zheng, Controlable fabrication of SnO_2-coated multiwalled carbon nanotubes by chemical vapor deposition, *Carbon*, **44**, 1166–1172 (2006).

[132] C.Y. Su, W.Y. Juang, Y.L. Chen, K.C. Leou and C.H. Tsai, The field emission characteristics of carbon nanotubes coated by boron nitride film, *Diamond Relat. Mater.*, **16**, 1393–1397 (2007).

[133] X. Pan, Z. Fan, W. Chen, Y. Ding, H. Luo and X. Bao, Enhanced ethanol production inside carbon-nanotube reactors containing catalytic particles, *Nature Mater.*, **6**, 507–511 (2007).

[134] J.-P. Tessonnier, L. Pesant, G. Ehret, M.J. Ledoux and C. Pham-Huu, Pd nanoparticles introduced inside multi-walled carbon nanotubes for selective hydrogenation of cinnamaldehyde into hydrocinnamaldehyde, *App. Catal. A*, **288**, 203–210 (2005).

[135] X.R. Ye, Y.H. Lin, C.M. Wang, M.H. Engelhard, Y. Wang and C.M. Wai, Supercritical fluid synthesis and characterization of catalytic metal nanoparticles on carbon nanotubes, **J.** *Mater. Chem.*, **14**, 908–913 (2004).

[136] E. Castillejos, P.-J. Debouttière, L. Roiban, A. Solhy, V. Martinez, Y. Kihn, O. Ersen, K. Philippot, B. Chaudret and P. Serp, An efficient strategy to drive nanoparticles into carbon nanotubes and the remarkable effect of confinement on their catalytic performance, *Angew. Chem. Internat. Ed.*, **48**, 2529–2533 (2009).

[137] J.-P. Tessonnier, O. Ersen, G. Weinberg, C. Pham-Huu, D.S. Su and R. Schlögl, Selective deposition of metal nanoparticles inside or outside multiwalled carbon nanotubes, *ACS Nano*, **3**, 2081–2089 (2009).

[138] R.M. Lago, S.C. Tsang, K.L. Lu, Y.K. Chen and M.L.H. Green, Filling carbon nanotubes with small palladium metal crystallites: the effect of surface acid groups, *Chem. Commun.*, 1355–1357 (1995).

[139] D.D. Perrin, in *Stability Constants of Metal Complexes*, IUPAC Chemical Series, **22**, Pergamon, New York (1979).

[140] Y. Fan, M. Burghard and K. Kern, Chemical defect decoration of carbon nanotubes, *Adv. Mater.*, **14**, 130–133 (2002).

[141] B. Kim and W.M. Sigmund, Functionalized multiwall carbon nanotube/gold nanoparticle composites, *Langmuir*, **20**, 8239–8242 (2004).

[142] X. Li, J. Niu, J. Zhang, H. Li and Z. Liu, Labeling the defects of single-walled carbon nanotubes using titanium dioxide nanoparticles, *J. Phys.Chem. B*, **107**, 2453–2458 (2003).

[143] F. Winter, G.L. Bezemer, C. van der Spek, J.D. Meeldijk, A.J. van Dillen, J.W. Geus and K.P. de Jong, TEM and XPS studies to reveal the presence of cobalt and palladium particles in the inner core of carbon nanofibers, *Carbon*, **43**, 327–332 (2005).

[144] A.J. Koster, U. Ziese, A.J. Verkleij, A.H. Janssen and K.P. de Jong, Three-dimensional transmission electron microscopy: A novel imaging and characterization technique with nanometer scale resolution for materials science, *J. Phys.Chem.*, **104**, 9368–9370 (2000).

[145] P.A. Midgeley and M. Weyland, 3D electron microscopy in the physical sciences: the development of Z-contrast and EFTEM tomography, *Ultramicrosopy*, **96**, 413–431 (2003).

[146] O. Ersen, J. Werckmann, M. Houlle, M.-J. Ledoux and C. Pham-Huu, Electron microscopy study of metal particles inside multiwalled carbon nanotubes, *Nano Lett.*, **7**, 1898–1907 (2007).

[147] J. Ma, C. Park, N.M. Rodriguez and R.T.K. Baker, Characteristics of copper particles supported on various types of graphite nanofibers, *J. Phys. Chem. B*, **105**, 11994–12002 (2001).

[148] J. Qiua, H. Zhang, X. Wang, H. Han, C. Liang and C. Li, Selective hydrogenation of cinnamaldehyde over carbon nanotube supported Pd-Ru catalyst, *React. Kinet. Catal. Lett.*, **88**, 269–275 (2006).

[149] R.V. Hull, L. Li, Y. Xing and C.C. Chusuei, Pt nanoparticle binding on multiwall carbon nanotubes, *Chem. Mater.*, **18**, 1780–1788 (2006).

[150] B.I. Rosario-Castro, E.J. Contès, M.E. Perez-Davis and C.R. Cabrera, Attachment of single wall carbon nanotubes on platinum surfaces by self assembling techniques, *Rev. Adv. Mater. Sci.*, **10**, 381–386 (2005).

[151] J.F. Watts and J. Wolstenholme, in *An Introduction to Surface Analysis by XPS and AES*, John Wiley & Sons, Inc., New York (2003).

[152] B.V. Crist, in *Handbooks of Monochromatic XPS Spectra, 1–5*, XPS International LLC, Mountain View, CA, USA (2004).

[153] J.T. Grant and D. Briggs, *Surface Analysis by Auger and X-ray Photoelectron Spectroscopy*, IM Publications (2003).

[154] U.K. Chichester, D. Koningsberger, F.B.M. van Zon, M. Vaarkamp and A. Munozpaez, *X-ray absorption fine structure for Catalysts and Surfaces*, Y. Iwasawa (ed.), Tokyo, 257 (1995).

[155] Y.-T. Kim, K. Ohshima, K. Higashimine, T. Uruga, M. Takata, H. Suematsu and T. Mitani, Fine size control of platinum on carbon nanotubes: from single atoms to clusters, *Ang. Chem., Int. Ed.*, **45**, 407–411 (2006).

[156] D.-Q. Yang, B. Hennequin and E. Sacher, XPS demonstration of - Interaction between benzyl mercaptan and multiwalled carbon nanotubes and their use in the adhesion of Pt nanoparticles, *Chem. Mater.*, **18**, 5033–5038 (2006).

[157] M. Vaarkamp, F.S. Modica and D.C. Konigsberger, Influence of hydrogen pretreatment on the structure of the metal-support interface in Pt/Zeolite catalysts, *J. Catal.*, **144**, 611–626 (1993).

[158] Y. Zhang, M.L. Toebes, A. van der Eerden, W.E. O'Grady, K.P. de Jong and D.C. Koningsberger, Metal particle size and structure of the metal-support interface of carbon-supported platinum catalysts as determined with EXAFS spectroscopy, *J. Phys. Chem. B*, **108**, 18509–18519 (2004).

[159] P. Zhang, X. Zhou, Y. Tang and T.K. Sham, Organosulfur-functionalized Au, Pd, and Au-Pd Nanoparticles on 1D silicon nanowire substrates: Preparation and EXAFS studies, *Langmuir*, **21**, 8502–8508 (2005).

[160] I. Coulthard and T.K. Sham, Novel preparation of noble metal nanostructures utilizing porous silicon, *Solid State Commun.*, **105**, 751–754 (1998).

[161] J. Zhou, X. Zhou, X. Sun, R. Li, M. Murphy, Z. Ding, X. Sun and T.-K. Sham, Interaction between Pt nanoparticles and carbon nanotubes – An X-ray absorption near edge structures (XANES) study, *Chem. Phys. Lett.*, **437**, 229–232 (2007).

[162] B.W. Reed and M. Sarikaya, TEM/EELS analysis of heat treated carbon nanotubes: experimental techniques, *J. Electron Microsc.*, **51**, S97–S105 (2002).

[163] M. Rühle, T. Seeger, P. Redlich, N. Grobert, M. Terrones, D.R.M. Walton and H.W. Kroto, Novel SiOx-coated carbon nanotubes, *J. Cer. Process. Res.*, **3**, 1–5 (2002).

[164] V. Stolojan, S.R.P. Silva, M.J. Goringe, R.L.D. Whitby, W.K. Hsu, D.R.M. Walton and H.W. Kroto, Dielectric properties of WS$_2$-coated multiwalled carbon nanotubes studied by energy-loss spectroscopic profiling, *Appl. Phys. Lett.*, **86**, 063112–063116 (2005).

[165] G. Singh, S. Priya, M.R. Hossu, S.R. Shah, S. Grover, A.R. Koymen and R.L. Mahajan, Synthesis, electrical and magnetic characterization of core–shell silicon carbo-nitride coated carbon nanotubes, *Mater. Lett.*, **63**, 2435–2438 (2009).

[166] Z. Zhu, N. Minami, S. Kazaoui and Y. Kim, π-Chromophore-functionalized SWCNTs by covalent bonding: substantial change in the optical spectra proving strong electronic interaction, *J. Mater. Chem.*, **14**, 1924–1926 (2004).

[167] F. Frehill, M.I. Panhuis, N.A. Young, W. Henry, J. Hjelm and J.G. Vos, Microscopy and spectroscopy of interactions between metallopolymers and carbon nanotubes, *J. Phys. Chem. B*, **109**, 13205–13209 (2005).

[168] R.R. Bacsa, C. Laurent, R. Morishima, H. Suzuki and M. Le Lay, Hydrogen storage in high surface area carbon nanotubes prepared by catalytic chemical vapor deposition, *J. Phys. Chem. B*, **108**, 12718–12723 (2004).

[169] J. Ma, C. Park, N.M. Rodriguez and R.T.K. Baker, Characteristics of copper particles supported on various types of graphite nanofibers, *J. Phys. Chem. B*, **105**, 11994–12002 (2001).

[170] J. Li and H.T. Ng, Carbon nanotube sensors, In *Encyclopedia of Nanoscience and Nanotechnology*, **1**, American Scientific Publishers, 591–601 (2004).

[171] N. Sinha, J. Ma and J.T.W. Yeow, Carbon nanotube-based sensors, *J. Nanosci. Nanotech.*, **6**, 573–590 (2006).

[172] M. Terrones, Science and technology of the twenty-first century: synthesis, properties, and applications of carbon nanotubes, *Annu. Rev. Mater. Res.*, **33**, 419–501 (2003).

[173] P. Boul, J. Liu, E. Mickelson, C. Huffman, L. Ericson, I. Chiang, K. Smith, D. Colbert, R. Hauge, J. Margrave and R. Smalley, Reversible sidewall functionalization of buckytubes, *Chem. Phys. Lett.*, **310**, 367–371 (1999).

[174] K. Besteman, J.O. Lee, F.G.M. Wiertz, H.A. Heering and C. Dekker, Enzyme-coated carbon nanotubes as single-molecule biosensors, *Nano Lett.*, **3**, 727–730 (2003).

[175] M. Shim, N. Wong, S. Kam, R.J. Chen, Y. Li and H. Dai, Functionalization of carbon nanotubes for biocompatibility and biomolecular recognition, *Nano Lett.*, **2**, 285–288 (2002).

[176] Y.X. Liang, Y.J. Chen and T.H. Wang, Low-resistance gas sensors fabricated from multiwalled carbon nanotubes coated with a thin tin oxide layer, *Appl. Phys. Lett.*, **85**, 666–669 (2004).

[177] Q. Zhao, M.B. Nardelli, W. Lu and J. Bernholc, Carbon nanotube-metal cluster composites: a new road to chemical sensors?, *Nano Lett.*, **5**, 847–851 (2005).

[178] J. Kong, M.G. Chapline and H. Dai, Functionalized carbon nanotubes for molecular hydrogen sensors, *Adv. Mater.*, **13**, 1384–1386 (2001).

[179] M. Yang, Y. Yang, Y. Liu, G. Shen and R. Yu, Platinum nanoparticles-doped sol–gel/carbon nanotubes composite electrochemical sensors and biosensors, *Biosens. Bioelectron.* **21**, 1125–1131 (2006).

[180] Y.L. Liu, H. Yang, U. Yang, Z. Liu, G. Shen and R. Yu, Gas sensing properties of tin dioxide coated onto multi-walled carbon nanotubes, *Thin Solid Films*, **497**, 355–360 (2006).

[181] C. Staii, A.T. Johnson Jr., M. Chen and A. Gelperin, DNA-Decorated carbon nanotubes for chemical sensing, *Nano Lett.*, **5**, 1774–1777 (2005).

[182] C. Park and R.T.K. Baker, Catalytic behavior of graphite nanofiber supported nickel particles. 3. The effect of chemical blocking on the performance of the system, *J. Phys. Chem. B*, **103**, 2453–2460 (1999).

[183] A. Chambers, T. Nemes, N.M. Rodriguez and R.T.K. Baker, Catalytic behavior of graphite nanofiber supported nickel particles. 1. Comparison with other support media, *J. Phys. Chem. B*, **102**, 2251–2261 (1998).

[184] C. Park and R.T.K. Baker, Catalytic behavior of graphite nanofiber supported nickel particles. 2. The influence of the nanofiber structure, *J. Phys. Chem. B*, **102**, 5168–5175 (1998).

[185] W. Xia, O.F.K. Schülter, C. Liang, M.W.E. van den Berg, M. Guraya and M. Muhler, The synthesis of structured Pd/C hydrogenation catalysts by the chemical vapor deposition of Pd(allyl)Cp onto functionalized carbon nanotubes anchored to vapor grown carbon micro-fibers, *Catal. Today*, **102–103**, 34–39 (2005).

[186] A. Corma, H. Garcia and A. Leyva, Catalytic activity of palladium supported on single wall carbon nanotubes compared to palladium supported on activated carbon: Study of the Heck and Suzuki couplings, aerobic alcohol oxidation and selective hydrogenation, *J. Mol. Catal.*, **230**, 97–105 (2005).

[187] C. Pham-Huu, N. Keller, L.J. Charbonniere, R. Ziessel and M.J. Ledoux, Carbon nanofiber supported palladium catalyst for liquid-phase reactions. An active and selective catalyst for hydrogenation of C=C bonds, *Chem. Commun.*, 1871–1873 (2000).

[188] I. Janowska, G. Winé, M.J. Ledoux and C. Pham-Huu, Structured silica reactor with aligned carbon nanotubes as catalyst support for liquid-phase reaction, *J. Mol. Catal.*, **267**, 92–97 (2007).

[189] M. Ruta, I. Yuranov, P.J. Dyson, G. Laurenczy and L. Kiwi-Minsker, Structured fiber supports for ionic liquid-phase catalysis used in gas-phase continuous hydrogenation, *J. Catal.*, **247**, 269–276 (2007).

[190] H. Ma, L. Wang, L. Chen, C. Dong, W. Yu, T. Huang and Y. Qian, Pt nanoparticles deposited over carbon nanotubes for selective hydrogenation of cinnamaldehyde, *Catal. Commun.*, **8**, 452–456 (2007).

[191] M.L. Toebes, Y. Zhang, J. Hájek, T. A. Nijhuis, J.H. Bitter, A.J. Van Dillen, D.Y. Murzin, D.C. Konigsberger and K.P. de Jong, Support effects in the hydrogenation of cinnamaldehyde over carbon nanofiber-supported platinum catalysts: characterization and catalysis, *J. Catal.*, **226**, 215–225 (2004).

[192] M.L. Toebes, T.A. Nijhuis, J. Hajek, J.H. Bitter, A.J. van Dillen, D.Y. Murzin and K.P. de Jong, Support effects in hydrogenation of cinnamaldehyde over carbon nanofiber-supported platinum catalysts: Kinetic modeling, *Chem. Eng. Sci.*, **60**, 5682–5695 (2005).

[193] M.L. Toebes, F.F. Prinsloo, J.H. Bitter, A.J. van Dillen and K.P. de Jong, Influence of oxygen-containing surface groups on the activity and selectivity of carbon nanofiber-supported ruthenium catalysts in the hydrogenation of cinnamaldehyde, *J. Catal.*, **214**, 78–87 (2003).

[194] V. Brotons, B. Coq and J-M. Planeix, Catalytic influence of bimetallic phases for the synthesis of single-walled carbon nanotubes, *J. Mol. Catal. A*, **116**, 397–403 (1997).

[195] Y. Li, G.H. Lai and R.X. Zhou, Carbon nanotubes supported Pt–Ni catalysts and their properties for the liquid phase hydrogenation of cinnamaldehyde to hydrocinnamaldehyde, *Appl. Surf. Sci.*, **253**, 4978–4984 (2007).

[196] H. Vu, F. Gonçalves, R. Philippe, E. Lamouroux, M. Corrias, Y. Kihn, D. Plee, P. Kalck and P. Serp, Bimetallic catalysis on carbon nanotubes for the selective hydrogenation of cinnamaldehyde, *J. Catal.*, **240**, 18–22 (2006).

[197] P. Gallezot and D. Richard, Selective hydrogenation of α,β-unsaturated aldehydes *Catal. Rev.*, **40**, 81–126 (1998).

[198] C. Ganley, F.S. Thomas, E.G. Seebauer and R.I. Masel, A priori catalytic activity correlations: The difficult case of hydrogen production from ammonia, *Catal. Lett.*, **96**, 117–122 (2004).

[199] A.S. Chellappa, C.M. Fischer and W.J. Thomson, Ammonia decomposition kinetics over Ni-Pt/Al$_2$O$_3$ for PEM fuel cell applications, *Appl. Catal. A: Gen.*, **227**, 231–240 (2002).

[200] T.V. Choudhary, C. Svadinaragana and D.W. Goodman, Catalytic ammonia decomposition: COx-free hydrogen production for fuel cell applications, *Catal. Lett.*, **72**, 197–201 (2001).

[201] S.J. Wang, S.F. Yin, L. Li, B.Q. Xu, C.F. Ng and C.T. Au, Investigation on modification of Ru/CNTs catalyst for the generation of CO$_x$-free hydrogen from ammonia, *Appl. Catal. B: Environn.*, **52**, 287–299 (2004).

[202] S.F. Yin, B.Q. Xu, X.P. Zhou and C.T. Au, A mini-review on ammonia decomposition catalysts for on-site generation of hydrogen for fuel cell applications, *Appl. Catal. A: Gen.*, **277**, 1–9 (2004).

[203] G. Leendert Bezemer, U. Falke, A.J. van Dillen and K.P. de Jong, Cobalt on carbon nanofiber catalysts: auspicious system for study of manganese promotion in Fischer–Tropsch catalysis, *Chem. Commun.*, 731–733 (2005).

[204] L. Guczi, G. Stefler, O. Geszti, Z. Koppány, Z. Kónya, É. Molnár, M. Urbán and I. Kiricsi, CO hydrogenation over cobalt and iron catalysts supported over multiwall carbon nanotubes: Effect of preparation, *J. Catal.*, **244**, 24–32 (2006).

[205] Y.-H. Chin, J. Hu, C. Cao, Y. Gao and Y. Wang, Preparation of a novel structured catalyst based on aligned carbon nanotube arrays for a microchannel Fischer-Tropsch synthesis reactor, *Catal. Today*, **110**, 47–52 (2005).

[206] J. Wang, H.B. Chen, H. He, Y. Cai, J.D. Lin, J. Yi, H.B. Zhang and D.W. Liao, New Rh-ZnO/carbon nanotubes catalyst for methanol Synthesis, *Chin. Chem. Lett.*, **13**, 1217–1220 (2002).

[207] R. Gao, C. D. Tan and R.T.K. Baker, Ethylene hydroformylation on graphite nanofiber supported rhodium catalysts, *Catal. Today*, **65**, 19–29 (2001).

[208] Y. Zhang, H.-B. Zhang, G.-D. Lin, P. Chen, Y.-Z. Yuan and a K.R. Tsai, Preparation, characterization and catalytic hydroformylation properties of carbon nanotubes-supported Rh–phosphine catalyst, *Appl. Catal. A*, **187**, 213–224 (1999).

[209] M. Alexandre, E. Martin, P. Dubois, M. Garcia-Marti and R. Jérôme, Use of metallocenes in the polymerization-filling technique with production of polyolefin-based composites, *Macromol. Rapid Commun.*, **21**, 931–936 (2000).

[210] M. Alexandre, E. Martin, P. Dubois, M. Garcia-Marti and R. Jérôme, Polymerization-filling technique: an efficient way to improve the mechanical properties of polyethylene composites, *Chem. Mater.*, **13**, 236–237 (2001).

[211] D. Bonduel, M. Mainil, M. Alexandre, F. Monteverde and P. Dubois, Supported coordination polymerization: a unique way to potent polyolefin carbon nanotube nanocomposites, *Chem. Commun.*, 781–783 (2005).

[212] W. Wang, P. Serp, P. Kalck and J.L. Faria, Photocatalytic degradation of phenol on MWCNT and titania composite catalysts prepared by a modified sol–gel method, *Appl. Catal. B*, **56**, 305–312 (2005).

[213] W. Wang, P. Serp, P. Kalck and J.L. Faria, Visible light photodegradation of phenol on MWCNT-TiO$_2$ composite catalysts prepared by a modified sol–gel method, *J. Mol. Catal.*, **235**, 194–199 (2005).

[214] Z. Kulesius and E. Valatka, Synthesis, characterization and photocatalytic activity of some TiO$_2$/carbon composites, *Polish J. Chem.*, **80**, 335–344 (2006).

[215] Y. Yu, J.C. Yu, J.G. Yu, Y.C. Kwok, Y.K. Che, J.C. Zhao, L. Ding, W.K. Ge and P.K. Wong, Enhancement of photocatalytic activity of mesoporous TiO$_2$ by using carbon nanotubes, *Appl. Catal. A*, **289**, 186–196 (2005).

[216] Y. Yu, J.C. Yu, C.Y. Chan, Y.K. Che, J.C. Zhao, L. Ding, W.K. Ge and P.K. Wong, Enhancement of adsorption and photocatalytic activity of TiO_2 by using carbon nanotubes for the treatment of azo dye, *Appl. Catal. B*, **61**, 1–11 (2005).

[217] Y. Ou, J. Lin, S. Fang and D. Liao, MWCNT–TiO_2: Ni composite catalyst: A new class of catalyst for photocatalytic H_2 evolution from water under visible light illumination, *Chem. Phys. Lett.*, **429**, 199–203 (2006).

[218] W. Vielstich, H.A. Gasteiger and A. Lamm, *Handbook of Fuel Cells-Fundamentals, Technology, and Applications*; John Wiley & Sons, Ltd, Chichester (2003).

[219] H. Tsuchiya and O. Kobayashi, Mass production cost of PEM fuel cell by learning curve, *Int. J. Hydrogen En.*, **29**, 985–990 (2004).

[220] Z.R Ismagilov, M.A Kerzhentsev, N.V. Shikina, A.S. Lisitsyn, L.B. Okhlopkova, C.N. Barnakov, M. Sakashita, S. Iijima and K. Tadokoro, Development of active catalysts for low Pt loading cathodes of PEMFC by surface tailoring of nanocarbon materials, *Catal. Today*, **102–103**, 58–66 (2005).

[221] X. Wang, W. Li, Z. Chen, M. Waje and Y. Yan, Durability investigation of carbon nanotube as catalyst support for protox exchange membrane fuel cell, *J. Power Sources*, **158**, 154–159 (2006).

[222] C.A. Bessel, K. Laubernds, N.M. Rodriguez and R.T.K. Baker, Graphite nanofibers as an electrode for fuel cell applications, *J. Phys. Chem. B*, **105**, 1115–1122 (2001).

[223] M. Tsuji, M. Kubokawa, R. Yano, N. Miyamae, T. Tsuji, M.S. Jun, S. Hong, S. Lim, S.H. Yoon and I. Mochida, Fast preparation of PtRu catalysts supported on carbon nanofibers by the microwave-polyol method and their application to fuel cells, *Langmuir*, **23**, 387–390 (2007).

[224] G. Mestl, N.I. Maksimova, N. Keller, V.V. Roddatis and R. Schlögl, Carbon nanofilaments in heterogeneous catalysis: An industrial application for new carbon materials, *Angew. Chem. Int. Ed.*, **40**, 2066–2068 (2001).

[225] D.S. Su, N. Maksimova, J.J. Delgado, N. Keller, G. Mestl, M.J. Ledoux and R. Schlögl, Nanocarbons in selective oxidative dehydrogenation reaction, *Catal. Today*, **102–103**, 110–114 (2005).

[226] K. Lee, J. Zhang, H. Wang and D. Wilkinson, Progress in the synthesis of carbon nanotube- and nanofiber-supported Pt electrocatalysts for PEM fuel cell catalysis, *J. Appl. Electrochem.*, **36**, 507–522 (2006).

[227] D. Villers, S.H. Sun, A.M. Serventi, J.P. Dodelet and S. Désilets, Characterization of Pt nanoparticles deposited onto carbon nanotubes grown on carbon paper and evaluation of this electrode for the reduction of oxygen, *J. Phys. Chem. B*, **110**, 25916–25925 (2006).

[228] W. Li, X Wang, Z. Chen, M. Waje and Y.Yan, Carbon nanotube film by filtration as cathode catalyst support for proton-exchange membrane fuel cell, *Langmuir*, **21**, 9386–9389 (2005).

[229] M. Saha, R. Li, X. Sun and S. Ye, 3-D composite electrodes for high performance PEM fuel cells composed of Pt supported on nitrogen-doped carbon nanotubes grown on carbon paper, *Electrochem. Commun.*, **11**, 438–441 (2009).

[230] J.M. Tang, K. Jensen, M. Waje, W. Li, P. Larsen, K. Pauley, Z. Chen, P. Ramesh, M.E. Itkis, Y. Yan and R.C. Haddon, High performance hydrogen fuel cells with ultralow Pt loading carbon nanotube thin film catalysts, *J. Phys. Chem. C*, **111**, 17901–17904 (2007).

[231] A.C. Dillon, K.M. Jones, T.A. Bekkedahl, C.H. Kiang, D.S. Bethune and M.J. Heben, Storage of hydrogen in single-walled carbon nanotubes, *Nature*, **386**, 377–379 (1997).

[232] M. Shiraishi, T. Takenobu, A. Yamada, M. Ata and H. Kataura, Hydrogen storage in single-walled carbon nanotube bundles and peapods, *Chem. Phys. Lett.*, **358**, 213–218 (2002).

[233] L. Schlapbach and A. Zuttel, Hydrogen-storage materials for mobile applications, *Nature*, **414**, 353–358 (2001).

[234] A. Zuttel, C. Nutzenadel, P. Sudan, P. Mauron,. C. Emmenegger, S. Rentsch,. L. Schlapbach, A. Weidenkaff and T. Kiyobayashi, Hydrogen sorption by carbon nanotubes and other carbon nanostructures, *J. Alloy Compd.*, **330–332**, 676–682 (2002).

[235] Y. Ye, C.C. Ahn, C. Witham, B. Fultz, J. Liu, A.C. Rinzler, D. Colbert, K.A. Smith and R.E. Smalley, Hydrogen adsorption and cohesive energy of single-walled carbon nanotubes, *Appl. Phys. Lett.*, **74**, 2307–2311 (1999).

[236] T. Yildirim and S. Ciraci, Titanium-decorated carbon nanotubes as a potential high-capacity hydrogen storage medium, *Phys. Rev. Lett.*, **94**, 175501–175505 (2005).

[237] J.W. Lee, H.S. Kim, J.Y. Lee and J.K. Kang, Hydrogen storage and desorption properties of Ni-dispersed carbon nanotubes, *Appl. Phys. Lett.*, **88**, 143126–143129 (2006).

[238] E. Yoo, L. Gao, T. Komatsu, N. Yagai, K. Arai, T. Yamazaki, K. Matsuishi, T. Matsumoto and J. Nakamura, Atomic hydrogen storage in carbon nanotubes promoted by metal catalysts, *J Phys Chem. B*, **108**, 18903–18907 (2004).

[239] A.L.M. Reddy and S. Ramaprabhu, Hydrogen storage properties of nanocrystalline Pt dispersed multi-walled carbon nanotubes, *Int. J. Hyd. En.*, **32**, 3998–4004 (2007).

[240] Y. Luo, P. Wang, L.-P. Ma and H.-M. Cheng, Enhanced hydrogen storage properties of MgH_2 co-catalyzed with NbF_5 and single-walled carbon nanotubes, *Scripta Mat.*, **56**, 765–768 (2007).

[241] K. Iyakutti, Y. Kawazoe, M. Rajarajeswari and V.J. Surya, Aluminum hydride coated single-walled carbon nanotube as a hydrogen storage medium, *Int. J. Hyd. En.*, **34**, 370–375 (2009).

[242] F. Balavoine, P.Schultz, C. Richard, V. Mallouh, T.W. Ebbesen and C. Mioskowski, Helical crystallization of proteins on carbon nanotubes: a first step towards the development of new biosensors, *Angew. Chem. Int. Ed.*, **38**, 1912–1915 (1999).

[243] Z. Liu, S. Tabakman, K. Welsher and H. Dai, Carbon nanotubes in biology and medicine: In vitro and in vivo detection, imaging and drug delivery, *Nano Res.*, **2**, 85–120 (2009).

[244] J. Ye, Y. Wen, W.D. Zhang, L.M. Gan, G.Q. Xu and F.-S. Sheu, Non enzymatic glucose detection using multi-walled carbon nanotube electrodes, *Electrochem. Commun.*, **6**, 66–70 (2004).

[245] G.L. Luque, M.C. Rodriguez and G.A. Rivas, Glucose biosensors based on the immobilization of copper oxide and glucose oxidase within a carbon paste matrix, *Talanta*, **66**, 467–471 (2005).

[246] I.T. Bae, E. Yeager, X. Xing and C.C. Liu, In situ infrared studies of glucose oxidation on platinum in an alkaline medium, *J. Electroanal. Chem.*, **309**, 131–145 (1991).

[247] X. Chu, D. Duan, G. Shen and R. Yu, Amperometric glucose biosensor based on electrodeposition of platinum nanoparticles onto covalently immobilized carbon nanotube electrode, *Talanta*, **71**, 2040–2047 (2007).

[248] K. Besteman, J.-O. Lee, F.G.M. Wiertz, H.A. Heering and C. Dekker, Enzyme-coated carbon nanotubes as single-molecule biosensors, *Nano Lett.*, **3**, 727–730 (2003).

[249] L.-Q. Rong, C. Yang, Q.-Y. Qian and X.-H. Xia, Study of the nonenzymatic glucose sensor based on highly dispersed Pt nanoparticles supported on carbon nanotubes, *Talanta*, **71**, 819–824 (2007).

[250] Y. Cheng, Y. Liu, J. Huang, K. Li, Y. Xian, W. Zhang and L. Jin, Amperometric tyrosinase biosensor based on Fe_3O_4 nanoparticles-coated carbon nanotubes nanocomposite for rapid detection of coliforms, *Electrochimica Acta*, **54**, 2588–2594 (2009).

[251] A. Bianco, K. Kostarelos and M. Prato, Applications of carbon nanotubes in drug delivery, *Curr. Opin. Chem. Biol.*, **9**, 674–679 (2005).

[252] W. Wu, S. Weckowski and C. Klumpp, Targeted delivery of amphotericin B to cells by using functionalized carbon nanotubes, *Angew. Chem.*, **44**, 6358–6362 (2005).

[253] R. Singh, D. Pantarotto, D. McCarthy, O. Chaloin, J. Hoebeke, C. D. Partidos, J.-P. Briand, M. Prato, A. Bianco and K. Kostarelos, Binding and condensation of plasmid DNA onto functionalized carbon nanotubes: toward the construction of nanotube-based gene delivery vectors, *J. Am. Chem. Soc.*, **127**, 4388–4396 (2005).

[254] D.A. Heller, E.S. Jeng, T.-K. Yeung, B.M. Martinez, A.E. Moll, J.B. Gastala and M.S. Strano, Optical detection of DNA conformational polymorphism on single walled carbon nanotubes, *Science*, **311**, 508–511 (2006).

[255] K. Welsher, Z. Liu, D. Daranciang and H. Dai, Selective probing and imaging of cells with single walled carbon nanotubes as near-infrared fluorescent molecules, *Nano Lett.*, **8**, 586–590 (2008).

[256] S.B. Zimmerman, The three-dimensional structure of DNA, *Ann. Rev. Biochem.*, **51**, 395–427 (1982).

[257] A. Herbert and A. Rich, The biology of left-handed Z-DNA, *J. Biol. Chem.*, **271**, 11595–11598 (1996).

[258] Y Wu, J.S. Hudson, Q. Lu, J.M. Moore, A.S. Mount, A.M. Rao, E. Alexov and P.C. Ke, Coating single-walled carbon nanotubes with phospholipids, *J. Phys. Chem. B*, **110**, 2475–2478 (2006).

[259] R. Qiao and P.C. Ke, Lipid-carbon nanotube self-assembly in aqueous solution, *J. Am. Chem. Soc.*, **128**, 13656–13657 (2006).

[260] R. Hurt, M. Monthioux and A. Kane, Toxicology of carbon nanomaterials: Status and trends, *Carbon*, **44**, 1028–1033 (2006).

[261] K. Donaldson and C.A. Poland, Nanotoxicology: new insights into nanotubes, *Nature Nanotech.*, **4**, 708–710 (2009).

[262] N. Ishigami, H. Ago, Y. Motoyama, M. Takasaki, M. Shinagawa, K. Takahashi, T. Ikuta and M. Tsuji, Micro reactor utilizing a vertically-aligned carbon nanotube array grown inside the channels, *Chem. Commun.*, 1626–1628 (2007).

5

Filled Carbon Nanotubes

5.1 Presentation of Chapter 5

Carbon (and related, i.e. –BN–, –BCN–, etc.) nanotubes have proven to be extraordinarily effective templates for the precise engineering of materials formation on a subnanometre scale. This is a consequence of firstly, the fact that the confining volume of nanotubes, specified by their internal van der Waals surface, corresponds to the scale either of discrete molecules or atomic wires just a few atomic distances in cross section and that secondly, this confining cylindrical volume demonstrates extraordinary mass-transport capabilities and makes possible the formation of unprecedented crystal structures. In addition, the intricately close and well protected situation of the filling material with respect to the encapsulating one (the nanotube) has raised promises for mutual interactions likely to generate new behaviours and properties. Because of this, filling carbon nanotubes (and related materials) has rapidly become an intensively explored field of research.

The most popular species to fill single-wall carbon nanotubes with are so far C_{60} fullerenes and related materials such as higher fullerenes (C_{70} and higher), endofullerenes, functionalized fullerenes, and so on. For this reason, a whole subchapter (Chapter 5b) will be specifically dedicated to these materials. On the other hand, filled nanotubes whose filling materials are not fullerene-related represent a wide variety of new materials that will be dealt in the first subchapter (Chapter 5a). In the latter, however, fullerene-filled nanotubes will be sometimes mentioned for specific purposes. Because new crystallographic and physical behaviours are the most expected for the most confined situations, both subchapters will focus on single-wall carbon nanotubes as more likely to provide the needed very narrow (~1 nm) inner cavities. Finally, the filling of nanotubes which are not made of pure carbon (e.g. boron nitride nanotubes, $C_x B_y N_z$ nanotubes) will be reported in Chapter 6.

Carbon Meta-Nanotubes: Synthesis, Properties and Applications, First Edition. Edited by Marc Monthioux.
© 2012 John Wiley & Sons, Ltd. Published 2012 by John Wiley & Sons, Ltd.

5a

Filled Carbon Nanotubes: (X@CNTs)

Jeremy Sloan[1] and Marc Monthioux[2]
[1]*Dept. of Physics, University of Warwick, UK*
[2]*CEMES, CNRS, University of Toulouse, France*

5a.1   Introduction

Whilst fullerene and related materials are the most inserted molecules in carbon nanotubes (CNTs), other molecular species including Zn-diphenylporphyrin (i.e. Zn-DPP) [1], *ortho*-carborane [2,3], metallocenes [4,5], octasiloxane [6] and squarylium dye [7], among others have been encapsulated within either single-, double- or multiple walled carbon nanotubes (SWCNTs, DWCNTs and MWCNTs, respectively) and directly imaged by either high resolution transmission electron microscopy (HRTEM) or related techniques. In tandem with these studies, one-dimensional (1D) nanowires of a wide variety of materials including multiplets of polymeric iodine chains [8,9], pure metals [10–15], metal oxides [16–18], metal carbides [19–21], $2 \times 2 \times \infty$ or $3 \times 3 \times \infty$ atomic layer thick KI crystals [22,23], polyhedral chains of lanthanide trihalides [24,25], twisted 1D Co_2I_4 double chains with tetrahedral Co^{2+} coordination [26], a new trigonal tubular form of HgTe [27], reduced coordination polymorphs of PbI_2 [28], alloys [29,30] and semiconductor liquid-metal heterojunctions [31] have also been encapsulated within CNTs. Further, ab initio theory has been used to either predict or explain the formation of some of the above and several other encapsulated species, for example, silver atoms [32], hydrogen molecules [33] and ice crystals [34,35].

Carbon Meta-Nanotubes: Synthesis, Properties and Applications, First Edition. Edited by Marc Monthioux.
© 2012 John Wiley & Sons, Ltd. Published 2012 by John Wiley & Sons, Ltd.

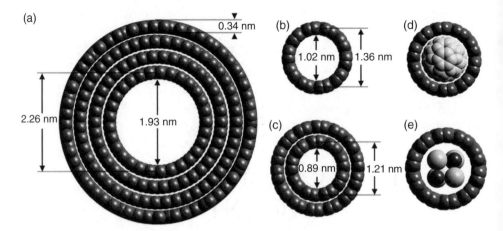

Figure 5a.1 *Structure models (a) a (29,0)@(38,0)@(47,0)@(48,13) MWCNT; (b) a (10,10) SWCNT; (c) a (9,9)@(14,14) DWCNT; (d) a C$_{60}$ molecule (white spheres) inside a (10,10) SWCNT (dark spheres); (e) a 2 × 2 KI crystal inside a (10,10) SWCNT (white spheres = iodine, black spheres = potassium).*

The extent to which a nanotube can accommodate a species may be considered, to a first approximation, to be a function of the overall diameter of the encapsulating nanotube, partly specified by the (*n,m*) structural conformation of the nanotube, which dictates the nanotube diameter [36] and by the twofold contribution of the 0.17 nm van der Waals radii of the wall carbons which constrains the internal volume still further [37]. In Figure 5a.1, we can see how these factors influence the available free internal cross-section of a selection of nanotubes, including a MWCNT, a SWCNT and a DWCNT. With regard to the size of a species which may be accommodated, this will be specified by the internal diameter of the innermost SWCNT of a multi-walled nanotube (i.e. as in the case of a MWCNT or a DWCNT) or of the SWCNT itself (which consists of one graphene tubule only). Thus the MWCNT in Figure 5a.1(a) has an internal cylinder of 1.93 nm in diameter specified by the innermost (29,0) SWCNT, while the corresponding dimension for the (10,10) SWCNT in Figure 5a.1(b) is just 1.02 nm. The inner tube in the DWCNT specified in Figure 5a.1(c) is a (9,9) SWCNT which has a corresponding internal diameter of 0.89 nm which is narrower than for the (10,10) SWCNT in Figure 5a.1(b). The size of the inner surfaces of the narrower SWNTs and DWNTs correspond to the external dimensions either of single molecules, as exemplified by the encapsulation C$_{60}$ molecule in Figure 5a.1(d) or of integral numbers of atomic layers of crystalline materials, as exemplified by the encapsulated 2 × 2 KI crystal shown in Figure 5a.1(e).

The dimensions of the nanotubes specified in Figure 5a.1 (and, by extension, nanotubes of both larger and smaller sizes) determine the available space within a nanotube for encapsulation although this does not tell the whole story. Beyond the simplistic consideration that a minimal inner diameter for the nanotube to fill could be merely required [38], we also have to consider the 'wetability' of the interior surface of the nanotube [39,40] in terms of the carbon and also in terms of the surface tension properties of the introduced liquid or solution and also whether or not the nanotubes are open at one or both ends (or contain sufficiently large voids in the graphene walls forming the nanotubes to permit filling). If we are considering a species introduced from the vapour phase, we need to think in terms of the diffusion characteristics of the gaseous species and the relative permeability of the nanotube in terms of wall and tip holes. In order to be more rigorous, we may also need to consider more subtle structural features of the encapsulating nanotubes, such as: (i) defects which may take the form of gross structural defects, for example, bends and internal caps; (ii) the effective pitch of the nanotube, specified by the relative (n,m) helical conformation of a particular nanotube; (iii) the effect of relative rigidity on the nanotube when the tubule comprises two or more shells, which can sometimes cause differences in crystallization behaviour; (iv) the possible influence of the wall corrugation on encapsulate behaviour which may play some role in terms of fixing species into place or providing a register which can either assist or inhibit crystallization (i.e. in the Burger's vector sense); (v) the physical size and shape of the filling material, especially if it is molecular; and finally (vi) probably the most significant interaction, that is, the electronic interaction between nanotube and filling which can be defined either in terms of a mutual interaction (i.e. between either a semiconducting or metallic nanotube and a filling with different electronic properties) or in terms of the effect that confinement can have on a particular material. Both can potentially cause significant effects on the band structure of the nanotube itself or on the corresponding electronic structure of the filling material as well.

The desire to synthesize meta-nanotubes consisting of encapsulated materials within carbon nanotubes (which we shall abbreviate here 'X@CNT' where X is the introduced material and CNT is the collective term for all kinds of carbon nanotubes) consists of a multiplicity of motivations, including, among others, (i) to 'see' and study low dimensional crystals, single molecules and even functional groups attached to the latter, protected from the environment by the carbon sheath, with their mobility hindered by the interaction with it; (ii) to observe the formation of low dimensional crystal structures and, in so doing, occasionally observe new ones; (iii) to modify the physical properties of the encapsulating nanotubes and of the encapsulated materials; (iv) to use the obtained species as a means for both testing and exploiting the very latest in nanoscale characterization methodologies; (v) to correlate structural and electronic behaviour on materials formed on such a scale as to be tractable to the most rigorous computational methods (including both molecular dynamics (MD) and density functional theory (DFT)); and, ultimately, (vi) to satisfy that most fundamental of solid state science objectives, in other words to synthesize useful new materials or devices with useful applications on the nanometre scale. The goal of this chapter is therefore to evaluate the progress made so far in realising these objectives and perhaps give some idea of the work still remaining to be done although hopefully the reader will be pleasantly surprised with the amount of progress that has already been made.

5a.2 Synthesis of X@CNTs

5a.2.1 A Glimpse at the Past

The story of filled nanotubes started in 1993, where four papers were published that reported filling of MWCNTs with various metals and compounds (PbO$_x$ [16], Y$_3$C and TiC [19] Bi$_2$O$_5$ [41] and Ni [42]) in attempts to obtain encapsulated inorganic nanowires.

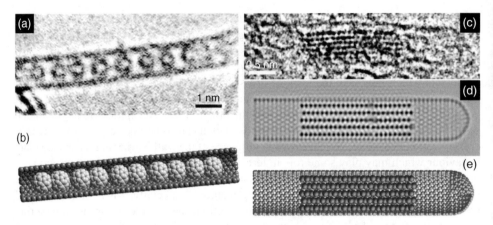

Figure 5a.2   *(a) The first HRTEM image of a C_{60}@SWCNT (peapod) nanocomposite (reproduced with permission from [46] Copyright (1998) Macmillan Publishers Ltd). (b) Corresponding structure model. (c) The first example of a metal@SWCNT (reprinted from [10] Copyright (1998) The Royal Society of Chemistry); in this case, the metal is Ru and was initially introduced as $RuCl_3$ and then was chemically reduced. (d) HRTEM image simulation. (e) Schematic structure model.*

However, inner diameters of MWCNTs [43] are comparatively large (in general in the range 5–50 nm) and the filling of the much smaller inner diameter SWCNTs (in general in the range 1–1.5 nm) that were discovered in 1993 [44,45] therefore appeared more challenging. It actually took five years following the reporting of filled MWCNTs for the first two definitive examples of filled SWCNTs to be reported (Figure 5a.2). One was the incidental discovery of the ability of fullerenes to diffuse and pack into SWCNTs, thereby forming the so-called 'peapods' [46]. The other, involving the hydrogen reduction of $RuCl_3$ introduced into pre-opened SWCNTs from solution was reported in the same year [10].

By analogy with the accepted notation for endofullerenes, SWCNTs filled with fullerenes were symbolized as C_{60}@SWCNTs [47], which rapidly became the overall notation for any kind of X@CNT, where X is the chemical symbol of the filling material (or, sometimes, an accepted abbreviation for a chemical compound) and CNT is the nanotube type, as described previously.

5a.2.2   The Expectations with Filling CNTs

In a similar fashion to the preparation of decorated nanotubes, in which the latter are essentially considered as a substrate, nanotubes can be considered with respect to their inner cavity, that is, an empty volume that can be used as a template or 'nanomould', and/or as a 'nanoreactor' (i.e. for the chemical transformation of an inserted material) for the synthesis of nanomaterials. Due to the immense length/cross-section aspect ratio of nanotubes, inserted materials are forced to adopt a nearly one-dimensional morphology, especially when considering SWCNTs whose internal cavities are consistently smaller than those of MWCNTs and whose aspect ratios are considerably longer. The host nanotubes may also act as nanocapsules protecting the contained materials from any reaction with the surrounding medium, typically (though not exclusively) oxidation by contact with the atmosphere, as well as dissolution of the filling material in aqueous or nonaqueous solvents. Nanotubes may therefore make possible the synthesis of nanowire-like materials that could never exist if not encapsulated. Indeed, metallic nanowires of 1 nm in diameter would rapidly turn into oxide nanowires or would even possibly collapse, as recently demonstrated by attempts to remove the carbon sheath from filled SWCNTs [48]. In both cases, chances are high that any peculiar property is lost. Generally speaking, growing phases within a very confined volume allows for the formation of new nanomaterials to be expected because of the possibility opened to enforce and stabilize new combinations of chemical elements, to enforce and stabilize novel crystal structures for regular chemical phases and to deform and stress lattices within regular structures.

Of course, new nanomaterials are mainly valuable if they come with new, altered or enhanced physical properties. The latter are expected from the stabilization of otherwise impossible new chemical compositions, structures, or morphologies. Typical examples are (i) the ballistic transport behaviour from the one-dimensional structure, which prevents electron scattering; (ii) the very unusual surface-atom to core-atom ratios, which may even reach an infinite value, (e.g. all the atoms of the structure can effectively be 'surface atoms'), and (iii) the protection by the carbon sheath from the surface adsorption of disturbing molecules. Typical anticipated and realized benefits regard nanowires made from magnetic elements or compounds and charge-transfer electronic interactions with the encapsulating graphene lattice and the guest materials.

Finally, should encapsulated phases not exhibit new structures, chemical composition or physical properties, encapsulation may provide external reactants peculiar access conditions to the filling materials, thereby controlling the behaviour of the latter by controlling the interaction kinetics with the surrounding medium. Of course, this could occur in a negative manner, for instance, regarding the filling material chemical reactivity that may be more or less inhibited in spite of the nanosize because of the presence of the carbon sheath. On the other hand, this modified behaviour regarding chemical reactivity may be useful to some applications, for instance, slowing down diffusion kinetics and/or chemical reactivity may be very valuable in fields such as chemical catalysis, drug or pesticide delivery, and so on.

Hence, encapsulating materials in CNTs is likely to promote new phases, new structures, new properties, and/or new behaviours. However, any of these features will be more likely as the tube cavity diameter is smaller. Filling nanotubes whose inner cavity is below ~2 nm wide, as is encountered in most of SWCNTs and DWCNTs, will therefore be preferably considered below. Detailed review papers specifically dealing with filling MWCNTs may be found in the early literature [49,50].

5a.2.3 Filling Parameters, Routes and Mechanisms

5a.2.3.1 Filling Strategy

Various ways of filling nanotubes may be considered, depending on the physical properties of the material being inserted, the cavity diameter to be filled, the cost issues (if for commercial purposes) and the capabilities of the methods themselves, with all parameters being more or less interdependent.

Materials can be inserted into CNT cavities as a solid, liquid, or vapour. Hence, relevant physical properties of the materials filling the nanotubes are primarily solubility, melting point, boiling point and of course, decomposition temperature, which should not be reached. When liquids are involved (i.e. as either a molten or dissolved material), a subsequent property of the utmost importance to consider is surface tension. It was experimentally observed that the surface tension (γ) of liquids should be below 100–200 mN m^{-1} in order to wet MWCNTs [40] while supporting the idea that the capillarity-driven mechanisms for filling nanotubes with liquids are obeying the Young-Laplace law and equations. Later on, this surface tension threshold was more precisely defined as 130–170 mN m^{-1} for SWCNTs [39]. In addition to surface tension, the viscosity of the liquid material is also interesting to consider as a parameter presumably important in the filling event, at least regarding the filling kinetics. In addition, it might also be interesting to know the vapour pressure of the material to be inserted to estimate, if needed, whether some vapour was introduced along with the liquid. These aspects will be discussed further in the following sections.

The nanotube inner diameter is obviously an important parameter because, in the first instance, it will determine the ultimate diameter of the basic entities (such as molecules, as a vapour or in solution, or the cross-sectional diameter of any included crystals) able to enter the nanotube cavity. On the other hand, it is also quite important when filling with liquids because it will contribute to the determination of the respective filling efficiency. Previous studies have actually proposed that a minimum inner diameter for the nanotube to fill is required [38]. Because this cut off value depends on the surface tension of the liquid involved, it does not correspond to a single value, hence, no 'magic' number for the nanotube diameter can be provided. For instance, threshold nanotube inner diameters were calculated to be 0.7 and 4 nm for molten V_2O_5 ($\gamma = 80$ mN m^{-1}) and molten $AgNO_3$, respectively [38,50].

In addition to the intrinsic cost of the starting materials, product recyclability and energy supply for thermal steps, an important feature for cost issues is the number of steps included in the filling process. The latter aspects may drive the selection of the filling process but may also drive the nanotube synthesis process because some may allow in situ filling, that is, these are able to fill nanotubes while they form.

5a.2.3.2   In Situ Filling Processes

Filling nanotubes in situ is interesting because it is a single-step synthesis procedure; it leaves the nanotube sheath intact and, depending on the synthesis process, closed (for arc-prepared nanotubes) or opened at one end only (CCVD-prepared nanotubes), thereby protecting as far as possible the encapsulated material from any contact with the surrounding post-synthesis atmosphere and it allows nanotubes to be filled with elements whose surface tension is high enough to prevent filling by wetting methods (see next).

In situ filling was one of the two early ways that foreign materials were introduced into MWCNT-type nanotubes simultaneously with their synthesis. Two techniques permit this: one is the electric arc process [51] and the other is the CCVD process (Figure 5a.3) [52,53]. The latter is rather limited because it most often corresponds to specific conditions for which the excess metal catalyst that is needed to grow the nanotubes is trapped and encapsulated inside them. Therefore, filling nanotubes this way is basically restricted to materials that are catalysts for carbon formation, typically transition metals such as Ni, Co, and Fe, although examples of other types of materials encapsulated this way can be found in the literature (e.g. Cu and Ge in [53]). Because catalyst metals for carbon are mainly ferromagnetic materials, it is actually quite a convenient way to produce ferromagnetic nanowires encapsulated in nanotubes [52,54,55] otherwise difficult to obtain (see next).

(a) (b) (c)

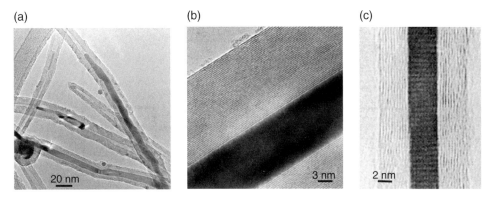

Figure 5a.3 *(a) Low-magnification TEM image of in situ filled MWCNTs obtained via a CCVD-related process. The filling material is pure α-Fe (bcc Fe crystals). (b) High resolution of one of the Fe@MWCNTs in (a). The perfection of graphenes from the surrounding carbon nanotube is shown. Lattice fringes of the contained iron crystal are not seen because of the high Z number of Fe that makes it barely transparent to electrons. The material was prepared by P. Watts (University of Sussex) following the procedure reported in [52]. (c) High resolution TEM image of an in situ filled MWCNT obtained via an electric arc plasma process. The filling material is a single crystal of pure chromium [57]. Modified with permission from [13] Copyright (2006) Cambridge University Press.*

The electric arc plasma process consists in drilling a coaxial hole within the graphite anode and filling it with the ground, desired element (or a mixture of it with graphite powder). Once the electric arc is run using conditions (current, voltage, pressure, and atmosphere) similar to that used to produce fullerenes, the MWCNTs that usually grow as a cathode deposit are found partially filled with the desired element. However, the efficiency and control of the filling process is lower than for the CCVD-related method, probably because of the huge temperature gradients that are typical of the plasma zone in arc-related processes, and that are also responsible for the heterogeneity in the diameter distribution of the MWCNTs synthesized that way. A specific requirement for achieving successful in situ filling in arc plasma processes is that some sulfur is present, added or naturally contained as an impurity (e.g. originating from the pitch used as a binder) within the graphite anode, otherwise the filling mechanism may fail [56,57]. As opposed to the CCVD-based method, the electric arc method is hardly able to form hybrid MWCNTs encapsulating transition metals such as Fe, Ni, or Co, although some rare exceptions may be found (e.g. with Co [51]). The reason is that as soon as such a metal is introduced into the system (i.e. by doping the graphite anode with it), it forms C-metal solid solution droplets that are encapsulated as carbides within the MWCNTs growing at the cathode, or that escape the plasma zone to subsequently promote the growth of (empty) SWCNTs from them. That is actually how SWCNTs were incidentally discovered, that is, as by-products of attempts to fill MWCNTs with transition metals (i.e. Fe [44] and Co [45]).

The CCVD method and the electric arc method conveniently complement each other because the former is mainly able to produce transition metal-filled MWCNTs, whereas the latter is not but is mainly able to produce MWCNTs filled with many other elements [51–57]. On the other hand, both methods exhibit severe limitations. One is that only single elements

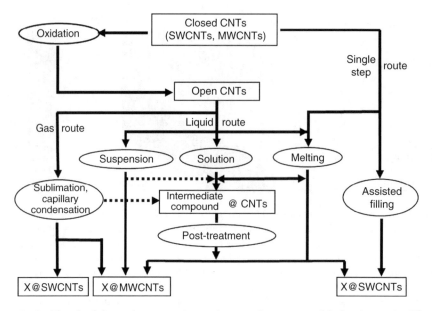

Figure 5a.4 *Sketch of the various experimental routes that are possible for the ex situ filling of carbon nanotubes. The rectangles indicate obtained materials while ovals indicate treatment steps. Modified with permission from [13] Copyright (2006) Cambridge University Press.*

or carbides, as stated above, are able to be inserted because of the requirement of being a carbon catalyst for the CCVD process or because of the high temperatures involved (for the electric arc plasma process) that do not allow oxides or salts to form. Typically, inserting multi-element compounds is not possible, nor is inserting labile materials. Another major limitation is that synthesizing filled SWCNTs this way is nearly not possible and when it is, it comes with very low yield, in the range of a few percent of the obtained product. Examples in the literature are very scarce and involve the electric arc process only. One involves the spontaneous formation of peapods obtained during arc experiments designed to grow SWCNTs [58,59] and the other is the encapsulation of Bi in SWCNTs grown with Co as catalyst [11], both Co and Bi having previously been introduced into the graphite anode, as described here.

The severe limitations related to in situ filling, specifically regarding SWCNTs, have therefore promoted the development of ex situ filling processes, as we shall see in the next section.

5a.2.3.3 Ex Situ Filling Processes

Ex situ filling is the most versatile route because it makes possible the ability to insert nearly any kind of material into nearly any kind of nanotube. This basic principle was acknowledged long ago when the first fillings of MWCNTs were achieved [20,49–51]. Figure 5a.4 summarizes the various ex situ filling pathways that will be described in the subsequent sections. Ex situ filling can be carried out following one-, two-, or three-step procedures, not mentioning a final cleaning step, which is needed for any of the ex situ methods to remove any extraneous material, by means of washing or dynamic vacuum heating (the latter may be preferred for avoiding a liquid phase step that often comes with residuals, and the filtration step that compacts the filled SWCNT material obtained).

5a.2.3.3.1 *Previous Opening of the Tubes*   Except for single-step filling procedures, and unless starting nanotubes are naturally opened because of the specificity of the nanotube synthesis process (e.g. templating), the first step for ex situ filling of nanotubes is dedicated to the prior opening of the latter, which can be achieved in two main ways, documented since the early days of filling MWCNTs [16,19,60]. These consist of (i) thermal treatments in an oxidizing gas phase (air or O_2) [41] or (ii) reaction with liquid reactants that are oxidizing for polyaromatic carbon materials, typically, acids such as concentrated nitric (preferred) or sulfuric acid [60] or a mixture of both. The latter can also include oxidants such as hydrogen peroxide, potassium permanganate, fluoric acid/bromine fluoride mixture, osmium tetroxide, and so on [61]. Gas or liquid phase oxidation procedures are efficient in opening MWCNTs or SWCNTs, although conditions have to be more severe for the former because of the higher number of graphene walls to penetrate. However, liquid phase oxidation tends to generate residuals that may more or less coat the nanotubes, thereby hindering subsequent treatments (including filling) or investigations (such as TEM). In this respect, gas phase oxidation is generally preferable and is the dominant technique for opening SWCNTs. For instance, subjecting raw SWCNTs obtained from arc process to a 380°C thermal treatment (including 1 h of isothermal) in air within an open tubular furnace allowing natural convection was proven in our laboratory to be able to create openings as wide as 0.7 nm or more in SWCNTs from the arc method with an overall weight loss in the range of 40%. This mass loss was not only from the nanotubes and varied according to the amount and type of carbon impurities and catalysts (if any) employed. Such a treatment allows SWCNTs to be subsequently filled with fullerenes with a filling efficiency that may reach ~90–95%. Interestingly, because purification procedures that are carried out on SWCNTs to remove carbon by-products (and catalyst remnants) are oxidizing treatments as well, purified SWCNTs available from commercial suppliers (e.g. Nanocarblab in Russia) may exhibit openings that are sufficient to allow them to be filled with high efficiency without needing any further treatment but the filling step.

Considering the comparatively high inertness of the graphene lattice toward chemical oxidation, opening occurs at the location of structural defects (typically five- or seven-membered rings) whatever the tube type. Pentagons are obviously found at the tips of nanotubes as a requirement to curve and close the graphenes involved, making nanotubes naturally able to be opened from the tips. However, a major difference regarding the behaviour of MWCNTs versus SWCNTs upon oxidation is that the former are able to open from the tips only, whereas SWCNTs are able to open from the tips and the side walls. Indeed, as discussed earlier [62], several experimental results such as TEM investigation, behaviour under electron irradiation, and titration of acid-treated SWCNTs have shown the presence of side defects to be very likely (typically, heptagon-pentagon ring pairs or double pairs, the latter corresponding to the so-called Dienes defects), with an average proportion of one chemically attackable site every 5 nm along a given SWCNT wall. Filling of SWCNTs by this route was further supported by calculations and modelling [63]. In contrast, even should every graphene tubule making up the composite wall of a MWCNT be defective, the chance for the defects from each graphene to superimpose at a given location of the wall to create a site for side opening is nil, except in the MWCNT tip region where pentagons obviously superimpose from reasons related to geometric constraints at the sites of the nanotube closure.

5a.2.3.3.2 Filling by Means of Gas Phase Processes Filling nanotubes via the gaseous phase consists in putting in contact the previously opened nanotubes (i.e. MWCNTs, DWCNTs or SWCNTs) and the vapour of the material to insert in them, for instance, in a Pyrex or quartz vessel that should be evacuated (primary or, preferably, secondary vacuum) before being sealed, and then heated up to the desired temperature which will typically be either the vaporization or sublimation temperature of the filling material, or slightly above. This is the way peapods (e.g. C_{60}@SWCNT) are usually prepared [64]. For a filling material such as fullerenes, which exhibit high electronic affinity with the graphene lattice (i.e. the nanotube surface) insertion into the SWCNT cavity was proven to be a surface diffusion-driven mechanism [64]. Hence, it is highly dependent on temperature and time, but weakly dependent on the partial pressure of filling material vapour. Therefore, operating at a not-too-high temperature is necessary to minimize temperature-induced stochastic movements of the filling molecules and to give them a sufficient residence time on the nanotube surface to enable them to find the entry ports. A long processing time, in the range of several hours to two days, is therefore necessary to achieve high filling rates (sometimes close to 100%) provided a sufficient amount of filling material is supplied. For other types of materials to insert that do not exhibit the same affinity for the graphene lattice (e.g. $ZrCl_4$ [65], Se [66], or Re_xO_y [67]), the filling mechanism is therefore slightly different and relates to capillary condensation. In the latter, a high partial pressure of the material to be inserted has to be developed at the vaporization/sublimation temperature or above, and the vapour condenses into the capillary porosity (i.e. the tube inner cavity) when the system is cooled down. High filling efficiencies, along with shorter processing times (e.g. with respect to peapod synthesis), are therefore expected for materials exhibiting high vapour pressure (e.g. Se [66]). Examples of quantitative filling are provided in Figure 5a.5.

The gas phase route has many advantages: its relative simplicity (basically two steps, i.e. opening and filling), its potentiality for a high filling rate and high purity/homogeneity of the filling material, and also the absence of the requirement that the material meet the threshold surface tension requirement (see the next section). It is applicable to SWCNTs and MWCNTs, although examples for the latter in literature are scarce [66,68,69]. Conversely, a major drawback of the method is the need for the filling material to exhibit vaporization or sublimation temperature below ca. 1000°C–1200°C, to minimize the chances for healing the nanotube openings and/or for possible reactions with carbon. Inserting compounds this way is limited as well because candidate materials such as oxides are often high temperature refractories or, on the contrary, compounds such as salts are barely able to vaporize or sublime without decomposing.

5a.2.3.3.3 Filling by Means of Liquid Phase Processes The major characteristic of the liquid phase methods is their close dependence on the physical interaction between the filling liquid and the encapsulating solid, typically via the Young-Laplace law for capillary wetting. As soon as a solvent is involved, as for the suspension and the solution methods (see next), wetting is not supposed to be a problem because surface tensions of the usual solvents are below 80 mN m^{-1}. However, viscosity is certainly another parameter assumed to play a leading role, yet it has been scarcely considered in the literature dealing with the topic.

The *suspension method* is dedicated to filling nanotubes with nanoparticles. Although examples (thus far involving MWCNTs only) are still limited because quite recent [70,71],

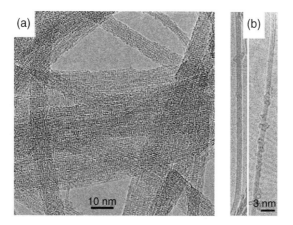

Figure 5a.5 *Examples of hybrid SWCNTs with high filling rates obtained via the gas phase route. (a) Low magnification of C$_{60}$@SWCNT (peapods) material prepared by the sublimation method (450°C, 24h). The periodically dashed contrast of the bundles is due to the overall presence of the fullerene molecules aligned within the SWCNT inner volume. High resolution images of such peapods are given in Chapter 5b (image credit: M. Berd, CEMES-CNRS). (b) Low resolution, high magnification images showing two examples of a full filling over lengths longer than 60nm with a sample of Se@SWCNTs prepared by capillary condensation (800°C, several hours). Modified with permission from [13] Copyright (2006) Cambridge University Press.*

filling rates can be fairly high. This is not very surprising considering the large diameters of the MWCNTs being filled in comparison to the small particle size of the introduced material (Figure 5a.6). In related experiments, open nanotubes were put into contact with a suspension of nanoparticles in a liquid at room temperature, the latter being evaporated during the experiment and/or afterward. Low volume concentrations of particles in the suspending fluid were used to maintain a low viscosity. The high filling efficiency is attributed to the combination of capillary forces [71] possibly added with the effect of the concomitant fluid evaporation [70], the latter being assumed to drive a continuous flow of fresh suspension from the outside to the inside of the tube.

The method is interesting because it is quite simple (two steps, opening and filling) and efficient but it comes with severe limitations related to geometrical constrains and the morphology of the filling material feedstock. Indeed, nanotubes to be filled will be limited to MWCNTs, unless clusters less than 1 nm large in diameter can be produced to fill pre-opened SWCNTs. One related example has actually been reported already in which nanohorns (nanohorns are short, tapering single-wall nanocapsules [72]) were partially filled with 1 nm or less Gd-containing clusters [73]. On a more positive note, obtaining encapsulated nanowires using this approach will be possible once nanoparticles will be inserted and aligned within CNTs whose inner diameter matches that of the nanoparticles, while the chemical nature of the nanoparticles would allow them to melt and coalesce upon annealing at a temperature range that does not damage the CNTs. Generally speaking, the trickiest part of the procedure is the need to be able to prepare nanoparticles of the desired materials before the filling step. This is not common knowledge yet. However, some are now commercially available (e.g. Sigma-Aldrich), even

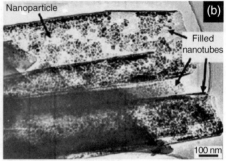

Figure 5a.6 *Examples of MWCNTs filled with nanoparticles via the suspension method (liquid route), taken from pioneering works. (a) Polystyrene nanobeads initially suspended in ethylene glycol (reprinted with permission from [53] Copyright (1996) Elsevier Ltd). (b) Fe$_3$O$_4$ nanoparticles initially suspended in water- or organic-based solvents, the whole making so-called 'ferrofluids' (reprinted with permission from [54] Copyright (1998) The Royal Society of Chemistry).*

as a ready-to-use suspension (e.g. so-called ferrofluids, as supplied by Ferrotec Corporation, Japan, for instance). At any rate, this field is just starting, and interesting developments are anticipated. Smaller diameter MWCNTs or large-diameter SWCNTs should be able to be used, annealing post-treatments should possibly be able to transform the packed nanoparticles into single nanorods in MWCNTs with narrow inner cavity, and so on.

The *solution method* is quite similar to the latter and consists of putting into contact a concentrated solution of the desired materials with the previously opened nanotubes. This generally requires considering soluble derivatives of the materials ultimately wanted to fill the nanotubes (MWCNT or SWCNT), such as salts (halides or nitrates, usually). It is the compulsory method to use when the desired filling material does not present the appropriate physical constants for inserting itself via the gas phase route (see Section 5a.2.3.3.2) or the molten phase method (see next), both of which should be preferred for their higher filling efficiency and fewer number of steps. Examples where the desired filling material is directly inserted into nanotubes via the solution method are actually seldom, and post-treatments (most often calcination as in [74] or reduction as in [10,11,75] but also other treatments such as photolysis or electron irradiation, as in [65,67]) are most often necessary to obtain the hybrid nanotube with the desired chemical composition (Figures 5a.7(a) and 5a.7(b)). This method was actually the very first one used to deliberately fill SWNTs with Ru from RuCl$_3$ [10] and has been widely used since then.

An interesting feature of the solution method is that it is can be operated at either room or colder temperatures. This allows nanotubes to be filled with thermally unstable materials (e.g. fullerenes grafted with organic functionalities [76,77] or N@C$_{60}$ endofullerenes [78]). A valuable alternative is to keep operating at mild temperature conditions (50°C) but use high pressure (150 bars) to be able to replace regular solvents (water, acids, chloroform, toluene, etc.) with supercritical CO$_2$ [79,80]. The latter is a very powerful solvent and exhibits a high penetrability in nanopores thanks to its very low viscosity and low surface tension. Although operating times as long as several days might be necessary [79], the method is to be considered as being applicable to a large range of compounds.

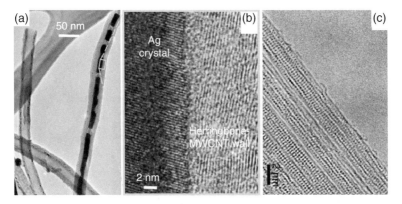

Figure 5a.7   (a) Example of CCVD MWCNTs partially filled with Ag via the solution method (liquid route): nanotubes were soaked in a water solution of silver nitrate upon stirring for 48 h at room temperature and then calcined at 300°C in vacuum. (b) High resolution detailed portion of the nanotube filled in (a), showing that each of the encapsulated Ag rods seen in (a) are monocrystals (reprinted with permission from [13] Copyright (2006) Cambridge University Press). Electric arc SWCNTs were also able to be filled using the same procedure [13]. (c) High resolution TEM image illustrating the high filling rate obtained for KI@SWCNT using the molten phase route combined with thermal cycling, by showing a bundle of SWCNTs, most of them being highly filled with elongated KI single crystals.

Finally, although the solution method is intrinsically multi-stepped to generally account for opening, filling and post-treating the nanotubes, it could also be made into a single step in some specific cases which allow raw nanotubes to be opened and then filled during the same process. For instance, when considering filling in solution, this requires the solution to be able to contain the dissolved filling material as well as to be oxidizing toward the nanotubes. This was achieved in [18] by soaking raw SWCNTs from arc within a super-saturated solution of CrO_3 in concentrated HCl. Both reacted to form CrO_2Cl_2, which is known for being highly oxidizing for polyaromatic carbon, including graphite [81], hence resulting in opening the SWCNTs. The excess CrO_3 was then able to enter the SWCNT cavity, thereby forming CrO_x@SWCNT hybrid nanotubes.

To summarize, the solution method is quite useful as an alternative when other routes (i.e. the gas phase route described in Section 5a.2.3.3.2 and the molten phase method described below) are not possible because of the inappropriate physical properties of the desired filling materials. Thanks to the large variety of compounds that can be found soluble in some solvent, the method appears widely applicable. Except for very specific cases, the method has to be multi-stepped (three steps), which can be a drawback. Nevertheless, the main limitation lies in the concomitant filling with the solvent molecules (along with the dissolved desired material), which cannot be avoided. As a common observation [64,66], this prevents high filling rates from being achieved.

As opposed to the solution method, the *molten phase method* allows high filling rates to be achieved, was among the first methods used to fill MWCNTs (with PbO [19]) and was subsequently demonstrated to be highly successful even for SWCNTs (Figure 5a.7(c) and Figure 5a.8). For instance, SWCNT filling rates as high as 70% were reported for KI [82]

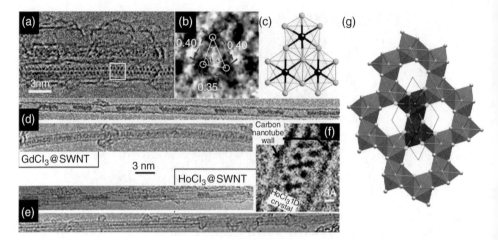

Figure 5a.8 *Examples of SWCNTs filled with lanthanide chlorides via the molten phase route: nanotubes were sealed under vacuum in a quartz ampoule along with the ground material to fill them with, and were then heat treated for 24 h. (a), (b) and (c) HRTEM image, detail and corresponding structure model of TbCl₃@SWCNT obtained at 588°C. (d) GdCl₃@ SWCNTs obtained at 700°C. (e) and (f) HRTEM images of HoCl₃@SWCNTs obtained at 820°C. Filling materials can form continuous nanowires (bottom region in (a) and bottom images in (d) and (e)) or short segments somewhat periodically displayed (top image in (d) and (e)). The crystal structure of these materials is very interesting and often corresponds to reduced fragments 'selected' from the bulk crystal structure, as shown in by the selection in (f). Likewise, the TbCl₃ fragment derived in (c) is a reduced coordination version of a fragment derived from the regular P6₃/m bulk structure shown in (g). Images (d) to (f) are reprinted with permission from [13] Copyright (2006) Cambridge University Press.*

although cumulating filling cycles were necessary to achieve it. This makes the method almost as efficient as the gas phase route obviously because, for both methods, only the desired filling material is entering the nanotube cavity.

The procedure is similar to that for the solution method of the gas phase route, that is putting together the previously opened (or not, see next paragraph) nanotubes and the material to fill the latter within a quartz vessel sealed under vacuum, and then heating up the whole. The temperature has of course to be above (~30–100°C higher) the melting temperature of the filling material and maintained for long times, in the range of 1–3 days (see [82–84], among others) although this may be shorter for some low viscosity materials. Such long times are necessary to account for the high viscosity that the molten materials may exhibit and that slows down the filling kinetics (specifically within SWCNTs), and also to account for cycling (Figure 5a.7(c)). Cycling does not require repeating the whole procedure and, provided a sufficient amount of the material to fill is supplied as a feedstock in the starting sealed ampoule, repeatedly operating within a limited temperature range (e.g. ± 30°C) bracketing the filling material melting temperature was shown to be enough [82]. Again, when high-melting-temperature materials, for example, metals, lanthanides, and so on, are the ultimate filling materials envisaged, compounds such as salts are the preferred materials for the filling steps because they usually exhibit lower melting points (see Table 5a.1). Hence,

Table 5a.1 *Surface tensions (and temperatures for state changes) of some elements and compounds (mostly halides and oxides) that have been investigated in the literature dealing with filling or wetting carbon nanotubes with molten materials (references to the respective data for a filling material are included within this section) (modified from [13]).*

Material	Surface tension (mN/m)	Melting temperature (°C)	Filling temperature (°C)
AgCl	113–173	560	660–560
AgBr	151	432	532–590
AgI	171	455	555
Al	860	660	
BaI_2	130	740	840
Bi_2O_3	200	825	
CaI_2	83	784	884
Cs	67	29	
CsI	69	627	727
Ga	710	30	
$GdCl_3$	92	609	659
Hg	490	−38	
K	117	336	
KCl	93	771	870
KI	70	681	781
$LaCl_3$	109	860	910
LiI	94	449	549
NaI	81	661	761
$NdCl_3$	102	784	834
Pb	470	327	
PbO	132	886	
Rb	77	39	
RbI	70	647	747
S	61	112	
Se	97	217	
Te	190	450	
UCl_4	27	590	690
V_2O_5	80	690	
$ZrCl_4$	1.3	437	487
HF	117		
HNO_3	43		

post-treatments for reducing the intermediate compounds into the desired filling materials are needed. However, interesting observations regarding the behaviour of one-dimensional crystals were already possible on such intermediate hybrid materials (see next).

A valuable alternative proposed by the Oxford Group as early as in 1999 [85] and which has been applied thoroughly since then in most of their filling experiments is to use the chemical activity of the halides toward polyaromatic carbons to get the nanotubes opened within the filling step, thereby making needless a preliminary, separate nanotube-opening step. However, this is mainly valid for SWCNTs only (and DWCNTs), for the reasons related to the number of walls already discussed above. Interestingly Grigorian et al. [86], in their early attempt to dope SWCNT bundles with iodine using the molten phase method, did not realize that this procedure could open SWCNTs. Had they demonstrated in their first paper that iodine was actually partly filling the SWCNT cavities as they showed later [8] this work would have been cited among the pioneering studies on filling SWCNTs with foreign materials.

To summarize, the molten phase method is the second method to be preferred for filling nanotubes because of the possibilities for high filling rates, simplicity (1–3 steps, depending on the goal and the material to fill), and versatility. Its main limitation is its requirement for selecting the materials to fill (or their compounds) among those exhibiting the convenient surface tension (i.e. < 130–170 mN m^{-1} [40]) when molten at the filling temperature.

Finally, the *assisted filling methods* are worth considering because they correspond to attempts to develop original and/or simpler and/or more efficient ways to synthesize filled nanotubes. Two examples will be reported here. One by Mittal et al. [88,89] who proposed to use solutions of halides (FeCl$_3$, MoCl$_5$, and I) in chloroform subsequently irradiated with UV at room temperature. Chlorine moieties are thereby created that are able to attack the SWCNTs, allowing the filling to proceed from the dissolved material. Filling rates were actually low but the procedure was not optimized and might have included cycling, for instance. The method should be applicable to, but also limited to, any compound soluble in chloroform or other UV-sensitive appropriate solvent. The method is dedicated to filling SWCNTs (and possibly DWCNTs) only, for the same reasons as developed above for the opening by molten halides. The other example was proposed by a Japanese consortium and consists of using collisions of accelerated atoms or molecules from a plasma generated between a grounded electrode and a stainless steel plate on which SWCNTs were previously deposited. Nanotubes were filled with fullerenes [90,91] and Cs [91] this way. Oxidized and therefore opened SWCNTs were used for purity purposes only, and the method should work with raw (i.e. closed) SWCNTs as well. The kinetic energies involved are in the range of ~150 eV (for alkali metals) and filling rates can be fairly high (~50% for C$_{60}$@SWCNTs) with respect to the short duration of the run (~1 h). However, damaging to the nanotubes by the energetic bombardment was suspected, yet the extent of this damage was not estimated. This method is likely to be dedicated to filling SWCNTs only, considering the tremendous energy that would be needed to enter a nanotube through a multi-graphene wall.

5a.2.4 Materials for Filling

Virtually every type of material (i.e. atoms, molecules, and phases) has now been inserted into SWNT cavities. This section will provide a thorough listing of various filled nanotubes whose preparation was reported in the literature but references dealing with virtual fillings via modelling will not be provided here and whether the filling claimed in the papers cited was actual will not be discussed either. Indeed, direct evidences for filling, such as TEM imaging, were not always provided in the related papers and the successful syntheses of filled nanotubes were sometimes reported on the mere basis of spectroscopic investigations, which may be mistaken. Because the literature has become particularly abundant, citing all of the related papers was impossible and citing selected references (for example, the pioneering papers) is preferred here.

5a.2.4.1 *Atoms (Isolated, or as Chains)*

Encapsulating atoms so that they remain isolated is difficult because of their relative instability and demonstrating the presence of isolated atoms in SWCNTs is experimentally challenging. When not forming elongated crystals as nanowires (see next), encapsulated

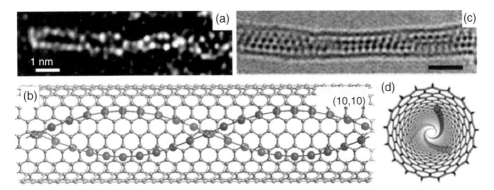

Figure 5a.9 *(a) The first example of SWCNTs filled with a double-helix chain of iodine atoms (I@SWCNT), as evidenced by means of high-resolution Z contrast TEM. (b) Side view of a model for the image in (a) considering a (10,10) tube as the containing SWCNT. (c) High- resolution TEM image of a I@SWCNT, identified as a triple-helix chain of iodine atoms. The bar is ~1.5 nm. (d) Top view of a model for the image in (c). Reprinted with permission from [87] Copyright (2000) American Physical Society (a and b) and from [9] Copyright (2007) American Chemical Society (c and d).*

atoms tend to gather as single atom wide chains, although examples of this are scarce. The first example was iodine [87], which was demonstrated to be able to adopt a beautiful helical chain structure, either single, or as double (Figure 5a.9) or even triple [9] assemblies within the same tube, presumably from commensurability situations between the I-I distance in the chain and some peculiar periodicities of the graphene lattice [87]. Based on this pioneering work, a similar structure was assumed for encapsulated Cs chains prepared later [91]. However, the formation of chains may be prevented and Cs atoms may remain as isolated, observable atoms if the latter are encapsulated along with large and low-reactivity molecules such as fullerenes, which sterically obstruct the filling pathway and do not favour the bonding or adsorption of the Cs atoms to their surface [92]. Isolated K atoms were observed similarly, that is, in SWCNTs concomitantly filled with both fullerenes and potassium [93,94].

5a.2.4.2 Molecules (Isolated, or as Chains)
Except for some rare examples whose motivation was more related to typology issues (e.g. filling BN- or C-MWCNTs with fullerenes [68,69]), filling nanotubes with molecules was and still does principally involve SWCNTs. For most of the following filling materials, the goal was generally to introduce electron donor or acceptor molecules, and thereby to tentatively modify the electronic structure of the subsequent hybrid nanotubes to obtain peculiar electronic behaviours via band gap modulations and charge transfers. The first molecules ever inserted into SWCNTs were C_{60} fullerenes [46,64] thereby forming the so-called nanopeapods [Figures 5a.2(a), and Chapter 5b]. Many fullerene-related filling materials have then followed, including higher fullerenes until C_{90} [95], endohedral fullerenes such as $N@C_{60}$ [78], $Sc_3N@C_{80}$ and $Er_xSc_{3-x}N@C_{80}$ [96], $Dy_3N@C_{80}$ [97], $Gd@C_{82}$ [95,98], $La@C_{82}$ and $La_2@C_{82}$ [99], $Dy@C_{82}$ [95,100], $Sm@C_{82}$ [101,102], $Sc_2@C_{84}$ [102], and $Gd_2@C_{92}$ [103,104], functionalized fullerenes such as [cyclopropa-C_{60}-dicarboxylic

acid diethyl ester (abbreviated as $C_{61}(COOEt)_2$@SWCNT [79]) and fullerenes grafted with a retinal chromophore [77].

Regarding the former category (i.e. higher fullerenes), C_{90} is the largest (empty) fullerene deliberately inserted in SWCNTs so far [95], while Gd@C_{92} endofullerenes were the largest such species encapsulated in this category [103,104]. Elongated capsules obtained from the in situ coalescence of encapsulated C_{60} upon various processes (see Chapter 5b) may also be considered higher fullerenes (for example, consider the formation of C_{120} from the coalescence of two C_{60} molecules). In this regard, the ultimate higher fullerene possibly encapsulated is an inner capped SWCNT, thereby making the whole become a DWCNT (that could also be abbreviated as SWCNT@SWCNT!).

Although fullerenoids are the most popular filling molecules, several other organic molecules have been recently inserted in SWCNTs, including metallocenes [4,5,12,96, 105–108], octasiloxane [6], *ortho*-carborane and related molecules [2,3], fulvalenes [109], Zn-diphenylporphyrin [1,110], Pt-porphyrin, rhodamin-6G, and chlorophyll [110], and squarylium dye [7].

In addition, some of the works deal with filling SWNTs with molecules that are gaseous at room temperature, as opposed to those listed above which exist in the solid phase. However, except for hydrogen, whose adsorption by SWCNTs, MWCNTs and carbon nanofibres was extensively investigated for reasons related to the perspective of a new era based on hydrogen economy [111], gaseous molecules investigated are still scarce, but include Xe [112], O_2, and N_2 [113], and so on. Even more critically than for any other filling material, filling SWCNTs (or MWCNTs) with gaseous molecules whose physisorption is the main interaction mechanism with the nanotubes makes highly questionable the actual location of the adsorbed molecules. Considering SWCNT bundles, the inner nanotube cavity is only one of four distinct adsorption sites identified from their various adsorption energy potential (see Figure 1.20 in Chapter 1, and [114]). In addition, the demonstration of the actual adsorption of some gaseous molecule onto or within any material, including CNT, is generally indirect, via so-called gravimetric or volumic methods, with which the only evidence for gas adsorption is obtained from the weight gain of the nanotube sample or the measured consumption of a known volume of feedstock gas, respectively. This opens wide the door for misinterpretations from various sources or errors independent from the tested material (e.g. leaks, H_2O traces, impurities, buoyancy effects, etc.), which are likely to contribute significantly (or even prevalently) to the demonstration of an adsorption phenomenon onto nanotubes. That is why the supposedly ~7 wt.% filling of gaseous H_2 in SWCNTs reported as early as 1997 [115] has to be considered with caution, first because it was later on demonstrated that the high amount of H_2 claimed for being trapped in the SWCNTs resulted from a misinterpretation [116], and also because there was no demonstration of the effective filling, that is, the insertion of H_2 molecule gas into the SWCNT inner cavity was not demonstrated, meaning that the H_2 molecule adsorption could have occurred in one of the other three adsorption sites mentioned above as well. It was demonstrated only far later that the inner cavity of the SWCNTs does contribute as an adsorption site for H_2 [117], yet the total amount of adsorbed hydrogen molecules remains low (less than 1 wt.%) at least at room temperature. The adsorption of permanent gases in nanotubes will not be considered further in this chapter.

Discrete molecular anions were also introduced and imaged within DWCNTs [118]. The type of anion observed is a Lindqvist ion which belongs to the family of inorganic poly-oxometalate (POM) ions which are effectively ordered oxide clusters built up of metal

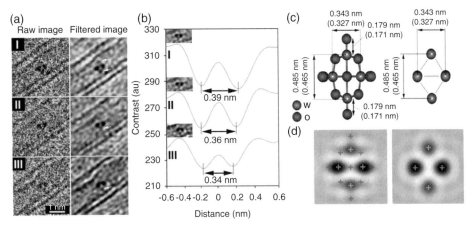

Figure 5a.10 *(a) A sequence of HRTEM images (left) obtained from a single [W_6O_{19}]$^{2-}$ anion locked into position within the capillary of a DWCNT shown with Fourier-filtered images (right). (b) Line profiles produced through equatorial spots corresponding to pairs of W atoms. (c) Structure models of the [W_6O_{19}]$^{2-}$ anion with oxygen included (left) and excluded (right). Selected bond distances are included for the unrelaxed anion (in brackets) and the relaxed anion (from MD). (d) HRTEM image simulations produced at ideal defocus from the structure models in (c) showing the effect of including oxygen in the image simulation calculation. When the oxygen atoms are included (left simulation), the centre of the spot contrast in the equatorial W atom pairs are distorted outwards towards the three terminal oxygen atoms on each side of the anion. When the oxygen atoms are not included (right simulation), there is relatively little distortion of the W positions (red crosses = O; green crosses = W). For a better understanding of the figure, please refer to the colour plate section. Reprinted with permission from [118] Copyright (2008) American Chemical Society.*

oxide polyhedra in various face, edge and corner sharing configurations. As a result of the corresponding oxidation states of the heavy metal cations and charge balancing with the surrounding O^{2-} atoms, these anions almost invariably have a net negative charge which is counterbalanced, typically, by organic counterions, for example [NBu_4]$^+$, or more complex counterions, such as dendrons. The resulting anions are nonspheroidal (cf. the fullerenes or endofullerenes) and are comparable in size to the inner diameter of the containing nano-tube. As a result of this external morphology, the anions can 'lock in' to position within sterically matched nanotube capillaries and this has the advantage that the structure of the anion can be studied with greater clarity than other more mobile species not rigidly fixed in position including, for example, the metallocenes, carborane, fulvalenes, and so on, mentioned above and also many of the fullerene and endofullerene examples mentioned in this section and Chapter 5b, in particular when the included species is not sterically matched with the encapsulating nanotube (i.e. consider the inclusion-nanotube spatial relationships in Figure 5a.1(b)–(e)). In the recent example, a Lindqvist ion of the form [W_6O_{19}]$^{2-}$ (i.e. a 'super-octahedra' comprised of six WO_6 octahedra fused together at their faces) was inserted into DWCNTs and then imaged by HRTEM (Figure 5a.10). As a result of the locking-in it was possible to study the separation between two atom columns each contain-ing just a pair of W atoms. Once the contribution of the HRTEM image contrast due to the oxygen atoms was taken into account, it was possible to verify that a small expansion had taken place in situ, possibly associated with an electronic interaction between the nanotube and the encapsulated anion. The nature of this interaction is currently the subject of ongoing theoretical investigations.

5a.2.4.3   Pure Elements (as Nanowires or Nanoparticles)

As opposed to encapsulated iodine or caesium atoms, which were demonstrated to gather as monoatomic chains (e.g. Figure 5a.9), other elements may form wires when encapsulated in SWCNTs, most often crystallized metals (e.g. Figures 5a.2(b) and 5a.8(c)). SWCNT- (or DWCNT-) encapsulated elements include Ru [10], Bi [1311], Co [15], Se [66], Ag [74,75,85,119,120], Au [75], Pt [75], Pd [75], and Fe [12,121,122] among others. The Bi and Se fillings are distinct from the other pure element fillings because they were inserted either from the liquid phase or the vapour phase, whereas the other fillings were inserted as intermediate compounds (e.g. salts) in a first step, and then formed in situ via post-treatments (e.g. thermal treatment and/or hydrogenation). The reasons why, upon encapsulation in SWCNTs, some atoms such as Bi and Se form several atom-large nanowires while other atoms such as I and Cs form single atom chains have not been fully discussed. The reasons do not lie in discrepancies in the respective, average diameters of SWCNTs, because the latter mainly originate from the regular electric arc method that provides SWCNT diameter distributions in the 1–2 nm range with a median diameter of ca. 1.4 nm. The reasons neither lie in the respective filling processes because that used for I (i.e. molten phase filling [85]) is very different from that used for Cs (ion bombardment [90]), whereas the former is similar to that used for Bi [11] and Se [14,66] among others. The existence of some commensurability between the atom chain periodicities and the surrounding, curved graphene lattice was first proposed in [8] that could provide the driving force for generating the helical, single atom chains as the configuration of lesser energy. This was, however, not confirmed by further studies [9] and the aspect has not been investigated for Cs chains. The commensurability-related explanation should therefore be reasonably considered as a mere possibility so far. For untwisted periodic crystallites formed within MWCNTs, they cannot be directly commensurate with the innermost SWCNT if the conformation of the latter is chiral (i.e. not zigzag or armchair) [49] and can only be truly commensurate if the crystal periodicity matches that of the encapsulating armchair or zigzag nanotube or is some integral multiple of them.

5a.2.4.4   Compounds (as Nanowires or Nanoparticles)

Filling SWCNTs with compounds is a very important topic likely to provide successful alternatives when filling with chemical elements is not possible, typically because of too high a melting point and/or surface tension of the considered elements at the molten state (see Table 5a.1). Considering compounds may also give access to filling via solutions, as an alternative when a low-temperature filling process is needed or when the elements considered are poorly soluble. Compounds are typically salts, among which halides are the most popular, for the several advantages they offer such as high versatility (many types of halides are possible, that is, involving Cl, I, Br, or F, thereby offering a wide range of melting temperatures and solubilities); suitable range of melting points (i.e. not too high, e.g. < 1000°C) in many cases; good solubility in usual solvents (chloroform, water, etc.); and high chemical reactivity toward carbon that allows the nanotube to be opened and then filled by the molten halides all at once.

One of the most active groups involved principally in filling SWCNTs with inorganic materials has been based in Oxford although this is now been carried out by many other groups around the world. First, based on previous works on filling MWCNTs [60], SWCNTs have been filled with many kinds of halides and other compounds since 1998 [10], mainly using the melting method, but also using the solution method and, in a few instances, the gas phase route. The following list mostly corresponds to the materials inserted in SWCNTs (or DWCNTs) as reported by this group in the references provided (including review papers [37,123–125]), and includes an increasing number of materials inserted by other groups also. The list is likely incomplete, but provides an idea of the extensive work carried out in the field: HgTe [27], MnTe2 [126], (Na/Tl/Cs/Ag)Cl [37,125,127], (Cd/Fe/Co/Pd)Cl$_2$ [37,61,125,127], Ln(La to Lu)Tb/Ru/Ho/Gd/Au/Ag/Y/Fe)Cl$_3$ [10,20,25,61,83,88,121,127, 129], Al$_2$Cl$_6$ [37], (Hf/Th/Zr/Pt)Cl$_4$ [65,75,124,127], MoCl$_5$ [88], WCl$_6$ [37], (KCl)$_x$(UCl$_4$)$_y$ [82,85], CsBr [37], AgCl$_x$Br$_y$I$_z$ [48,82,85,130,131], (Li/Na/Cs/K/Rb/Ag/Cu)I [18,23,48, 123,131,132,133], (Ca/Cd/Co/Sr/Ba/Pb/Hg)I$_2$ [26,28, 123,127,134] (Te/Sn)I$_4$ and Al$_2$I$_6$ [37,129].

From 2001, oxides were also directly inserted into SWNTs, starting with CrO$_3$ [18,119] and Sb$_2$O$_3$ [135,136] and then followed by others such as PbO [84], Re$_x$O$_y$ [67], and UO$_2$ [137] (although the latter was obtained from the primary filling of uranyl acetate) and hydroxides (KOH and CsOH [137]), sometimes with a fairly high filling rate (e.g. 80–90% for PbO). Although oxides can be interesting alternatives as intermediate compounds or can exhibit interesting intrinsic properties on their own (e.g. CrO$_3$ is electrically conductive), they will never become as popular as halides because oxides with high solubility in harmless solvents or reasonable melting temperatures are not many. Other popular compounds that were attempted for filling SWCNTs are typically nitrates, for example, silver nitrate [75,119,120], bismuth nitrate [11], uranyl nitrate [137], and so on.

Figures 5a.7 and 5a.8 provide examples of filling achieved with or via such compounds, respectively. In the case of Figure 5a.7, elemental Ag was initially introduced as the corresponding nitrate dissolved in water, while GdCl$_3$ and HoCl$_3$ (Figure 5a.8) were inserted from the corresponding molten salt.

5a.2.5 Filling Mechanisms

Data regarding physical properties of hybrid SWCNTs have now started to be reported, which is the most exciting part of this research field (see Section 5a.3). This should not suggest that synthesis mechanisms are well understood yet, specifically regarding filling via liquid routes although the work by Dujardin et al. [39,40] clearly indicates that the surface tension of the liquid used to fill nanotubes is as a key determinant for successful filling. However, because the pressure difference that makes a liquid enter and proceed within a tube is proportional to the liquid surface tension and inversely proportional to the tube diameter, according to a relation obtained by combining Young-Laplace's laws and Jurin's law [138], narrow tubes should fill over a longer length range than large tubes, meaning higher filling efficiency. However, this is not consistent with the conclusions from early works dealing with filling nanotubes with liquids, while comparing MWCNTs with a narrow and wide inner cavity, respectively. Indeed, until 1999, the filling of SWCNTs was estimated to be inefficient [39,139], and Ugarte et al. [38] actually stated that 'there appears to be a preference for the wider cavities to fill compared with the narrower cavities.' This suggested that the laws (such as Young-Laplace's laws) driving regular (macro)wetting phenomena that were established for sub-millimetre capillaries could not apply to capillaries in the nanometre range ('nanowetting'). This was further supported by early predictions [140] who calculated that increasing the tubule radius decreases the incarceration energy and the overlap repulsion activation barrier at the mouth of the tube, from which it was deduced that material encapsulation in narrow SWCNTs via liquid or gas routes should be not be favoured with respect to larger MWCNTs.

However, actual experiments carried out on SWNTs were found to contradict such early statements based on MWCNTs. We did observe that filling efficiency appeared often better for regular SWCNTs (inner cavity ~1.4 nm on average) than for MWCNTs (inner cavity ~30–40 nm on average, herringbone type) for molten materials such as $GdCl_3$ and CoI_2. Likewise, condensation of Se was found to occur first in the narrowest nanotubes [66] and filling with Bi via the gas and liquid routes was found to be more efficient in SWCNTs than in MWCNTs 'as a result from stronger capillary forces' [11]. Generally speaking, SWCNT filling rates as high as ~70% (PbI_2 [28] and KI [82]) or even 80–90% (for PbO [84]) were observed and we can therefore conclude that nanowetting is a reality.

However, chances are high that discrepancies may occur in nanowetting with respect to regular macrowetting. For instance, it is reasonable to admit that there should be a minimal tube diameter below which filling is no longer possible. In macrowetting, molecules from the liquid located at the liquid/solid interface go slower than molecules from the liquid core, somehow inducing a convection-like effect at the progression front where friction-affected molecules are continuously replaced by unaffected molecules, which thereby feed the liquid/solid interface as the liquid progresses within the tube. In nanowetting, there should be a nanotube diameter below which the friction forces may affect all of the molecules of the liquid because they all can be considered as surface molecules, thereby preventing convection-like effects from occurring, thereby hindering filling. However, giving such a threshold value cannot be given because it also depends on the surface tension of the liquid material used to fill the nanotubes, for example, 4 nm for molten Al_2O_3 and 0.7 nm for molten V_2O_5 [38]. This might explain why extensive filling studies on SWCNTs have demonstrated that there is no direct relation between surface tension and filling efficiency [82]. In addition, as opposed to macrowetting, gravity is certainly not to consider in nanowetting because of the tiny volumes of liquid involved that make related weights negligible with respect to capillary forces. On the contrary, the role of viscosity, which is independent from surface tension (Hg provides a good example of a low viscosity liquid, 1.53 cP, with high surface tension of 490 mN m^{-1}), is likely to be enhanced in nanowetting. In macrowetting, viscosity is considered to only affect the wetting dynamics. The reason is, whatever the tube diameter, friction forces increasingly oppose the wetting forces because the tube/liquid contact surface where the friction forces take place increases as the liquid proceeds in the tube, while the gas/liquid/solid contact line where the capillary forces take place remains constant. This may ultimately cause the liquid progression to stop.

It is likely that viscosity and/or nanowetting are not the only criteria to consider for controlling the filling of nanotubes. The size of the atoms or molecules in the liquid, for instance, certainly matter for nanotube cavities below ~1 nm, specifically considering the solution method where the molecules to fill the nanotube with are surrounded by solvating molecules. In addition, the intrinsic saturating vapour pressure value of the desired filling materials is likely to be important when using the molten phase method. Despite the filling temperature being maintained below the vaporization temperature of the filling material, some vapour pressure actually develops in the sealed ampoule. This pressure is likely to help the molten material to enter the SWNT cavity, with the consequence that materials with higher saturating vapour pressure should provide higher filling rates (considering that other parameters are equal). Such an aspect has not yet been investigated. Finally, local conditions may play an important role as well, as probably the only reason why, within the same filling experiment, some nanotubes are found extensively filled, whereas neighbouring ones are found empty, yet obviously wide open (Figure 5a.7(a)).

5a.3 Behaviours and Properties

As reminded above, the synthesis of X@CNTs, specifically involving SWCNTs, has a rather young history and examples of experimental measurements showing the peculiar properties of filled SWCNTs are still relatively few. Likewise, little theoretical work to tentatively predict the electronic properties of X@CNTs (excepting peapods) has been carried out thus far and expectations are mainly based on the anticipated possibility to modulate the SWCNT band gap thanks to the filling material, or to modulate the band gap of the latter thanks to the enforced structural deformations. Yet a variety of filled SWCNTs have been modelled (e.g. filled with materials as varied as KI [141], $(Sc,Ti,V)_8C_{12}$ carbohedrenes [142], water [143], DNA [144], Cu [145], Ag and CrO_3 [32], acetylene [146], Ge [147]), a majority of theoretical works mostly concerns transition metal filled SWCNTs, more specifically Fe, and aims at predicting the related magnetic behaviour with respect to the iron wire structure [148,149–152].

5a.3.1 Peculiar in-Tube Behaviour (Diffusion, Coalescence, Crystallization)

Thanks to the unidirectional feature of the inner cavity in carbon nanotubes, specifically in SWCNTs, a variety of peculiar in-tube behaviours of the filling materials can be observed, which occur either spontaneously upon filling, or under the effect of an external stress (annealing, irradiation…).

Several interesting examples can be found with peapods, whose some are given below, that will be further commented in Chapter 5b. When subjected to an electron beam in a TEM, fullerenes encapsulated in SWCNTs have actually shown a series of peculiar behaviours including diffusion [153], dimerization [46,153,154] and coalescence [154]. Such behaviours were specifically permitted by the encapsulated situation of the fullerenes because no such behaviour was observed for fullerenes in bulk fullerite (i.e. nonencapsulated) when electron irradiated. The former (fullerene diffusion) requires the filling of the SWCNTs to be partial only, in order to leave the room for observing the phenomenon. It is illustrated in Figure 5a.11. Another example can be found in [13].

On the contrary, the two latter (dimerization and coalescence) require the fullerenes to be densely packed. They are illustrated in Figure 5a.12. Briefly, irradiated fullerenes tend to first get closer to each other in pairs, assuming dimerization (Figure 5a.12(b)), and then to coalesce into elongated, somewhat distorted capsules (which may be regarded as higher, elongated fullerenes).

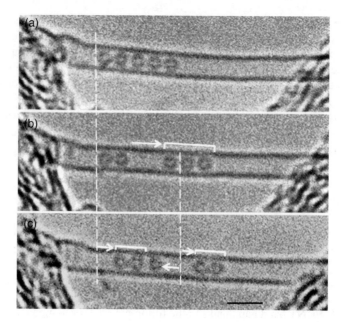

Figure 5a.11   *Sequence of high resolution TEM images (100 kV, ~10 seconds between images) illustrating the irradiation-promoted diffusion of the encapsulated fullerenes within a SWCNT (the 'nanoshuttle'). (a) Starting situation, showing an ensemble of five fullerenes. (b) The five fullerenes suddenly split into an ensemble of three fullerenes which move to the right, leaving behind an ensemble of two fullerenes. (c) One fullerene from the ensemble of three moves back to join the former ensemble of two and leaves another ensemble of two. Meanwhile, both the ensembles of two have slightly moved to the right. All the moves are presumed to be ionization-driven, explaining why they are random and either repulsive or attractive (depending on whether electrons are captured by or taken from the fullerenes). Modified with permission from [103] Copyright (2003) American Physical Society.*

Chapter 5b will show that fullerene coalescence may also occur upon several other kinds of treatments, including thermal annealing [103,155], combinations of thermal annealing and electron irradiation [56,156,157] and photon (laser) irradiation [158,159]. Thermal annealing goes in the direction of thermodynamic stability and ultimately can result in the destruction of the tubular morphology then transform the whole peapod material into a graphite-like material. On the other hand, irradiation combines thermal energy supply and ballistic damaging which ultimately can be destructive to the tube. Hence, depending on the amount of energy supplied and the way it is supplied (e.g. by temperature, or situations of resonance upon laser irradiation), the coalescence process can produce elongated capsules, an inner SWCNT (thereby making a DWCNT), a polyaromatic material, or amorphous carbon.

Low-dimensional crystals formed within SWCNTs do not undergo quite the same types of structural transformations which are described in the previous section for peapods. Related dynamic behaviours upon electron irradiation have however been reported for other materials encapsulated in SWCNTs such as clusterization of $ZrCl_4$ crystals [65] and the observed steady rotation of Re_xO_y clusters [67]. The former (clusterization) is distinct

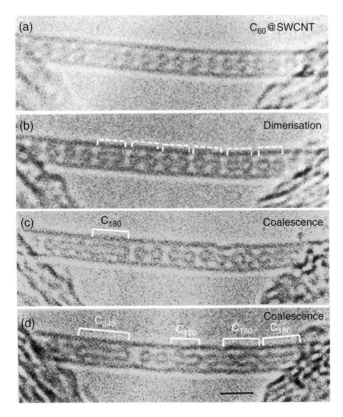

Figure 5a.12 *Sequence of high resolution TEM images (100 kV, ~300 s between images) showing the progressive coalescence of C_{60} molecules to higher fullerenes as elongated, distorted capsules under the effect of electron irradiation, starting from a regular C_{60}@SWCNT peapod. (a) Starting situation, exhibiting a well periodic display of the fullerenes. (b) Dimerization occurs first, making the C_{60} get closer by pairs. (c) Coalescence starts, yet not uniformly. (d) Coalescence proceeds, which may make the formerly formed elongated capsules merge as well. Modified with permission from [154] Copyright (2000) Elsevier Ltd.*

from the progressive agglomeration that was described for fullerenes, as instead we see an effective breaking down of a local crystal structure, whereas the latter (cluster rotation) is commonly observed for both types of encapsulates formed within SWCNTs.

In addition, the mere fact that encapsulating crystals in SWCNTs enforces a nanowire morphology that is generally unusual for most of the filling materials tested thus far is intrinsically a peculiar behaviour, as demonstrated by attempts to remove the carbon sheath, which resulted in the loss of the nanowire morphology for several halide crystals [48]. Moreover, a second major behaviour that is specific of the encapsulation in SWCNTs is to frequently generate often profound structural modification for a material with respect to the same material as it is observed in the bulk state. Structural crystallographic peculiarities actually include preferred orientation with respect to the nanowire elongation axis, systematically reduced coordination, lattice expansion and/or contraction, and, occasionally,

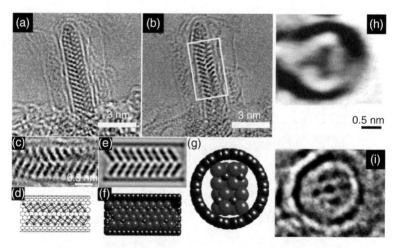

Figure 5a.13 *(a) and (b) High-resolution TEM images of a tip of a SWCNT filled with PbI₂ for two different focus values of the objective lens. (c) Enlarged image of the area framed in (b). (e) Corresponding calculated image, based on the model in (d). Although bulk PbI₂ structure is a two-dimensional layered type, encapsulation in SWCNTs enforces it to become a one-dimensional polyhedral chain. (f) and (g) Ball and stick models of the structure modelled in (d), as a side and a cross-section views respectively. (h) High-resolution TEM image of a cross-section of a SWCNT filled with CrOₓ. The filling material structure exhibits a three-fold symmetry in which each lobe is assumed to be the projection of a chain of CrO₄ tetrahedra (reprinted with permission from [18] Copyright (2001) Elsevier Ltd). (i) High-resolution TEM image of a cross-section of a SWCNT filled with HoCl₃. The filling material structure exhibits a somewhat five-fold symmetry, yet is deformed. Based on the observed structural similarities with NdCl₃ [24], each dark spot is assumed to a projection of a single 1D chain of HoClₓ polyhedra. For a better understanding of the figure, please refer to the colour plate section.*

new crystal structures. Preferred orientation may originate from the need to maintain the stoichiometry as far as possible, in the case of binary or higher order compounds [123]. Reduced coordination is obviously a consequence of the sterically confined space in which the encapsulated crystals have to grow, which may also be the cause for the lattice expansion in the radial direction as observed for CrO_x [18] and KI [22]. Lattice distances along the wire axis are less likely to overcome distortions because the confinement is relatively nil in this specific direction but numerous examples of a lattice contraction in this specific direction can be found, for example, KI, [23] Sb_2O_3 [135,136] and even fullerenes because the 1 nm C_{60}-C_{60} distance in fullerite was found to become 0.97 nm for C_{60} chains in peapods [160]. Such lattice contraction behaviours correspond to the Poisson effect, which is well known in bulk material mechanics. Such lattice distortions are significant, and may reach 14% or more. Despite the crystallization of binary halides may adopt different structures (three-dimensional, layered, chain and molecular in bulk state) [161], it is clear that the limited space available in SWCNTs (and DWCNTs) will not allow all of them to develop (specifically for the three-dimensional and the layered structures), depending on the radius of the atoms involved (Figure 5a.13 (a)–(d)). Interestingly, a recent study [162] showed that in wide inorganic nanotubes based on WS_2, not only does the layered iodide PbI_2 introduced from the liquid phase, revert to its bulk structure type but this distorts in order to allow the encapsulated iodide to form core-shell nanotube structures, forming in effect tubes within tubes, inside the WS_2 capillaries. It is interesting to speculate that similar structural behaviour might be observed for wide nanotubes based on carbon although this is yet to be reported.

For a given compound, the structure type may change with respect to the available inner space within the nanotube capillaries [28,37,124,134]. Ultimately, structural constraints can be such that new structures are created, for example, for $HoCl_3$ [13], CrO_3 [18], $NdCl_3$ [24], CoI_2 [26], or BaI_2 [134], giving rise to peculiar features such as unusual symmetry (Figures 5a.13(h),(i)). Finally, it was observed that the ability of the filling material to crystallize may depend on the available inner space on the one hand, and on the deformability of the tube wall on the other hand. Indeed, small diameter SWCNTs were found to accommodate the contained crystal structure (CoI_2 [26] or PbI_2 [28]) to such an extent that the host SWCNTs exhibit oval cross-sections. Accordingly, PbI_2 was found to be amorphous when filling DWCNTs with similar small diameter, presumably because the deformability of the tube wall is much decreased, hence not allowing the crystal structure to develop.

Molecular dynamics (MD) simulations which are based on interatomic and intermolecular forces rather than specific electronic interactions are also proving to be very useful tool for both interpreting and predicting filling behaviour. Indeed, this has been prominently applied to ab initio predictions of the crystallization of water within nanotubes [34,35]. Wilson and Madden [141] have used such an approach to model the crystal growth behaviour of KI in variable diameter SWCNTs using such an approach. The interactions between the ionic K^+ and I^- species were described by standard Born–Mayer pair potentials with the assumption that the ions retained their formal integer charges. Interactions between the ions and the walls of nanotubes of varying diameters were described by Lennard–Jones potentials, with parameters derived from potentials for the interactions of isoelectronic inert gas atoms with the carbon atoms of the graphene surface. Time-resolved and minimum energy simulations predicted the filling of (10,10) SWCNTs with thermodynamically ordered arrays of 2×2 KI crystals starting from an open SWCNT immersed in molten KI. Minimum energy simulations also predict lattice distortions that were consistent with those observed experimentally. Wilson subsequently adopted a similar approach to produce in effect a 'phase diagram' of minimum energy versus observed structure type (Figure 5a.14(b)) [163]. As a result, a phase diagram could be constructed which predicts, in an ab initio fashion, the predominant structure type for the typical median range of nanotube diameters which form for SWCNTs. For diameter ranges of 1.1–1.3 nm and 1.6–1.8 nm, effectively $2 \times 2 \times \infty$ and $3 \times 3 \times \infty$ rock salt-type 1D KI crystal structures are predicted. Outside these ranges, other 1D crystal structures are expected. In the lowest diameter range studied (i.e. 1.0–1.1 nm), a so-called T3 structure is predicted, which is a twisted crystal structure in which the cross-section of the crystal consists of a three-membered ring (i.e. –I–K–K– or –I–I–K–). Similarly, in the ranges 1.3–1.35 nm and 1.47–1.55 nm, T5 and T7 twisted 1D crystal structures with five– (e.g. –I–K–I–K–I–, etc.) and seven– (e.g. –I–K–I–K–I–K–I–, etc.) membered rings formed in cross-section, respectively. In one range, from 1.35–1.47 nm, a so-called 'hex.' structure is formed in which the encapsulated structure can best be described in terms of a stacked 1D array of alternating six-membered shells of the form –I–K–I–K–I–K– and –K–I–K–I–K–I–.

Even more remarkably, this technique was applied by the same researcher to predict the formation of inorganic nanotubes within SWCNTs (i.e. a tube within tubes) [164]. Similar approaches were also adopted with regard to predicting both ice phase transitions within carbon nanotubes and also the formation of ice tunnels and crystals within the same (see Figure 5a.15 (a)–(d)) [34,35]. In an extension to the KI modelling work, both chiral and achiral inorganic nanotubes form within SWCNTs according to the ambient nanotube diameter (Figure 5a.15(e) and (f)). What made this work particularly remarkable is that it successfully predicted the formation of a later observed nanostructure, HgTe in SWCNTs (see Figure 5a.15(g)-(n)) [27].

5a.3.2 Electronic Properties (Transport, Magnetism and Others)

The structural features described above provide good prospects that some hybrid SWCNTs should exhibit quite peculiar physical properties, specifically regarding transport and magnetism. As far as properties are concerned, the distinction should be made between (i) the peculiar behaviours of the filling materials specifically resulting from their encapsulation; (ii) the properties intrinsically from the filling materials that are naturally transferred to the whole hybrid nanotube; (iii) and the peculiar properties that may result from the nanotube/filling material interactions. With respect to filled MWCNTs, case (iii) is not to be expected because the interaction, if any, can barely affect more than the innermost graphene and the properties of the multi-graphene wall are dominated by that of the polyaromatic layer stacks. Case (i) is not impossible but seldom because, as already pointed out in the *Introduction* section to the chapter, the inner cavity is often too large for the confining conditions to be significantly stressful to the filling materials, leaving little chance for peculiar properties. Case (ii) is actually more likely and one example of it is the filling with ferromagnetic material, making the whole hybrid CNTs ferromagnetic as well.

Hence, the investigations of filled CNT properties have focused on filled SWCNTs rather than filled MWCNTs. Unfortunately, such examples are still relatively few. Probable reasons are that although high filling rates can be achieved that are suitable for structural investigation, the whole material may remain too imperfect (e.g. impurity content) to investigate them as a bulk in a reliable manner. On the other hand, investigating isolated, single X@CNTs requires nanolithography facilities and specifically-designed measurement devices that are not of common and easy access yet. In addition to possessing the related technological knowhow, explaining and understanding the phenomena observed requires one to be able to characterize the very same X@CNT tested to ascertain the extent of filling and the structural features of the encapsulating SWNT, which is definitely not routine.

Again, most of the work published thus far deals with peapods and related materials and has addressed issues related to thermal properties [165], transport and charge transfers (between the encapsulated materials and the containing SWCNT) [95,100,165–172]. Those aspects will be presented in Chapter 5b.

Less abundant in literature, other interesting and promising observations have been made on SWCNTs filled with materials other than fullerene-related. As early as in 1998, the electrical resistivity of I@SWNT was found to be significantly decreased with respect to pristine SWNTs, with some evidence for charge transfer [86]. However, it is difficult to ascertain that this result accounted for the contribution of encapsulated iodine, because iodine *intercalation* (i.e. of I atoms in the interstitial spaces between hexagonally packed SWCNTs within bundles and also in the spaces between bundles) certainly occurred as well. It is not a trivial issue to question whether I doping in interstitial sites and within capillaries contribute similarly because both sites do not necessarily 'see' the same graphene curvature and different electronic interactions may therefore result. Interactions between foreign materials other than iodine for which the filling position is ascertained and the encapsulating SWCNT is, however, demonstrated, resulting in, for example, charge transfer (with Ag and CrO_x [32], AgCl [128]), reduction of band gap (with $MnTe_2$ [126]), energy transfer modifying the optical adsorption range (with squarylium dye) [7].

At this stage, it is certainly useful to consider first principles ab initio density functional theory (DFT) calculations and, in this regard, some very useful work is starting to be done. In the case of filled SWCNTs formed between $M(\eta\text{-}C_5H_5)_2$ where M = an early transition metal (i.e. Fe, Co), first principles density functional pseudo-potential calculations were used to investigate the nature of interactions between 'flat' graphene, SWCNT, and the intercalated metallocene complex [173]. A further variable considered was also the respective diameter and structural conformation of the encapsulating nanotube. The obtained results demonstrated that the composites were stabilized by weak π-stacking and CH $\leftrightarrow$ π interactions, and, in the case of the $Co(\eta\text{-}C_5H_5)_2$@SWCNT composites there is an additional electrostatic contribution as a result of charge transfer from $Co(\eta\text{-}C_5H_5)_2$ to the nanotube. The extent of this charge transfer was rationalized in terms of the electronic structures of the two fragments, or more specifically, the relative positions of the metallocene highest occupied molecular orbital (HOMO) and the conduction band of the nanotube in the electronic structure of the composite. The study concluded that control over the electronic properties of specific SWCNTs, which have been selected from the bulk according to the nanotube diameter and/or chirality, might be achievable via the non-covalent modification of SWCNTs by neutral and charge transfer molecules. Experimentally, this kind of transfer had previously been observed by a combination of optical absorption and Raman spectroscopy [5] so this type of synergy between theory and experiment is clearly valuable.

In this regard, modelling has certainly shown to be useful to understand hybrid SWCNTs well and to support experimental measurements. Such a consistency is claimed for Ag- or CrO_3-filled SWCNTs, whose modelling confirms to the donor and acceptor roles, respectively, of the filling materials toward the containing SWCNTs in agreement with the resonant Raman data [32]. Electron transfer was also found in bulk ferrocene@SWCNT material upon photoexcitation [105]. However, the recent work by Carter et al. [27] can be considered the first full demonstration of the modification of the electronic behaviour (namely, the change from semimetal to semiconductor) from the geometric constrains imposed on the encapsulated crystal in a HgTe@SWNT material. It was subsequently noted [174] that the observed new structural form of HgTe in which Hg is in trigonal coordination and Te is in half-octahedral coordination [27,174] is effectively isostructural with the new form of inorganic nanotube predicted by Wilson (see Figure 5a.15 (e)–(n)) [164]. Recent DFT work by Yam et al. [175] Sceats et al. [176] and also Christ and Sadeghpour [177] has also been applied to the KI@SWNT system and all demonstrate a clear tendency for, in particular, the 2×2 inclusion to promote metallic behaviour in the encapsulating nanotube principally as a result of charge transfer. The recent apparent observation of Lindqvist ion expansion in DWCNTs [118] also offers intriguing possibilities for studying the structural response of polyoxometalates to charge transfer between the filling material and the containing nanotube.

The magnetic behaviour of filled SWCNTs is also a fast-growing topic, although theoretical works [145,149–151] are still more numerous than experimental ones [12,121,122,178].

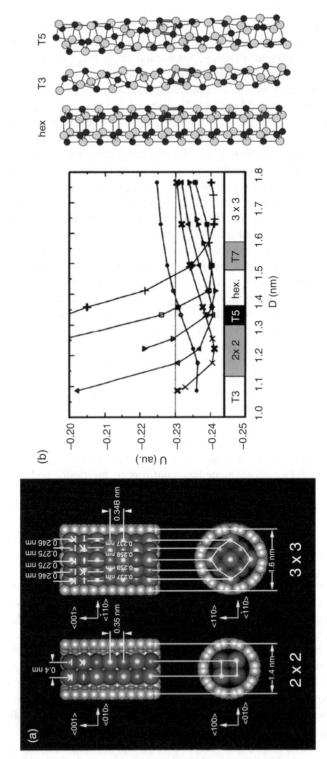

Figure 5a.14 (a) Structure models of 2 × 2 and 3 × 3 layer KI@SWCNT hybrid nanotubes determined experimentally. (b) Minimum energy (U) phase diagram depicted for structure types of KI predicted to fill SWCNTs of varying diameter (D). T3, T5 and T7 are helical KI 1D crystals with three-, five- and seven-membered rings in cross-section, respectively. For a better understanding of the figure, please refer to the colour plate section. Reprinted with permission from [163] Copyright (2002) Elsevier Ltd.

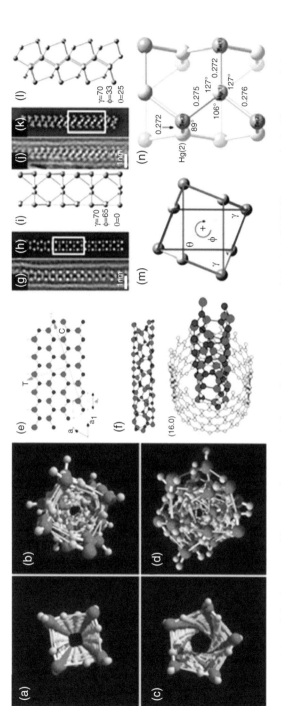

Figure 5a.15 (a) and (b) Square and hexagonal ice tunnels formed within (14,14) and (15,15) SWCNTs respectively. The containing SWCNTs are not shown for clarity. (c) and (d) show the corresponding liquid phases, respectively (reprinted with permission from [34] Copyright (2001) Macmillan Publishers Ltd). (e) to (f): The construction of an inorganic nanotube from a corresponding hexagonal planar structure. The red and blue circles represent the two ionic species. C_h is the chiral vector along which the structure must be folded in order to form a (3,2) hexagonal nanotube. The bottom sketch in (f) shows a perspective view of a (3,2) inorganic nanotube within a (16,0) SWCNT. This nanostructure turned out to be nearly identical to a similar nanostructure later reported within a SWCNT by HgTe [27]. (g), (h) and (i) show a reconstructed phase image, corresponding image simulation, and structure model respectively, of one projection of a twisted, chiral HgTe crystal in a SWCNT. (j), (k) and (l) show similarly a phase image, simulation, and model from a second twisted HgTe crystal in a separate carbon nanotube. (m) and (n) show the end on views of the essential structure motif and corresponding DFT-optimized model corresponding to the materials shown above them respectively. There is a close correspondence between this model and the model shown in (f) (reprinted with permission from [164] Copyright (2004) Elsevier Ltd). For a better understanding of the figure, please refer to the colour plate section.

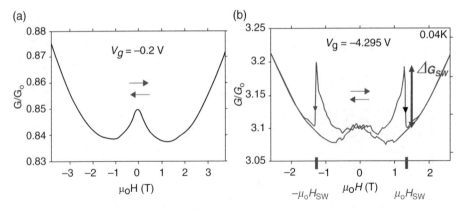

Figure 5a.16 *Variation of conductance versus cycling magnetic field measured on individual Co@SWCNT bundles contacted (with Pd) following the transistor-type device, for an angle of 25° between the direction of the outer magnetic field and the elongation axis of the contacted bundle, and at a temperature of 40 mK. (a) 90% of the devices do not show a hysteretic behaviour, as illustrated, whatever the gate voltage. (b) 10% of the devices show a hysteretic behaviour, as illustrated. Conductance curves indeed exhibit a hysteresis with two conductance jumps. They correspond to the magnetic field value at which the magnetization direction of a nanomagnet (presumably one, or several but aligned, encapsulated cobalt wire segment(s) reverses (reprinted with permission from [179] Copyright (2011) American Chemical Society). For a better understanding of the figure, please refer to the colour plate section.*

The latter examples showed that, for iron, the ferromagnetic property is maintained in spite of its nanowire morphology with a nanometre-sized diameter, as a good example of an intrinsic property of the encapsulated material that is transferred to the whole hybrid material. Interactions between the magnetic filling material and the encapsulated SWCNT are also possible, as recently demonstrated with Co@SWCNT materials contacted according to a transistor-type device, which showed the occurrence of magneto-resistance effects when operated at temperatures close to absolute zero in cycling magnetic field conditions (Figure 5a.16) [179]. It is also worth noting that, in the same material, the main direction of magnetization was found to be perpendicular or oblique to the elongation axis (as opposed to what was found for large Co wires encapsulated in MWCNTs [180]), as a direct consequence of the high proportion of surface atoms in the encapsulated Co nanowire whose radial dimension is minimized due to the narrow diameter of the containing SWCNT [179].

5a.4 Applications (Demonstrated or Expected)

5a.4.1 Applications that Make Use of Mass Transport Properties

5a.4.1.1 Filtration
Current studies on mass transport are principally concerned with using carbon nanotubes as a filter for processing gases and molecules (i.e. either in the gas phase or in solution) through permeable membranes in which embedded nanotubes provide the pores through

which these species are filtered [181]. As indicated in previous sections, gas/liquid transport is the principal mechanism whereby materials are introduced into nanotubes. In one sense, however, this section does not really belong here as we are actually more interested in the properties of the kind of nanocomposites that filled nanotubes actually are, rather than in nanotube pipes through which liquids or gases transit and which obviously are not nanocomposites. But these studies are worth noting in this context because they are useful in terms of giving further guidance as to the limitations of nanotube inclusion as a function of tube diameter because this is clearly both a sterically controlled phenomenon and also one which is closely controlled by the physical smoothness (or potential energy surface) of the host nanotubes. In practical terms, we should also note that this field has undergone considerable development in recent years.

Various researchers have undertaken significant investigations into the transport properties of various gases and solutes through SWCNT membranes. Johnson and Scholl (and coworkers) for example [181], have studied processes such as the absorption and diffusion of CO_2 and N_2 through SWCNT membranes [182] and also the rapid transport of H_2 and CH_4 through SWCNTs [183] Similarly, Hinds et al. [184] have produced results on the transport of N_2 gas and $Ru(NH_3)_6^+$ in solution through aligned MWCNT membranes and Holt et al. [185] have studied the diffusion of a variety of gaseous species including H_2, He, N_2, O_2, Ar, CO_2 and the hydrocarbons gases including CH_4 (smallest) and C_4H_8 (largest) through aligned DWCNT and MWCNT membranes. The significant theoretical work is probably that of Johnson's and Scholl's team which reveals that the inner potential energy surface of carbon nanotubes, to a very good approximation, makes them behave like atomically flat cylinders that easily permit the flow of molecules through their inner capillary channels. Interestingly, they also comment: 'We have considered only defect free nanotubes in our calculations. The presence of defects in the nanotubes (heteroatoms, holes, etc.) may have a profound effect on molecular diffusion by adding corrugation to the molecule-solid potential energy surface.' In fact, nanotubes themselves are corrugated even without this consideration, and it is enough to look at STM studies such as in [188] (and references therein) to see this. Also, none of the works cited above but one [185], addresses the limiting case where the diffusing species is sterically similar (i.e. almost the same size) to the internal volume of the nanotube; nearly all the molecular species addressed (i.e. CO_2, N_2, H_2, CH_4 and $Ru(NH_3)_6^+$ etc.) are significantly smaller than the internal van der Waals surface and are therefore 'free' to diffuse through the nanotubes. The notable exception is the work by Holt et al. [185] which describes the transport of a variety of gaseous species performed with sterically matched gold colloids and one large molecular species $Ru^{2+}(bipyr)_3$ which were used to estimate the diameter of the pores of the nanotubes making the 'holes' in their permeable membranes. As the authors themselves stated, these studies provided them with an estimate of the average pore size (for example, their membranes fabricated with aligned DWNTs were estimated to have pore sizes between 1.3–2 nm based on the fact that they exclude most gold colloids with a diameter of 2 nm: +/− 0.4 nm) but this study provides some very interesting insights into how nanotubes falling within a certain diameter range can accommodate species of a particular size.

In summary, such studies provide two useful additions to a consideration of the properties and applications of filled nanotubes: (i) they give us very clear indications as to the steric properties of nanotubes and therefore give us semiquantitative information on how they are able to discriminate and/or accommodate species of a particular size to fill them with; (ii) in the potentially important area of drug or substance delivery (see next), such diffusion studies give us important indicators as to how effective nanotubes are likely to be in this role.

5a.4.1.2 Targeted Molecular Species Delivery (Biomedical Applications and Others)

Further to the foregoing discussion, several researchers have proposed filled carbon nanotubes as vehicles for the delivery of bio-active agents for medical applications, because of the overall biocompatibility of graphene. The related works deal with the potential of nanotubes to be used in effect as:

1) Nanocontainers, to instil in the tissue then activate by induction, as with for example, iron [181,189,190] for magnetotherapy, which is used for the selective heating and destruction of diseased tissue (by the application of an inductive field). The filled nanotubes can also be opened after instillation to release their contained agent, for example, when filled with anticancer drugs such as carboplatin [191,192] and cisplatin [193] to initialize their medical virtue. In both cases, however, the molecular species to deliver does not really transit through the nanotubes but is delivered along with the nanotubes which are merely used as a vehicle. This makes that this application better pertains to the next Section 5a.4.2 but was placed here in order to merge all biomedical applications. Another biomedical application that uses CNTs as nanocontainers is to instil wherever is needed in the tissues contrast agents for medical imaging such as magnetic resonance imaging (MRI); when encapsulated in CNTs, those contrast agents, whose most popular ones are currently gadolinium-based compounds, have better chance of not interacting with the body as well as better chances of being eliminated from it. Gd-filled SWCNTs [194] as well as Fe_2O_3-filled SWCNTs [195] were successfully tested, with a dramatically enhanced imaging contrast resulting from the absence of any organic component and the enrichment of the magnetic atom (or compound) concentration.
2) Microcatheters [196] or nanopipettes [197–199], in which nanotubes are merely used as nanopipes. In truth, the latter was properly demonstrated only very recently [199] whereas most of studies until now have concentrated instead on the thorny issue of the biocompatibility of carbon nanotubes which have a tendency to promote in vivo cell coagulation (in this respect, the review presented in [196] is a typical example).
3) Applications related to that above should follow, such as nanoprinting: connecting a nanopipette to a reservoir of molecules whose diameter is small enough to diffuse freely within the nanotube cavity, should allow for the controlled and patterned deposition of the latter onto a surface. Using charge-bearing molecules and applying a potential difference between the printing head and the deposition surface might be necessary to actually extract the molecules from the CNT needle.

5a.4.2 Applications Arising as a Result of Filling

5a.4.2.1 Field Emission

One of the potentially most realisable applications of carbon nanotubes has been their use as field emitters. This is in part due to the comparative ease with which 2D arrays of carbon nanotubes can be grown on a flat or on a curved substrate. Because the nanotubes can be formed from catalyst metal islands whose positional and size distribution can be precisely

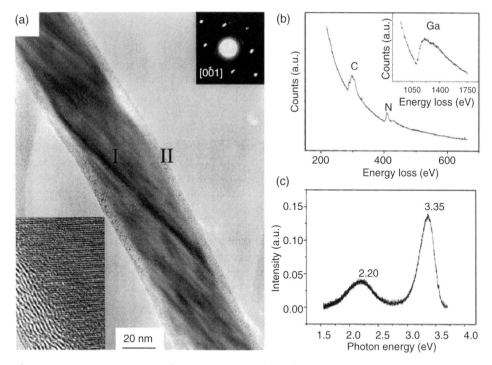

Figure 5a.17 *(a) TEM image of a GaN@MWCNT hybrid nanotube; bottom-left inset shows the well crystallized structure of the filling GaN (whereas the nanotexture of the containing MWCNT wall is rather low); top-right inset shows that the GaN filling actually makes a single crystal. (b) EELS spectrum showing that the chemical composition is only C (from the MWCNT) and N, and Ga (inset). (c) Photoluminescence spectrum of the GaN nanowires encapsulated in CNTs. Reprinted with permission from [200].*

controlled, producing carefully grown arrays is comparatively trivial. In addition, the atomically thin cross-sectional dimensions of nanotubes combined with their high aspect ratio and semiconducting/metallic character (i.e. according to conformation) makes them ideal candidates as field emitters in micron scale devices. An obvious enhancement to this application could arise if filling the nanotubes were to produce some obvious benefit with regard to enhancing the field emission properties. There are relatively few works describing the enhancement of such field emission apart from a brief communication by Liao et al. [200] describing the stable field emission from GaN filled nanotubes (Figure 5a.17) synthesized by a combination of catalytic growth and plasma enhanced deposition [201]. A close study was carried out by Domrachev et al. [202] who have also investigated the MOCVD synthesis and field emission characteristics of Ge-filled carbon nanotubes, but the containing nanotubes were made of a 'diamond-like' carbon material instead of genuine graphene-based carbon. At least one study has also described the field emission characteristics of an array of nanotubes filled with iron oxide [203] but, as with the GaN studies, this work has concentrated on filled MWCNTs rather than filled SWCNTs presumably due to the comparative ease of synthesis and also due to the relative mechanical stability of MWCNTs compared to SWCNTs during the electron emission process.

5a.4.2.2 Detectors

An area of considerable future potential may arise from the ability of nanotubes formed at the lowest dimension (i.e. SWCNTs and DWCNTs) to change or alter the crystallographic state of the filling material, as we have described above for a variety of filling materials, for example, CrO_x [18], BaI_2 [134], CoI_2 [26], and HgTe [27] in SWCNTs, and PbI_2 in DWCNTs [28]. In particular with the semimetal HgTe (with a band gap of -0.3 eV) it has been possible to show that encapsulating this material within SWNTs forces this structure into a new graphene-like crystallographic form which is now semiconducting (with a band gap of +1.3 eV) [27]. Similarly, PbI_2 is usually a wide but variable band gap semiconductor in the bulk (with a band gap of 2.2–2.6 eV) which is useful in the X- and γ-ray detection domain. The variability of the band gap is generally attributed to polytypism (2H, 4H and 12R stacking, etc.) which can be induced by kinetic heating of the halide in the bulk but which can only be healed with difficulty by careful annealing. Such a polytypism is sterically forbidden when the material is constrained within the capillaries of SWCNTs and DWCNTs which will similarly have the effect of constraining the band gap (and possibly also tuning it as a function of the encapsulating nanotube). These properties may lead to the development of these materials in detector applications.

5a.4.2.3 Catalyst Support

Other areas of interest have included the use of filled nanotubes as catalyst materials [60,204] or as catalyst support [205]. In the latter case, the nanotube is not only used, as usual, as a high surface area substrate for nanosized catalyst particles located at the nanotube outer surface (as reported in Chapter 4) but the synergetic confinement inside the MWCNTs of both the catalyst (Rh) and the fluids ($CO + H_2$) to convert (into ethanol) was shown to result into a formation rate exceeding by one order of magnitude that of the similar reaction carried out with the Rh nanoparticles located outside, instead of inside, the MWCNTs. This is quite promising for industrial applications, once cost issues related with preparing the nanotubes and fill them will be ruled out.

5a.4.2.4 Electrochemical Energy Storage and Production

Their use in electrochemical energy storage and production, in point of fact, was among the earliest suggested applications for filled nanotubes [60,204]. But related demonstrations are seldom, specifically when involving SWCNTs as the containing nanotubes. It is therefore worth noting the first demonstration of it, which also was the first demonstrated application for X@SWCNT materials, which involves CrO_x@ SWCNTs as a material to build electrodes for symmetric supercapacitor devices capable of unprecedented charging speeds, thanks to Faradaic reactions that occur between the nanosized CrO_x crystals and the acidic electrolyte [206]. In this study, the material was not optimized for reaching high capacitance values, a work still to be carried out but Galvanometry curves was still showing the desired square-like profile even at scanning speed of $1 V s^{-1}$ (Figure 5a.18).

5a.4.2.5 Future Electronics

Finally, one of the most intriguing applications put forward for filled nanotubes, particularly those containing endofullerenes, is the idea that they may be used in single spin devices which may partly arise from the polarizing electronic properties of the encapsulating

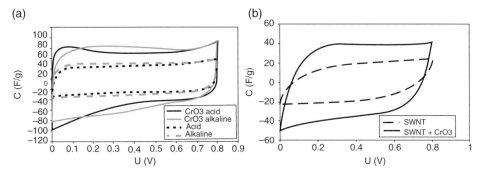

Figure 5a.18 *CrO$_x$@SWCNT materials were used as electrodes for a symmetric electrochemistry cell and the supercapacitor performances were measured by Galvanometry in either acidic or alkaline electrolyte, and compared to the same using pristine SWCNTs. The voltage was cycled from 0 to 0.8V at various scan rates (from 0.002 to 1V s^{-1}) while the capacitance was measured. (a) Whatever the electrolyte or the scan rate, using CrO$_x$@SWCNT (solid lines, labelled 'CrO$_3$') instead of pristine SWCNT (dashed lines) electrodes results in a capacitance gain. The irregular shape of the CrO$_x$@SWCNT-related curves indicates the occurrence of Faradaic reactions, which are believed to be responsible for pseudo-capacitance effects. The example shown is for a scan rate of 20mV s^{-1}. (b) Remarkably, the desired square-like shape of the galvanometric curve is maintained with the CrO$_x$@SWCNT electrodes at a scan rate as high as 1V s^{-1}, whereas it is not with pristine SWCNT electrodes (fish-like shape of the curve). Reprinted with permission from [206] Copyright (2007) Elsevier Ltd.*

nanotubes [207] but also partly arising from the fact that SWCNTs are one of the few species capable of isolating discrete single molecules containing single isolated spins (e.g. N@C$_{60}$) [208]. Progress in this area is slow but, such materials could one day form the basis of a solid state quantum computer [209].

Acknowledgements

J. S. is indebted to the Royal Society for the previous award of a University Research Fellowship and also to the Warwick Centre for Analytical Science (EPSRC funded Grant EP/F034210/1). M.M. is indebted to NATO, Imra-Europe, Toyota Motor-Corporation, the "Hierarchised Nanomaterials" CNRS program, and the "ACI Nanoscience" government program for their financial support.

References

[1] H. Kataura, Y. Maniwa, M. Abe, A. Fujiwara, T. Kodama, K. Kikuchi, H. Imahori, Y. Misaki, S. Suzuki and Y. Achiba, Optical properties of fullerene and non-fullerene peapods, *Appl. Phys. A*, **74**, 349–354 (2002).

[2] D.A. Morgan, J. Sloan and M.L.H. Green, Direct imaging of *o*-carborane molecules within single walled carbon nanotubes, *Chem. Commun.*, 2442–2443 (2002).

[3] M. Koshino, T. Tanaka, N. Solin, K. Suenaga, H. Isobe and E. Nakamura, Imaging of single organic molecules in motion, *Science*, **316**, 853–855 (2007).

[4] L. Guan, Z. Shi, M. Li and Z. Gu, Ferrocene filled single walled carbon nanotubes, *Carbon*, **43**, 2780–2785 (2005).

[5] L.-J. Li, A.N. Khlobystov, J.G. Wiltshire, G.A.D. Briggs and R.J. Nicholas, Diameter-selective encapsulation of metallocenes in single-walled carbon nanotubes, *Nature Mater.*, **4**, 481–485 (2005).

[6] A.N. Khlobystov, D.A. Britz and G.A.D. Briggs, Molecules in carbon nanotubes, *Accts. Chem. Res.*, **38**, 901–909 (2005).

[7] K. Yanagi, K. Iakoubovskii, H. Matsui, H. Matsuzaki, H. Okamoto, Y. Miyata, Y. Maniwa, S. Kazaoui, N. Minami and H. Kataura, Photosensitive function of encapsulated dye in carbon nanotubes, *J. Am. Chem. Soc.*, **129**, 4992–4997 (2007).

[8] X. Fan, E.C. Dickey, P.C. Eklund, K.A. Williams, L. Grigorian, R. Buczko, S.T. Pantelides and S.J. Pennycook, Reversible intercalation of charged iodine chains into carbon nanotubes, *Phys. Rev. Lett.*, **84**, 4621–4624 (2000).

[9] L. Guan, K. Suenaga, S. Zujin, Z. Gu and S. Iijima, Polymorphic structures of iodine and their phase transition in confined nanospace, *Nano Lett.*, **7**, 1532–1535 (2007).

[10] J. Sloan, J. Hammer, M. Zwiefka-Sibley and M.L.H. Green, The opening and filling of single walled carbon nanotubes (SWTs), *Chem. Commun.*, 347–348 (1998).

[11] C.H. Kiang, J.-S. Choi, T.T. Tran and A.D. Bacher, Molecular nanowires of 1 nm diameter from capillary filling of single-walled carbon nanotubes, *J. Phys. Chem. B*, **103**, 7449–7451 (1999).

[12] Y.F. Li, R. Hatakeyama, T. Kaneko, T. Izumida, T. Okada and T. Kato, Electrical properties of ferromagnetic semiconducting single-walled carbon nanotubes, *Appl. Phys. Lett.*, **89**, 083117/1–2 (2006).

[13] M. Monthioux, J.-P. Cleuziou and E. Flahaut, Hybrid carbon nanotubes: Strategy, progress, and perspectives, *J. Mater. Res.*, **21**, 2774–2793 (2006).

[14] J. Chancolon, F. Archaimbault, S. Bonnamy, A. Traverse, L. Olivi and G. Vlaic, Confinement of selenium inside carbon nanotubes. Structural characterization by X-ray diffraction and X-ray absorption spectroscopy, *J. Non-Cryst. Sol.*, **352**, 99–108 (2006).

[15] J.-P. Cleuziou, W. Wernsdorfer, T. Ondarçuhu and M. Monthioux, Magneto-transport in carbon nanotube-templated 1D cobalt-crystal, *Proc. The World Conference on Carbon "CARBON'10"*, July 11–16, Clemson, South Carolina, USA, Extend. Abstr. #626 (CD-Rom) (2010).

[16] P.M. Ajayan and S. Iijima, Capillarity-induced filling of carbon nanotubes, *Nature*, **361**, 333–334 (1993).

[17] P.M. Ajayan, O. Stephan, P. Redlich and C. Colliex, Carbon nanotubes as removable templates for metal oxide nanocomposites and nanostructures, *Nature*, **375**, 564–567 (1995).

[18] J. Mittal, M. Monthioux, H. Allouche and O. Stephan, Room temperature filling of single-wall carbon nanotubes with chromium oxide in open air. *Chem. Phys. Lett.*, **339**, 311–318 (2001).

[19] S. Seraphin, D. Zhou, J. Jiao, J.C. Withers and R. Loufty, Yttrium carbide in nanotubes, *Nature*, **362**, 503 (1993).

[20] C. Guerret-Piécourt, Y. Le Bouar, A. Loiseau and H. Pascard, Relation between metal electronic structure and morphology of metal compounds inside carbon nanotubes, *Nature*, **372**, 761–765 (1994).

[21] M. Liu and J.M. Cowley, Encapsulation of lanthanum carbide in carbon nanotubes and carbon nanoparticles, *Carbon*, **33**, 225–232 (1995).

[22] J. Sloan, M.C. Novotny, S.R. Bailey, G. Brown, C. Xu, V.C. Williams, S. Friedrichs, E. Flahaut, R.L. Callendar, A.P.E. York, K.S. Coleman, M L.H. Green, R.E. Dunin-Borkowski and J.L. Hutchison, Two layer 4:4 coordinated KI crystals grown within single walled carbon nanotubes, *Chem. Phys. Lett.*, **29**, 61–65 (2000).

[23] R.R. Meyer, J. Sloan, A.I. Kirkland, R.E. Dunin-Borkowski, M.C. Novotny, S.R. Bailey, J.L. Hutchison and M.L.H. Green, Discrete atom imaging of one-dimensional crystals formed within single-walled carbon nanotubes, *Science*, **289**, 1324–1326 (2000).

[24] C. Xu, J. Sloan, G. Brown, S.R. Bailey, V.C. Williams, S. Friedrichs, K.S. Coleman, J.L. Hutchison, R.E. Dunin-Borkowski and M.L.H. Green, 1D lanthanide halide crystals inserted into single-walled carbon nanotubes, *Chem. Commun.*, 2427–2428 (2000).

[25] B.C. Satishkumar, A. Taubert and D.E. Luzzi, Filling single wall carbon nanotubes with d- and f-metal chloride and metal nanowires, *J. Nanosci. Nanotech.*, **3**, 159–163 (2003).

[26] E. Philp, J. Sloan, A.I. Kirkland, R.R. Meyer, S. Friedrichs, J.L. Hutchison, and M.L.H. Green, An encapsulated helical one dimensional cobalt iodide nanostructure, *Nature Mater.*, **2**, 788–791 (2003).

[27] R. Carter, J. Sloan, A. Vlandas, M.LH. Green, A.I. Kirkland, R.R. Meyer, J.L. Hutchison, P.J.D. Lindan, G. Lin and J. Harding, Correlation of structural and electronic properties in a new low dimensional form of mercury telluride, *Phys. Rev. Lett.*, **96**, 215501.1–215501.4 (2006).

[28] E. Flahaut, J. Sloan, S. Friedrichs, A.I. Kirkland, K.S. Coleman, V.C. Williams, N. Hanson, J.L. Hutchison and M.L.H. Green, Crystallisation of 2H and 4H PbI_2 in carbon nanotubes of varying diameters and morphologies, *Chem. Mater.*, **18**, 2059–2069 (2006).

[29] W.K. Hsu, S. Trasobares, H. Terrones, M. Terrones, N. Grobert, Y.Q. Zhu, W.Z. Li, J.P. Hare, R. Escudero, H.W. Kroto and D.R.M. Walton, Electrolytic formation of carbon-sheathed mixed Sn-Pb nanowires, *Chem. Mater.*, **11**, 1747–1751 (1998).

[30] N. Grobert, M. Mayne, R.R.M. Walton, H.W. Kroto, M. Terrones, R. Kamalakaran, T. Seeger, M. Ruhle, H. Terrones, J. Sloan, R.E. Dunin-Borkowski and J.L. Hutchison, Alloy nanowires: Invar inside carbon nanotubes, *Chem. Commun.*, 471–472 (2001).

[31] J.Q. Hu, Y. Bando, J.H. Zhan, C.Z. Li and D. Golberg, Mg_3N_2-Ga: Nanoscale semiconductor-liquid metal heterojunctions inside graphitic carbon nanotubes, *Adv. Mater.*, **19**, 1342–1346 (2007).

[32] S.B. Fagan, A.G. Souza Filho, P. Corio, J. Mendes Filho and M.S. Dresselhaus, Electronic properties of Ag- and CrO_3-filled carbon nanotubes, *Chem. Phys. Lett.*, **406**, 54–59 (2005).

[33] J.A. Alonso, J.S. Arellano, L.M. Molina, A. Rubio and M.J. Lopez, Interaction of molecular and atomic hydrogen with single-wall carbon nanotubes, *IEEE Trans. Nanotech.*, **3**, 304–310 (2004).

[34] K. Koga, G.T. Gao, H. Tanaka and X.C. Zeng, Formation of ordered ice nanotubes inside carbon nanotubes, *Nature*, **412**, 802–805 (2001).

[35] Y. Maniwa, M. Abe, H. Kataura, S. Suzuki, Y. Achiba, H. Kira and K. Matsuda, Phase transition in confined water inside carbon nanotubes, *J. Phys. Soc. Jpn.*, **71**, 2863–2866 (2002).

[36] M.S. Dresselhaus, G. Dresselhaus and R. Saito, Physics of carbon nanotubes, *Carbon*, **33**, 883–891 (1995).

[37] J. Sloan, A.I. Kirkland, J.L. Hutchison and M.L.H. Green, Integral atomic layer architectures of 1D crystals inserted into single walled carbon nanotubes, *Chem. Commun.*, 1319–1332 (2002).

[38] D. Ugarte, A. Chatelain and W.A. de Heer, Nanocapillarity and chemistry in carbon nanotubes, *Science*, **274**, 1897–1899 (1996).

[39] E. Dujardin, T.W. Ebbesen, A. Krishnan and M.M.J. Treacy, Wetting of single shell carbon nanotubes, *Adv. Mater.*, **10**, 1472–1475 (1998).

[40] E. Dujardin, T.W. Ebbesen, H. Hiura and K. Taginaki, Capillarity and wetting of carbon nanotubes, *Science*, **265**, 1850–1852 (1994).

[41] P.M. Ajayan, T.W. Ebbesen, T. Ichihashi, S. Iijima, K. Tanigaki and H. Hiura, Opening carbon nanotubes with oxygen and implications for filling, *Nature*, **362**, 522–525 (1993).

[42] Y. Saito and T. Yoshikawa, Bamboo-shaped carbon tube filled partially with nickel, *J. Cryst. Growth*, **134**, 154–156 (1993).

[43] S. Iijima, Helical microtubules of graphite carbon, *Nature*, **354**, 56–58 (1991).

[44] S. Iijima and T. Ichihashi, Single-shell carbon nanotubes of 1–nm diameter. *Nature*, **363**, 603–605 (1993).

[45] D.S. Bethune, C.H. Kiang, M.S. De Vries, G. Gorman, R. Savoy, J. Vasquez and R. Breyers, Cobalt catalysed growth of carbon nanotubes with single-atomic-layer walls, *Nature*, **363**, 605–607 (1993).

[46] B.W. Smith, M. Monthioux and D.E. Luzzi, Encapsulated C_{60} in carbon nanotubes, *Nature*, **396**, 323–324 (1998).

[47] B. Burteaux, A. Claye, B.W. Smith, M. Monthioux, D.E. Luzzi and J.E. Fischer, Abundance of encapsulated C_{60} in single-wall carbon nanotubes, *Chem. Phys. Lett.*, **310**, 21–24 (1999).

[48] J.S. Bendall, A. Ilie, M.E. Welland, J. Sloan and M.L.H. Green, Thermal stability and reactivity of metal halide filled single walled carbon nanotubes, *J. Phys. Chem.*, **110**, 6569–6573 (2006).

[49] J. Sloan, J. Cook, M.L.H. Green, J.L. Hutchison and R. Tenne, Crystallization inside fullerene-related structures, *J. Mater. Chem.*, **7**, 1089–1095 (1997).

[50] D. Ugarte, T. Stöckli, J.M. Bonard, A. Châtelain and W.A. de Heer, Filling carbon nanotubes. *Appl. Phys. A*, **67**, 101–105 (1998).

[51] A. Loiseau and H. Pascard, Synthesis of long carbon nanotubes filled with Se, S, Sb, and Ge by the arc method, *Chem. Phys. Lett.*, **256**, 246 (1996).

[52] P.C.P. Watts, W.K. Hsu, V. Kotzeva and G.Z. Chen, Fe-filled carbon nanotube-polystyrene: RCL composites, *Chem. Phys. Lett.*, **366**, 42–50 (2002).

[53] J.Y. Dai, J.M. Lauerhaas, A.A. Setlur and R.P.H. Chang, Synthesis of carbon-encapsulated nanowires using polycyclic aromatic hydrocarbon precursors, *Chem. Phys. Lett.*, **258**, 547–553 (1996).

[54] C.N.R. Rao, R. Sen, B.C. Satishkumar and A. Govindaraj, Large aligned-nanotube bundles from ferrocene pyrolysis, *Chem. Commun.*, 1525–1526 (1998).

[55] A. Leonhardt, M. Ritschel, R. Kozhuharova, A. Graff, T. Mühl, R. Huhle, I. Mönch, D. Elefant and C.M. Schneider, Synthesis and properties of filled carbon nanotubes, *Diamond Relat. Mater.*, **12**, 790–793 (2003).

[56] N. Demoncy, O. Stéphan, N. Brun, C. Colliex, A. Loiseau and H. Pascard, Filling carbon nanotubes with metals by the arc discharge method: The key role of sulphur, *Eur. Phys. J. B*, **4**, 147–157 (1998).

[57] A. Loiseau and F. Willaime, Filled and mixed nanotubes: From TEM studies to the growth mechanism within a phase-diagram approach, *Appl. Surf. Sci.*, **164**, 227–240 (2000).

[58] Y. Zhang, S. Iijima, Z. Shi and Z. Gu, Defects in arc-discharge produced single-walled carbon nanotubes, *Philos. Mag. Lett.*, **79**, 473–479 (1999).

[59] J. Sloan, R.E. Dunin-Borkowski, J.L. Hutchison, K.S. Coleman, V.C. Williams, J.B. Claridge, A.P.E. York, C. Xu, S.R. Bailey, G. Brown, S. Friedrichs and M.L.H. Green, The size distribution, imaging and obstructing properties of C_{60} and higher fullerenes formed within arc-grown single walled carbon nanotubes, *Chem. Phys. Lett.*, **316**, 191–198 (2000).

[60] S.C. Tsang, Y.K. Chen, P.J.F. Harris and M.L.H. Green, A simple chemical method of opening and filling carbon nanotubes, *Nature*, **372**, 159–162 (1994).

[61] B.C. Satishkumar, A. Govindaraj, J. Mofokeng, G.N. Subbanna and C.N.R. Rao, Novel experiments with carbon nanotubes: Opening, filling, closing, and functionalizing nanotubes, *J. Phys. B*, **29**, 4925–4934 (1996).

[62] M. Monthioux, Filling single wall carbon nanotubes, *Carbon*, **40**, 1809–1823 (2002).

[63] S. Berber, Y.-K. Kwon and D. Tománek, Microscopic formation mechanism of nanotube peapods, *Phys. Rev. Lett.*, **88**, 185502/1–4 (2002).

[64] B.W. Smith and D.E. Luzzi, Formation mechanism of fullerene peapods and coaxial tubes: A path to large scale synthesis, *Chem. Phys. Lett.*, **321**, 169–174 (2000).

[65] G. Brown, S. Bailey, J. Sloan, C. Xu, S. Friedrichs, E. Flahaut, K.S. Coleman, J.L. Hutchison, R.E. Dunin-Borkowski and M.L.H. Green, Electron beam induced in situ clusterization of 1D $ZrCl_4$ chains within single-walled carbon nanotubes, *Chem. Commun.*, 845–846 (2001).

[66] J. Chancolon, F. Archaimbault, A. Pineau and S. Bonnamy, Filling of carbon nanotubes with selenium by vapor phase process, *J. Nanosci. Nanotechnol.*, **6**, 1–5 (2006).

[67] P.M.F.J. Costa, J. Sloan, T. Rutherford and M.L.H. Green, Encapsulation of Re_xO_y clusters within single-walled carbon nanotubes and their *in tubulo* reduction and sintering to Re metal, *Chem. Mater.*, **17**, 6579–6582 (2005).

[68] W. Mickelson, S. Aloni, W.-Q. Han, J. Cumings and A. Zettl, Packing C_{60} in boron nitride nanotubes, *Science*, **300**, 467–469 (2003).

[69] T. Fröhlich, P. Scharff, W. Schliefke, H. Romanus, V. Gupta, C. Siegmund, O. Ambacher and L. Spiess, Insertion of C_{60} into multi-wall carbon nanotubes: A synthesis of C_{60}@MWCNT, *Carbon*, **42**, 2759–2762 (2004).

[70] B.M. Kim, S. Qian and H.H. Bau, Filling carbon nanotubes with particles, *Nano Lett.*, **5**, 873–878 (2005).

[71] G. Korneva, H. Ye, Y. Gogotsi, D. Halverson, G. Friedman, J.-C. Bradley and K.G. Kornev, Carbon nanotubes loaded with magnetic nanoparticles, *Nano Lett.*, **5**, 879–884 (2005).

[72] S. Iijima, M. Yudasaka, R. Yamada, S. Bandow, K. Suenaga, F. Kokai and K. Takahashi, Nano-aggregates of single-walled graphitic carbon nano-horns, *Chem. Phys. Lett.*, **309**, 165–170 (1999).

[73] A. Hashimoto, H. Yorimitsu, K. Ajima, K. Suenaga, H. Isobe, J. Miyawaki, M. Yudasaka, S. Iijima and E. Nakamura, Selective deposition of a gadolinium (III) cluster in a hole opening of single-wall carbon nanohorn, *Proc. Natl. Acad. Sci.*, **101**, 8527–8530 (2004).

[74] Z.L. Zhang, B. Li, Z.J. Shi, Z.N. Gu, Z.Q. Xue and L.M. Peng, Filling of single-walled carbon nanotubes with silver, *J. Mater. Res.*, **15**, 2658–2661 (2000).

[75] A. Govindaraj, B.C. Satishkumar, M. Nath and C.N.R. Rao, Metal nanowires and intercalated metal layers in single-walled carbon nanotube bundles, *Chem. Mater.*, **12**, 202–204 (2000).

[76] Z. Liu, M. Koshino, K. Suenaga, A. Mrzel, H. Kataura and S. Iijima, Transmission electron microscopy imaging of individual functional groups of fullerene derivatives, *Phys. Rev. Lett.*, **96**, 088304/1–4 (2006).

[77] Z. Liu, K. Yanagi, K. Suenaga, H. Kataura and S. Iijima, Imaging the dynamic behaviour of individual retinal chromophores confined inside carbon nanotubes, *Nature Nanotechnol.*, **2**, 422–425 (2007).

[78] F. Simon, H. Kuzmany, H. Rauf, T. Pichler, J. Bernardi, H. Peterlik, L. Korecz, F. Fülöp and A. Jànossy, Low temperature fullerene encapsulation in single wall carbon nanotubes: Synthesis of N@C_{60}@SWCNT, *Chem. Phys. Lett.*, **383**, 362–367 (2004).

[79] D.A. Britz, A.N. Khlobystov, J. Wang, A.S. O'Neil, M. Poliakoff, A. Ardavan and G.A.D. Briggs, Selective host-guest interaction of single-walled carbon nanotubes with functionalized fullerenes, *Chem. Commun.*, 176–177 (2004).

[80] D.A. Britz, A.N. Khlobystov, K. Porfyrakis, A. Ardavan and G.A.D. Briggs, Filling of fullerene oxide in supercritical CO_2, *Chem. Commun.*, 37–38 (2005).

[81] J. Mittal, H. Konno and M. Inagaki, Synthesis of GICs of Cr^{VI} compound using CrO_3 and HCl at room temperature, *Synth. Met.*, **96**, 103–108 (1998).

[82] G. Brown, S.R. Bailey, M. Novotny, R. Carter, E. Flahaut, K.S. Coleman, J.L. Hutchison, M.L.H. Green and J. Sloan, High-yield incorporation and washing properties of halides incorporated into single walled carbon nanotubes, *Appl. Phys. A*, **76**, 1–6 (2003).

[83] B.C. Satishkumar, A. Taubert and D.E. Luzzi, Filling single wall carbon nanotubes with d- and f-metal chloride and metal nanowires, *J. Nanosci. Nanotechnol.*, **3**, 159–163 (2003).

[84] M. Hulman, H. Kuzmany, P.M.F.J. Costa, S. Friedrichs and M.L.H. Green, Light-induced instability of PbO-filled single wall carbon nanotubes, *Appl. Phys. Lett.*, **85**, 2068–2070 (2004).

[85] J. Sloan, D.M. Wright, H.G. Woo, S. Bailey, G. Brown, A.P.E. York, K.S. Coleman, J.L. Hutchison and M.L.H. Green, Capillarity and silver nanowire formation observed in single walled carbon nanotubes, *Chem. Commun.*, 699–700 (1999).

[86] L. Grigorian, K.A. Williams, S. Fang, G.U. Sumanasekera, A.L. Loper, E.C. Dickey, S.J. Pennycook and P.C. Eklund, Reversible intercalation of charged iodine chains into carbon nanotube ropes, *Phys. Rev. Lett.*, **80**, 5560–5563 (1998).

[87] X. Fan, E.C. Dickey, P.C. Eklund, K.A. Williams, L. Grigorian, R. Buczko, S.T. Pantelides and S.J. Pennycook, Atomic arrangement of iodine atoms inside single-walled carbon nanotubes, *Phys. Rev. Lett.*, **84**, 4621–4624 (2000).

[88] J. Mittal, M. Monthioux and H. Allouche, Synthesis of SWNT based hybrid nanomaterials from photolysis-enhanced chemical processes, in *Proc. 25th Biennial Conference on Carbon*, Lexington, KY, USA, Novel/14.2 (2001).

[89] J. Mittal, M. Monthioux, V. Serin and J.-P. Cleuziou, UV photolysis: An alternative for the synthesis of hybrid carbon nanotubes, in *Chinese-French Workshop on Carbon Materials*, September 30-October 1, Orléans, France (2005).

[90] G.-H. Jeong, R. Hatakeyama, T. Hirata, K. Tohji, K. Motomiya, N. Sato and Y. Kawazoe, Structural deformation of single walled carbon nanotubes and fullerene encapsulation due to magnetized-plasma ion irradiation, *Appl. Phys. Lett.*, **79**, 4213–4215 (2001).

[91] G-H. Jeong, R. Hatakeyama, T. Hirata, K. Tohji, K. Motomiya, T. Yaguchi and Y. Kawazoe, Formation and structural observation of cesium encapsulated in single-walled carbon nanotubes, *Chem. Commun.*, 152–153 (2003).

[92] B.-Y. Sun, Y. Sato, K. Suenaga, T. Okazaki, N. Kishi, T. Sugai, S. Bandow, S. Iijima and H. Shinohara, Entrapping of exohedral metallofullerenes in carbon nanotubes: $(CsC_{60})_n$@ SWNT nanopeapods, *J. Am. Chem. Soc.*, **127**, 17972–17973 (2005).

[93] T. Pichler, A. Kukovecz, H. Kuzmany, H. Kataura and Y. Achiba, Quasicontinuous electron and hole doping of C_{60} peapods, *Phys. Rev. B*, **67**, 125416.1–125416.7 (2003).

[94] M. Kalbac, L. Kavan, M. Zukalova and L. Dunsch, Two positions of potassium in chemically doped C_{60} peapods: an in situ spectroelectrochemical study, *J. Phys. Chem. B*, **108**, 6275–6280 (2004).

[95] T. Shimada, Y. Ohno, T. Okazaki, T. Sugai, K. Suenaga, S. Kishimoto, T. Mizutani, T. Inoue, R. Taniguchi, N. Fukui, H. Okubo and H. Shinohara, Transport properties of C_{78}, C_{90} and Dy@ C_{82} fullerenes-nanopeapods by field effect transistors, *Physica E*, **21**, 1089–1092 (2004).

[96] D.E. Luzzi, B.W. Smith, R. Russo, B.C. Satishkumar, F. Stercel and N. Nemes, Encapsulation of metallofullerenes and metallocenes in carbon nanotubes, in *Electronic Properties of Molecular Nanostructures*, H. Kuzmany, J. Fink, M. Mehring and S. Roth (eds.), American Institute of Physics Conference Proceedings Series, **591**, 622–626 (2001).

[97] H. Shiozawa, H. Rauf, T. Pichler, M. Knupfer, M. Kalbac, S. Yang, L. Dunsch, B. Büchner, D. Batchelor and H. Kataura, Effective valency of Dy ions in $Dy_3N@C_{80}$ metallofullerenes in peapods, in *Electronic Properties of Molecular Nanostructures*, H. Kuzmany, J. Fink, M. Mehring and S. Roth (eds.), American Institute of Physics Conference Proceedings Series, **786**, 325–328 (2002).

[98] K. Suenaga, M. Tence, C. Mory, C. Colliex, H. Kato, T. Okazaki, H. Shinohara, K. Hirahara, S. Bandow and S. Iijima, Element selective single atom imaging, *Science*, **290**, 2280–2282 (2000).

[99] K. Suenaga, K. Hirahara, S. Bandow, S. Iijima, T. Okazaki, H. Kato and H. Shinohara, Core level spectroscopy on the valence state of encaged metal in metallofullerenes-peapods, in *Electronic Properties of Molecular Nanostructures*, H. Kuzmany, J. Fink, M. Mehring and S. Roth (eds.), American Institute of Physics Conference Proceedings Series, **591**, 256–260 (2001).

[100] P.W. Chiu, G. Gu, G.T. Kim, G. Philipp, S. Roth, S.F. Yang and S. Yang, Temperature-induced change from *p* to *n* conduction in metallofullerene nanotube peapods, *Appl. Phys. Lett.*, **79**, 3845–3847 (2001).

[101] T. Okazaki, K. Suenaga, K. Hirahara, S. Bandow, S. Iijima and H. Shinohara, Real time reaction dynamics in carbon nanotubes, *J. Amer. Chem. Soc.*, **123**, 9673–9674 (2001).

[102] T. Okazaki, K. Suenaga, K. Hirahara, S. Bandow, S. Iijima and H. Shinohara, Electronic and geometric structures of metallofullerenes peapods, *Physica B*, **323**, 97–99 (2002).

[103] K. Suenaga, T. Okazaki, C.-R. Wang, S. Bandow, H. Shinohara and S. Iijima, Direct imaging of $Sc_2@C_{84}$ molecules encapsulated inside single-wall carbon nanotubes by high resolution electron microscopy with atomic sensitivity, *Phys. Rev. Lett.*, **90**, 055506/1–4 (2003).

[104] K. Suenaga, R. Taniguchi, T. Shimada, T. Okazaki, H. Shinohara and S. Iijima, Evidence for the intramolecular motion of Gd atoms in a $Gd_2@C_{92}$ nanopeapod, *Nano Lett.*, **3**, 1395–1398 (2003).

[105] D.M. Guldi, M. Marcaccio, D. Paolucci, F. Paolucci, N. Tagmatarchis, D. Tasis, E. Vasquez and M. Prato, Single wall carbon nanotube-ferrocene nanohybrids: Observing intramolecular electron transfer in functionalized SWNTs, *Angew. Chem. Int. Ed.*, **42**, 4206–4209 (2003).

[106] H. Qiu, Z. Shi, Z. Gu and J. Qiu, Controllable preparation of triple-walled carbon nanotubes and their growth mechanism, *Chem. Commun.*, 1092–1094 (2007).

[107] L. Guan, K. Suenaga and S. Iijima, Smallest carbon nanotube assigned with atomic resolution accuracy, *Nano Lett.*, **8**, 459–462 (2008).

[108] H. Shiozawa, T. Pichler, A. Grüneis, R. Pfeiffer, H. Kuzmany, Z. Liu, K. Suenaga and H.Kataura, A Catalytic reaction inside a single-walled carbon nanotube, *Adv. Mater.* **20**, 1443–1449 (2008).

[109] T. Takenobu, T. Takano, M. Shiraishi, Y. Murakami, M. Ata, H. Kataura, Y. Achiba and Y. Iwasa, Stable and controlled amphoteric doping by encapsulation of organic molecules inside carbon nanotubes, *Nature Mater.*, **2**, 683–688 (2003).

[110] H. Kataura, Y. Maniwa, T. Kodama, K. Kikuchi, Y. Susuki, Y. Achiba, K. Sugiura, S. Okubo and K. Tsukagoshi, One dimensional system in carbon nanotubes, in *Electronic Properties of Molecular Nanostructures*, H. Kuzmany, J. Fink, M. Mehring and S. Roth (eds.), American Institute of Physics Conference Proceedings, **685**, 349–353 (2003).

[111] D.M. Kammen, T.E. Lipman, A.B. Lovins, P.A. Lehman, J.M. Eiler, T.K. Tromp, R.-L. Shia, M. Allen and Y.L. Yung, Assessing the future hydrogen economy, *Science*, **302**, 226–229 (2003).

[112] A. Kuznetsova, J.T. Yates Jr., J. Li and R.E. Smalley, Physical adsorption of xenon in open single walled carbon nanotubes: Observation of a quasi-one-dimensional confined Xe phase, *J. Chem. Phys.*, **112**, 9590–9598 (2000).

[113] A. Fujiwara, K. Ishii, H. Suematsu, H. Kataura, Y. Maniwa, S. Susuki and Y. Achiba, Gas adsorption in the inside and outside of single-walled carbon nanotubes, *Chem. Phys. Lett.*, **336**, 205–211 (2001).

[114] K.A. Williams and P.C. Eklund, Monte Carlo simulations of H_2 physisorption in finite-diameter carbon nanotube ropes, *Chem. Phys. Lett.*, **320**, 352–358 (2000).

[115] A.C. Dillon, K.M. Jones, T.A. Bekkedahl, C.H. Kiang, D.S. Bethune and M. J. Heben, Storage of hydrogen in single walled carbon nanotubes, *Nature*, **386**, 377–379 (1997).

[116] M. Hirscher, M. Becher, M. Haluska, U. Dettlaff-Weglikowska, A. Quintel, G.S. Duesberg, Y.M. Choi, P. Downes, M. Hulman, S. Roth, I. Stepanek and P. Bernier, Hydrogen storage in sonicated carbon materials, *Appl. Phys. Mater. Sci. Process*, **2**, 129–132 (2001).

[117] S. Farhat, B. Weinberger, F.D. Lamari, T. Izouyar, L. Noé and M. Monthioux, Performance of carbon arc-discharge nanotubes to hydrogen energy storage, *J. Nanosci. Nanotechnol.*, **7**, 3537–3542 (2007).

[118] J. Sloan, G. Matthewman, C. Dyer-Smith, A.-Y. Sung, Z. Liu, K. Suenaga, A.I. Kirkland and E. Flahaut, Direct imaging of the structure, relaxation and sterically constrained motion of encapsulated tungsten polyoxometalate Lindqvist ions within carbon nanotubes, *ACS Nano*, **2**, 966–976 (2008).

[119] P. Corio, A.P. Santos, P.S. Santos, M.L.A. Temperini, V.W. Brar, M.A. Pimenta and M. S. Dresselhaus, Characterization of single wall carbon nanotubes filled with silver and with chromium compounds, *Chem. Phys. Lett.*, **383**, 475–480 (2004).

[120] E. Borowiak-Palen, M.H. Rummeli, E. Mendoza, S.J. Henley, D.C. Cox, C.H.P. Poa, V. Stolojan, T. Gemming, T. Pichler and S.R.P. Silva, Silver intercalated carbon nanotubes, in *Electronic Properties of Molecular Nanostructures*, H. Kuzmany, J. Fink, M. Mehring and S. Roth (eds.), American Institute of Physics Proceedings Series, **786**, 236–239 (2005).

[121] E. Borowiak-Palen, E. Mendoza, A. Bachmatiuk, M.H. Rummeli, T. Gemming, J. Nogues, V. Skumryev, R.J. Kalenczuk, T. Pichler and S.R.P. Silva, Iron filled single-wall carbon nanotubes: A novel ferromagnetic medium, *Chem. Phys. Lett.*, **421**, 129–133 (2006).

[122] J. Jorge, E. Flahaut, F. Gonzalez-Jimenez, G. Gonzalez, J. Gonzalez, E. Belandria, J.-M. Broto and B. Raquet, Preparation and characterization of α-Fe nanowires located inside double wall carbon nanotubes, *Chem. Phys. Lett.*, **457**, 347–351 (2008).

[123] J. Sloan, A.I. Kirkland, J.L. Hutchison and M.L.H. Green, Aspects of crystal growth within carbon nanotubes, *C. R. Phys.*, **4**, 1063–1074 (2003).

[124] J. Sloan, S. Friedrichs, R.R. Meyer, A.I. Kirkland, J.L. Hutchison and M.L.H. Green: Structural changes induced in nanocrystals of binary compounds confined within single walled carbon nanotubes: A brief review, *Inorg. Chim. Acta*, **330**, 1–12 (2002).

[125] J. Sloan and M.L.H. Green, Synthesis and characterisation of materials incorporated within carbon nanotubes, in *Fullerenes: Chemistry, Physics and Technology*, K.M. Kadish and R.S. Ruoff (eds.), John Wiley & Sons, Inc., New York, 826–828 (2000).

[126] L.-J. Li, T.-W. Lin, J. Doig, I.B. Mortimer, J.G. Wiltshire, R.A. Taylor, J. Sloan, M.L.H. Green and R.J. Nicholas, Crystal-encapsulation-induced band-structure change in single-walled carbon nanotubes: Photoluminescence and Raman spectra, *Phys. Rev. B*, **74**, 245418/1–5 (2006).

[127] E. Flahaut, J. Sloan, K.S. Coleman and M.L.H. Green, Synthesis of 1D p-block halide crystals within single walled carbon nanotubes, in *Electronic Properties of Molecular Nanostructures*, H. Kuzmany, J. Fink, M. Mehring and S. Roth (eds), American Institute of Physics Proceedings Series, **591**, 283–286 (2001).

[128] G. Chen, J. Qiu and H. Qiu, Filling double-walled carbon nanotubes with AgCl nanowires, *Scripta Mater.*, **58**, 457–460 (2008).

[129] J. Sloan, S. Friedrichs, E. Flahaut, G. Brown, S.R. Bailey, K.S. Coleman, C. Xu, M.L.H. Green, J.L. Hutchison, A.I. Kirkland and R.R. Meyer, The characterization of subnanometer scale structures within single walled carbon nanotubes, in *Electronic Properties of Molecular Nanostructures*, H. Kuzmany, J. Fink, M. Mehring and S. Roth (eds.), American Institute of Physics Proceedings Series, **591**, 277–282 (2001).

[130] J. Sloan, M. Terrones, S. Nufer, S. Friedrichs, S.R. Bailey, H.G. Woo, M. Rühle, J.L. Hutchison and M.L.H. Green, Metastable one-dimensional $AgCl_{1-x}I_x$ solid-solution wurzite 'tunnel' crystals formed within single walled carbon nanotubes, *J. Am. Chem. Soc.*, **124**, 2116–2117 (2002).

[131] J. Sloan, M. Terrones, S. Nufer, S. Friedrichs, S.R. Bailey, H-G. Woo, M. Rühle, J.L. Hutchison and M.L.H. Green, Spatially resolved EELS applied to the study of a one-dimensional solid solution of $AgCl_{1-x}I_x$ formed within single-wall carbon nanotubes, in *Electronic Properties of Molecular Nanostructures*, H. Kuzmany, J. Fink, M. Mehring and S. Roth (eds.), American Institute of Physics Proceedings Series, **633**, 135–139 (2002).

[132] P.M.F.J. Costa, S. Friedrichs, J. Sloan and M.L.H. Green, Imaging lattice defects and distortions in alkali-metal iodides encapsulated within double-walled carbon nanotubes, *Chem. Mater.* **17**, 3122–3129 (2005).

[133] M.V. Chernysheva, A.A. Eliseev, A.V. Lukashin, Y.D. Tretyakov, S.V. Savilov, N.A. Kiselev, O.M. Zhigalina, A.S. Kumskov, A.V. Krestinin and J.L. Hutchison, Filling of single-walled carbon nanotubes by CuI nanocrystals via capillary technique, *Physica E*, **37**, 62–65 (2007).

[134] J. Sloan, S.J. Grosvenor, S. Friedrichs, A. Kirkland, J.L. Hutchison and M.L.H. Green, A one-dimensional BaI_2 chain with five-and six-coordination, formed within a single-walled carbon nanotube, *Angew. Chem., Int. Ed.*, **41**, 1156–1159 (2002).

[135] S. Friedrichs, R.R. Meyer, J. Sloan, A.I. Kirkland, J.L. Hutchison and M.L.H. Green, Complete characterization of a Sb_2O_3/(21,-8)SWNT inclusion composite, *Chem. Commun.*, 929–930 (2001).

[136] S. Friedrichs, J. Sloan, M.L.H. Green, J.L. Hutchison, R.R. Meyer and A.I. Kirkland, Simultaneous determination of inclusion crystallography and nanotube conformation for a Sb_2O_3/single-walled nanotube composite. *Phys. Rev. B*, **64**, 045406/1–8 (2001).

[137] N. Thamavaranukup, H.A. Höppe, L. Ruiz-Gonzalez, P.M.F.J. Costa, J. Sloan, A. Kirkland and M.L.H. Green, Single-walled carbon nanotubes filled with M OH (M = K, Cs) and then washed and refilled with clusters and molecules, *Chem. Commun.*, 1686–1697 (2004).

[138] P.G. Gennes, F. Brochard-Wyart and D. Quéré, *Capillarity and Wetting Phenomena: Drops, Bubbles, Pearls, Waves*, Springer, New York (2004).

[139] M. Terrones, N. Grobert, W.K. Hsu, Y.Q. Zhu, W.B. Hu, H. Terrones, J.P. Hare, H.W. Kroto and D.R. Walton, Advances in the creation of filled nanotubes and novel nanowires, *MRS Bull.*, **24**, 43–49 (1999).

[140] M.R. Pederson and J.Q. Broughton, Nanocapillarity in fullerene tubules, *Phys. Rev. Lett.*, **69**, 2689–2692 (1992).

[141] M. Wilson and P.A. Madden, Growth of ionic crystals in carbon nanotubes, *J. Am. Chem. Soc.*, **123**, 2101–2102 (2001).

[142] A.A. Sofronov, V.V. Ivanovskaya, Y.N. Makurin and A.L. Ivanovskii, New one-dimensional crystals of $(Sc,Ti,V)_8C_{12}$ metallocarbohedrenes in carbon and boron–nitrogen (12,0) nanotubes: quantum chemical simulation of the electronic structure, *Chem. Phys. Lett.*, **351**, 35–41 (2002).

[143] R.J. Mashl, S. Joseph, N.R. Aluru and E. Jakobsson, Anomalously immobilized water: A new water phase induced by confinement in nanotubes, *Nano Lett.*, **3**, 589–592 (2003).

[144] H. Gao, Y. Kong and D. Cui, Spontaneous insertion of DNA oligonucleotides into carbon nanotubes, *Nano Lett.*, **3**, 471–473 (2003).

[145] Y. Guo, Y. Kong, W. Guo and H. Gao, Structural transition of copper nanowires confined in single-walled carbon nanotubes, *J. Comput. Theor. Nanosci.*, **1**, 93–98 (2004).

[146] G. Kim, Y. Kim and J. Ihm, Encapsulation and polymerization of acetylene molecules inside a carbon nanotube, *Chem. Phys. Lett.*, **415**, 279–282 (2005).

[147] J. Zhang, X.Q. Zhang, H. Li and K.M. Liew, The structures and electrical transport properties of germanium nanowires encapsulated in carbon nanotubes, *Appl. Phys.*, **102**, 073709/1–5 (2007).

[148] C.-K. Yang, J. Zhao and J.P. Lu, Magnetism of transition-metal/carbon-nanotube hybrid structures, *Phys. Rev. Lett.*, **90**, 257203/1–4 (2003).

[149] Y.-J. Kang, J. Choi, C-Y. Moon and K.J. Chang, Electronic and magnetic properties of single-wall carbon nanotubes filled with iron atoms, *Phys. Rev. B*, **71**, 115441.1–115441.7 (2005).

[150] M. Weissmann, G. García, M. Kiwi and R. Ramírez, Theoretical study of carbon-coated iron nanowires, *Phys. Rev. B*, **70**, 201401/1–4 (2004).

[151] M. Weissmann, G. García, M. Kiwi, R. Ramírez and C-C. Fu, Theoretical study of iron-filled carbon nanotubes, *Phys. Rev. B*, **73**, 125435/1–8 (2004).

[152] N. Fujima and T. Oda, Structures and magnetic properties of iron chains encapsulated in tubal carbon nanocapsules, *Phys. Rev. B*, **71**, 115412/1–9 (2005).

[153] B.W. Smith, M. Monthioux and D.E. Luzzi, Carbon nanotube encapsulated fullerenes: A unique class of hybrid materials, *Chem. Phys. Lett*, **315**, 31–36 (1999).

[154] D.E. Luzzi and B.W. Smith, Carbon cage structures in single wall carbon nanotubes: A new class of materials, *Carbon*, **38**, 1751–1756 (2000).

[155] S. Bandow, M. Takizawa, K. Hirahara, Y. Yudasaka and S. Iijima, Raman scattering study of double-wall carbon nanotubes derived from the chains of fullerenes in single-wall carbon nanotubes, *Chem. Phys. Lett.*, **337**, 48–54 (2001).

[156] Y. Sakurabayashi, M. Monthioux, K. Kishita, Y. Suzuki, T. Kondo and M. Le Lay, Tailoring double wall carbon nanotubes, in *Electronic Properties of Molecular Nanostructures*, H. Kuzmany, J. Fink, M. Mehring and S. Roth (eds.), American Institute of Physics Conference Proceedings, **685**, 302–305 (2003).

[157] C. Arrondo, M. Monthioux, Y. Kishita and M. Le Lay, In situ coalescence of aligned C_{60} in peapods, in *Electronic Properties of Molecular Nanostructures*, H. Kuzmany, J. Fink, M. Mehring and S. Roth (eds.), American Institute of Physics Conference Proceedings, **786**, 329–332 (2005).

[158] C. Kramberger, A. Waske, K. Biedermann, T. Pichler, T. Gemming, B. Büchner and H. Kataura, Tailoring carbon nanostructures via temperature and laser irradiation, *Chem. Phys. Lett.*, **407**, 254–259 (2005).

[159] M. Kalbac, L. Kavan, L. Juha, S. Civis, M. Zukalova, M. Bittner, P. Kubat, V. Vorlicek and L. Dunsch, Transformation of fullerene peapods to double-walled carbon nanotubes induced by UV radiation, *Carbon*, **43**, 1610–1616 (2005).

[160] M. Yoon, S. Berber and D. Tománek, Energetics and packing of fullerenes in nanotube peapods, *Phys. Rev. B*, **71**, 155406/1–4 (2005).

[161] A.F. Wells, in *Structural Inorganic Chemistry*, Oxford University Press, Oxford (1990).

[162] R. Kreizman, S.-Y. Hong, J. Sloan, R. Popovitz-Biro, A. Albu-Yaron, G. Belén Ballesteros, B.G. Davis, M.L.H. Green and R. Tenne, Core-shell $PbI_2@WS_2$ inorganic nanotubes from capillary wetting, *Angew. Chem., Int. Ed.*, **48**, 1230–1233 (2009).

[163] M. Wilson, Structure and phase stability of novel 'twisted' crystal structures in carbon nanotubes, *Chem. Phys. Lett.*, **366**, 504–509 (2002).

[164] M. Wilson, The formation of low-dimensional chiral inorganic nanotubes by filling single-walled carbon nanotubes, *Chem. Phys. Lett.*, **397**, 340–343 (2004).

[165] J. Vavro, M.C. Llaguno, B.C. Satishkumar, D.E. Luzzi and J.E. Fischer, Electrical and thermal properties of C_{60}-filled single-wall carbon nanotubes, *Appl. Phys. Lett.*, **80**, 1450–1452 (2002).

[166] K. Hirahara, K. Suenaga, S. Bandow, H. Kato, T. Okazaki, H. Shinohara and S. Iijima, One-dimensional metallofullerene crystal generated inside single-walled carbon nanotubes, *Phys. Rev. Lett.*, **85**, 5384–5387 (2000).

[167] H. Hongo, F. Nihey, M. Yudasaka, T. Ichihashi and S. Iijima, Transport properties of single-wall carbon nanotubes with encapsulated C_{60}, *Physica B*, **323**, 244–245 (2002).

[168] P.W. Chiu, S.F. Yang, S.H. Yang, G. Gu and S. Roth, Temperature dependence of conductance character in nanotube peapods, *Appl. Phys. A*, **76**, 463–467 (2003).

[169] Y. Cho, S. Han, G. Kim, H. Lee and J. Ihm, Orbital hybridization and charge transfer in carbon nanopeapods, *Phys. Rev. Lett.*, **90**, 106402/1–4 (2003).

[170] A. Rochefort, Electronic and transport properties of carbon nanotube peapods, *Phys. Rev. B*, **67**, 115401/1–3 (2003).

[171] P. Utko, J. Nygard, M. Monthioux and L. Noé, Sub-Kelvin transport spectroscopy of fullerene peapod quantum dots, *Appl. Phys. Lett.*, **89**, 233118/1–3 (2006).

[172] A. Eliasen, J. Paaske, K. Flensberg, S. Smerat, M. Leijnse, M.R. Wegewijs, H.I. Jorgensen, M. Monthioux and J. Nygard, Transport via coupled states in a C_{60} peapod quantum dot, *Phys. Rev. B*, **81**, 155431/1–5 (2010).

[173] E.L. Sceats and J.C. Green, Noncovalent interactions between organometallic metallocene complexes and single-walled carbon nanotubes, *J. Chem. Phys.*, **125**, 154704/1–12 (2006).

[174] J. Sloan, R. Carter, R.R. Meyer, A. Vlandas, A.I. Kirkland, P.J.D. Lindan, G. Lin, J. Harding and J.L. Hutchison, Structural correlation of band-gap modifications induced in mercury

telluride by dimensional constraint in single walled carbon nanotubes, *Phys. Stat. Sol. (b)*, **243**, 3257–3262 (2006).

[175] C. Yam, C. Ma, X. Wang and G. Chen, Electronic structure and charge distribution of potassium iodide intercalated single-walled carbon nanotubes, *Appl. Phys. Lett.*, **85**, 4484–4486 (2004).

[176] E.L. Sceats, J.C. Green and S. Reich, Theoretical study of the molecular and electronic structure of one-dimensional crystals of potassium iodide and composites formed upon intercalation in single-walled carbon nanotubes, *Phys. Rev. B*, **73**, 125441/1–11 (2006).

[177] K.V. Christ and H.R. Sadeghpour, Energy dispersion in graphene and carbon nanotubes and molecular encapsulation in nanotubes, *Phys. Rev. B*, **75**, 195418/1–7 (2007).

[178] C.Z. Loebick, M. Majewska, F. Ren, G.L. Haller and L.D. Pfefferle, Fabrication of discrete nanosized cobalt particles encapsulated inside single-walled carbon nanotubes, *J. Phys. Chem. C*, **114**, 11092–11097 (2010).

[179] J.-P. Cleuziou, W. Wernsdorfer, T. Ondarçuhu and M. Monthioux, Electrical detection of individual magnetic nanoparticles encapsulated in carbon nanotubes, *ACS Nano*, **5**, 2348–2355 (2011).

[180] R. Kozhuharova, M. Ritschel, D. Elefant, A. Graff, A. Leonhardt, I. Mönch, T. Mühl, S. Groudeva-Zotova, C.M. Schneider, Well-aligned Co-filled carbon nanotubes: preparation and magnetic properties, *Appl. Surf. Sci.*, **238**, 355–359 (2004).

[181] D.S. Scholl and J.K. Johnson, Making high flux membranes with carbon nanotubes, *Science*, **312**, 1003–1004 (2006).

[182] A.I. Skoulidas, D.S. Scholl and J.K. Johnson, Adsorption and diffusion of carbon dioxide and nitrogen through single-walled carbon nanotube membranes, *J. Chem. Phys.*, **124**, 054708/1–7 (2006).

[183] S. Kim, J.R. Jinscheck, H. Chen, D.S. Johnson and E. Marand, Scalable fabrication of carbon nanotube/polymer composite membranes for high flux gas transport, *Nano Lett.*, **7**, 2806–2811 (2006).

[184] B.J. Hinds, N. Chopra, T. Rantell, R. Andrews, V. Gavalas and L.G. Bachas, Aligned multiwalled carbon nanotube membranes, *Science*, **303**, 62–65 (2004).

[185] K. Holt, H.G. Park, Y. Wang, M. Stadermann, A.B. Artyukhin, C.P. Grigoropoulos, A. Noy and O. Bakajin, *Science*, **312**, 1034–1037 (2006).

[186] A.I. Skoulidas, D.M. Ackerman, D.S. Scholl and J.K. Johnson, Rapid transport of gases in carbon nanotubes, *Phys. Rev. Lett.*, **89**, 185901/1–4 (2002).

[187] H. Chen, J.K. Johnson and D.S. Scholl, Transport of diffusion is rapid in flexible carbon nanotubes, *J. Phys. Chem. B*, **110**, 1971–1975 (2006).

[188] M. Ouyang, J.-L. Huang and C.M. Lieber, Fundamental electronic properties and applications of single-walled carbon nanotubes, *Accts. Chem. Res.*, **35**, 1018–1025 (2002).

[189] A. Leonhardt, I. Mönch, A. Meye, S. Hampel and B. Büchner, Synthesis of ferromagnetic filled carbon nanotubes and their biomedical applications, *Adv. Sci. Tech.*, **49**, 74–78 (2006).

[190] F. Yang, J. Hu, D. Yang, J. Long, G. Luo, C. Jin, X. Yu, J. Xu, C. Wang, Q. Ni and D. Fu, Pilot study of targeting magnetic carbon nanotubes to lymph nodes, *Nanomedicine*, **4**, 317–330 (2009).

[191] S. Hampel, D. Kunze, D. Haase, K. Krämer, M. Rauschenbach, M. Ritschel, A. Leonhardt, J. Thomas, S. Oswald, V. Hoffmann and B. Büchner, Carbon nanotubes filled with a chemotherapeutic agent: a nanocarrier mediates inhibition of tumor cell growth, *Nanomedicine*, **3**, 175–180 (2008).

[192] M. Arlt, D. Haase, S. Hampel, S. Oswald, A. Bachmatiuk, R. Klingeler, R. Schulze, M. Ritschel, A. Leonhardt, S. Fuessel, B. Büchner, K. Kraemer and M.P. Wirth, Delivery of carboplatin by carbon-based nanocontainers mediates increased cancer cell death, *Nanotechnology*, **21**, 335101/1–9 (2010).

[193] C. Tripisciano, S. Costa, R.J. Kalenczuk and E. Borowiak-Palen, Cisplatin filled multiwalled carbon nanotubes – a novel molecular hybrid of anticancer drug container, *The Eur. Phys. J. B*, **75**, 141–146 (2010).

[194] B. Sitharaman, K.R. Kissell, K.B. Hartman, L.A. Tran, A. Baikalov, I. Rusakova, Y. Sun, H.A. Khant, S.J. Ludtke, W. Chiu, S. Laus, É. Tóth, L. Helm, A.E. Merbach and L.J. Wilson,

Superparamagnetic gadonanotubes are high-performance MRI contrast agents, *Chem. Commun.*, 3915–3917 (2005).

[195] J.H. Choi, F.T. Nguyen, P.W. Barone, D.A. Heller, A.E. Moll, D. Patel, S.A. Boppart and M.S. Strano, Multimodal biomedical imaging with asymmetric single-walled carbon nanotube/iron oxide nanoparticle complexes, *Nano Lett.*, **7**, 861–867 (2007).

[196] S. Koyama, H. Haniu, K. Osaka, H. Koyama, N. Kuroiwa, M. Endo, Y.A. Kim and T. Hayashi, Medical application of carbon-nanotube-filled nanocomposites: The microcatheter, *Small*, **2**, 1406–1411 (2006).

[197] P. Kohli and C.R. Martin, Template-synthesized nanotubes for bionanotechnology and medicine, *J. Drug Deliv. Sci. Tech.*, **15**, 49–57 (2005).

[198] H.J. Hwang, K.H. Byun and J.W. Kang, Carbon nanotubes as nanopipette: modelling and simulations, *Physica E*, **23**, 208–216 (2004).

[199] R. Singhal, Z. Orynbayeva, R.V.K. Sundaram, J.J. Niu, S. Bhattacharyya, E.A. Vitol, M.G. Schrlau, E.S. Papazoglou, G. Friedman and Y. Gogotsi, Multifunctional carbon-nanotube cellular endoscopes, *Nature Nanotechnol.*, **6**, 57–64 (2011).

[200] L. Liao, J.C. Li, C. Liu, Z. Xu, W.L. Wang, S. Liu, X.D. Bai and E.G. Wang, Field emission of GaN-filled carbon nanotubes: High and stable emission current, *J. Nanosci. Nanotech.*, **7**, 1080–1083 (2007).

[201] C.Y. Zhi, D.Y. Zhong and E.G. Wang, GaN-filled carbon nanotubes: synthesis and photoluminescence, *Chem. Phys. Lett.*, **381**, 715–719 (2003).

[202] G.A. Domrachev, A.M. Ob'edkov, B.S. Kaverin, A.A. Zaitsev, S.N. Titova, A.I. Kirillov, A.S. Strahkov, S.Y. Ketkov, E.G. Domracheva, K.B. Zhogova, M.V. Kruglova, D.O. Filatov, S.S. Bukalov, L.A. Mikhalitsyn and L.A. Leites, MOCVD synthesis of germanium filled "diamond-like" carbon nanotubes from organo-germanium precursors and their field emission properties, *Chem. Vap. Dep.*, **12**, 357–363 (2006).

[203] C. Yang, L.G. Yu, M.S. Wang, Q.F. Zhang and J.L. Wu, Low-Field Emission from iron oxide-filled carbon nanotube arrays, *Chin. Phys. Lett.*, **22**, 911–914 (2005).

[204] G. Che, B.B. Lakshmi, C.R. Martin and E.R. Fisher, Metal-nanocluster-filled carbon nanotubes: Catalytic properties and possible applications in electrochemical energy storage and production, *Langmuir*, **15**, 750–758 (1998).

[205] X. Pan, Z. Fan, W. Chen, Y. Ding, H. Luo and X. Bao, Enhanced ethanol production inside carbon-nanotube reactors containing catalytic particles, *Nature Mater.*, **6**, 507–511 (2007).

[206] G. Lota, E. Frackowiak, J. Mittal and M. Monthioux, High performance supercapacitor from hybrid-nanotube-based electrodes, *Chem. Phys. Lett.*, **483**, 73–77 (2007).

[207] D.Y. Lu, Y. Li, S.V. Rotkin, U. Ravaioli and K. Schulten, Finite-size effect and wall polarization in a carbon nanotube channel, *Nano Lett.*, **4**, 2383–2387 (2004).

[208] B. Corzilius, A. Gembus, N. Weiden, K.-P. Dinse and K. Hata, EPR characterization of catalyst-free SWNT and N@C_{60}-based peapods, *Phys. Stat. Solid.*, **243**, 3273–3276 (2006).

[209] S.C. Benjamin, A. Ardavan, G.A.D. Briggs, D.A. Britz, D. Gunlycke, J. Jefferson, M.A.G. Jones, D.F. Leigh, B.W. Lovett, A.N. Khlobystov, S.A. Lyon, S.A. Morton, K. Porfyrakis, M.R. Sambrook and A.M. Tyryshkin, Towards a fullerene-based quantum computer, *J. Phys. Cond. Mat.*, **18**, S867–S883 (2006).

5b

Fullerenes inside Carbon Nanotubes: The Peapods

Ferenc Simon[1,2] *and Marc Monthioux*[3]

[1]*Institute of Physics, Budapest University of Technology and Economics, Hungary*
[2]*Fakultät für Physik Universität Wien, Austria*
[3]*CEMES, CNRS, University of Toulouse, France*

5b.1   Introduction

A particularly compelling possibility for filling the inner space of the SWCNTs is that with fullerenes. The existence of such a structure had been long speculated prior to its actual discovery. The speculations were fueled by the capability of fullerenes themselves of retaining atoms or molecules inside [1]. This structure is called endohedral fullerene, referring to the icosahedral C_{60} symmetry [2]. Such materials are denoted as $A@C_x$ where A marks the encapsulated atom or molecule and $x = 60, 70\dots$ is the index of the fullerene.

C_{60} fullerenes encapsulated inside SWCNTs were discovered by Smith, Monthioux, and Luzzi at the University of Pennsylvania using high resolution transmission electron microscopy [3]. This discovery soon led to the birth of a vivid new field within the nanotube research, the field of the *peapods*, as the structure was baptized. It not only provided a system of inner beauty for the field but it also led to a number of breakthroughs in the study of optical, electronic, and transport properties of the SWCNTs and to a number of applications.

The almost artistic beauty of the peapod structure is shown in Figure 5b.1. The interest in this material originates from the fact that it combines two fundamental carbon molecules: the zero-dimensional fullerene and the one-dimensional SWCNT, which both generate two

Carbon Meta-Nanotubes: Synthesis, Properties and Applications, First Edition. Edited by Marc Monthioux.
© 2012 John Wiley & Sons, Ltd. Published 2012 by John Wiley & Sons, Ltd.

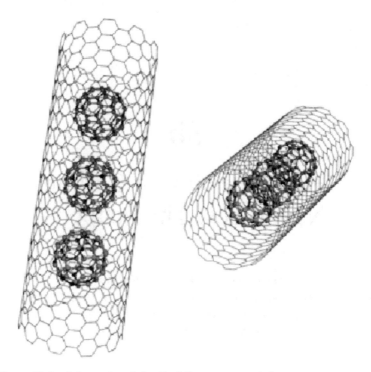

Figure 5b.1 *Schematics of the C$_{60}$ fullerene peapods from two viewpoints.*

of the carbon allotropes when periodically assembled, namely 'fullerite' and 'SWCNT bundles', respectively. The latter is sometimes referred to as 'nanotubulite'. The fate of these two materials is further linked by the discovery that the encapsulated fullerenes merge together to form an inner SWCNT shell upon electron irradiation or heating [4,5].

Here, we review the literature of the peapods covering all aspects from discovery up until applications. Some previous reviews are available in Refs. [6,7]. This chapter is organized as follows: in Section 5b.2, the history and circumstances of the peapod discovery is discussed, followed by a classification of the different peapods in Section 5b.3. In Section 5b.4, the synthesis methods, theory of the synthesis and the peapod stability as well as the behavior of peapods under electron and light radiation and thermal treatment is discussed. The structural, electronic, optical, vibrational, and magnetic properties of peapods are discussed in Section 5b.5, followed by a summary of their realized and expected applications in Section 5b.6.

5b.2 The Discovery of Fullerene Peapods

It is well documented that the point of discovery of carbon nanotubes is a debated issue [8] but it is widely accepted that the discovery of multi-wall carbon nanotubes in the cathode deposit of fullerene synthesis experiments by the electric arc method by Sumio Iijima [9]

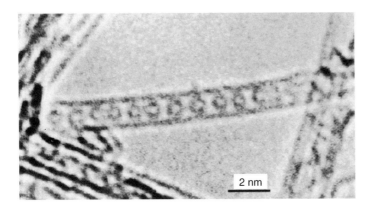

Figure 5b.2 *The first image showing a row of encapsulated fullerenes. The center-to-center spacing of about 1 nm was found to be consistent with the van der Waals separation expected for C_{60}s. Micrograph reproduced with permission from [3] Copyright (1998) Macmillan Publishers Ltd.*

attracted the attention of the scientific community to the topic. In contrast, the discovery of SWCNTs is much clearer as it was reported by two subsequent papers in *Nature* by Iijima and Ichihashi [10] and by Bethune and coworkers [11]. Similarly, the discovery of the peapod structure is well identified as it was reported in Nature in 1998 by Smith *et al.* [3]. The possibility for the existence of the peapod structure had been speculated previously but its detection had not been reported. The reason for the speculation came from the fact that C_{60} has a van der Waals diameter of 0.7 nm [2] which could be optimally accommodated inside SWCNTs with diameters of 1.4 nm given an approximately 0.35 nm van der Waals distance (its exact value is 0.335 nm in graphite) between the SWCNT walls and the fullerene cages. An energetic preference for this structure is given by the nature of the molecular orbitals: both the C_{60} exterior and the SWCNT interior has π orbitals which are similar (although not identical) to those found in graphite. The second reason why the possible presence of C_{60} fullerenes inside SWCNTs had been speculated was that the SWCNT synthesis was known to produce a sizeable amount of C_{60} as a side product apart from the desired carbon nanotubes. In fact, the original SWCNT synthesis was a modified version of the so-called Krätschmer-Huffman arc-discharge process [12] which was developed to yield fullerenes. Much as the expectations and predictions were put forward, fullerenes inside SWCNT remained elusive.

In Figure 5b.2, we show the first high resolution transmission electron micrograph from Ref. [3] which shows an array of spherical molecules encapsulated inside SWCNTs. The center-to-center separation of the molecules was found to be around 1 nm which is consistent with the van der Waals separation expected for a linear array of C_{60} molecules given the above-discussed 0.7 nm fullerene cage diameter and the van der Waals distance. A similar experimental value is found for a three-dimensional fullerene crystal (fullerite), where the center-to-center separation is 1.00 nm [2]. The samples used in Ref. [3] were purified by the HNO_3 acid refluxing method followed by a vacuum annealing up to 1100°C, a method developed by Rinzler and coworkers [13]. The presence of the spherical C_{60} molecules is further corroborated by studying cross-sectional images, which we show

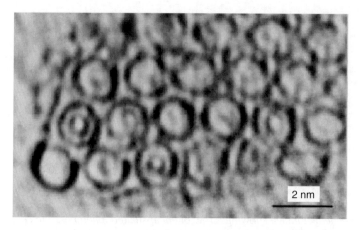

Figure 5b.3 *First cross-sectional image of the C_{60} fullerene peapods. The smaller circular features inside the outer shells of the SWCNTs were identified as the encapsulated C_{60} fullerenes. Reprinted with permission from [3] Copyright (1998) Macmillan Publishers Ltd.*

in Figure 5b.3 from Ref. [3]. Images of this kind show the presence of smaller, circular features inside the SWCNTs which exclude that the structure in Figure 5b.2 would result from a fullerene contamination above or below the studied nanotube. The fact that the circular features are observed in some of the nanotubes only, even though all the tubes are at the same focus depth, excludes that they originate from Fresnel fringes which are common artifacts in HR-TEM. These observations led to the unambiguous identification of this structure which was originally named *nanoscopic peapod* of which later the name *peapod* stuck.

The coalescence of C_{60} was reported already in Ref. [3] to larger 'super-fullerenes' due to the lower binding energy C_{60}, as compared to the SWCNTs. It is discussed next that this observation eventually led to the discovery of the growth of an inner nanotube from the encapsulated fullerenes and to the so-called peapod-based double-wall carbon nanotubes.

The observation in Ref. [3] was first confirmed independently (although on a somewhat different system) by Kwon *et al.* [14], who studied multi-wall carbon nanotubes synthesized from diamond powder and observed the C_{60}s inside the smallest voids. They also proposed a quantum information storage and processing application for the structure, which we discuss further below. The first spectroscopic evidence that the encapsulated molecule is indeed C_{60} came also from the U-Penn group [15]. They performed a moderate etching of the peapod material in acids in order to 'liberate' the encapsulated C_{60}s, which were then removed by rinsing in toluene, which is a common fullerene solvent. The resulting solution was studied by UV-vis spectroscopy which showed unambiguous evidence for the presence of C_{60} and it also allowed quantifying the amount of encapsulated fullerenes.

The mechanism of peapod synthesis was also discussed in Ref. [15]. Two scenarios were considered: production of peapods during the SWCNT synthesis and the production of peapods during the post-treatment purification process. The first scenario was excluded on the basis that no peapod structure was observed for the as-prepared SWCNT soot. In addition, the catalytic synthesis of SWCNTs, yields a very low amount of tubes and fullerenes per consumed carbon, making the simultaneous synthesis of SWCNTs and fullerenes (the latter being embedded inside the former) very improbable. It was argued that the second possibility is more probable since during the acid refluxing some fullerenes become mobile and have sufficient time to find an opening on a nanotube. The peapod structure itself is energetically favored compared to an isolated fullerene and nanotube, as we discuss next. It is interesting to note here, that this originally suggested mechanism corresponds to peapod formation in solution as compared to the latter developed vapor method. However, the solution method was somewhat forgotten just to be rediscovered latter [16–19].

Another peapod synthesis method was proposed by Smith and Luzzi, who showed [5] that effective peapod formation happens for acid purified SWCNTs if these are heated to 450°C irrespective whether the annealing is performed in a closed or in an open environment. At this temperature, C_{60}s, which are stuck to SWCNT walls from the outside, can diffuse along tube walls until an opening is found through which the buckyballs are sucked to the inside. A third synthesis method was described in the work of Kataura and coworkers [20] who heated C_{60} at 650°C in a closed ampoule together with SWCNTs. At this temperature, C_{60} molecules contribute to the gas phase and the peapod formation proceeds in this gas or vapor phase.

It is worth discussing in retrospect the circumstances which led to the discovery of the peapod structure:

1) The use of a relatively low electron energy, 100 kV, by the U-Penn group for the HR-TEM microscopy. At that time (1998), state-of-the art HR-TEM instruments used up to 1 MeV electron beam energies. However, beam energies above 100 kV destroy the encapsulated peapods very rapidly, which prevents to acquire high quality micrographs.
2) The synthesis method, using the Co/Ni catalyst process with either arc discharge or laser ablation, which provides fullerenes and the approximately 1.4 nm diameter tubes, which are ideal for the encapsulation. Other methods such as, for example, CVD growth give diameters which are less optimal for encapsulation and provide virtually no fullerenes as side-products.
3) The purified nature of the samples since unpurified samples do not contain peapods, the fullerenes being found outside the tubes in the SWCNT soot.
4) The actual purification protocol, that is, refluxing in acids followed by vacuum annealing up to 1100°C. The acid refluxing induces openings in the nanotube walls, while only moderately oxidizing or destroying the fullerenes. At the same time, it also provides a medium in which fullerenes can travel until an opening in the nanotube walls is found. The temperature and the duration of the annealing is also critical as a long annealing time at temperature above ~1000°C leads to the coalescence of fullerenes and to the formation of inner nanotubes [5].

5b.3 Classification of Peapods

The peapods can be classified according to the variations that are possible for their two constituents, the host nanotubes and the encapsulated fullerenes. Classification is given

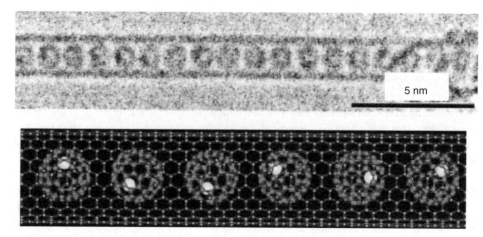

Figure 5b.4 *HR-TEM micrograph and a schematic representation (not to scale) of Gd@C$_{82}$ metallofullerene peapods. Dark spots seen on the fullerene cages correspond to the heavy Gd atoms. For a better understanding of the figure, please refer to the colour plate section. Reprinted with permission from [26] Copyright (2000) American Physical Society.*

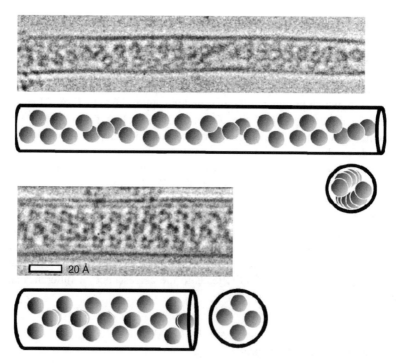

Figure 5b.5 *Two HR-TEM observations for the 'silo' arrangement of fullerene peapods inside carbon nanotubes. The arrangement depends on the diameter of the host nanotube. Reprinted with permission from [23] Copyright (2004) American Physical Society.*

according to: (i) the nanotube type; (ii) the fullerene type (with several subclasses) and (iii) the type of arrangement of the fullerenes.

We first give a classification according to the host nanotube:

1) Normal SWCNT peapods such as those reported in 1998 in Ref. [3].
2) Multi-wall CNT peapods [21,22] synthesized first in 2003. As a subcase of this class are double-wall carbon nanotube peapods [23,24].
3) Noncarbon nanotubes, for example, multi-wall boron nitride nanotube peapods, first reported in 2003 [25].

The classification according to the fullerene type is as follows:

1) Pristine fullerenes, for example, C_{60}, C_{70}, and so on, Ref. [3].
2) Endohedral fullerenes, that is, endohedral metallofullerenes [26,27] and the unique $N@C_{60}$ fullerene [28] whose first encapsulation was reported in Ref. [18]. In Figure 5b.4, we show a HR-TEM image of heterofullerene peapods from Ref. [26].
3) Functionalized fullerenes [29].
4) Heterofullerenes, for example, the $C_{59}N$ azafullerene [30,31].
5) ^{13}C isotope enriched fullerene peapods, reported in Ref. [32].

According to the arrangement type, the encapsulated fullerenes can form a linear chain [3] or can organize themselves in a 'silo' configuration [23,25,33] in larger diameter nanotubes as we show in Figure 5b.5.

5b.4   Synthesis and Behavior of Fullerene Peapods

5b.4.1   Synthesis of Peapods

5b.4.1.1   *Mass Scale Synthesis*
Clarifying the peapod formation mechanism in Ref. [15] by ascertaining that the fullerenes are originally outside the SWCNT, led to the optimization of the synthesis method. Smith and Luzzi reported [5] that adding a droplet of C_{60} suspended in dimethylformamide (DMF) to HR-TEM grids and subsequently heating the samples in vacuum in situ at 450°C leads to an abundant filling of SWCNTs with C_{60}, as shown in Figure 5b.6.

The choice of the temperature is very delicate: C_{60} is known to sublime, that is, to enter to the gas phase above 350°C with a vapor pressure increasing with the temperature. However, the finding in Ref. [15] shows that C_{60}s, which are stuck to SWCNT surface, can travel at this temperature along the tube walls until they find an opening then enter the SWCNT cavity to become encapsulated. However, 450°C is low enough not to remove the fullerenes away from the tubes.

This method of encapsulating additional fullerenes (i.e., which are not present as a side-product in the soot) into the SWCNTs opened the way to prepare other peapods such as of higher fullerenes, metallofullerenes, endohedral fullerenes, and so on. The C_{70} fullerenes were also present albeit with a low concentration in the SWCNT soot along with C_{60} therefore C_{70} peapods were also produced in the original report [3]. This, along with the highly developed nature of the fullerene synthesis and chemistry, opened the way for a new level of SWCNT functionalization.

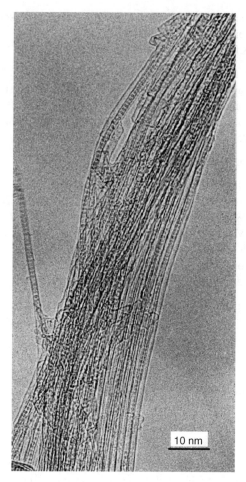

Figure 5b.6 *Ropes of SWCNT peapods showing an abundant filling with fullerenes. C_{60}s suspended in DMF was added to the sample and it was annealed at 450°C for 1 h in vacuum. Reprinted with permission from [5] Copyright (2000) Elsevier Ltd.*

5b.4.1.2   The Gas Phase Synthesis

A further step toward the mass scale synthesis of peapods, which eventually led to their spectroscopic investigation, was reported by Kataura *et al.* [20]. They performed the filling on a purified and opened buckypaper of SWCNTs. Opening was a side effect of purification done by refluxing in H_2O_2 and washing in HCl. The buckypaper samples were prepared by filtering and sealed together with an abundant amount of fullerene powder in a quartz ampoule and heated to 650°C for 1–2 h. At this temperature, the fullerenes have a high vapor pressure which enables them to enter through the SWCNT openings. This synthesis method is therefore termed as the gas phase synthesis method. The nonencapsulated fullerenes were removed from the SWCNTs by sonication in toluene (which is an effective solvent for C_{60}). Encapsulated fullerenes are energetically favored inside the SWCNTs and

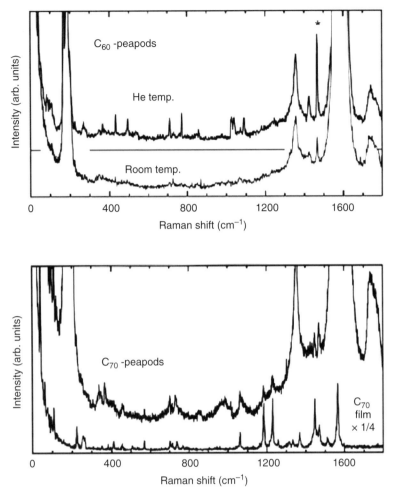

Figure 5b.7 *Raman spectra of C_{60} and C_{70} peapods with a 488 nm (2.54 eV) laser excitation. The upper panel also shows that the narrow C_{60} appear stronger at lower temperature. Asterisk shows the pentagonal pinch mode (PPM) of the peapod at 1466 cm^{-1}. Reprinted with permission from [20] Copyright (2001) Elsevier Ltd.*

therefore are not removed meanwhile (exception from this is discussed below for very large diameter nanotubes according to Ref. [33]). The toluene-sonicated samples were again filtered which yielded the final buckypaper peapod sample.

The possibility to synthesize macroscopic amounts of peapod samples in the form of buckypapers opened the way for a variety yet unavailable spectroscopic and structural investigations, which are discussed further below. In Figure 5b.7, we show the first bulk spectroscopic study on the peapods using Raman spectroscopy.

A number of narrow lines, which are originating from the encapsulated fullerenes (C_{60} and C_{70}) were observed. Interestingly, the 488 nm laser line used in the study is close to the optical transition energy for the fullerene HOMO-LUMO gap [2], therefore the fullerene

Raman modes are greatly enhanced. Without this resonance condition, the fullerene modes are unobservable. The Raman spectra provides clear-cut experimental evidence for the encapsulated nature of the fullerenes as a downshift of the strongest pentagonal pinch mode (PPM) with the $A_g(2)$ symmetry (asterisk in Figure 5b.7) is observed.

A parallel work by the Iijima group clarified the role of and the method for opening the SWCNTs [26]. They found that a heat treatment at 400–450°C for 10–30 min in air opens very effectively the tubes, whereas the acid treatment alone does not produce sufficient number of openings. The treatment in H_2O_2 by Kataura *et al.* essentially provided the same effect as the nascent oxygen acts similarly to the oxidation in air. There is certainly a balance between the number of SWCNT openings and the amount of unwanted defects created and nanotube material consumed. This optimal treatment (duration, temperature) depends on the nanotube material and morphology (long or short tubes, highly bundled or more individual) it therefore has to be optimized for each sample [34].

The removal of nonencapsulated fullerenes from the outside of the nanotubes was simplified as it can be performed directly on the nanotube samples by heating them in dynamic vacuum in a quartz tube above 650°C for 1–2 h [18]. Then, fullerenes enter the vapor phase in the middle of the furnace and diffuse to the colder parts of the quartz tube toward the pump where they deposit.

In summary, the peapod synthesis with the vapor method can be performed in three steps on the SWCNT soot: (i) heating in air at 400–450°C for 10–30 min; (ii) sealing the SWCNT material in a quartz (or glass) ampoule with a torch along with 1–2 times larger mass of fullerenes and heating at 550–650°C for 1–12 h (the short duration applies to the high temperature); and (iii) placing the resulting material in a quartz tube connected to a vacuum pump (vacuum levels of 10^{-4} mbar suffice throughout) and heating to 650°C for 1 h.

The first step may be omitted when the starting SWCNT material has been previously purified via oxidizing methods, with purification conditions severe enough to generate openings in the SWCNT walls.

5b.4.1.3 Low Temperature Synthesis

The vapor synthesis method is applicable to pristine fullerenes such as C_{60}, C_{70} and to metallofullerenes [26] which are stable at the 450–650°C required. However, it cannot be applied to temperature-sensitive fullerenes derivatives such as the endohedral $N@C_{60}$ (decays above 200°C [35]) and functionalized fullerene derivatives [36]. The interest in preparing peapods from such fullerenes emerged in 2002–2003 when four groups independently described low temperature synthesis methods to obtain high yield peapods [16–19]. The activity was partly triggered by the suggested use of $N@C_{60}$ peapods for quantum computing by Harneit [37].

The Iijima group reported a method termed 'nanoextraction' [16] in which C_{60} and the SWCNTs are placed into ethanol for one day. This method was demonstrated for 2 nm diameter nanotubes, which is larger than the usually considered 1.4 nm nanotubes. 1.4 nm SWCNTs were considered in a similar method proposed by Monthioux and Noé [17] who used a saturated solution of C_{60} in toluene in which SWCNTs were soaked upon sonication at room temperature. They pointed out that optimum filling conditions correspond to mid-sonication conditions, because of the damaging of SWCNTs induced by the dynamic shocks of the solvated fullerenes onto the SWCNT walls.

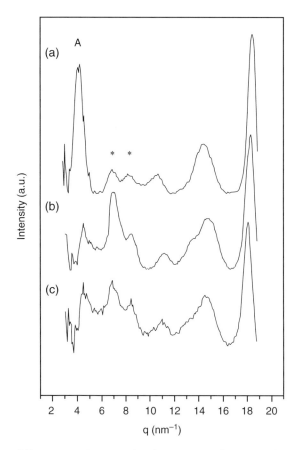

Figure 5b.8 *X-ray diffractogram (reprinted with permission from [18] Copyright (2004) Elsevier Ltd) showing that C$_{60}$ peapods prepared with the vapor (b) and solvent (c) methods are structurally identical on a macroscopic scale. The diffractogram for the starting, empty SWCNT is shown (a). Note that the Bragg peaks corresponding to the hexagonal lattice of SWCNT bundles are changed upon the fullerene encapsulation as the structure factor of the one-dimensional fullerene lattice modulates these (peak A is reduced, those marked with an asterisk are increased).*

Simon *et al.* [18] described a method in which the two peapod constituents are placed into refluxing n-hexane (at 69°C). The boiling of n-hexane itself provides a continuous stirring which reduces the required reaction time to 2 h. Khlobystov *et al.* reported the use of supercritical carbon-dioxide as a medium and a reaction time of 10 days. The relative ease, efficiency, and short duration of the n-hexane refluxing method made it more popular than the alternatives and it was also used to encapsulate functionalized fullerenes [38]. Samples prepared with the low temperature (also termed the *solvent method*) do not differ in any physical properties from those prepared with the vapor method. It is best shown with X-ray diffraction (Figure 5b.8), which provides a bulk structural characterization of the ordered, one-dimensional peapod filling. Were the filling only partial, or the fullerenes not forming a perfect 1D array, the X-ray diffractogram would not be clearly visible.

5b.4.1.4 Quantification of Peapod Filling

Characterization of peapod filling efficiency is important to optimize their synthesis routes, to apply bulk spectroscopic methods on them, and also for the applications. In this respect, microscopic methods such as HR-TEM are very time-consuming as a large number of nanotubes have to be sampled. The most straightforward method to estimate the peapod filling ratio is by weight uptake. Raman spectroscopy and X-ray diffractometry are viable candidates for this characterization, and these are macroscopic methods. However, Raman scattering intensity does not directly sense the number of fullerenes or the peapod filling fraction as it depends on a number of factors such as the optical density of the peapod material and the Raman resonance condition [39]. X-ray diffractometry on the other hand requires a peapod material where both the SWCNTs are well ordered in the bundle structure and also the fullerenes form a perfect one-dimensional lattice. However, X-ray diffraction can in principle provide a direct measure of the filling fraction.

Two further methods provide direct counting of carbon atoms on the fullerenes and on the SWCNTs: electron spectroscopy (which measures some characteristic carbon core levels) and nuclear magnetic resonance which is only sensitive to carbon and with which SWCNT and the encapsulated peapod signals can be discriminated.

We discuss the filling characterization methods in this order. An accurate, calibrated measurement was performed by the Luzzi group [40]. It was found that under optimal circumstances, up to 90% filling fraction can be achieved. Here, filling fraction means what amount of the available inner volume of the SWCNTs is occupied by the fullerenes. This was also correlated with the HR-TEM measurements and a good agreement was observed. Kataura and Maniwa studied the X-ray diffractogram of C_{60} peapods. The presence of peapods is to modulate the Bragg peaks of the hexagonal bundle arrangement through the form factor of the one-dimensional C_{60} chain. The effect of this modulation is shown in Figure 5b.8, and the details of the X-ray analysis are discussed in Section 5b.5.1. This analysis allowed for a quantitative determination of the filling fraction in Refs. [41,42]. A more general treatment of the X-ray diffraction for peapods was given by the Launois group [43]. It also considered the effect of the tube diameter distribution, SWCNT alignment, and crystallinity on the determination of the peapod filling.

The group of Pichler performed a direct carbon counting of the peapods separately for the carbons on the fullerenes and on the SWCNTs (or on other carbonaceous phases) [44]. Electron energy loss spectroscopy (EELS) was used to study the carbon core level excitation for the peapods. In this method, a core electron is excited to an unoccupied state; for the case of carbon, the $1s$ electrons are excited to the unoccupied $2p$-state-related π^* state with an energy onset of 285.4 eV. In Figure 5b.9, we show the EELS data for the peapod as well as for an unfilled SWCNT reference sample. The peapod spectrum contains a shoulder around 284.5 eV and an extra peak at 288.3 eV, which were assigned as spectral features related to the encapsulated fullerenes. The reference spectrum is subtracted from the peapod EELS spectrum after multiplying with a scaling factor such that the difference resembles most to the well-known spectrum of C_{60}s. The difference is also shown in Figure 5b.9.

The number of carbon atoms on the encapsulated C_{60}s relative to the host nanotubes is obtained by integrating the C $1s$ spectrum of the difference normalized by the integral of the reference sample. This allowed essentially counting the carbon atoms which make the encapsulated fullerenes. The carbon counting provided by EELS was used to calibrate the Raman spectra [45], which is a more available method for the peapod filling efficiency

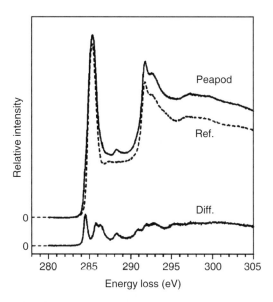

Figure 5b.9 *C 1s core level excitation spectra of C_{60} peapods, of the reference material, and of the difference between the two after an appropriate scaled subtraction was performed (see text). Note that the difference spectrum is dominated by the exceptionally narrow C_{60} transitions. Reprinted with permission from [44] Copyright (2002) American Physical Society.*

characterization. Benchmarks were provided for the filling efficiency as a function of the observed Raman modes at well-defined laser energies. For instance, for a SWCNT sample with 1.4 nm mean diameter and 0.1 nm Gaussian diameter variance, the maximum volume that could be filled with C_{60}s is 94% of the total inner volume of the tubes (SWCNTs smaller than 1.2 nm cannot be filled). Considering the best visible C_{60} $A_g(2)$ mode and the SWCNT G modes, the Raman intensity ratio is $A_g(2)/G = 4.6 \cdot 10^{-3}$ at 488 nm laser excitation for such a sample if all the available volume is occupied by fullerenes.

Another, bulk carbon counting method is based on encapsulating ^{13}C enriched fullerenes inside the SWCNTs, and performing nuclear magnetic resonance, NMR, on them [32]. Since NMR is only sensitive to the ^{13}C isotope, the encapsulated isotope-enriched fullerenes (with enrichments up to 90%) are detected with a large contrast as compared to the host nanotubes, which contain ^{13}C with a natural, 1.1%, abundance only. The NMR signal intensity provides an absolute measurement of the detected number of ^{13}C nuclei, therefore calibrating it with a NMR standard with known number of ^{13}C nuclei directly yields the number of carbon atoms encapsulated inside the SWCNTs.

5b.4.1.5 Theory of Peapod Synthesis, Stability and Arrangement

The theoretical understanding of the peapod formation and energetic stability is important for the successful synthesis and to explore similar systems. In addition, the description of the peapod stability is an interesting test-bed for the theoretical methods in a true nanosized environment. It is a well-known problem in the first principles based calculation methods that the van der Waals interaction is difficult to handle accurately. Therefore the description

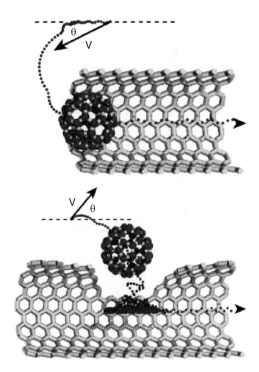

Figure 5b.10 *Schematics of the fullerene encapsulation inside a (10,10) nanotube for two different scenarios: through a tube-end (upper panel) and through a tube-wall opening (lower panel). The dotted lines indicate possible C_{60} trajectories, characterized by the launch velocity y and launch angle θ with respect to the tube axis. Reprinted with permission from [46] Copyright (2002) American Physical Society.*

of the peapod system, which is controlled by the van der Waals interaction, led to the development of the theoretical methods as well.

The first theoretical description for the peapod synthesis was reported by the Tomanek group [46]. Molecular dynamics calculations were performed using a parameterized interaction Hamiltonian which had been developed to describe the wall-wall interaction in MWCNTs. The authors found that the fullerene encapsulation happens in two steps: first a fullerene outside the SWCNTs is physisorbed on the surface with an energy gain of about 0.07 eV. Then, at the elevated temperatures of the encapsulation (above 400°C), the fullerene can freely diffuse along the nanotube until an opening is found through which it can enter the nanotube with a further energy gain (discussed next). It was found that the openings themselves do not form an activation barrier for the encapsulation. Of the two possible scenarios, that is tube-wall or tube-end openings, it was found that the encapsulation is more probable through the tube-wall opening. The two scenarios are shown schematically in Figure 5b.10. A simple explanation of this difference is that, for a tube-end encapsulation, the fullerene is required to make a 'U-turn', which is a less probable trajectory at the synthesis temperature.

The energetic stability of the peapod structure motivated a number of theoretical studies to calculate the binding energy (Figure 5b.11).

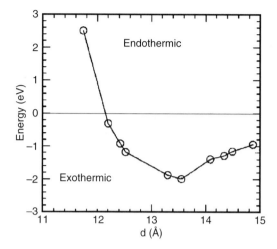

Figure 5b.11 *Binding energy of C_{60} in various peapods as a function of the host outer tube diameter, d. The calculation was performed with the density functional theory in the local density approximation. The negative binding energy corresponds to the energetically stable, exothermic encapsulation. Reprinted with permission from [47] Copyright (2004) American Physical Society.*

Table 5b.1 *Summary of the available theoretical results on the energetics of peapod encapsulation. The calculation method, the binding energy for the (10,10) host outer tube, E_b, and the minimal host tube diameter d_{min} for which the encapsulation is exothermal are given. DFT-LDA stands for the Density Functional Theory method with the Local Density Approximation.*

Ref.	Method	E_b [eV]	d_{min} [nm]
[49]	DFT-LDA	1.39	1.23
[86]	DFT-LDA	0.51	1.28
[140]	Brenner potential force model	3.26	1.174
[141]	Molecular force model	3.69	1.18
[142]	DFT-LDA	2.1	1.22
[89]	pseudopotential DFT	0.9	–
[143]	total energy functional	0.4	1.32

It was observed experimentally [5] that the encapsulated peapods cannot escape the host nanotubes even at elevated temperatures (above 1000°C) and they form an inner nanotube, which leads to the double-wall carbon nanotube structure. Therefore the encapsulation is essentially an irreversible process with a relatively large binding energy, yet it was found that releasing encapsulated C_{60}s was as easier as the host tube diameter is larger [33,48].

In Table 5b.1, we summarize the theoretical reports on this matter. The applied calculation method, the largest binding energy observed, and the diameter of the smallest outer tube

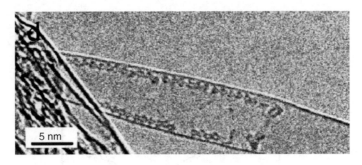

Figure 5b.12 *HR-TEM micrograph of a flattened large diameter SWCNT containing encapsulated C_{60} chains along the tube-like channels on both sides. Reprinted with permission from [48] Copyright (2007) Elsevier Ltd.*

where the encapsulation is still energetically preferred, are given. Although the parameters scatter somewhat, some general conclusions can be drawn from these works. First, the encapsulation is energetically preferred if the host outer tube has a diameter above ~1.2 nm, but the incorporability of C_{60}s inside SWCNTs decreases for diameters beyond ~2 nm [48]. Next, the typical binding energy observed for a (10,10) outer tube is 1–2 eV per C_{60} molecule. This corresponds to a binding energy of about 5 meV per carbon atom (considering the carbon atoms of the host tube as well) [49]. This is much smaller than that found in graphite (23 meV per atom) which reflects the trivial fact that the coordination of most of the carbon atoms on the fullerene is such that they do not face optimally the surrounding outer tube, thus the van der Waals energy gain is not optimal.

Once the energetic stability of C_{60} upon encapsulation was understood theoretically, interest turned toward understanding configuration as a function of the host tube diameter, d, and also for other fullerenes. We discuss the related results in this order in the following. The encapsulated fullerenes occupy a high symmetry position inside the host nanotubes and the relative orientation of the fullerenes with respect to outer tube is incommensurable, that is, it is ill-defined except for some special cases (e.g., C_{60} inside a (10,10) nanotube) when a commensurable structure can be achieved. As it is discussed below, this results in a low energy barrier for the C_{60} rotation [50]. Provided the fullerenes are situated perfectly along the nanotube axis (which is realized for $d \leq 1.5$ nm), their mutual arrangement of the neighboring C_{60}s can take several well-defined arrangements; facing pentagons, facing hexagons, or facing double bonds, which were studied by Michel *et al.* [51]. It was found that for $d \leq 1.4$ nm the facing pentagon structure is energetically preferred. For $d \geq 1.5$ nm, the fullerenes are displaced from the tube axis thus no mutually ordered structure of the C_{60} can be identified and the structure takes the so-called silo arrangement such as that shown in Figure 5b.5 after Ref. [23].

It is worth noting that considering SWCNTs with diameters above d ~7 nm cannot compare to other SWCNTs as such large nanotubes are flattened. Due to this specific morphology that those nanotubes are enforced to adopt upon energetical constraints, encapsulated fullerenes are found to align along the nanotube edges, revealing that they are actually trapped in the tube-like channels left on both sides of the flattened part [48]. We show this situation in Figure 5b.12.

There exists a transition region, for ~ 1.4 nm $< d \leq 1.5$ nm where the facing hexagons and facing double bond configurations occur. These are energetically so close that no distinction could be made for the preferred alignment. The silo structure in larger diameter nanotubes was understood on the basis of the arrangement taken by fullerenes for the optimal van der Waals energy [47,52]. It is energetically preferred for the fullerenes to displace themselves from the middle of the nanotube, so that the fullerene-tube wall separation is the optimal van der Waals distance. Since all fullerenes prefer to stick to the host nanotube walls on different sides, they are essentially located around the tube center, giving rise to the structure shown in Figure 5b.5. We discuss below that apart from the local HR-TEM evidence, X-ray diffractometry shows a smaller fullerene lattice constant along the tube axis for the silo arrangement [33].

Okada *et al.* studied the configuration for C_{70} as a function of the host diameter nanotubes [53]. It was found that a lying configuration (i.e., when the longer axis of C_{70} is parallel with the nanotube axis) is realized for tube diameters below 1.41 nm and the standing configuration (i.e., when the shorter C_{70} axis is parallel to the tube axis) is realized for larger diameter host nanotubes. Verberck and Michel [54] found that the lying structure is realized for $d \leq 1.4$ nm and the standing for $d \geq 1.44$ nm and a transition region in between with a mixture of the two configurations. It is discussed next that this structure is indeed observed experimentally [55,56]. The Launois group found in X-ray studies [56] that structural change from the lying to standing configuration happens for $d \approx 1.42$ nm in good agreement with the theoretical prediction [54].

5b.4.2 Behavior of Peapods under Various Treatments

5b.4.2.1 *Behavior under Electron Irradiation*

The behavior of encapsulated peapods under the electron beam of the HR-TEM was already studied in the first report on peapods [3]. It was found that exposure to the 100 kV electron beam causes the encapsulated fullerenes to coalesce into smaller tubular structures with a capped ends. This behavior was interpreted with the energy per carbon atom which is higher for a fullerene than for a carbon nanotube. Obviously, the lieu where the π-electrons are shared with the outer tube that results in the so-called van der Waals interaction is limited to a single circle per fullerene molecule, while it corresponds to a genuine cylindrical surface for coalesced fullerenes. In this sense, coalescence into the smaller tubular structure is energetically preferred and the electron beam irradiation provides sufficient energy to overcome an activation barrier. The exposure effect has an important technical implication: the peapod structure can be best studied for relatively low electron beam energies in HR-TEM. Many state-of-the-art HR-TEM instruments operate with beam energy above 200 keV, which cannot be exploited when studying peapods.

The electron beam-induced coalescence of the encapsulated fullerenes was subsequently studied in more detail [4]. The resulting structure was termed 'coaxial tubes' (CATs) and was found to be the smallest multi-wall carbon nanotubes for a while. It was shown that the diameter of an inner nanotube is 0.7 nm, which is inside a 1.4 nm outer tube. Therefore the wall-wall separation is close to the ideal van der Waals distance, which is 0.335 nm in graphite. It is more common to call this structure 'double-wall carbon nanotubes' (DWCNTs), a structure which had been described prior to the discovery of peapods [57]. However, electron irradiation also goes with a ballistic interaction with electrons, inducing atom displacements and ultimately, knocking off of carbon atoms from the lattices (yet in an extent that depends on the electron energy). The resulting coalesced structure is highly defective, and barely long enough for being called a genuine 'inner nanotube'. Hence, the destructive and constructive effects of the interacting electrons compete, ultimately possibly resulting in the amorphization of the whole irradiated material.

5b.4.2.2 Behavior upon Annealing: The Double-Wall Carbon Nanotubes

The ability to obtain the inner nanotubes under the electron beam exposure motivated to study their temperature-induced formation. For this purpose, Smith and Luzzi developed the method of preparing a macroscopic amount of peapods as we described earlier [5]. This sample was subjected to a 1200°C annealing in dynamic vacuum for 24 h. In the resulting material, an abundant occurrence of double-wall carbon nanotubes was observed, which confirmed that the temperature excitation can be also used to overcome the activation barrier for the inner tube formation.

The annealing-induced coalescence is potentially capable of producing double-wall carbon nanotubes in macroscopic amounts. The attention to study this material with macroscopic spectroscopy methods was drawn by the report of Bandow *et al.* [58], who showed that Raman spectroscopy can be performed on the DWCNTs and that the

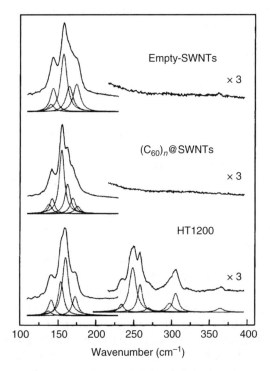

Figure 5b.13 Raman spectra of empty nanotubes, C_{60} peapods, and annealed peapods (at 1200°C) with a 515 nm laser excitation. A deconvolution into different components is shown with dashed curves. Note the peaks which appear for the annealed peapod sample in the 230–370 cm^{-1} spectral range, which correspond to the RBMs of the inner tubes. Reprinted with permission from [58] Copyright (2001) Elsevier Ltd.

characteristic Raman modes, which correspond to the inner tubes, are observed. We show the Raman spectra for empty nanotubes, C_{60} peapods, and for annealed peapods in Figure 5b.13. The annealing was performed at 1200°C, which is sufficient to induce the coalescence of all the encapsulated fullerenes and to form the inner tubes.

Raman spectroscopy is particularly useful to characterize the growth of carbon nanotubes in a diameter-selective manner, as the so-called radial breathing mode (RBM), which is unique to SWCNTs (and DWCNTs), is sensitive to the nanotube diameter. There is a reciprocal relation between the energy of Raman mode and the diameter: $v_{RBM} \approx C_1/d$ [59]. The value of the C_1 constant depends on the environment of the tubes and $C_1 \approx 235\,cm^{-1}\,nm^{-1}$ is an accepted value. Clearly, new Raman modes in the 230–370 cm^{-1} range arise upon annealing the peapod material which correspond to small diameter nanotubes with $d = 1$ to 0.65 nm and these modes were identified as the radial breathing modes of the small diameter inner tubes.

5b.4.2.3   Alternatives to Mere Electron Irradiation or Annealing
An interesting alternative to the thermally induced coalescence is the local heating using a high power focused laser. Kramberger *et al.* [60] found that peapods suspended on TEM grids can be made white glowing in vacuum using a 1064 nm laser of an FT-Raman spectrometer if it is focused onto a 1 µm^2 spot with 500 mW power. This in situ heating also allowed for Raman monitoring of the inner tube growth process. The widespread transformation of the peapods into genuine DWCNTs was, however, not confirmed by clear TEM images.

A similar approach was used by Kalbac *et al.* [61]. They also claimed the successful formation of DWCNTs, yet with a higher yield when starting from C_{70}@SWCNT than from C_{60}@SWCNT due to the higher sensitivity to photolysis of the former. Since the irradiation was carried out with a pulsed laser in order to prevent thermal load of the material, the coalescence mechanism is explained via multiphoton photolysis of the encapsulated fullerenes, without passing via the intermediate event of [2+2] cycloaddition-driven fullerene polymerization (see Section 5b.4.2.2). The outer tubes of the resulting DWCNTs are found larger and less defective than upon coalescence by thermal annealing. This might be surprising as temperature is always a structuring process for graphene-based materials, provided the atmosphere is nonreactive. However, it is worth noting that no TEM image of the resulting DWCNTs was shown in the paper.

Puech *et al.* [62] also carried out UV light laser irradiation experiments, yet with a different goal as above since the purpose was to induce local heating (i.e., convenient for temperature-sensitive substrates such as Si-doped wafer-supported devices) and check whether reaching temperatures high enough for inducing the formation of DWCNTs was possible. Using nonpulsed UV-light laser made possible to record the Raman spectrum evolution simultaneously. Starting from empty SWCNTs or CVD-prepared DWCNTs as reference materials [63] and using the downshift of the G^+ and G^- bands as local temperature probe, it was shown that temperatures as high as 800–1000°C could be reached for a laser power of 280 mW, that is, able to initiate the fullerene coalescence mechanisms. However, deep changes (merging of the G^+ and G^- bands, and sudden upshift of the resulting G band) occurred on irradiated peapods as early as for a laser power of 140 mW (corresponding to a temperature of ~480°C), which was explained by oxidation phenomena locally promoted by the contained-fullerenes, thanks to a some partial pressure of oxygen as an impurity in the low pressure argon atmosphere.

Achieving the peapod-to-DWCNT transformation via laser irradiation seems to be possible both photo-induced mechanisms, that is, photolysis or thermal annealing, but further work is still needed to support this statement further.

Another alternative was proposed by Monthioux *et al.* who studied the coalescence process combining electron beam irradiation with in situ heating [64,65]. First, the two extremal handling was studied, that is, no heating and long 200 keV electron beam irradiation time and 1200°C heating without exposure to irradiation. The heating alone produced structurally perfect double-wall carbon nanotubes, whereas the elongated irradiation alone produced a large number of defects on the outer and inner walls in addition to promoting the coalescence of the encapsulated fullerenes. It was found that the combination of the two methods, that is annealing at 200°C while irradiating produced inner wall to outer wall distances of 0.5–0.55 nm which is far beyond of that anticipated from the van der Waals distance of ~0.35 nm.

Then, by applying various couples of temperature (in the range 480–700°C) and electron irradiation (energy in the range 150–300 kV, in addition to various electron dose and flow values) conditions, it was found that some were able to successfully achieve the peapod-to-DWCNT transformation whereas others were resulting in the peapod destruction, with all the intermediate features also obtained depending on the conditions. The most interesting observation was that, based on TEM investigation statistics, the diameter of the resulting inner tube or capsules as well as the inner-outer shell distance were also depending on the temperature/irradiation condition couple used. The leading parameter was found to be the host (outer) nanotube diameter, whose increase relates linearly to an increase of both the inter-tube distance (see Figure 5b.14) and the inner tube diameter.

This indicates that, in such conditions, the coalescence mechanisms are driven by a compromise between two contradictory energetic requirements, that is, tentatively enforcing the regular 0.34 nm van der Waals distance between the inner and outer graphenes, and tentatively maintaining the initial 0.7 nm diameter originating from the starting fullerenes. With respect to this overall mechanism, temperature appeared to be an important parameter to control the dimensional features of DWCNTs. Typically, for a given electron energy, both the resulting inner tube and outer tube diameters appeared smaller (and the intertube distance larger) after 490°C than after 700°C heat-treatment (see Figure 5b.14), and both resulting inner tube and outer tube diameters were larger after 700°C heat treatment than for the starting peapods. On the other hand, none of the electron beam features were found to have a significant influence on the DWCNT features, but possibly on the electron dose. This suggested distinct coalescence regimes according to the temperature range. At high temperature (700°C), thermal effects may prevail, resulting in larger (inner and outer) tube diameters possibly via a tube fattening mechanism which involves the creation of Stone-Wales defects, their subsequent splitting into two 5–7 ring pairs, and the migration of the two 5–7 pairs apart from each other [66]. At low temperature (490°C), irradiation effects such as knocking-off of atoms and hindering of the fattening mechanism just mentioned may prevail, resulting in a relative shrinkage of both inner and outer tube diameter. This effect is more pronounced for inner tubes than for outer tubes, which probably relates to the fact that the former are building-up whereas the latter pre-exist.

5b.4.2.4 Coalescence Mechanisms

The possibility to form single-wall carbon nanotube inside another nanotube is interesting for both the fundamental understanding of the coalescence process and also for practical

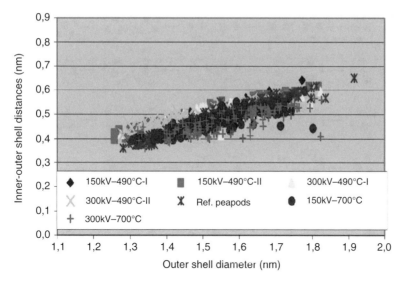

Figure 5b.14 *Plot of the inner-outer shell distance with respect to the outer diameter for various temperature/electron energy (1V = 1eV) condition couples. The trend for 490°C experiments, regardless of the voltage value (specifically for the set of experiments II, that is, a pink square for 150kV, and a light blue cross for 300kV) correspond to larger inner-outer shell distances, as opposed to the trend for 700°C experiments (brown dots for 150kV, and dark 'plus' sign for 300kV). Reprinted with permission from [65] Copyright (2005) American Institute of Physics.*

reasons. To date, the growth of SWCNTs requires the use of metal catalysts which results in inevitable contaminations and defects. In contrast, the peapod-derived SWCNTs are formed in a catalyst-free 'nano clean-room' where the outer wall as a template enforces the building of a highly curved tubular structure. As we mentioned above, the clear preference of the sp^2-like bonds on the carbon nanotube and the absence of any pentagons (which are energetically not preferred) make them energetically more stable than the peapods. However, the details of the coalescence process are somewhat subtle.

The first attempt to describe the formation of the inner tubes was made by the Tomanek's group [67]. They found a particularly low energy route (which is therefore a very probable growth channel), which starts with a $(2+2)$ cycloadditional bond on neighboring C_{60}s. This requires that the fullerenes can take a position where their doubly bonded carbon atoms are facing each other as the cycloadditional bond is only possible upon breaking up this bond. Once the $(2+2)$ bond is formed a series of transformations can take place where the bonds are rotating but bond breaking is avoided. This process is the so-called Stone-Wales transformations [68], which were first proposed to describe the evolution of nonoptimal fullerene cages toward the energetically stable ones. The series of such transformations, which can lead to cycloadditionally bonded fullerenes to a fully merged C_{120} molecule, is shown in Figure 5b.15 from Ref. [67].

This mechanism has the advantage that it only involves the relatively low energy Stone-Wales transformations. In principle it predicts that the inner tube formation results in

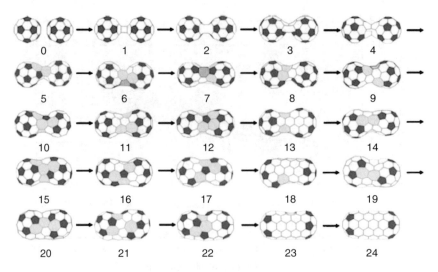

Figure 5b.15   *Schematics of the pathway for the 2 $C_{60} \rightarrow C_{120}$ fusion. The minimum possible Stone-Wales transformations were found using a geometrical search. Note that the starting point of the fusion is the formation of a (2 + 2) cycloadditional bond between the adjacent C_{60}s. Reprinted with permission from [67] Copyright (2004) American Physical Society.*

perfect (5,5) inner tubes. However, it is observed experimentally that not only there is more than one type of inner tube that is formed, it is also observed that the inner tube diameter increases until that of the outer tube until the van der Waals separation between the walls is attained. Therefore, additional (probably also Stone-Wales related) transformations are required to produce all the inner tubes which are observed experimentally. Nevertheless, this growth model suggests that the relatively small diameter (5,5) tube is formed first which is followed by its enlargement toward larger diameter tubes.

This mechanism found experimental support in the Raman studies of Bandow *et al.* [69] who monitored the development of the inner tube spectra during a careful, relatively low temperature annealing process. Annealing temperatures as low as 800°C were used, making that the inner tube formation proceeded over weeks. As mentioned above, the Raman spectra are directly sensitive to the diameter of the grown inner tubes. It was observed that the small diameter inner tubes grow earlier which is followed by their enlargement to the larger tubes that have diameters which is optimal inside the outer tube.

A proposed alternative mechanism for the growth of inner tubes from peapods suggests the complete decay into small carbon fragments which then join to form the inner tubes. This was based on the information that the source of carbon is unimportant for the growth of inner tubes: it can also proceed from, for example, toluene [70] and ferrocene [71]. At present it is believed that both kinds of mechanisms are at play: for fullerene peapods probably the Stone-Wales route prevails. For the encapsulated non-fullerenes, the molecules decay into small carbon fragments which readily combine to form the inner tube and there is no substantial diffusion along the carbon axis.

It was shown experimentally with the help of ^{13}C isotope labeling that, for C_{60} peapods, there is no diffusion of the carbon atoms [72]. A mixture of ^{13}C labeled and natural C_{60} (1.1% ^{13}C content) was encapsulated into SWCNTs and transformed to DWCNTs. It was found that the Raman RBM lines of the inner tubes are substantially broadened due to the inhomogeneous distribution of the ^{13}C nuclei. This could be explained by assuming that the ^{13}C and the ^{12}C do not mix with each other thus ^{13}C rich and poor regions alternate along the inner tube axis. This experimental result also supports the inner tube formation model proposed in Ref. [67].

5b.4.2.5 Behavior upon Alkali Doping or Pressure: The Polymerization of the Peas

The C_{60} fullerenes possess a number of different polymer phases, such as that induced by charging [73] and two-dimensional [74], and three-dimensional [75] polymers induced by pressure. This motivated the search for similar polymers *inside* the host outer tubes. The motivation arose from the rich nature of the physics encountered for the fullerene polymers which includes, for example, low-dimensional correlated states [76] or superionic conductivity [77]. First principles calculations showed [78] that the structures of the different C_{60} polymer phases are dictated by the charging. For instance C_{60}^- gives rise to a one-dimensional (2+2) cycloadditional polymer, C_{60}^{3-} prefers a single-bonded one-dimensional polymer, C_{60}^{4-} results in a single bonded two-dimensional polymer structure, and for C_{60}^{6-} a single bonded one-dimensional polymer is preferred. The three-dimensional polymer structure is formed in neutral C_{60} upon high pressure and temperature treatment.

For peapods, the C_{60}–C_{60} coordination can take positions [51] which allow the (2+2) bond formation and also a single bonded polymer structure (when single carbon atoms are facing each other). In contrast, the (2+2) configuration is hindered for C_{70} in the lying configuration since the pentagons are facing each other with no double bonds [54]. Pichler *et al.* studied the behavior of C_{60} peapods upon alkali doping [79]. It was found that a one-dimensional, single-bonded, metallic polymer structure is formed from the peapods when the encapsulated fullerenes are charged to C_{60}^{6-}. This structure is consistent with the theoretical prediction by Pekker *et al.* [78].

The pressure induced polymerization of peapods was studied using X-ray scattering [80] and it was shown that 25 GPa (250 kbar) causes an irreversible polymerization. The C_{60}–C_{60} distance was reduced from 0.956 nm to 0.845 nm. The role of the temperature in promoting the polymerization was also studied [81]. It was shown that a simultaneous annealing at 1000°C reduces the pressure required for the polymerization down to 4 GPa. In contrast, for C_{70} no polymerization was observed [82] using a pressure of 2.5 GPa and annealing at 550°C, even though crystalline C_{70} polymerizes with the application of 2 GPa and 300°C [83]. This was interpreted that as the confinement inside the nanotubes prevents the C_{70} molecules from taking positions which is required for the bond formation.

5b.5 Properties of Peapods

5b.5.1 Structural Properties

The compelling one-dimensional structure of the peapods was recognized immediately in the first publication [3] and is shown in Figure 5b.2. It indicated a perfect linear chain arrangement with an approximate C_{60} center-to-center separation of 1 nm. More detailed

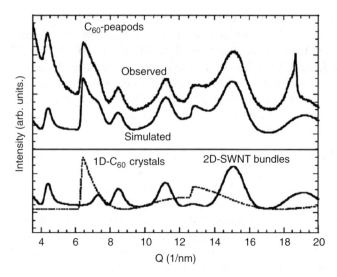

Figure 5b.16   *Observed and simulated XRD profiles of C$_{60}$ peapods. The lower panel shows the diffraction pattern for the hexagonal bundle structure of empty SWCNT and the form factor of the one-dimensional C$_{60}$ chain. Reprinted with permission from [84] Copyright (2003) American Physical Society.*

structural information was provided by electron an X-ray diffraction (XRD), upon the availability of peapods with a large fullerene filling [5]. The advantage of electron diffraction is that it can be performed within the TEM and also on microscopic amounts of sample. The first report on the peapod structure using electron diffraction was performed on the Gd@C$_{82}$ metallofullerene peapod [26]. While we do not intend to consider any priority question, we are aware that the Luzzi group also performed electron diffraction studies on the peapods around the same time.

XRD performed on pristine peapods was mentioned by Kataura *et al.* in [41], but the details of the measurement and a careful analysis of the XRD data were actually published first in Ref. [42]. In Figure 5b.16, we show the measured XRD profile for C$_{60}$ peapods along with a simulation from Ref. [84].

The XRD profile for peapods is a product of the Bragg peaks for the hexagonal bundle structure of SWCNTs and the form factor of the fullerenes. The latter is a one-dimensional chain of C$_{60}$s whose diffraction profile is also shown in Figure 5b.16. The effect of the encapsulated C$_{60}$s is most apparent for the first two Bragg peaks of the unfilled SWCNT around Q=4.5 and 7.5 nm^{-1} (Q is the scattering wave-vector); the first peak is larger in the unfilled material whereas it is smaller in the filled one. This change is due to the multiplying effect of the form factor of the fullerene chain, which is small below Q=6 nm^{-1} with a strong peak around 6.5 nm^{-1}.

The good agreement between the experimental and simulated XRD profiles means that the encapsulated fullerenes indeed form a perfect one-dimensional lattice with a long range structural coherence, that is, there is not a significant number of chain breaking defects or voids. On the other hand, the validity of the simulated one-dimensional C$_{60}$ form factor means that there is no correlation between fullerene chains in neighboring tubes due to the

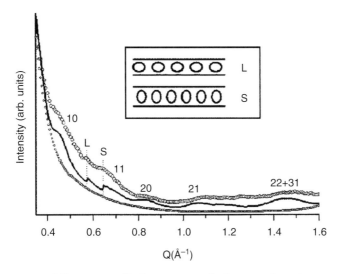

Figure 5b.17 *X-ray diffraction profile for C$_{70}$ peapods (large circles) along with a simulated diffractogram (solid curve) and the diffuse background (small circles). Miller indices indicate the Bragg peaks of the SWCNT lattice and L and S indicate the diffraction positions of the lying and standing configurations, respectively. Inset shows the schematics of the two configurations. Reprinted with permission from [56] Copyright (2007) American Physical Society.*

'shielding' of the host nanotube. The fullerene center-to-center distance was found to be 0.96(1) nm, in good agreement with the earlier TEM observations. This also confirms that the encapsulated fullerenes form a one-dimensional van der Waals crystal as this separation agrees with that in crystalline fullerite and is significantly larger than the lattice constant of a fullerene polymer (0.92 nm) [2].

The most rigorous treatment of the description of the peapod XRD profile was given by the group of Launois [43]. The influences of tube diameter distribution, different filling rates, and bundles with various sizes on the XRD profile were considered. We discussed above that the encapsulated fullerenes form a perfect one-dimensional chain only in nanotubes with sufficiently small diameter whereas, for larger diameter nanotubes, the so-called silo crystal arrangement is preferred. As a first approximation, the one-dimensional lattice constant of the fullerenes is reduced. This effect was reported using XRD in Ref. [33], that is, as an observed upshift of the C$_{60}$ related peak at 6.5nm^{-1}.

The structure of C$_{70}$ peapods was first studied by Kataura *et al.* using electron diffraction and XRD in Refs. [41,42]. Clear signatures of a double peak was observed with both techniques corresponding to lattice constants of 1.0 and 1.1 nm. This observation evidences that two types of arrangements are observed for the C$_{70}$ peas, the so-called standing and lying. The energetic stability for both structures was shown (also discussed in Section 5b.4.1.5) to depend on the host tube diameter, a value of ~1.45 nm being a boundary between the two [53].

In Figure 5b.17, we show the XRD image for C$_{70}$ peapods from Ref. [56] along with a simulated diffractogram assuming a superposition of the two different configurations. Two peaks can be identified at 5.7 and 6.3 nm^{-1} which correspond to one-dimensional lattices with lattice constants of 1.1 and 1.0 nm, respectively. The presence of separated peaks is the evidence that the two configurations exist in separate nanotubes and no experimental signature for the presence of a mixed configuration was found. The relative intensity (i.e., the structural abundance) of the two kinds of configuration is about 1:1. The host SWCNTs used in Ref. [56] had a mean diameter of 1.4 nm which corresponds to the earlier mentioned boundary between the two configurations, which results in the equal abundance of the two types of C_{70} arrangements.

5b.5.2 Peapod Band Structure from Theory and Experiment

Before going into the details, we will summarize in short the present knowledge on the electronic structure of the C_{60} peapods. There is little interaction between the encapsulated fullerenes and the host SWCNT. The fullerene LUMO (lowest unoccupied molecular orbital) and the band derived from the LUMO is situated about 0.5 eV above the Fermi energy of metallic or semiconducting nanotubes (the mid-gap energy is understood as Fermi energy for the latter). This meets accidentally the first Van Hove singularity of the semiconducting nanotubes but, for neutral SWCNTs, this state is unfilled therefore the encapsulated fullerenes remain neutral. The early literature on the peapod electronic properties is reviewed in Ref. [85].

Describing the peapod band structure starts with the brief introduction to the band structure of pristine SWCNTs given in Chapter 1. Then we discuss the molecular orbitals of C_{60} and the molecular bands in the solid molecular crystal, that is, the fullerite. The HOMO (highest occupied molecular orbital) of C_{60} has an h_{1u} symmetry and is five-fold degenerated thus it is fully occupied by 10 electrons. This is separated by a gap of about 2.5 eV from the triply degenerated LUMO with the t_{1u} symmetry.

The first theoretical work addressing the band structure of C_{60} peapods appeared in 2001 [86]. Using DFT-LDA calculations, it was found that for the C_{60} peapods with a (10,10) armchair host tube, the fullerene t_{1u} level is located near the Fermi energy of the SWCNTs. This would cause that the fullerene chain becomes a metal. Instead of the usual two linear bands, this material was claimed to have four conduction bands, two of which are fullerene related with a smaller Fermi velocity due to the weaker metallicity of this component. This result was reproduced later in another DFT-LDA calculation by Dubay and Kresse [47] and was extended to semiconducting SWCNTs. For the latter, the fullerene t_{1u} state was claimed to be near the middle of the gap.

However, our present knowledge and in particular the experimental results do not support the first principles results. This in fact shows that for such a delicate van der Waals system as the peapods, otherwise powerful methods fail. Liu *et al.* performed electron energy loss spectroscopy (EELS) studies on C_{60} peapods [44]. They found that the electronic and optical properties of the encapsulated C_{60}s are very similar to those of solid fcc fullerite. This excludes the possibility that the t_{1u} of the encapsulated C_{60}s is occupied in contrast to the prediction of the first principles result. This means that the fullerene t_{1u} band is located above the Fermi energy (or mid-gap for semiconducting tubes) level of the SWCNTs and is thus unfilled.

The EELS result, which is a spectroscopic method on the bulk scale, was clearly confirmed by a local, microscopic method, scanning tunneling spectroscopy (STS). Hornbaker *et al.* performed STM and STS on the C_{60} peapods [87]. For usual metals with broad conduction bands, this method provides a topographic information on the studied surface

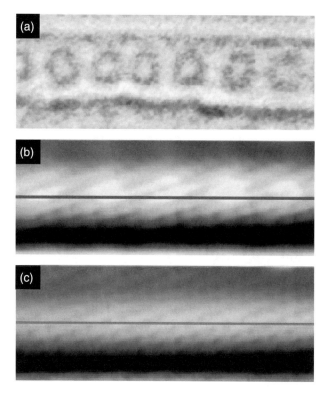

Figure 5b.18 *Structural and STM image of a C_{60} peapod. A representative HR-TEM image (a), STM image with 1.5 V bias and 700 pA (b) and -1.5 V and 700 pA (c). Bright areas indicate a topographically higher surface, the amplitude of the height modulation is 0.04 nm. Note that only the positively biased STM image indicates the presence of the fullerenes. Reproduced from [87] Copyright (2002) AAAS.*

irrespective of the applied bias voltage. For peapods, a strong sensitivity on the bias was observed. For negative sample bias, that is, when the occupied states are monitored, no signature of the embedded fullerenes is observed. This drastically changes for positive sample bias, that is, when the unoccupied states are monitored, when a periodic modulation of the apparent sample height is observed along the nanotube axis for constant current operation. We show this result in Figure 5b.18. The topographically higher surface translates to larger density of states for the unoccupied states in the positive bias measurement. This means that the unoccupied fullerene t_{1u} band is located near the unoccupied SWCNT states, which confirms the result of the EELS and also contradicts the first principles prediction. We note that the STS measurement was performed on a semiconducting host SWCNT and to our knowledge no data for metallic host tubes were reported.

In a similar STS study, the effect of encapsulated metallofullerenes was investigated [88]. It was found that the encapsulation of Gd@C_{82} reduces the band gap in semiconducting SWCNTs from about 0.5 eV to 0.1 eV. This is mainly the result of the additional charges transferred from Gd (Gd^{3+} when ionized).

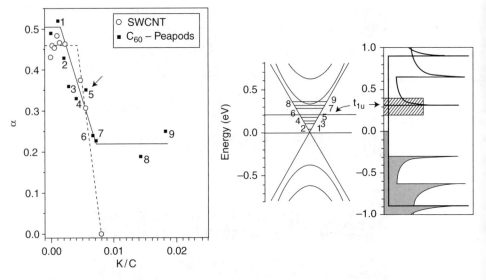

Figure 5b.19 *The power law scaling factor of the DOS, α, as a function of K-doping for pristine SWCNTs and peapods (left panel) and schematic band dispersion and DOS (right panel). Numbers in the panels show the doping stages. Grey and black lines denote the DOS for metallic and semiconducting SWCNTs, respectively. Note that for C_{60} peapods, the TLL state is preserved for higher doping level as the exponent remains finite, however, its value indicates a two-channel TLL state. Reprinted with permission from [90] Copyright (2005) American Physical Society.*

In an improved first principles calculation, Louie's group [89] qualitatively confirmed the validity of the above description. Using ab initio pseudopotential density functional approach with a linear combination of localized orbitals basis, it was found that the Fermi energy of SWCNTs lies indeed between the HOMO and LUMO levels of the encapsulated C_{60}s, thus the latter molecules are neutral and do not contribute to the conduction.

Another way of studying the unoccupied states is to dope them with donor atoms and to monitor the changes to the density of states. The Pichler group studied [90] the electronic structure of C_{60} peapods by means of shifting the Fermi energy upon K-doping using photoemission spectroscopy. Previously, the method was successfully used to evidence a crossover from the Tomonaga-Luttinger liquid (TLL) phase [91] to the Fermi liquid (FL) [92]. It was shown for pristine metallic SWCNTs using photoelectron-spectroscopy (PES) [93], that strong electron-electron correlations combined with the one-dimensionality gives rise to the TLL state. This is observable as a power law behavior in the density of states near the Fermi level (DOS $\propto (E_F - E)^{\alpha}$, where E_F is the Fermi energy), rather than the Fermi-edge behavior, which characterizes usual metals. However, the Fermi edge and thus the FL state is recovered upon K-doping when all the SWCNTs are rendered metallic due to the electron doping [92], and the system loses its one-dimensional character. The TLL to FL crossover happens when the dopant electrons start occupying the valence band of the otherwise semiconducting nanotubes.

For C_{60} peapods, the situation is different and is a fingerprint of the presence of the encapsulated fullerenes [90]. In Figure 5b.19, we show the power law scaling factor, α, for

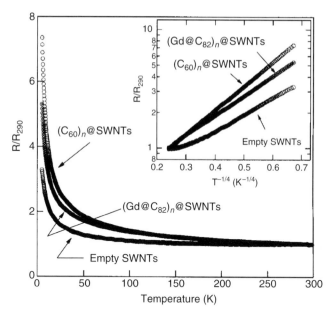

Figure 5b.20 *Temperature dependence of the resistance for Buckypapers of (Gd@ C_{82})@SWCNT, C_{60}@SWCNT, and unfilled SWCNT. Inset is a semilogarithmic plot against $T^{-1/4}$. Reprinted with permission from [26] Copyright (2000) American Physical Society.*

the pristine SWCNT and for the peapod material as a function of the K-doping [92]. For the pristine material, the FL state (i.e., $\alpha=0$) is recovered for a K/C ratio of $8\cdot10^{-4}$. This point corresponds to the charging of the valence band of the semiconducting nanotubes. For peapods, the fullerene t_{1u} level is situated around the valence band of the semiconducting SWCNTs (see the right panel in Figure 5b.19). In real space, this band corresponds to the states on the one-dimensional chain of the fullerenes. Therefore, upon doping, this lower lying band gets doped earlier, forming a perfect one-dimensional conductor. Inside metallic SWCNTs, the presence of a one-dimensional metal inside gives rise to an additional channel of TLL, thus their ensemble appears to be a two-channel TLL, which results in a reduced (but non-zero) α. Therefore, the phase diagram of the K-doped peapods is such that for low doping it is a TLL, for high doping it is a FL, and for intermediate doping it is a two-channel TLL.

5b.5.3 Transport Properties

The first transport studies were performed on metallofullerene (Gd@C_{82}) peapods, the sample being in the form of a buckypaper [26]. The temperature dependent resistance for a metallofullerene and C_{60} peapods, and for an empty SWCNT is shown in Figure 5b.20.

It was found that the resistance follows a $\ln R \propto T^{-1/4}$ temperature dependence in the 5–300 K range, which is characteristic for a three-dimensional variable range hopping behavior [94]. The peapod films were found to have a higher resistance than the empty SWCNT sample, which was assigned to the electrostatic scattering or due to disorder induced by the presence of the fullerenes. This temperature dependence was confirmed for a thin film of C_{60} peapods [95], which confirms the presence of disorder.

Transport measurements on individual SWCNT peapods were also performed [96,97]. The advantage of such studies is that the individual behavior can be better studied and the measurements are not influenced by the tube-tube contacts which are inevitable for the bucky-paper measurements. The power low-like nature of the low temperature conductivity was confirmed in the individual peapod measurements [96]. Below 30 K, it was found that the peapods behave as a regular array of individual quantum dots [97]. This was assigned to the local changes induced by the presence of the encapsulated fullerenes.

5b.5.4 Optical Properties

The optical properties of SWCNTs are clearly unique among the common materials. First, the optical transitions are extremely well defined and intense due to the presence of van Hove singularities in the density of states for both the occupied and unoccupied states. Second, the optical transition energies strongly depend on the nanotube type (metallic or semiconducting) and helicity (or diameter). Depending the SWCNT helicity, optical transition energies ranging from the near infrared (1 eV) to UV (3 eV or higher) can be found. For a given SWCNT helicity, the optical transition energies form almost a mathematical series. The important question related to the peapods is whether the optical properties are significantly affected by the encapsulation. The answer in brief is no, although some minor changes specific to peapods were observed.

The optical properties were studied by the Pichler group using optical absorption spectroscopy and EELS [44]. In Figure 5b.21, we show the optical absorption spectra for empty SWCNTs and for the peapod sample. A small downshift of 14 meV was observed, which can be thought of as a fingerprint for the presence of the fullerenes. It was suggested that the downshift is either due to a small increase of the SWCNT diameter, or due to a change in the intertube interaction. The earlier possibility is related to the fact that the

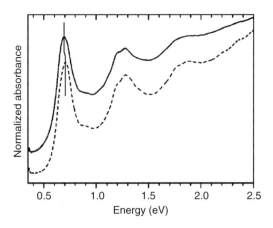

Figure 5b.21 *Optical absorption spectra of peapods (solid curve) and empty SWCNTs (dashed curve). The vertical guideline indicates the small downshift in the peapod sample for the first semiconducting optical transition. Reprinted with permission from [44] Copyright (2002) American Physical Society.*

optical transition energy is inversely proportional to the tube diameter. The second possibility is related to the fact that upon encapsulating fullerenes, the inner van der Waals interaction could make the tube-tube interaction weaker on the outside.

The present understanding of the optical properties is more complicated than a single electron picture with a band-band transition. It is known that the optical excitations are of excitonic nature, that is, the excited electron and hole pair forms a bound state with a sizeable binding energy [98]. It is also known that the dielectric environment plays an important role in the exciton binding energy due to dielectric screening effects [99], which could also account for the observed shift. Our conclusion is that it cannot be decided at present what the real source of the observed optical transition energy downshift is.

Concluding the optical properties, we mention another yet unsolved problem. It is known (although being a negative result, it is unpublished) that the infrared vibrations of the encapsulated C_{60} has not been observed. C_{60} has four infrared active vibrations in the mid-infrared (IR) range below $1429\,cm^{-1}$ [2]. It was suggested that these are invisible since, for a SWCNT assembly, these lie below the plasma edge thus they are screened. However, optical measurements did not show a clear plasma edge for SWCNTs and rather show the evidence for a low energy ($10\,cm^{-1}$) gap instead [100].

5b.5.5 Vibrational Properties

Studying the vibrational properties of peapods is restricted to Raman spectroscopy due to the above mentioned anomalous absence of the IR lines of the encapsulated fullerenes. We discussed the first Raman study in Section 5b.4.1.2. and in Figure 5b.7. from Ref. [20]. It was shortly followed by a detailed Raman study by Pichler *et al.* involving several laser energies and low temperatures [79]. C_{60} has four IR and 10 Raman active modes [2]. Of the latter two are totally symmetric, the $A_g(1)$ and the $A_g(2)$ at 496 and $1469\,cm^{-1}$, respectively. The totally symmetric nature means that these modes are nondegenerate, *that* is, they cannot split even upon lowering the icosahedral symmetry of C_{60}. The other nonsymmetric modes are the $H_g(1)$ ($272\,cm^{-1}$), $H_g(2)$ ($433\,cm^{-1}$), $H_g(3)$ ($709\,cm^{-1}$), $H_g(4)$ ($772\,cm^{-1}$), $H_g(5)$ ($1099\,cm^{-1}$), $H_g(6)$ ($1252\,cm^{-1}$), $H_g(7)$ ($1425\,cm^{-1}$), $H_g(8)$ ($1575\,cm^{-1}$). Among all the Raman modes, the $A_g(2)$ (also called pentagonal pinch mode, or PPM) is the strongest.

Two properties prove that the C_{60} vibrations are indeed observed in the peapod Raman spectra. First, most of the modes can be identified, even though the SWCNT modes are also present and overlap with the C_{60} modes. As we show in Figure 5b.22, the $H_g(2)$, $A_g(1)$, and the $A_g(2)$ can be undoubtedly identified. The other property is the Raman resonance behavior. It is known that the Raman signal is enhanced by several orders of magnitude when the exciting laser energy matches an optical transition of the studied molecule [101]. C_{60} has a HOMO-LUMO gap of around $2.5\,eV$ where a relatively strong optical transition takes place. In the Raman spectra, a clear enhancement of the fullerene vibrational mode intensity (best seen for the PPM) is observed.

The C_{60} PPM shows a characteristic splitting for the peapods. The two components are at 1466 and $1474\,cm^{-1}$. In fact, the splitting itself is very surprising. As mentioned above, the totally symmetric $A_g(2)$ mode should not be split even upon symmetry lowering. This means that the observed splitting is an inhomogeneous effect, that is, the two components come from different fullerene molecules, which are located in different areas of the sample. The optical transition energy seems to be different for the two components, as the stronger,

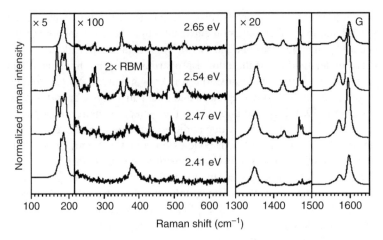

Figure 5b.22   *Energy dependent Raman spectra at 20 K for C_{60} peapod samples. The spectra are scaled relative to the Raman G mode. The lines between 360 and 390 cm^{-1} originate from the overtone of the SWCNT RBMs. Reprinted with permission from [79] Copyright (2001) American Physical Society.*

low frequency one is in resonance with the 2.54 eV laser, whereas the weaker, high frequency component is stronger with the 2.47 eV laser. Two further Raman experiments led to the clarification of the origin of the split components.

Pfeiffer *et al.* [102] studied the temperature dependence of the split $A_g(2)$ components. It was found that the high frequency one reversibly disappears with increasing temperature (above 500 K). This suggests that the high frequency mode vanishes due to the activated rotation of the molecule. Simon *et al.* [33] studied the Raman spectra of the PPM when C_{60} is encapsulated inside SWCNTs with different diameters and the result is shown in Figure 5b.23. The Raman spectrum of molecular C_{60} is shown for reference. For a host SWCNT with a mean diameter of 1.34 nm, the characteristic double PPM peaks are found with the Raman shifts discussed above. For a somewhat larger diameter host SWCNT sample (1.4 nm) the higher PPM peak is absent but the lower frequency component remains at the original position (1466 cm^{-1}). For even larger diameter host SWCNTs (for which the so-called silo arrangement occurs), the single PPM mode is found at the position where it appears in crystalline C_{60}.

These observations can be interpreted in conjunction with the theoretical work of Michel *et al.* [51], who found that two different C_{60} configurations are realized depending on the diameter of the host tube. For $d \leq 1.4$ nm, C_{60}s form a structure with facing pentagons whereas the fullerenes rotate freely for larger diameter host SWCNTs. This leads to the conclusion that the mode with the higher Raman shift originates from the fullerenes with facing pentagons and the mode with the low Raman shift originates from the rotating C_{60}. This explains the observation in Ref. [102] that the high frequency mode disappears at higher temperatures as fullerenes formerly arranged with facing pentagons start to rotate. The 3 cm^{-1} downshift of the low frequency mode with respect to the mode in crystalline fullerenes disappears for the very large diameter host tubes as the tube-fullerene interaction therein is smaller.

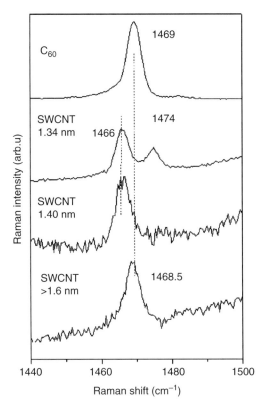

Figure 5b.23 *Dependence of the C_{60} PPM on the diameter of the host SWCNT using a 488 nm laser excitation (2.54 eV). Labels indicate the position of the split components and the mean diameter of the host SWCNT is also given. Note that (i) the high frequency mode is absent for the 1.4 nm SWCNTs, and (ii) the PPM is found at the position of crystalline C_{60} for very large diameter SWCNT host. Reprinted with permission from [33] Copyright (2007) Elsevier Ltd.*

The behavior of the PPM mode upon charge transfer was studied using K vapor [79,103] and electrochemical doping [104]. It was found that the PPM downshifts of 6 cm^{-1} per transferred electron. This charging-induced downshift is identical to that found for the crystalline fullerenes [2].

5b.5.6 Magnetic Properties

The magnetism of peapods originates from encapsulated magnetic fullerenes. Among the available experimental methods, we discuss magnetic resonance only, that is, electron spin resonance (ESR) and nuclear magnetic resonance (NMR). Other magnetism related experiments such as, for example, bulk magnetometry, are not usable for SWCNTs due to the inevitable presence of magnetic catalyst particles which gives a nonseparable contribution to the net magnetism [105]. In contrast, magnetic resonance spectroscopy enables to separate the magnetism of the catalytic particles from the encapsulated fullerenes due to its spectroscopic nature.

Two different kinds of magnetic fullerenes could be encapsulated, those carrying electron magnetism such as, for example, $N@C_{60}$, $C_{59}N$, and other fullerene nitroxide radicals, and those carrying a nuclear spin on the ^{13}C nuclei. We discuss below (Section 5b.6), that efforts to encapsulate magnetic fullerenes into SWCNTs is motivated by a proposal from Harneit [37] that this system would be suitable for quantum computing.

The first encapsulation of a magnetic fullerene was performed in 2001 [26] in a now seminal paper, which is discussed herein in many aspects (HR-TEM on metallofullerenes and transport). To our knowledge the magnetic properties of the encapsulated $Gd@C_{82}$ metallofullerene has not been studied in its peapod form, although the electron spin resonance for the pristine metallofullerene was investigated [106].

The first encapsulation of a C_{60}-derived magnetic fullerene inside SWCNTs was reported by Simon *et al.* [18]. The endohedral $N@C_{60}$ was encapsulated with the low temperature peapod synthesis method, which is discussed in Section 5b.4. In $N@C_{60}$ [28], the nitrogen is in an atomic state with an $S=3/2$ high spin state due to Hund's rules. The $I=1$ nuclear spin state of the most abundant ^{14}N nuclei gives rise to a characteristic hyperfine splitting into three components with a uniquely large hyperfine coupling constant. This is the result of the compressed nitrogen orbitals inside the fullerene cage, which thus increases the overlap of the electron wave functions with the nucleus. The unique magnitude of the splitting allowed the unambiguous identification of the encapsulated molecule. However, an inhomogeneous broadening was observed whose origin remains unclear.

The major limitation related to the $N@C_{60}$ magnetic fullerene is that it is prepared with N ion bombardment of C_{60} fullerenes with a very low yield (tens to hundreds of ppm in C_{60}) and concentrating it requires a costly repetition of high-performance liquid chromatography [107] which relies on the longer retention time of the heavier $N@C_{60}$ for achieving 100% concentration.

Another magnetic fullerene is the azafullerene $C_{59}N$. In this molecule, nitrogen replaces a carbon atom whose extra electron is unpaired and the monomer form of the molecule is a free radical. The material can be chemically prepared [108,109], however the unpaired electrons readily form a bond and the material occurs as $(C_{59}N)_2$ dimers which can be converted to $C_{59}N-X$ (where X denotes a side-group). However, both of these azafullerene compounds are ESR silent as the extra electron is located in the bond. This also means that encapsulating $(C_{59}N)_2$ dimers into the SWCNTs does not result in a magnetic peapod.

$C_{59}N$ could be only temporarily produced in light [110] or thermally excited [111] conditions. It was shown that the $C_{59}N$ free radical could be also stabilized if it is embedded in the C_{60} matrix, that is, when it forms a solid solution [112] and is denoted as $C_{59}N:C_{60}$. This motivated to attempt a similar stabilization inside SWCNTs. First, $C_{59}N-X$ was encapsulated inside the SWCNTs along with C_{60} [30]. The C_{60} was intended to separate the azafullerene molecules from each other. Then, a heat treatment around 600°C removes the -X side-group leaving behind the azafullerene free radical with high concentration [31]. This material is denoted as $C_{59}N:C_{60}@SWCNT$.

In Figure 5b.24, we show the ESR spectrum for the azafullerene peapod from Ref. [31]. For reference, the ESR spectrum for the crystalline $C_{59}N:C_{60}$ solid solution is also shown. It was recognized previously that the latter ESR spectrum consists of a triplet component and a singlet line component [113]. The triplet signal corresponds to electron spins which reside on the $C_{59}N$ molecule and interact strongly with the ^{14}N nucleus. However, the singlet component comes from electron spins which are transferred nominally to a neighboring C_{60} when a

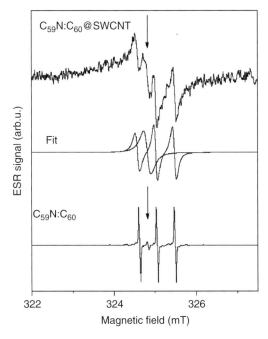

Figure 5b.24 *Room temperature ESR spectra of the azafullerene peapod (upmost panel) and of the azafullerene-fullerene solid solution (lowest panel). The middle panel shows deconvolution into a triplet and a single line components. The latter is indicated with an arrow. Reprinted with permission from [31] Copyright (2006) American Physical Society.*

$C_{59}N$–C_{60} heterodimer molecule is formed. The latter is the ground state of the $C_{59}N$:C_{60} system and the excited state is the rotating $C_{59}N$ monomer, surrounded by neutral C_{60}s.

For the azafullerene peapod, several properties of the encapsulated azafullerene change with respect to the crystalline form. These changes are all characteristic for the encapsulation. First, the rotation of the $C_{59}N$ inside the SWCNT is strongly anisotropic with a preferred axis of rotation parallel to the nanotube axis. Second, the binding energy of the heterodimer is larger, probably due to the more compact fullerene-azafullerene arrangement. Third, a reversible charge transfer from the azafullerene towards the host nanotube was observed above ~350 K. Measurement of the spin-lattice relaxation time on the heterodimer at low temperatures allowed measuring the density of states on the host nanotubes.

Following the azafullerene peapod synthesis, nitroxide spin-labeled fullerenes were also prepared [114,115]. The advantage of the nitroxide spin labels is that these are (i) air stable, and (ii) do not tend to dimerize, thus these are stable as a peapod even without being separated, as a fully linear spin chain.

Nuclear magnetism of fullerenes is sizeable for highly ^{13}C enrichment only since natural abundance of ^{13}C in carbon is only 1.1%. The first encapsulation of ^{13}C enriched C_{60} inside SWCNTs was reported in Ref. [32]. NMR spectroscopy was also performed but the attention was focused on the properties of the peapod derived inner tube. These NMR investigations led to an evidence for the Tomonaga-Luttinger liquid phase for the inner nanotubes [116,117]. The rotational dynamics of the encapsulated C_{60}s was studied by Maniwa *et al.* [50]. It was found that C_{60} rotates quasi-freely with rotational correlation times of 5–10 ps at 300 K. Down to 30 K, the large angle molecular jumps persist but there is no phase transition in the rotational behavior which is related to the one-dimensional nature of the arrangement.

5b.6   Applications of Peapods

We split the discussion of peapod applications into already existing (or demonstrated) on the one hand and expected (or theoretically envisioned) on the other hand. We note that 'existing' or 'demonstrated' does not mean a device functioning on an industrial scale but rather an application which is demonstrated already at least in laboratory conditions. To our knowledge, there is no application of peapods which has gone to the real industrial use level yet. The expected applications are those which have been only theoretically speculated.

5b.6.1   Demonstrated Applications

5b.6.1.1   Transistor
SWCNTs are ideal building blocks for nanoelectronics given their small size and their high current-carrying capability. There are three obstacles for the applicability though: (i) the inability of patterned growth and contacting; (ii) the random chirality (and thus conductivity) distribution of the tubes; (iii) the relatively large, above 1 eV, direct gap of the semiconducting tubes. Modern electronics tends to use small gap semiconductors for increased speed and reduced dissipation. The band gap problem was overcome using fullerenes. The Shinohara group showed that the band gap can be reduced to 0.1 eV by means of encapsulating charge donor metallofullerenes inside the SWCNTs [88].

Later, the transistor behavior of this metallofullerene peapod system was also proven. In Figure 5b.25, we show the HR-TEM image of such a peapod material along with the transistor characteristics, from Ref. [118]. The device (shown in the Figure 5b.25 inset) shows conduction for both polarities of the gate bias (gate-source voltage), V_{GS}. For comparison, a C_{60}-peapod-based transistor device was also investigated and a unipolar p-type behaviour was found. This is related to the strong electron acceptor nature of C_{60} and the behavior of the metallofullerene peapod is related to the fact that the metallofullerene can behave either as an acceptor or as a donor. The latter is due to the three valence electrons of Gd which are removed with relative ease. This experiment showed the potential of peapods in enabling a control of the electronic properties of the SWCNT hosts which can lead to a variety of applications.

Evidence for single-electron transistor behavior was found for C_{60} peapods by the group of Roth [119]. The conductance through a gate-biased peapod was studied as a function of temperature. At room temperature, the result of the Shinohara group was reproduced, that is, that the conductance is p-type, being finite for negative bias and vanishing for positive bias. At 1.8 K, however, the conductance is reduced for both polarities, but surprisingly the positive bias conductance shows (random) oscillations as a function of the gate potential. This oscillatory behavior was associated with the presence of the C_{60}. Indeed, the molecule modifies the Fermi surface in its vicinity [87,88] and when some fullerenes are separated

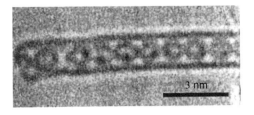

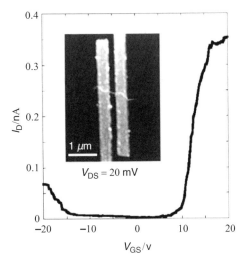

Figure 5b.25 *HR-TEM micrograph of Gd@C$_{82}$ metallofullerene peapods (upper panel) and the current voltage characteristics of the device shown in the inset. Note the darker spots in the HR-TEM image which are the heavy Gd ions. Also note the finite current for both positive and negative V$_{GS}$ which is characteristic for an ambipolar transistor behavior. Reprinted with permission from [118] Copyright (2002) American Institute of Physics.*

from each other by empty tube segments, a quantum dot is created. Transport in the nanotube thus proceeds through the quantum dots in the form of single electrons.

5b.6.1.2 Solar Cell

Solar cells are in the focus of applied research as these could enable sustainable development for the energy industry. Solar cells consist of an active medium which provides an electron-hole pair upon light excitation which are separated from each other with an electrode. The electrode has to be transparent and highly conducting. A well-known problem is that the most common electrode material, ITO (indium tin oxide), is expensive and it has a limited supply of indium (projected to run out in 20 years). Viable and promising alternatives to ITO are conducting thin films of SWCNTs as proposed by the Rinzler group [120]. The use of SWCNTs for solar cell purposes received another boost by the observation of Li *et al.* [121] that fullerenes encapsulated inside SWCNTs act as the active electron-hole producing medium. This means that peapods alone can act as solar cells as they have the active medium as well as the electrode naturally integrated. Although this effect requires further studies and in particular benchmarks for the quantum efficiency for the light induced electron transfer, it shows a high application potential of peapods for solar cell purposes.

5b.6.1.3 Drug Delivery

Drug delivery is one of the promising areas of application of carbon nanotubes, the activities being reviewed in [122]. This application potential stems from (i) the biocompatibility of the carbonaceous materials, (ii) the possibility to attach the active molecules to the nanotubes, as the SWCNTs have a high specific surface area, and (iii) the possibility to track the carbon nanotubes in vivo, for example, by fluorescence spectroscopy [123]. In addition to functionalization, the inner space of the SWCNTs provides another possibility for drug delivery, either using functionalized fullerenes or other nonfullerene peapods.

A potential hindrance to use encapsulated materials for drug delivery purposes is how to control the removal at the target. As we discussed above, the encapsulated fullerenes are energetically stable inside the host SWCNTs allowing the largest binding energy, typically the SWCNTs with d ~1.4nm such as the (10,10). The encapsulation is so stable, that the fullerenes do not escape at any temperature and rather form an inner tube. Clearly, the removal of the fullerenes could be attempted for larger diameter host nanotubes. In Ref. [33] the controlled removal of encapsulated fullerenes was presented. First, C_{60}s were encapsulated inside larger diameter SWCNTs, where the so-called silo (i.e., a non-linear, zigzag like) arrangement is realized. The encapsulation was performed in methanol and the removal proceeded in dichlorobenzene. Although the mechanism of this observation is not fully clear yet it is probably related to the different solubility of fullerenes in these two solvents: C_{60} has a three orders of magnitude lower solubility in methanol than in dichlorobenzene in units of mg ml^{-1} [2]. This suggests that the fullerenes prefer entering the nanotubes when in methanol and prefer leaving the tubes when the latter are surrounded by the dichlorobenzene. Although neither of these solvents is biocompatible, it was speculated that, for example, water-soluble fullerenes (with an attached bioactive group) would enable drug-delivery.

A promising alternative to the fullerene peapods was presented recently by the group of Borowiak-Palen [124]. Instead of the fullerene, an anticancer drug was directly encapsulated inside SWCNTs. Its effect was verified for an in vitro environment. We suggest herein, that a combination of external functionalization of the SWCNTs and encapsulated drugs could be a viable route. The external functionalization would be responsible for the targeted nature of the delivery, whereas the encapsulated molecule would be the drug itself.

5b.6.2 Expected Applications

5b.6.2.1 Photonic Crystals

Photonic crystals refer to materials where the periodic array of their constituents give rise to a periodic modulation of the dielectric constant. This results in unique optical properties such as, for example, well-defined gaps in the transmittance in a manner analogous to the band gap behavior of solids. Peapods were suggested to be used as photonic crystals [125] thanks to the periodicity of the encapsulated fullerenes which is well-defined. It was found that the periodicity and the dielectric properties of C_{60} fullerenes give rise to an optical gap around 7 nm which means that soft X-rays with this wavelength could be trapped inside the peapods. This means that if an effective medium, which emits radiation with this wavelength, is placed within the peapod structure, the emitted radiation would be amplified such as in a laser, which could provide a way to construct a soft X-ray laser.

5b.6.2.2 'Nanogun'

The effect of charging the outer walls of the host SWCNT on the motion of a neutral C_{60} was studied in Ref. [126]. It was found that the motion of the inner C_{60} is different depending on the sign of the outer tube charging. If the outer tube is positively charged, it drives out the molecule with speeds over $1\,km\,s^{-1}$, just like a 'nanogun', while a negatively charged tube can drive the molecule into oscillation inside it and can absorb inwards a neutral molecule in the vicinity of its open end, like a 'nanomanipulator'. This asymmetric behavior is related to the different affinity of C_{60} toward charging: it is an effective electron acceptor but it is a bad electron donor. The validity of this proposal lies in the ability to charge the host outer tube almost independently from the encapsulated fullerenes. Pichler *et al.* [103] observed that upon intercalating K (electron doping) or $FeCl_3$ (hole doping) into peapods, the host SWCNT walls are charged at low intercalation levels and that there is no charge transfer to the encapsulated fullerenes.

5b.6.2.3 Peapod 'Shuttle' Memory

Early dynamic observations of the behavior of encapsulated fullerenes by TEM showed that they were able to move back-and-forth in the SWCNT cavity, presumably upon random ionization effects due to electron irradiation in the TEM [4] (see Figure 5a.11 in Chapter 5a). The first paper confirming the existence of C_{60} inside carbon shells via modeling by Kwon *et al.* discussed the possibility to use these as 'shuttle' memories. It was shown that C_{60} can move relatively easily inside a carbon shell while retaining energy minima around the ends of the outer shell. The small but well defined energy minima would enable to use the C_{60} as an information carrier, the position being the different bit values. The name 'shuttle' describes the fact that the encapsulated C_{60} switches between the different bit values by a rapid shuttling. In Figure 5b.26, we show the schematics of the proposed device from Ref. [14] along with a HRTEM observation of a peapod-like structure which motivated the shuttle memory concept.

In addition to the issue of actually making such a short single-wall capsules deliberately (which could possibly be solved by, e.g., considering nanohorns) a potential problem of the shuttle memory device is to provide the driving force for the memory switch. Charging the C_{60} with K atoms was proposed [14,127], which would enable electrostatic control, but another possibility (to our knowledge proposed herein for the first time) is to encapsulate $C_{59}N$ azafullerenes, which can be ionized with ease [109].

Alternative shuttle memory devices were proposed such as those based on small, encapsulated K ion clusters [128] and also using BN nanotubes as host material for the fullerenes [129]. Another interesting proposal is to form a short inner tube segment from the encapsulated fullerenes and to use this as information carrier unit inside a host nanotube [130]. Whereas these suggestions are definitely interesting, we are not aware of any experimental realization of a similar device, neither of any attempt to achieve control of a single encapsulated fullerene.

5b.6.2.4 High Frequency Oscillator

Quartz crystal oscillators are common in information technology to provide synchronizing pulses. However, typical crystal sizes are millimeters, which prevent their direct integration. In addition, energy dissipation becomes an increasing problem for such crystals at the

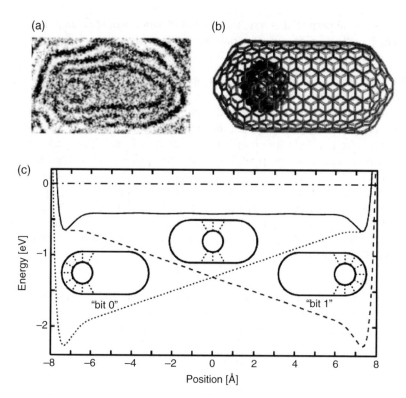

Figure 5b.26 *(a) HR-TEM micrograph of a fullerene encapsulated inside a multiple-wall shell of sp² carbons. (b) The schematics of the proposed 'shuttle memory' device. (c) The energetics of the C_{60} for different configurations. Reprinted with permission from [14] Copyright (1999) American Physical Society.*

required high frequencies (several 10 GHz for state-of-the-art computing). Recently, interest turned toward utilizing carbonaceous structures for oscillator purposes motivated by the demonstration of very low friction in coaxial multiwall carbon nanotubes by the Zettl group [131]. First, double-wall carbon nanotube structures were considered [132] and studied using molecular dynamics calculations. It was shown that a (9,0) nanotube inside an (18,0) would behave as a 38 GHz nanooscillator. A peapod-based oscillator was proposed in Ref. [133] and a fundamental eigenfrequency of 50 GHz was found for the C_{60} motion inside a 5 nm long section of a (10,10) SWCNT. In a recent work, the evidence for nanoelectromechanical coupling in peapods has suggested them as NEMS resonators, with an operation frequency set by the vibrational motion of the encapsulated fullerenes in the 100 GHz range. The condition for the encapsulated fullerenes is not to be closely packed so that they can arrange into short (e.g., involving 10 C_{60}s) chains [134]. The major obstacle for such a resonator is how to provide the continuous driving force for the motion. It is difficult to control externally the motion of C_{60} thus the $C_{59}N^+$ ion was suggested as a better candidate for such a purpose [133,135]. This molecule has a sizeable electric dipole moment which allows initiating its motion with an electric field impulse. It was also shown that the tube length allows controlling the frequency of the proposed oscillator [135].

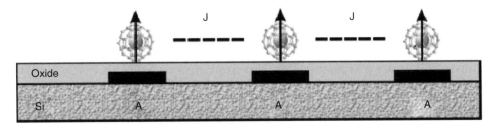

Figure 5b.27 *Concept of the N@C$_{60}$ fullerene based quantum computer suggested by Harneit. The spin qubits are addressed with ESR pulses with selectivity provided by a local magnetic field from the A gate. The qubit-qubit coupling, J, is provided by the dipole-dipole coupling between the spins. Reprinted with permission from [37] Copyright (2002) American Physical Society.*

We note, that neutral C$_{59}$N monomers were successfully encapsulated inside the SWCNTs [30,31]. Preparation of a sample containing short SWCNT segments filled with single C$_{59}$Ns could be the next step toward the realization of such an oscillator. Ionization could be achieved by UV light or soft X-ray irradiation as the extra electron on the C$_{59}$N is relatively weakly bound [109].

5b.6.2.5 Quantum Computing

Quantum information processing can be divided into quantum computing (QC) and quantum information storage [136]. Both areas require the availability of a true quantum variable or qubit (such as, e.g., an atomic state, a supercurrent, photon polarization, or a spin). For information storage, a high level of decoupling of the quantum variable from the environment is desired. For information processing, a number of the quantum variables has to interact in order to form an entangled state on which the calculation proceeds. There are two potential challenges in quantum computing: upscalability and decoherence. Practical calculations require a large number of qubits, which however increases complexity and also the decoherence due to the increased level of interactions, thus these two objectives are somewhat exclusive. QC has been demonstrated on a 7 qubit system using nuclear magnetic resonance [137] with already an enormous complexity. A possible alternative for an upscalable solid state qubit system with reasonable complexity was proposed by Kane [138], which relies on Si nuclei arranged in a linear fashion.

Building upon this suggestion, Harneit [37] suggested to prepare a linear chain of spin qubits for QC. It was proposed that a single qubit would contain a N@C$_{60}$ endohedral fullerene [28] and the concept of the proposal is shown in Figure 5b.27. In this molecule, the atomic nitrogen is remarkably decoupled from its environment due to the protecting shielding of the fullerene cage. The utility of this molecule for quantum computing has been demonstrated by Mehring *et al.* [139].

Although not suggested explicitly in the original paper, it became evident that the peapod structure would be an ideal realization for the linear arrangement of the endohedral fullerenes. This motivated the low temperature peapod synthesis efforts (described in Section 5b.4.1.3) as the N@C$_{60}$ molecule is unstable above 200°C [35]. Although, the QC field itself is continuously progressing, there is a long way until practical application, nevertheless the proposal to use peapods for this purpose remains an interesting and compelling suggestion.

Acknowledgements

F.S. thanks the hospitality of the group of Prof. Thomas Pichler during the preparation of this chapter. This work was supported by the New Széchenyi Plan Nr. TÁMOP-4.2.1/B-09/1/ KMR-2010-0002, by the Austrian Science Funds (FWF) project Nr. P21333–N20, by the Deutsche Forschungsgemeinschaft (DFG) project Nr. 440/15 and by the European Research Council Grant Nr. ERC259374-Sylo. M.M. is indebted to NATO, Imra-Europe, Toyota Motor-Corporation, and the "Hierarchised Nanomaterials" CNRS program for their financial support.

References

[1] R.D. Johnson, M.S. Devries, J. Salem, D.S. Bethune and C.S. Yannoni, Electron-paramagnetic resonance studies of lanthanum-containing C_{82}, *Nature*, **355**, 239–240 (1992).

[2] M.S. Dresselhaus, G. Dresselhaus and P.C. Ecklund, *Science of Fullerenes and Carbon Nanotubes*, Academic Press, New York (1996).

[3] B.W. Smith, M. Monthioux and D.E. Luzzi, Encapsulated C_{60} in carbon nanotubes, *Nature*, **396**, 323–324 (1998).

[4] B.W. Smith, M. Monthioux and D.E. Luzzi, Carbon nanotube encapsulated fullerenes: a unique class of hybrid materials, *Chem. Phys. Lett.*, **315**, 31–36 (1999).

[5] B.W. Smith and D.E. Luzzi, Formation mechanism of fullerene peapods and coaxial tubes: a path to large scale synthesis, *Chem. Phys. Lett.*, **321**, 169–174 (2000).

[6] M. Monthioux, Filling single-wall carbon nanotubes, *Carbon*, **40**, 1809–1823 (2002).

[7] M. Monthioux, E. Flahaut and J.-P. Cleuziou, Hybrid carbon nanotubes: Strategy, progress, and perspectives, *J. Mat. Res.*, **21**, 2774–2793 (2006).

[8] M. Monthioux and V.L. Kuznetsov, Who should be given the credit for the discovery of carbon nanotubes?, *Carbon*, **44**, 1621–1624 (2006).

[9] S. Iijima, Helical microtubules of graphitic carbon, *Nature*, **354**, 56–58 (1991).

[10] S. Iijima and T. Ichihashi, Single-shell carbon nanotubes of 1-nm diameter, *Nature*, **363**, 603–605 (1993).

[11] D.S. Bethune, C.H. Kiang, M.S. DeVries, G. Gorman, R. Savoy and R. Beyers, Cobalt catalyzed growth of carbon nanotubes with single-atomiClayer walls, *Nature*, **363**, 605 (1993).

[12] W. Krätschmer, L.D. Lamb, K. Fostiropoulos and D.R. Huffmann, Solid C_{60}: a new form of carbon, *Nature*, **347**, 354 (1990).

[13] A. Rinzler, J. Liu, H. Dai, P. Nikolaev, C.B. Huffman, F.J. Rodriguez-Macias, P.J. Boul, A.H. Lu, D. Heymann, D.T. Colbert, R.S. Lee, J.E. Fischer, A.M. Rao, P.C. Eklund and R.E. Smalley, Large-scale purification of single-wall carbon nanotubes: process, product, and characterization, *Appl. Phys. A*, **67**, 29–37 (1998).

[14] Y.K. Kwon, D. Tomanek and S. Iijima, "Bucky shuttle" memory device: Synthetic approach and molecular dynamics simulations, *Phys. Rev. Lett.*, **82**, 1470–1473 (1999).

[15] B. Burteaux, A. Claye, B.W. Smith, M. Monthioux, D.E. Luzzi and J.E. Fischer, Abundance of encapsulated C_{60} in single-wall carbon nanotubes, *Chem. Phys. Lett.*, **310**, 21–24 (1999).

[16] M. Yudasaka, K. Ajima, K. Suenaga, T. Ichihashi, A. Hashimoto and S. Iijima, Nanoextraction and nano-condensation for C_{60} incorporation into single-wall carbon nanotubes in liquid phases, *Chem. Phys. Lett.*, **380**, 42–46 (2003).

[17] M. Monthioux and L. Noé, Liquid phase synthesis of "peapods" at room temperature. In *International Conference on Carbon "Carbon'04"*, Ext. Abstract (CD-ROM), Oral/C105 (2004).

[18] F. Simon, H. Kuzmany, H. Rauf, T. Pichler, J. Bernardi, H. Peterlik, L. Korecz, F. Fülöp and A. Janossy, Low temperature fullerene encapsulation in single wall carbon nanotubes: synthesis of $N@C_{60}@SWCNT$, *Chem. Phys. Lett.*, **383**, 362–367 (2004).

[19] A.N. Khlobystov, D.A. Britz, J.W. Wang, S.A. O'Neil, M. Poliakoff and G.A.D Briggs, Low temperature assembly of fullerene arrays in single-walled carbon nanotubes using supercritical fluids, *J. Mat. Chem.*, **14**, 2852–2857 (2004).

[20] H. Kataura, Y. Maniwa, T. Kodama, K. Kikuchi, K. Hirahara, K. Suenaga, S. Iijima, S. Suzuki, Y. Achiba and W. Krätschmer, High-yield fullerene encapsulation in single-wall carbon nanotubes, *Synthetic Met.*, **121**, 1195–1196 (2001).

[21] D. Kondo, A. Kawabata, M. Horibe, M. Nihei and Y Awano, Vertically aligned peapod formation of position-controlled multi-walled carbon nanotubes (MWNTs), *Supperlatt. Microstruct.*, **34**, 389–394 (2003).

[22] T. Frohlich, P. Scharff, W. Schliefke, H. Romanus, V. Gupta, C. Siegmund, O. Ambacher and L. Spiess, Insertion of C_{60} into multi-wall carbon nanotubes – a synthesis of C_{60}@MWCNT, *Carbon*, **42**, 2759–2762 (2004).

[23] A.N. Khlobystov, D.A. Britz, A. Ardavan and G.A.D. Briggs, Observation of ordered phases of fullerenes in carbon nanotubes, *Phys. Rev. Lett.*, **92**, 245507.1/4 (2004).

[24] G. Ning, N. Kishi, H. Okimoto, M. Shiraishi, Y. Kato, R. Kitaura, T. Sugai, S. Aoyagi, E. Nishibori, M. Sakata and H. Shinohara, Synthesis, enhanced stability and structural imaging of C_{60} and C_{70} double-wall carbon nanotube peapods, *Chem. Phys. Lett.*, **441**, 94–99 (2007).

[25] W. Mickelson, S. Aloni, W.Q. Han, J. Cumings and A. Zettl, Packing C_{60} in boron nitride nanotubes, *Science*, **300**, 467–469 (2003).

[26] K. Hirahara, K. Suenaga, S. Bandow, H. Kato, T. Okazaki, T. Shinohara and S. Iijima, One-dimensional metallofullerene crystal generated inside single-walled carbon nanotubes, *Phys. Rev. Lett.*, **85**, 5384–5387 (2000).

[27] B.W. Smith, D.E. Luzzi and Y. Achiba, Tumbling atoms and evidence for charge transfer in La_{-2}@C_{80}@SWNT, *Chem. Phys. Lett.*, **331**, 137–142 (2000).

[28] T. Almeida Murphy, T. Pawlik, A. Weidinger, M. Hhne, R. Alcala and J.-M. Spaeth, Observation of atom-like nitrogen in nitrogen-implanted solid C_{60}, *Phys. Rev. Lett.*, **77**, 1075 (1996).

[29] D.A. Britz, A.N. Khlobystov, J.W. Wang, A.S. O'Neil, M. Poliakoff, A. Ardavan and G.A. Briggs, Selective host-guest interaction of single-walled carbon nanotubes with functionalized fullerenes, *Chem. Comm.*, 176–177, (2004).

[30] F. Simon, H. Kuzmany, J. Bernardi, F. Hauke and A. Hirsch, Encapsulating $C_{59}N$ azafullerene derivatives inside single-wall carbon nanotubes, *Carbon*, **44**, 1958–1962 (2006).

[31] F. Simon, H. Kuzmany, B. Nafradi, T. Fehér, L. Forro, L. Fülöp, A. Janossy, A. Rockenbauer, L. Korecz, F. Hauke and A. Hirsch, Magnetic fullerenes inside single-wall carbon nanotubes, *Phys. Rev. Lett.*, **97**, 136801.1/4 (2006).

[32] F. Simon, C. Kramberger, R. Pfeiffer, H. Kuzmany, V. Zolyomi, J. Kürti, P.M. Singer and H. Alloul, Isotope engineering of carbon nanotube systems, *Phys. Rev. Lett.*, **95**, 017401.1/4 (2005).

[33] F. Simon, H. Peterlik, R. Pfeiffer, J. Bernardi and H. Kuzmany, Fullerene release from the inside of carbon nanotubes: A possible route toward drug delivery, *Chem. Phys. Lett.*, **445**, 288–292 (2007).

[34] F. Hasi, F. Simon and H. Kuzmany, Reversible hole engineering for single-wall carbon nanotubes, *J. Nanosci. Nanotechn.*, **5**, 1785–1791(2005).

[35] M. Waiblinger, K. Lips, W. Harneit, A. Weidinger, E. Dietel and A. Hirsch, Thermal stability of the endohedral fullerenes N@C_{60}, N@C_{70}, and P@C_{60}, *Phys. Rev. B*, **64**, 159901.1/4 (2001).

[36] M. Prato, Fullerene chemistry for materials science applications, *J. Mat. Chem.*, **7**, 1097–1109 (1997).

[37] W. Harneit, Fullerene-based electron-spin quantum computer, *Phys. Rev. A*, **65**, 032322 (2002).

[38] A. Mrzel, A. Hassanien, Z. Liu, K. Suenaga, Y. Miyata, K. Yanagi and H. Kataura, Effective, fast, and low temperature encapsulation of fullerene derivatives in single wall carbon nanotubes. *Surf. Sci.*, **601**, 5116–5120 (2007).

[39] H. Kuzmany, *Solid-State Spectroscopy: An Introduction*, Springer-Verlag, Berlin (1998).

[40] B.W. Smith, R.M. Russo, S.B. Chikkannanavar and D.E. Luzzi, High-yield synthesis and one-dimensional structure of C_{60} encapsulated in single-wall carbon nanotubes, *J. Appl. Phys.*, **91**, 9333–9340 (2002).

[41] H. Kataura, Y. Maniwa, T. Kodama, K. Kikuchi, K. Hirahara, S. Iijima, S. Suzuki, W. Krätschmer and Y. Achiba. Fullerene-peapods: Synthesis, structure, and Raman spectroscopy, In *Electronic*

Properties of Molecular Nanostructures, H. Kuzmany, J. Fink, M. Mehring and S. Roth (eds.), *AIP Conf. Proc.*, **591**, 251–255 (2001).

[42] H. Kataura, Y. Maniwa, M. Abe, A. Fujiwara, T. Kodama, K. Kikuchi, J. Imahori, Y. Misaki, S. Suzuki and Y. Achiba, Optical properties of fullerene and non-fullerene peapods, *Appl. Phys. A*, **74**, 349–354 (2002).

[43] J. Cambedouzou, V. Pichot, S. Rols, P. Launois, P. Petit, R. Klement, H. Kataura and R. Almairac, On the diffraction pattern of C_{60} peapods, *Eur. Phys. J. B*, **42**, 31–45 (2004).

[44] X. Liu, T. Pichler, M. Knupfer, M.S. Golden, J. Fink, H. Kataura, Y. Achiba, K. Hirahara and S. Iijima, Filling factors, structural, and electronic properties of C_{60} molecules in singlewall carbon nanotubes, *Phys. Rev. B*, **65**, 045419.1/6 (2002).

[45] H. Kuzmany, R. Pfeiffer, C. Kramberger, T. Pichler, X. Liu, M. Knupfer, J. Fink, H. Kataura, Y. Achiba, B.W. Smith and D.E. Luzzi, Analysis of the concentration of C_{60} fullerenes in single wall carbon nanotubes, *Appl. Phys. A*, **76**, 449–455 (2003).

[46] S. Berber, Y.-K. Kwon and D. Tomanek, Microscopic formation mechanism of nanotube peapods, *Phys. Rev. Lett.*, **88**, 185502 (2002).

[47] O. Dubay and G. Kresse, Density functional calculations for C_{60} peapods, *Phys. Rev. B*, **70**, 165424 (2004).

[48] J. Fan, M. Yudasaka, R. Yuge, D.N. Futaba, K. Hata and S. Iijima, Efficiency of C_{60} incorporation in and release from single-wall carbon nanotubes depending on their diameters, *Carbon*, **45**, 722–726 (2007).

[49] M. Otani, S. Okada and A. Oshiyama, Energetics and electronic structures of one-dimensional fullerene chains encapsulated in zigzag nanotubes, *Phys. Rev. B*, **68**, 125424.1/8 (2003).

[50] K. Matsuda, Y. Maniwa and H. Kataura, Highly rotational C_{60} dynamics inside single-walled carbon nanotubes: NMR observations, *Phys. Rev. B*, **77**, 075421/1–6 (2008).

[51] K.H. Michel, B. Verberck and A.V. Nikolaev, Anisotropic packing and one-dimensional fluctuations of C_{60} molecules in carbon nanotubes, *Phys. Rev. Lett.*, **95**, 185506/1–4 (2005).

[52] S. Okada, Energetics of carbon peapods: Elliptical deformation of nanotubes and aggregation of encapsulated C_{60}, *Phys. Rev. B*, **77**, 235419/1–7 (2008).

[53] S. Okada, M. Otani and A. Oshiyama, Energetics and electronic structure of C_{70}-peapods and one-dimensional chains of C_{70}, *New J. Phys.*, **5**, 122–125 (2003).

[54] B. Verberck and K.H. Michel, Nanotube field and orientational properties of C_{70} molecules in carbon nanotubes, *Phys. Rev. B*, **75**, 045419/1–14 (2007).

[55] K. Hirahara, S. Bandow, K. Suenaga, H. Kato, T. Okazaki, T. Shinohara and S. Iijima, Electron diffraction study of one-dimensional crystals of fullerenes, *Phys. Rev. B*, **64**, 115420.1/5 (2001).

[56] M. Chorro, A. Delhey, L. Noé, M. Monthioux and P. Launois, Orientation of C_{70} molecules in peapods as a function of the nanotube diameter, *Phys. Rev. B*, **75**, 035416/1–11 (2007).

[57] M.S. Dresselhaus, G. Dresselhaus and R. Saito, Physics of carbon nanotubes, *Carbon*, **33**, 883–891 (1995).

[58] S. Bandow, M. Takizawa, K. Hirahara, M. Yudasaka and S. Iijima, Raman scattering study of double-wall carbon nanotubes derived from the chains of fullerenes in single-wall carbon nanotubes, *Chem. Phys. Lett.*, **337**, 48–54 (2001).

[59] J. Kürti, G. Kresse and H. Kuzmany, First-principles calculations of the radial breathing mode of single-wall carbon nanotubes, *Phys. Rev. B*, **58**, R8869–R8872 (1998).

[60] C. Kramberger, A. Waske, K. Biedermann, T. Pichler, T. Gemming, B. Buchner and H. Kataura, Tailoring carbon nanostructures via temperature and laser irradiation, *Chem. Phys. Lett.*, **407**, 254–259 (2005).

[61] M. Kalbac, L. Kavan, L. Juha, S. Civis, M. Zukalova, M. Bittner, P. Kubat, V. Vorlicek and L. Dunsch, Transformation of fullerene peapods to double-walled carbon nanotubes induced by UV radiation, *Carbon*, **43**, 1610–1616 (2005).

[62] P. Puech, F. Puccianti, R. Bacsa, C. Arrondo, M. Monthioux, W. Bacsa, V. Paillard, A. Bassil and F. Barde, Thermal transfer in SWNTs and peapods under UV-irradiation, *Phys. Stat. Sol. B*, **244**, 4064–4068 (2007).

[63] P. Puech, F. Puccianti, R. Bacsa, C. Arrondo, V. Paillard, A. Bassil, M. Monthioux, E. Flahaut, F. Barde and W. Bacsa, Ultraviolet photon absorption in single- and double-wall carbon nano-

tubes and peapods: Heating-induced phonon line broadening, wall coupling, and transformation, *Phys. Rev. B*, **76**, 054118/1–4 (2007).

[64] Y. Sakurabayashi, M. Monthioux, K. Kishita, Y. Suzuki, T. Kondo and M. Le Lay, Tailoring double-wall carbon nanotubes?, In *Electronic Properties of Molecular Nanostructures*, H. Kuzmany, J. Fink, M. Mehring and S. Roth (eds.), *AIP Conf. Proc.*, **685**, 302–305 (2003).

[65] C. Arrondo, M. Monthioux, K. Kishita and M. Le Lay, In-situ coalescence of aligned C_{60} molecules in peapods, In *Electronic Properties of Molecular Nanostructures*, H. Kuzmany, J. Fink, M. Mehring and S. Roth (eds.), *AIP Conf. Proc.*, **786**, 329–332 (2005).

[66] F. Ding, Z. Xu, B.I. Yakobson, R.J. Young, I.A. Kinloch, S. Cui, L. Deng, P. Puech, and M. Monthioux, Formation mechanism of peapod-derived double-walled carbon nanotubes, *Phys. Rev. B*, **82**, 041403/1–4 (2010).

[67] S.W. Han, M. Yoon, S. Berber, N. Park, E. Osawa, J. Ihm and D. Tomanek, Microscopic mechanism of fullerene fusion, *Phys. Rev. B*, **70**, 113402.1/4 (2004).

[68] A.J. Stone and D.J. Wales, Theoretical-studies of icosahedral C_{60} and some related species, *Chem. Phys. Lett.*, **128**, 501–503 (1986).

[69] S. Bandow, T. Hiraoka, T. Yumura, K. Hirahara, H. Shinohara and S. Iijima, Raman scattering study on fullerene derived intermediates formed within single-wall carbon nanotube, *Chem. Phys. Lett.*, **384**, 320–325 (2004).

[70] F. Simon and H. Kuzmany, Growth of single-wall carbon nanotubes from ^{13}C isotope labeled organic solvents inside single wall carbon nanotube hosts, *Chem. Phys. Lett.*, **425**, 85–88, (2006).

[71] H. Shiozawa, T. Pichler, A. Grüneis, R. Pfeiffer, H. Kuzmany, Z. Liu, K. Suenaga and H. Kataura, A catalytic reaction inside a single-walled carbon nanotube, *Adv. Mater.*, **20**, 1443–1449 (2008).

[72] V. Zolyomi, F. Simon, A. Rusznyak, R. Pfeiffer, H. Peterlik, H. Kuzmany and J. Kürti, Inhomogeneity of ^{13}C isotope distribution in isotope engineered carbon nanotubes: Experiment and theory, *Phys. Rev. B*, **75**, 195419.1/8 (2007).

[73] P.W. Stephens, G. Bortel, G. Faigel, M. Tegze, A. Janossy, S. Pekker, G. Oszlanyi and L. Forro, Polymeric fullerene chains in RbC_{60} and KC_{60}, *Nature*, **370**, 636–639 (1994).

[74] G. Oszlanyi, G. Baumgartner, G. Faigel and L. Forro, Na_4C_{60}: An alkali intercalated twodimensional polymer, *Phys. Rev. Lett.*, **78**, 4438–4441 (1997).

[75] Y. Iwasa, T. Arima, R.M. Fleming, T. Siegrist, O. Zhou, R.C. Haddon, L.J. Rothberg, K.B. Lyons, H.L. Carter, A.F. Hebard, R. Tycko, G. Dabbagh, J.J. Krajewski, G.A. Thomas, and T. Yagi, New phases of C_{60} synthesized at high-pressure, *Science*, **264**, 1570–1572 (1994).

[76] O. Chauvet, G. Oszlanyi, L. Forro, P.W. Stephens, M. Tegze, G. Faigel and A. Janossy, Quasi-one-dimensional electronic structure in orthorhombic RbC_{60}, *Phys. Rev. Lett.*, **72**, 2721–2724 (1994).

[77] M. Ricco, M. Belli, M. Mazzani, D. Pontiroli, D. Quintavalle, A. Janossy and G. Csanyi, Superionic conductivity in the Li_4C_{60} fulleride polymer, *Phys. Rev. Lett.*, **102**, 145901/1–4 (2009).

[78] S. Pekker, G. Oszlanyi and G. Faigel, Structure and stability of covalently bonded polyfulleride ions in A(x)C(60) salts, *Chem. Phys. Lett.*, **282**, 435–441 (1998).

[79] T. Pichler, H. Kuzmany, H. Kataura and Y. Achiba, Metallic polymers of C_{60} inside single-walled carbon nanotubes, *Phys. Rev. Lett.*, **87**, 267401.1/4 (2001).

[80] S. Kawasaki, T. Hara, T. Yokomae, F. Okino, H. Touhara, H. Kataura, T. Watanuki and Y. Ohishi, Pressure-polymerization of C_{60} molecules in a carbon nanotube, *Chem. Phys. Lett.*, **418**, 260–263 (2006).

[81] M. Chorro, S. Rols, J. Cambedouzou, L. Alvarez, R. Almairac, J.-L. Sauvajol, J.-L. Hodeau, L. Marques, M. Mezouar and H. Kataura, Structural properties of carbon peapods under extreme conditions studied using in situ x-ray diffraction, *Phys. Rev. B*, **74**, 205425/1–5 (2006).

[82] M. Chorro, J. Cambedouzou, A. Iwasiewicz-Wabnig, L. Noé, S. Rols, M. Monthioux, B. Sundqvist and P. Launois, Discriminated structural behaviour of C_{60} and C_{70} peapods under extreme conditions, *Eur. Phys. Lett.*, **79**, 56003.1/5 (2007).

[83] A.V. Soldatov, G. Roth, A. Dzyabchenko, D. Johnels, S. Lebedkin, C. Meingast, B. Sundqvist, M. Haluska and H. Kuzmany, Topochemical polymerization of C_{70} controlled by monomer crystal packing, *Science*, **293**, 680–683 (2001).

[84] M. Abe, H. Kataura, H. Kira, T. Kodama, S. Suzuki, Y. Achiba, K. Kato, M. Takata, A. Fujiwara, K. Matsuda and Y. Maniwa Structural transformation from single-wall to double-wall carbon nanotube bundles, *Phys. Rev. B*, **68**, R041405 (2003).

[85] I.V. Krive, R.I. Shekhter and M. Jonson, Carbon "peapods" - a new tunable nanoscale graphitic structure (Review), *Low Temp. Phys.*, **32**, 887–905 (2006).

[86] S. Okada, S. Saito and A. Oshiyama, Energetics and electronic structures of encapsulated C_{60} in a carbon nanotube, *Phys. Rev. Lett.*, **86**, 3835–3838 (2001).

[87] D.J. Hornbaker, S.J. Kahng, S. Misra, B.W. Smith, A.T. Johnson, E.J. Mele, D.E. Luzzi and A. Yazdani, Mapping the one-dimensional electronic states of nanotube peapod structures, *Science*, **295**, 828–831 (2002).

[88] J. Lee, H. Kim, S.J. Kahng, G. Kim, Y.W. Son, J. Ihm, H. Kato, Z.W. Wang, T. Okazaki, H. Shinohara and Y. Kuk, Bandgap modulation of carbon nanotubes by encapsulated metallofullerenes, *Nature*, **415**, 1005–1008 (2002).

[89] Y.G. Yoon, M.S.C. Mazzoni and S.G. Louie, Quantum conductance of carbon nanotube peapods, *Appl. Phys. Lett.*, **83**, 5217–5219 (2003).

[90] H. Rauf, H. Shiozawa, T. Pichler, M. Knupfer, B. Buchner and H. Kataura, Influence of the C_{60} filling on the nature of the metallic ground state in intercalated peapods, *Phys. Rev. B*, **72**, 245411/1–8 (2005).

[91] R. Egger and A.O. Gogolin, Effective low-energy theory for correlated carbon nanotubes, *Phys. Rev. Lett.*, **79**, 5082–5085 (1997).

[92] H. Rauf, T. Pichler, M. Knupfer, J. Fink and H. Kataura, Transition from a Tomonaga-Luttinger liquid to a Fermi liquid in potassium-intercalated bundles of single-wall carbon nanotubes, *Phys. Rev. Lett.*, **93**, 096805.1/4 (2004).

[93] H. Ishii, H. Kataura, H. Shiozawa, H. Yoshioka, H. Otsubo, Y. Takayama, T. Miyahara, S. Suzuki, Y. Achiba, T. Nakatake, T. Narimura, M. Higashiguchi, K. Shimada, H. Namatame and M. Taniguchi, Direct observation of Tomonaga-Luttinger-liquid state in carbon nanotubes at low temperatures, *Nature*, **426**, 540–544 (2003).

[94] J.M. Ziman, *Principles of the Theory of Solids*, Cambridge University Press, Cambridge (1986).

[95] H. Hongo, F. Nihey, M. Yudasaka, T. Ichihashi and S. Iijima, Transport properties of single wall carbon nanotubes with encapsulated C_{60}, *Phys. B Cond. Matt.*, **323**, 244–245 (2002). In *Tsukuba Symposium on Carbon Nanotube in Commemoration of the 10th Anniversary of its Discovery*, Tsukuba, Japan, Oct. 03–05 (2001).

[96] P. Utko, J. Nygard, M. Monthioux and L. Noé, Sub-Kelvin transport spectroscopy of fullerene peapod quantum dots, *Appl. Phys. Lett.*, **89**, 233118/1–3 (2006).

[97] A. Eliasen, J. Paaske, K. Flensberg, S. Smerat, M. Leijnse, M.R. Wegewijs, H.I. Jorgensen, M. Monthioux and J. Nygard, Transport via coupled states in a C_{60} peapod quantum dot, *Phys. Rev. B*, **81**, 155431/1–5 (2010).

[98] F. Wang, G. Dukovic, L.E. Brus and T.F. Heinz, The optical resonances in carbon nanotubes arise from excitons, *Science*, **308**, 838–841 (2005).

[99] C.D. Spataru, S. Ismail-Beigi, L.X. Benedict and S.G. Louie, Excitonic effects and optical spectra of single-walled carbon nanotubes. *Phys. Rev. Lett.*, **92**, 077402.1/4 (2004).

[100] F. Borondics, K. Kamaras, M. Nikolou, D.B. Tanner, Z.H. Chen and A.G. Rinzler, Charge dynamics in transparent single-walled carbon nanotube films from optical transmission measurements, *Phys. Rev. B*, **74**, 045431.1/6 (2006).

[101] R.M. Martin and L.M. Falicov, In *Resonance Raman Scattering*, Springer, Berlin, 79 (1983).

[102] R. Pfeiffer, H. Kuzmany, T. Pichler, H. Kataura, Y. Achiba, M. Melle-Franco and F. Zerbetto, Electronic and mechanical coupling between guest and host in carbon peapods, *Phys. Rev. B*, **69**, 035404 (2004).

[103] T. Pichler, A. Kukovecz, H. Kuzmany, H. Kataura and Y. Achiba, Quasi continuous electron and hole doping of C_{60} peapods, *Phys. Rev. B*, **67**, 125416/1–7 (2003).

[104] L. Kavan, L. Dunsch and H. Kataura, In situ Vis-NIR and Raman spectroelectrochemistry at fullerene peapods, *Chem. Phys. Lett.*, **361**, 79–85 (2002).

[105] F. Simon, D. Quintavalle, A. Janossy, B. Nafradi, L. Forro, H. Kuzmany, F. Hauke, A. Hirsch, J. Mende and M. Mehring, Metallic bundles of single-wall carbon nanotubes probed by electron spin resonance, *Phys. Stat. Sol. B*, **244**, 3885–3889 (2007).

[106] K. Furukawa, S. Okubo, H. Kato, H. Shinohara and T. Kato, High-field/high-frequency ESR study of Gd@C_{82}-I, *J. Phys. Chem. A*, **107**, 10933–10937 (2003).

[107] K.-P. Dinse, EPR investigation of atoms in chemical traps, *Phys. Chem. Chem. Phys.*, **4**, 5442–5447 (2002).

[108] K. Hasharoni, C. Bellavia-Lund, M. Keshavarz-K, G. Srdanov and F. Wudl, Light-induced ESR studies of the heterofullerene dimers, *J. Am. Chem. Soc.*, **119**, 11128–11129 (1997).

[109] J.C. Hummelen, C. Bellavia-Lund and F. Wudl, In *Heterofullerenes*, Springer, Berlin, Heidelberg, **199**, 93 (1999).

[110] A. Gruss, K.-P. Dinse, A. Hirsch, B. Nuber and U. Reuther, Photolysis of $(C_{59}N)_2$ studied by time-resolved EPR, *J. Am. Chem. Soc.*, **119**, 8728–8729 (1997).

[111] F. Simon, D. Arcon, N. Tagmatarchis, S. Garaj, L. Forro and K. Prassides, ESR signal in azafullerene $(C_{59}N)_2$ induced by thermal hemolysis, *J. Phys. Chem. A.*, **103**, 6969–6971 (1999).

[112] F. Fülöp, A. Rockenbauer, F. Simon, S. Pekker, L. Korecz, S. Garaj and A. Janossy, Azafullerene $C_{59}N$, a stable free radical substituent in crystalline C_{60}, *Chem. Phys. Lett.*, **334**, 233–237 (2001).

[113] A. Rockenbauer, G. Csanyi, F. Fülöp, S. Garaj, L. Korecz, R. Lukacs, F. Simon, L. Forro, S. Pekker and A. Janossy, Electron delocalization and dimerization in solid $C_{59}N$-doped C_{60} fullerene. *Phys. Rev. Lett.*, **94**, 066603 (2005).

[114] S. Campestrini, C. Corvaja, M. De Nardi, C. Ducati, L. Franco, M. Maggini, M. Meneghetti, E. Menna and G. Ruaro, Investigation of the inner environment of carbon nanotubes with a fullerene-nitroxide probe, *Small*, **4**, 350–356 (2008).

[115] S. Toth, D. Quintavalle, B. Nafradi, L. Forro, L. Korecz, A. Rockenbauer, T. Kalai, K. Hideg and F. Simon, Stability and electronic properties of magnetic peapods, *Phys. Stat. Sol. B*, **245**, 2034–2037 (2008).

[116] P.M. Singer, P. Wzietek, H. Alloul, F. Simon and H. Kuzmany, NMR evidence for gapped spin excitations in metallic carbon nanotubes, *Phys. Rev. Lett.*, **95**, 236403.1/4 (2005).

[117] B. Dora, M. Gulacsi, F. Simon and H. Kuzmany, Spin gap and Luttinger liquid description of the NMR relaxation in carbon nanotubes, *Phys. Rev. Lett.*, **99**, 166402 (2007).

[118] T. Shimada, T. Okazaki, R. Taniguchi, T. Sugai, H. Shinohara, K. Suenaga, Y. Ohno, S. Mizuno, S. Kishimoto and T. Mizutani, Ambipolar field-effect transistor behavior of Gd@C_{82} metallo-fullerene peapods, *Appl. Phys. Lett.*, **81**, 4067–4069 (2002).

[119] H.Y. Yu, D.S. Lee, S.H. Lee, S.S. Kim, S.W. Lee, Y.W. Park, U. Dettlaff-Weglikowskaand and S. Roth, Single-electron transistor mediated by C_{60} insertion inside a carbon nanotube, *Appl. Phys. Lett.*, **87**, 163118/1–3 (2005).

[120] Z.C. Wu, Z.H. Chen, X. Du, J.M. Logan, J. Sippel, M. Nikolou, K. Kamaras, J.R. Reynolds, D.B. Tanner, A.F. Hebard and A.G. Rinzler, Transparent, conductive carbon nanotube films, *Science*, **305**, 1273–1276 (2004).

[121] Y.F. Li, T. Kaneko and R. Hatakeyama, Photoinduced electron transfer in C_{60} encapsulated single-walled carbon nanotube, *Appl. Phys. Lett.*, **92**, 183115/1–3 (2008).

[122] A. Bianco, K. Kostarelos and M. Prato, Applications of carbon nanotubes in drug delivery, *Curr. Op. Chem. Biol.*, **9**, 674–679 (2005).

[123] S.M. Bachilo, M.S. Strano, C. Kittrell, R.H. Hauge, R.E. Smalley and R.B. Weisman, Structure-assigned optical spectra of single-walled carbon nanotubes, *Science*, **298**, 2361–2366 (2002).

[124] C. Tripisciano, K. Kraemer, A. Taylor and E. Borowiak-Palen, Single-wall carbon nanotubes based anticancer drug delivery system, *Chem. Phys. Lett.*, **478**, 200–205 (2009).

[125] S.L. He and J.Q. Shen, Nanoscale lasers based on carbon peapods, *Chin. Phys. Lett.*, **23**, 211–213 (2006).

[126] Y. Dai, C. Tang and W. Guo, Simulation studies of a "nanogun" based on carbon nanotubes, *Nano Res.*, **1**, 176–183 (2008).

[127] J.W. Kang and H.J. Hwang, Fullerene shuttle memory device: Classical molecular dynamics study, *J. Phys. Soc. Jap.*, **73**, 1077–1081 (2004).

[128] J.W. Kang and H.J. Hwang, Nano-memory-element applications of carbon nanocapsule encapsulating potassium ions: Molecular dynamics study, *J. Kor. Phys. Soc.*, **44**, 879–883 (2004).

[129] W.Y. Choi, J.W. Kang and H.J. Hwang, Bucky shuttle memory system based on boronnitride nanopeapod, *Phys. E*, **23**, 135–140 (2004).

[130] J.W. Kang and H.J. Hwang, 'Carbon nanotube shuttle' memory device, *Carbon*, **42**, 3018–3021 (2004).

[131] J. Cumings and A. Zettl, Low-friction nanoscale linear bearing realized from multiwall carbon nanotubes, *Science*, **289**, 602–604 (2000).

[132] S.B. Legoas, V.R. Coluci, S.F. Braga, P.Z. Coura, S.O. Dantas and D.S. Galvao, Moleculardynamics simulations of carbon nanotubes as gigahertz oscillators, *Phys. Rev. Lett.*, **90**, 055504/1–4 (2003).

[133] H. Su, W.A. Goddard, III and Y. Zhao, Dynamic friction force in a carbon peapod oscillator, *Nanotechnol.*, **17**, 5691–5695 (2006).

[134] P. Utko, R. Ferone, I.V. Krive, R.I. Shekhter, M. Jonson, M. Monthioux, L. Noé and J. Nygard, Nanoelectromechanical coupling in fullerene peapods probed by resonant electrical transport experiments, *Nature Comm.*, **1**, 37/1–6 (2010).

[135] M. Wang and C.M. Li, An oscillator in a carbon peapod controllable by an external electric field: a molecular dynamics study, *Nanotechnol.*, **21**, 035704/1–6 (2010).

[136] D.P. Divincenzo, Quantum computation, *Science*, **270**, 255–261 (1995).

[137] I.L. Chuang, L.M.K. Vandersypen, X.L. Zhou, D.W. Leung and S. Lloyd, Experimental realization of a quantum algorithm, *Nature*, **393**, 143–146 (1998).

[138] B.E. Kane, A silicon-based nuclear spin quantum computer, *Nature*, **393**, 133–137 (1998).

[139] M. Mehring, W. Scherer and A. Weidinger, Pseudoentanglement of spin states in the multilevel ^{15}N@C$_{60}$ system, *Phys. Rev. Lett.*, **93**, 206603/1–4 (2004).

Overall Conclusion for Chapter 5

Regarding filled CNTs, most of the expectations for the discovery of unprecedented properties – and subsequent promises for a variety of applications – involve SWCNTs instead of MWCNTs as the containing nanotubes. In contrast with pessimistic early predictions, the narrow cavity of SWCNTs can actually be filled with liquid and gaseous materials with high filling rate efficiency. For filling methods involving liquids, this means that, in first approximation, wetting nanosized capillaries obeys the same laws (such as Laplace-Young's laws) as wetting of submillimeter capillaries. However, it is now ascertained that the surface tension of the filling materials at liquid state is not the only relevant parameter to consider in nanowetting when the latter involves narrow (~1 nm and below) nanotubes. A large panel of methods is now available to fill SWCNTs with a large variety of materials (elements or compounds) with an efficiency that unfortunately still largely depends on empirically determined filling conditions. Filled SWCNTs are now currently investigated worldwide with respect to a variety of properties and behaviors, with the reasonable hope that breakthrough reports may be anticipated in the close future, as it has started to happen.

In this broad picture, although they are the first example, evidenced as early as in 1998 of a hybrid structure involving SWCNTs filled with individual molecules, fullerene-filled single-wall carbon nanotubes; *peapods* are still the most studied filled nanotubes. This is because they are: (i) fascinating materials, built from the combination of two of the most recently discovered carbon molecular structures; (ii) easy to prepare, with large yield if necessary; (iii) beautiful model objects for revealing unprecedented physical behaviors and new properties; (iv) useful training objects for practicing nanoengineering; and (v) full of promise for the synthesis of a variety of novel materials. It is not ascertained yet whether practical applications will actually come out once, as those aspects are still investigated at a point where cost issues are not yet considered. If they will, however, the first application area will probably be electronics. The next decade will tell.

6

Heterogeneous Nanotubes: (X*CNTs, X*BNNTs)

Dmitri Golberg[1,2] Mauricio Terrones[3,4]
[1]International Center for Materials Nanoarchitectonics (MANA),
National Institute for Materials Science (NIMS), Japan
[2]University of Tsukuba, Japan
[3]Department of Physics, Department of Materials Science and Engineering and Materials
Research Institute, The Pennsylvania State University, USA
[4]Research Center for Exotic Nanocarbons (JST), Shinshu University, Japan

6.1 Overall Introduction

Heterogeneous nanotubes (or heteronanotubes) are the nanotubes in which carbon atoms in a honeycomb graphene lattice are fully or partially substituted with heteroatoms, typically nitrogen and/or boron (see Introduction). One of the major consequences of such substitution is the significant modification of the tube electronic properties and atomic structure. This, in turn, leads to the appearance of new electrical, magnetic, optical, thermal, chemical and mechanical properties in the resultant ternary B-N-C or binary B-N tubular systems. Other benefits for such heteroatomic nanotubes are the expected improvements in oxidation resistance, for example, the BN phase is known to be much more oxidation-proof than C. In addition, an enhancement of tubular surface reactivity occurs, thanks to the presence of nitrogen or boron atoms. This chapter is organized in three major sections related to heterogeneous nanotubes: firstly,

Carbon Meta-Nanotubes: Synthesis, Properties and Applications, First Edition. Edited by Marc Monthioux.
© 2012 John Wiley & Sons, Ltd. Published 2012 by John Wiley & Sons, Ltd.

the synthesis, analysis and properties of pure BN tubes are reviewed, subsequently, the generation of $B_xC_yN_z$ nanotubes and their properties are discussed, and finally the production, characterization and applications of N- and B-substituted carbon nanotubes are presented.

6.2   Pure BN Nanotubes

6.2.1   Introduction

A boron nitride nanotube (BNNT) can be considered as a structural analog of a carbon nanotube (CNT) in which carbon atoms are entirely substituted with alternating B and N atoms (Figures 6.1a and 6.1b). The existence of homo-elemental bonds, like B-B and N-N, is not favorable in BN due to energetic constrains. In this respect, a BN nanostructure is quite different from that of C. In the C system the odd-member atomic rings (pentagons) create defective sites introducing positive curvature, while in the BN system such defect would prefer to have an even number of constituting atoms (e.g., squares or octagons; Figure 6.1c and d).

Bulk BN crystallizes in several layered forms [1] which are suitable for wrapping into nanotubes, similarly to graphite conversion into CNTs. Hexagonal BN and rhombohedral BN possess analogous lattice parameters (a = 2.50 Å and c = 6.66 Å), but have a different six-membered BN unit stacking along the c-axis. From the electronic standpoint, BN is a wide bandgap semiconductor with an energy gap varying in the range of 4.5–6 eV. BNNTs were first theoretically predicted in 1994 [2, 3] and then experimentally synthesized in 1995 [4]. Strictly speaking, 14 years prior to this discovery, Ishii *et al.* [5] reported on the formation of hexagonal BN whiskers that in the modern terminology would be called bamboo-like BNNTs, but the latter authors neither paid attention to their discovery nor used such a term at that time. In the pioneering theoretical paper by Rubio *et al.* [2] using tight-binding approximation it was demonstrated that a single-layered BNNT would behave as a semiconductor with a bandgap typically larger than 2 eV. The authors concluded that the larger the diameter of a tube, the larger the bandgap. Subsequently, Blase *et al.* [3], using more sophisticated theoretical approaches involving local density approximations (LDA), concluded that a single-walled BNNT should be insulating with a bandgap of 5.5 eV. Strikingly and quite oppositely to the CNT case [6], the bandgap of a wrapped BN sheet was found to be almost independent of the tube diameter (at least for the realistic range of diameters), helicity, and the number of concentric tubes in case of multi-walled BNNTs. These characteristics make BNNTs very convenient to work with, as each tube in a given bunch behaves similarly. It is interesting that there have been computations indicating that very narrow – and unrealistic – BNNTs (diameters < 2 Å) become semiconductors with a bandgap values approaching ~1 eV.

For a long time, BNNTs have not been easily accessible to the experimentalists. In fact, the publication activity in this area was only 5% of that of the CNT systems. The theory-oriented papers have markedly prevailed over the experimental ones [7–60]. This could be due to the fact that the synthesis of BNNTs is a challenging work compared to the relative ease in making kilogram quantities of their C cousins.

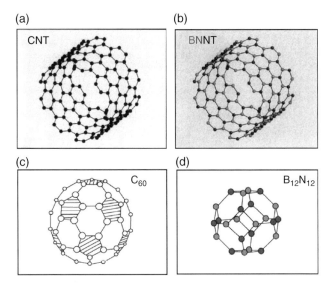

Figure 6.1 *Structural models of (a) carbon and (b) boron nitride nanotubes. (c) corresponding C and (d) BN closed atomic cages optimally sealed. The latter reveals a striking difference in the defective atomic ring nature, that is, a pentagon (odd-member ring defect) for the C system and a square (even-number ring defect) for the BN system. For a better understanding of the figure, please refer to the colour plate section.*

6.2.2 Synthesis of BN Nanotubes

6.2.2.1 Arc-Discharge

Historically, an arc-discharge technique was the first that resulted in the fabrication of multi-walled BNNTs [4]. A BN material could not be used as an electrode because of its electrical insulating nature, but in 1995, Chopra *et al.* [4] reported on a successful arc-discharge experiment by using a tungsten anode filled with hexagonal BN powder and a cathode made of copper. The tubes contained metallic particles (most probably made of W) at the tip-ends. Only multi-walled tubes were observed at that time. Subsequently, next year, the Loiseau's team [61] reported on the production of BN tubes having different number of walls (including a scarce amount of single-walled tubes) using an arc between HfB_2 electrodes in a nitrogen atmosphere. At this time it was noticed that the tip caps of BNNTs had always flat motifs. This was assigned to the presence of three B_2N_2 units (four-member atomic rings, see Figure 6.1) giving rise to the overall octahedral symmetry of a cap. At the same time, Terrones *et al.* [62] claimed on the formation of multi-walled BNNTs through arching Ta-BN electrodes in a N_2 atmosphere. The nanotube caps were found to be either flat or contain metallic particles. In 2000, Cumings and Zettl [63] reported on the production of large amounts of preferably double-walled BNNTs through arcing B electrodes containing 1 wt.% of Ni or Co in N_2 atmosphere.

6.2.2.2 Laser Ablation

The laser ablation was for the first time applied in combination with superhigh N_2 pressure by Golberg *et al.* [64] in 1996. Cubic or hexagonal BN targets were placed in a diamond

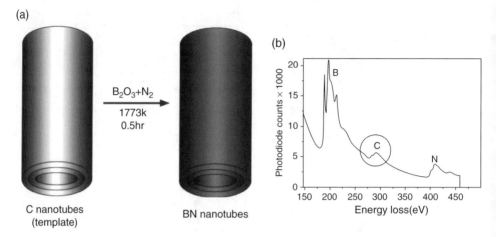

Figure 6.2 *(a) A scheme of the substitution reaction starting from a CNT template towards a BNNT. (b) A typical electron-energy-loss (EELS) spectrum of a product revealing an unwanted C remaining phase in it. The C residue can be up to 10–20 at.%.*

anvil cell filled with liquid nitrogen and heated by a continuous CO_2 laser beam to ~4700°C. During the process the BN targets melted. Several dozens of multi-walled BNNTs were found to protrude from a melted amorphous-like residue under high-resolution transmission electron microscope (HRTEM) observations. As a rule, the nanotubes were rather short, only several dozens of nanometers and exhibited a very limited number of layers, typically 3–6. Further developments of the laser ablation technique were made by Yu *et al.* [65] in 1998. Here, the laser ablated targets consisted of mixed BN and nanosized Ni and Co particles. The experiments were conducted in an inert atmosphere of Ar, N_2 or He in an oven heated to 1200°C. The procedure itself mimicked a highly effective laser ablation pathway for making single-walled CNTs developed at Rice University, by Smalley's group [66]. The production of single-, double- and triple-layered BNNTs were reported, albeit at very low yields not suitable for a macro-scale usage. Subsequently, a continuous CO_2 laser ablation technique was utilized by a French group on a rotating catalyst free BN target [67]. The researchers were able to detect a considerable amount of single-walled BNNTs in a product. The tubes were bundled in a way similar to bundles of single-walled CNTs. Later, a continuous laser ablation of BN targets was reproduced by Laude *et al.* [68], who, in the contrary, did not find single-walled NTs, but multi-walled BNNTs (typically exhibiting two or three concentric shells).

6.2.2.3 CNT Substitution Reactions
In 1998, Han *et al.* [69] discovered a so-called 'substitution reaction' synthesis method for producing BNNTs from CNTs (used as templates). These templates were annealed in B_2O_3 under N_2 or NH_3 atmospheres. During the reaction C atoms were oxidized by a highly reactive B_2O_3 vapor. The templates started to convert into BN, as BN islands were gradually substituted for C units. The whole tubular shape was preserved and the diameters of starting templates and resultant tubes correlated. The scheme of this process is depicted in Figure 6.2.

Interestingly enough, it was found that BNNT tubular layers produced in such a way displayed the preferential zigzag orientation. Golberg *et al.* [70, 71] assigned this to the fact that the zigzag graphene edges are more reactive than the armchair ones, and once a reaction, which is fully oxidation-driven, starts at a given zigzag edge, the forming BNNTs inherit this pre-existing orientation. Therefore, a CNT serves not only as a template but also as a seed for the preferred zigzag BNNTs. The disadvantage of this technique was the fact that it was not possible to get rid of C completely at realistic reaction temperatures. The C species partially remained, especially in the intermediate BNNT sections, as documented by an electron-energy loss spectrum (EELS) of a substitution reaction product in Figure 6.2b. This can be explained by the fact that a substitution reaction occurs more efficiently on the surfaces, which are well exposed to the reactive gases. And apparently, this is not the case for the C shells embedded inside the multi-walled CNT templates. These layers are hardly accessible to the B_2O_3 vapor. The substitution rate was found to be proportional to the induction furnace temperature at which the reaction took place. As the reaction increased above 1500°C, the amount of fully substituted C domains was notably increased. However, a drawback of a high-temperature regime was that the CNT template morphology itself became unstable. The templates entirely deteriorated if the temperature was too high (above 1700°C). A very narrow temperature range window, 1500–1550°C was found to be the most effective. Golberg *et al.* [72] found that the CNT (template) opening by sublimating highly reactive metal oxide vapors may be an efficient way to increase the ratio of a pure BNNT fraction. In this case, the CNT conversion into BN may occur simultaneously from outside-in and inside-out, thus increasing the overall efficiency of the process.

The pure BNNT fraction may also be obtained during a post-treatment of not fully substituted B-C-N NTs. Oxidation studies of $B_xC_yN_z$ NTs have been performed using thermogravimetric analysis (TGA) by Han *et al.* [73]. Their results may be summarized as follows: (i) $T_{room} \rightarrow 550°C$: the NTs remain stable; (ii) $550°C \rightarrow 675°C$ the excess of C is removed; (iii) $675°C \rightarrow 800°C$ the total mass of tubes stabilizes; and (iv) $800°C \rightarrow 1000°C$ BN layers start to oxidize and transform to B_2O_3. It was finally concluded that the best conditions to perform the entire transformation of not fully converted B-C-N tubes into pure BN tubular phase is an oxidation in air at 700°C over 30 min.

In 1999 and 2000, Golberg *et al.* [74,75] tried to utilize the similar substitution methodology to fabricate a single-walled BNNT starting from single-walled CNT templates. However, the success of this reaction in the latter case was found to be rather limited. The authors observed all types of doped single-walled C tubes, namely, B-doped and B/N-doped NTs having various ratios of B and N contents, but only a very marginal fraction of pure single-walled BNNTs were found to exist. This again points out at the problem of the C layer accessibility to the reactant gases. It is noteworthy that pure SWCNTs are packed in bundles and it is very difficult to separate them during a substitution reaction. The temperature range of the effective substitution reactions was found to be even narrower than that in the case of multi-walled CNT templates due to a lower stability of SWCNTs to the thermal and oxidation treatments.

6.2.2.4 *Chemical Vapor Deposition (with Catalyst)*

Lourie *et al.* [76] succeeded for the first time in fabricating multi-walled BNNTs when using a borazin precursor ($B_3N_3H_6$) in a CVD chamber with integrated NiB and Ni_2B powders as catalysts. The reaction was carried out at 1000–1100°C over 30 min. Subsequently, Ma *et al.* [77] used another precursor, namely, $B_4N_3O_2H$, and multi-walled BNNTs were produced using the CVD approach. The yield of CVD-grown multi-walled BNNTs was reported to be of hundreds of milligrams in a single experimental run. The tubes frequently revealed flat or flag-like tip-end terminations. The production of single-layered BNNTs using a CVD pathway still remains a challenge. Some encouraging data have recently been obtained by Wang *et al.* [78], who used a strategy effectively exploited for the successful preparation of single-walled B-C-N NTs [79]. The process presumed the usage of MgO-supported Fe-Mo bimetallic catalysts, CH_4, B_2H_6 and ethylenediamine as the precursors in a horizontal electrical furnace with several temperature zones.

Low-temperature CVD on Fe-particle functionalized substrates at ~600°C resulted in multi-walled BNNT aligned bundles [80]. Each bundle consisted of several dozens of multi-shelled BNNTs protruding perpendicular to the substrates. The authors were aware of possible doping effects during the growth, as the measured optical bandgaps of the BNNTs were found to be slightly lower that those in pristine BNNTs.

6.2.2.5   Ball Milling

In 1999, Chen *et al.* [81,82] of the Australian National University invented an alternative way of making relatively large quantities of a product containing multi-walled, mostly bamboo-like BNNTs with the ill-defined inner channel. The method relies on the ball milling of elemental B in ammonia gas followed by its thermal annealing at 1000–1200°C under N_2 or Ar atmospheres. Boron powder can also be replaced by BN powder. The authors suggested that the bamboo-like BNNTs had been formed through a solid-state process. This involved deposition from a vapor and surface self-diffusion mechanisms. The continuous efforts in the understanding of ball milling BNNT growth mechanism resulted in the claim that the most likely mechanism is the root-growth process catalyzed by the Fe nanoparticles coming from the ball mills [83, 84]. However, more recently Ji *et al.* [85] have stated that it is possible to grow BNNTs by annealing of a ball-milled B powder in NH_3 atmosphere without Fe particles. By all means the ball milling process seems to be relatively cheap and promising for the production of multi-walled BNNTs and/or fibers, especially those possessing bamboo-like morphology, as evidenced by the launch of a pilot company [86] selling the ball milled BN products at a price of $600–800 per gram and affiliated with the Australian National University and Chen's group.

Regarding the growth of BNNTs using this technique, Chadderton and Chen [87] have proposed that the thermally activated process of surface self-diffusion in the milled powders is the key factor for growing BNNTs and other nanomaterials. These authors believe that an iron particle always located at the tip of the growing tube is responsible for the BN nucleation by a catalytic capillary process. This results in the 'epitaxial' growth of the tube involving preferential planes of Fe and the BN (001) plane [88].

In this context, Velázquez-Salazar *et al.* [89] reported the synthesis and detailed characterization of BN bamboo-like NTs. The authors demonstrated that: (i) the ball-milling time induces considerable damage to the h-BN powder after 60 h, which is less than that used by Chen and collaborators; (ii) BNNTs always contain a metallic nanoparticle at one end, whereas the other end is closed and does not include a particle of any type; (iii) a metallic particle (e.g., Fe) appears to be responsible for BN agglomeration and subsequent NT growth; (iv) no epitaxial relationship exists between the crystallographic planes of a Fe nanoparticle and the BN (001) planes; and (v) the catalytic nanoparticle is not monocrystalline but an agglomeration of various Fe nanocrystals that somehow controls the diameter of the

BN bamboo-type NTs during growth. All these results reveal that the most likely growth mechanism accounting for the formation of bamboo-like BNNTs is the root growth process catalyzed by Fe nanoparticles. Figure 6.3 depicts SEM images and the mechanism proposed by Velázquez-Salazar *et al.* [89].

6.2.2.6   Floating Catalyst Method

In 2002, Tang *et al.* [90] from the National Institute for Materials Science, Japan, discovered that it is possible to greatly improve the yield of straight and highly crystalline BNNTs containing no carbon within an induction furnace set-up similar to that used for the substitution reaction process. At present this method gives the best results as far as non-bamboo well-ordered highly pure BNNTs are concerned. The scheme of the nanotube production is illustrated in Figure 6.4.

A mixture of MgO and B powders is prepared and placed into a graphite crucible within an induction furnace under ammonia or nitrogen flows (Figure 6.4a). The powders are quickly heated to 1300°C, and after a sequence of chemical reactions:

$$MgO + B \rightarrow Mg + B_2O_2; B_2O_2 + NH_3 \rightarrow BN + H_2O;$$

multi-walled BNNTs form inside the furnace and crystallized in its different parts. The typical yield of a product in a single experimental run may reach several hundreds of milligrams. During the reaction, the quasiliquid Mg particles are supposed to serve as a floating catalyst for the BNNT growth (Figure 6.4b). The reactions can be effectively improved if other metal oxides with low sublimation points are added into the starting mixtures, for example, SnO or FeO [91]. This efficient synthesis method is suggested to be highly valuable for a short-term commercialization. The BNNT products produced in such way exhibit a characteristic white color appearance implying that they are extremely pure (Figure 6.4c). Nowadays, this sort of BNNTs is widely used for numerous laboratory tests on BNNTs within different research groups throughout the world.

6.2.2.7   Other Promising Methods

Among other valuable methods: thermal annealing of rhombohedral B powders and *h*-BN mixtures at 1200°C in Li vapor [92], plasma-jet method [93], plasma-assisted laser vapor deposition [94], pyrolysis of melamine diborate ($C_3H_6N_6$ $2H_3BO_3$) without any metal catalysts [95], and direct atom deposition on clean metal substrates [96] should be mentioned. However, the available yields of BNNTs produced using these routes could not be regarded as high enough for possible commercialization.

6.2.2.8   Post-Synthesis Purification

In a recent report Chen *et al.* [97] described a four-step BNNT purification procedure that involved: (i) dispersion of large aggregates by ultrasonic treatment in ethanol; (ii) removal of metallic particles by selective chemical leaching using HCl; (iii) dissolution of BN particles by selective oxidation in air; and (iv) hot water washing and filtering to remove B_2O_3 residues. An alternative work reported by Zhi *et al.* [98] relied on strong interaction between the BNNT surfaces and conjugated polymers, for example, poly(m-phenylenevinylene)- co-(2,5-dioctoxy-p-phenylene) or PmPV. In this case, a polymer selectively wraps the BNNTs and can be then used to differentiate between BNNT and BN impurities, for example, nanoparticles, ribbons, or amorphous-like residues. This purification process was performed in three steps: (i) a BNNT powder was washed in HNO_3 in order

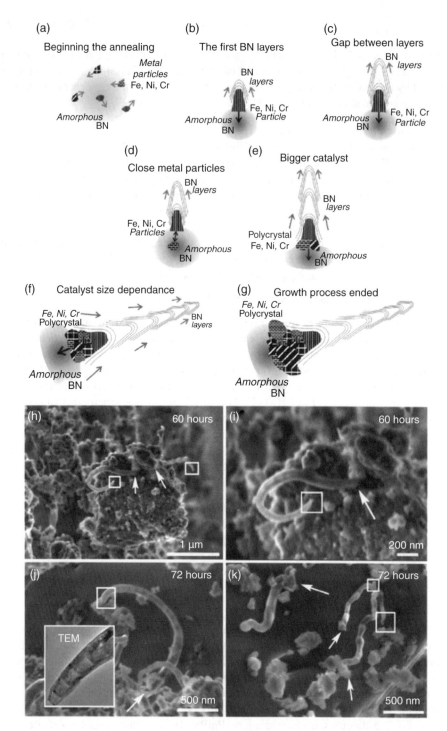

Figure 6.3 *Schematic diagram of BNNT growth during the annealing part of a ball milling process: (a) small Fe-based alloy nanoparticles are spread in the BN amorphous material after ball-milling; (b) as the temperature increases, BN material starts to migrate and precipitate layers, which adopt the shape of small Fe-based alloy nanoparticles;*

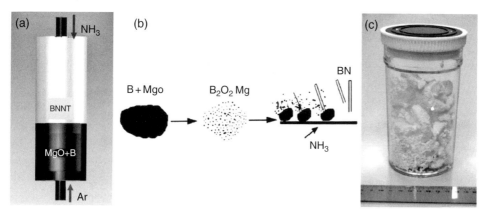

Figure 6.4 *Floating catalyst synthesis of pure BNNT products as developed at the National Institute for Materials Science (NIMS), Tsukuba, Japan: (a) experimental set-up; (b) scheme of the process; (c) resultant pure BNNT material.*

to remove catalytic particles, such as Mg or MgO; (ii) the BNNTs and the conjugated polymer were mixed in chloroform (this makes BNNTs soluble); the mixture was then centrifuged, accompanied by the removal of insoluble BN materials; (iii) the final annealing of the material in air at 700°C removed the polymer and extracted pure BNNTs. The nanotube phase normally occupies 100% of the product volume after purification using this technique.

6.2.3 Morphology and Structure of BN Nanotubes

When compared to pure carbon multi-walled CNTs, BNNTs exhibit various important structural differences. Primarily these differences are due to the specific B-N chemistry, for example, charge transfer, ionic-like features, defective atomic rings, etc. As mentioned earlier, the B-B and N-N bonds are not energetically stable when compared to B-N bonds. This gives a unique structural trend — the energetically favourable structural defects in the BN-system are even-membered atomic rings (not odd ones as in the C system). This leads to stable 90° angle on two-dimensional projections of the graphene-like hexagonal BN lattice, as frequently observed under an electron microscope for BNNT samples [61, 62].

Figure 6.3 (cont'd) *(c) sliding of the first cone-shaped cuplike structure occurs as a result of the strains created inside the cup, leaving a gap below the tip; (d) repetition of the process resulting in agglomeration of additional Fe-based alloy nanoparticles, which start to attach to other metal particles; (e) following the coalescence of various metallic particles, the seeding particle increases in size, resulting in an enlargement of the precipitated BN cups; (f) the metal agglomeration process continues, as well as the sliding of the cups, resulting in NT growth; (g) Fe-based particles reach a critical size that inhibits the formation of additional BN layers and the growth process is interrupted. (h-i) SEM images showing BNNTs produced after annealing amorphous BN powders obtained from h-BN powders ball-milled for 60 h; (j-k) for 72 h. Note that these filaments are always attached to a larger particle (indicated by white arrows). A TEM image of a typical BNNT end is also shown in the inset of (j). Squares denote the thin end of the tubular BN structure. Reprinted with permission from [89] Copyright (2005) Elsevier Ltd.*

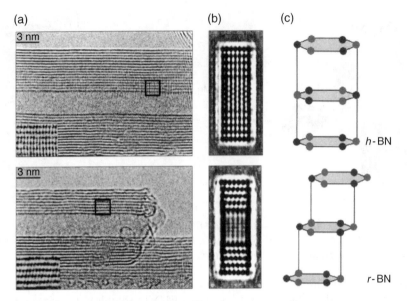

Figure 6.5 *(a) Experimental and (b) computer-simulated HRTEM images highlighting two possible types of layer stacking in multi-walled zigzag BNNTs. The insets display the magnified portions of the wall fragments in which hexagonal type (a) or rhombohedral type (b) of atomic column arrangements are visible. (c) The corresponding structural models. Reprinted with permission from [102] Copyright (2000) American Institute of Physics.*

The whole BN tubular system is described by the overall octahedral symmetry (not icosahedral as in the case of carbon system). For example, BNNT caps frequently look flat rather than conical as in the case of CNTs. Of course, the appearance of a particular defect is determined by the interplay between the chemical and elastic factors/constraints. In fact, the even-membered rings lead to high deformations of a graphene-like sheet and, hence, to larger elastic energies. As a result, conical features, such as flag-like terminations, or conical caps have also been observed by various researchers in BNNTs [99–101]. Another important trend is a sort of preferred stacking order between shells of multi-walled BNNTs due to the so-called 'lip-lip' interactions: the B atoms prefer to have N atoms, but not homo-elemental B ones, in the consecutive tubular shells. The same is true for the N atoms. Either hexagonal or rhombohedral types of stacking could exist in BNNTs, as shown in Figure 6.5, commensurate with hexagonal or rhombohedral boron nitrides [102]. Both orders result in perfectly straight, needle-like BNNT shapes [102, 103].

A careful study using the Rietveld phase analysis of a typical BN tubular product showed that a ratio of hexagonal to rhombohedral ordering types within nested BN shells is 73.6%/21.2% [104]. Clearly, such stacking would be not possible in cylindrical tubes (but very locally) because of a disorder caused by various lengths of the shell peripheries. Therefore, in many occasions it appears energetically preferable to have faceted BNNT cross-sections rather than cylinder-like, such facets are joined through defect lines, for example, dislocations or stacking faults. Polygonized cross-sections in BNNTs were indeed documented through the preparation of cross-sectional BNNT samples

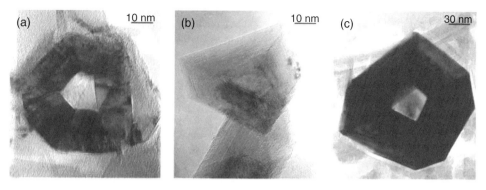

Figure 6.6 *(a) to (c): TEM images of multi-walled BN nanotube cross-sections revealing polygonal morphologies. The tubes were produced using a floating catalyst method. Reprinted with permission from [105] Copyright (2007) American Chemical Society.*

[105], as shown in Figure 6.6. It is noteworthy that such cross-sections are rather rare in C systems, albeit they were also detected.

The existence of a sort of a stacking trend in BNNTs also led to the shell helicity selection criteria, as shown in Figure 6.7. In the C system all layers of a given multi-walled nanotube could be freely rotated one upon another and with respect to the growth axis, thus imposing various helicities. In contrast, in the BN case, once a tubular layer is crystallized in some particular fashion (either zigzag, armchair or a certain type of helix), the consecutive layers try to accommodate themselves due to the B-N interactions persisting throughout the whole tube. Therefore in many cases, researchers have documented a very uniform chirality in BN tubes [106–113]. There has been a dominant preference for the zigzag type ordering, especially for BNNT products synthesized at high temperature (see Figure 6.7). However, the other high-symmetry case (armchair), has also been found.

The preferential BNNT zigzag atomic configuration has been observed in various BN tubular products by different researchers. Golberg *et al.* showed that such assembly is natural for BNNTs produced by BN target laser heating under super-high N_2 pressure [64], and for BNNTs obtained via high-temperature chemical syntheses [106–109]. Preferential zigzag arrangement of BNNTs was later observed by Bourgeois *et al.* [101] in BN tubular fibers synthesized by heating BN powders. Other scientists observed armchair and zigzag configurations being nearly equally in population [108, 109]. Theoretical predictions based on the molecular dynamics calculations performed by Blase *et al.* [114] favored the growth of armchair BN tubes rather than zigzag ones. The authors found that armchair edges tend to entirely eliminate dangling bonds, creating square-like defects and closing the BNNTs, whereas the zigzag edges possess dangling bonds that may close and reopen. It is noted that the significant difference between experimental synthesis temperatures, ~4700°C (expected during laser ablation at superhigh pressure [106]) and 1500–1700°C [107], and that used for Blase's calculations, ~2700°C, and the fact that because of computational limitations only a single-wall BNNT could be considered, should be taken into account for explaining this discrepancy. In addition, later on, Menon and Srivastava [115] calculated that for the most frequently observed

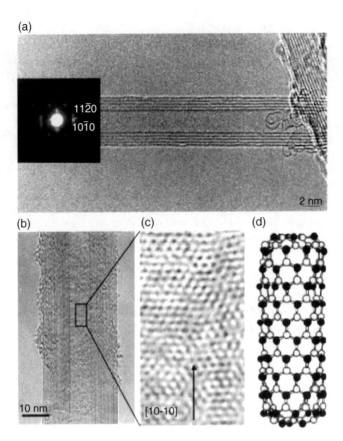

Figure 6.7 *(a) and (b) HRTEM images of multi-walled BNNTs; (c) enlarged image of (b) showing the spatially-resolved structure of part of the tube wall when oriented perpendicular to the electron beam (i.e., the superimposed graphene-like BN layers are seen face on); (d) structural model of the most common zigzag BNNT. The inset in (a) demonstrates a representative electron diffraction pattern from an individual BNNT; this is a characteristic of the zigzag tubular shells. The standard diffraction spots are marked. Reprinted with permission from [106] Copyright (1999) American Institute of Physics (a) and from (77) Copyright (2001) American Chemical Society (b and c).*

flat BNNT caps the zigzag morphology appeared to be energetically preferable. The latter conclusion is in line with the majority of experimental HRTEM and electron diffraction data obtained by different research groups, including those of the present authors. In order to explain these experimental-theoretical discrepancies in BN tube chiralities, one should take into account the fact that chiralities should be strongly related to the methods used and processing temperatures. However, further studies are needed. For example, a typical electron diffraction pattern recorded on a multi-walled BNNT is shown in Figure 6.7a, which reveals a zigzag orientation of all five tubular shells. The existence of preferred growth orientation for all shells frequently leads to nicely visualized patterns from the top and back tube wall portions recorded in an

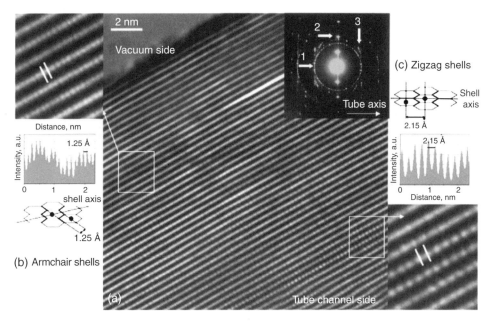

Figure 6.8 *(a) A HRTEM image of a multi-walled BNNT wall fragment revealing various atomic column separations in the innermost and outermost tube parts; the one in (b) corresponds to armchair layers; whereas the one in (c) to zigzag layers. The inset in (a) shows an electron diffraction pattern possessing common characteristics of a polygonal cross-section nanotube, as marked with the arrow '3'. Arrows '1' and '2' point at the features normally seen in standard cylindrical cross-section layers: the group 1 of reflections originates from the front and back nanotube portions; group 2 – for the side portions of a tube. Reprinted with permission from [116] Copyright (2007) Springer Science + Business Media.*

electron microscope under high-resolution, as shown in Figure 6.7b-d. Such patterns are not often seen in pure C system because of possible misorientations between consecutive layers. It is worth noting that the zigzag atomic arrangement in B-substituted CNTs (B*CNTs) was also found to be dominant. The zigzag BNNTs were repeatedly observed by Terauchi *et al.* [92], Ma *et al.* [77], Demczyk *et al.* [112], and Lee *et al.* [67] in BNNTs prepared via different synthesis routes. Thanks to the ~1 Å resolution now reached by modern electron microscope instruments, it was found for example that a typical BN nanotube material produced under the floating catalyst method may have various helicities within the different tubular shells in multi-walled BNNTs [116], as displayed in Figure 6.8. However, several authors agree on a characteristic grouping of experimentally determined helicities of all concentric shells of a given multi-walled nanotube around one single crystallographic orientation. This was presumed to be typical for BNNTs, as opposed to CNTs.

This leads to the specific appearance of diffraction patterns possessing the whole range of zigzag and armchair-like reflections and, in addition, the *101* reflections forbidden in cylindrical, concentric tubes (Figure 6.8a, inset, marked set 3). In line with a general theory of the electron diffraction [117–119], such patterns are the fingerprints of

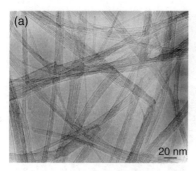
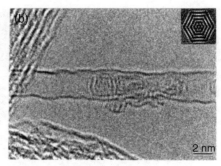

Figure 6.9 *(a) Multi-walled BNNT ropes; (b) a single-walled BNNT with an encapsulated octahedral BN fullerene; both structures were produced using a substitution reaction technique (reprinted with permission from (74) Copyright (1999) Elsevier Ltd).*

the tubule polygonization. Indeed, polygonization allows the stacked graphene-like BN layers to be coherently related according to the ABAB Bernal (hexagonal) or the ABC (rhombohedral) stacking, whose scattering characteristics is to generate reflections typical of 3D periodicity (*hkl* reflections with $l \neq 0$ other than *00l*).

The formation of defects in BNNTs could also be induced by the electron irradiation inside an electron microscope, which to date remains the best tool for the detailed analysis of NT structures. Zobelli *et al.* [46] demonstrated that divacancies had been created under such electron irradiation. Clustering of such vacancies was energetically stable. In turn, these clusters led to the change in chirality of a given tube, or could even deform it. Golberg *et al.* [102] experimentally observed a change in an individual BN tube helicity from armchair to zigzag in a kinked three-layered BN tube having the defects at the kink. These defects do not significantly modify the electronic bandgap but introduce acceptor states in it. Therefore, it is suggested that even defective BN tubes should remain electrically insulating. By contrast, BN tube flattening may lead to the dramatic changes in the conductivity. Kim *et al.* [120] computed that the bandgap of pristine (no chemical doping) individual single-walled BNNT should be reduced from ca. 5 eV to ca. 2 eV in a collapsed or entirely flattened tube. Such effect was calculated to take place only in zigzag tubes, but not in the armchair ones.

It is important to note other BNNT morphologies, for example, the multi-tube ropes [107] and single-walled (SW) BNNTs [67,74,75], as presented in Figure 6.9. The ropes may contain dozens of nanotubes packed together in a honeycomb-like fashion due to van der Waals interactions in a way analogous to pure SWCNT bundles. However, SWBNNTs have been rarely reported. Such an individual SWBNNT is shown in Figure 6.9b. The tube was produced using the substitution reaction method. It had an encapsulated nested octahedral BN fullerene similar to the peapod (see Chapter 5) structure in CNTs [121].

As we already mentioned, a laser ablation method seems to be capable of producing notable yields of such SWBNNTs but further process optimization and clear criteria for distinguishing single-walled structures from double-layered within the densely packed bundles should be obtained.

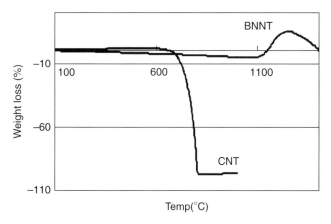

Figure 6.10 *Comparative thermogravimetry curves recorded in air for multi-walled CNTs prepared by CVD and BNNTs fabricated using a floating catalyst method.*

6.2.4 Properties of BN Nanotubes

6.2.4.1 Thermal Properties

BNNTs inherited a high thermal/oxidation stability of layered BN phases often used for crucible materials working at temperatures of 1000°C or even higher. The thermal and oxidation stabilities of BNNTs significantly exceed that of standard CNTs [122,123]. Figure 6.10 shows comparative thermogravimetry (TG) in air of CVD-grown multi-walled CNTs and BNNTs [122].

Oxidation of arc-produced CNTs starts at ~600°C (as revealed by the corresponding dramatic weight losses for the TG curve in Figure 6.10) whereas the degradation of BNNT occurs at much higher temperature (~1100°C). A control of the BNNT oxidation process reveals that B_2O_3 starts to form above ~1000°C and, then, it sublimes above 1200°C. Therefore, the use of BNNTs is preferred for specific device applications at high temperatures and in chemically active, and/or hazardous environments. Thermal and chemical stability of BNNTs may also be of importance for the performance of NT-based field-emitters in flat panel displays and in field-emission (FE) tips for scanning tunnelling (STM) and atom force (AFM) microscopes.

It has also been theoretically predicted that both CNTs and BNNTs are excellent thermal conductors. It has indeed been verified that CNT possesses high thermal conductivity values. Kim *et al.* [124] measured that the thermal conductivity of CNTs is 3000 W mK^{-1}, the value which rivals that of diamond. Values even higher than 3000 W mK^{-1} (e.g., up to 6000 W mK^{-1}) have been theoretically predicted by Tomanek's group [11]. Theoretical calculations demonstrated that the phonon scattering mechanism of BNNTs is similar to CNTs. However, BNNTs may have some alternations in a mean free path of electrons. Therefore, there have been intuitive expectations that the thermal conductivity of BNNTs may be comparable to that of CNTs.

However, thermal conductivity *(k)* measurements conducted on BNNTs and isotopically [11]B-enriched BNNTs have demonstrated that isotopically pure BN exhibits a value in the range of 350 W mK^{-1} [125]. The data was obtained at room temperature for BNNTs having a 30–40 nm outer diameter. Hence, a dramatic dependence of k on the isotopic disorder was found with a room temperature enhancement in k of 50%, the largest for any material known.

Recently, Zettl's group [126] has demonstrated that BNNTs (as well as CNTs) mass-loaded externally and inhomogeneously with heavy molecules, for example, $C_9H_{16}Pt$, possess asymmetric axial thermal conductance with a greater heat flow in the direction of decreasing mass density. The authors proposed that solitons may be accounted for the observed phenomenon. BNNT thermal rectifiers were proposed based on these results. The devices were suggested to have potential applications for fabricating nanoscale calorimeters, microelectronic processors, macroscopic refrigerators and energy-saving buildings.

6.2.4.2 Wetting Properties

For many years it was assumed that the wetting properties of graphene-based CNTs and BNNTs should markedly differ. However, recently Yum and Yu [127] have measured the wetting properties of BNNTs and found that these two types of tubes have very similar properties. The authors applied the standard Wilhelmy method while analyzing the interactions between multi-walled BNNTs and various liquids, including bromonaphtalene (nonpolar liquid), poly(ethylene glycol), glycerol and water (polar liquids). The measured contact angles of BNNTs were found to be comparable to those of CNTs, but slightly larger. For water, either partial wetting or non-wetting behaviors were observed. The total surface tension of BNNTs was estimated to be 27 mN m^{-1}, which is comparable to CNTs (27.8 mN m^{-1}) or surface-untreated, micron-sized, polyaromatic carbon fibers (31.5 mN m^{-1}). In addition, molecular-dynamic simulations have revealed that BNNTs do not offer any significant functional advantages over their CNTs when used in fluid-conduit applications [37].

6.2.4.3 Electrical Properties

Despite the fact that structurally BNNTs are very similar to CNTs, from the electronic standpoint they are totally different. Due to a partially ionic origin, a BNNT bandgap of ~5.5 eV is almost independent of tube morphology (see above), and normally BNNTs are insulating. However, flattening deformation [120] or decrease in tube diameter down to 0.4–1 nm or less [128], or application of a transverse electric field (Stark effect) [28,48] may dramatically reduce the bandgap, as revealed by several theoretical calculations. For example, using the local density approximation (LDA) for a single-walled armchair (22,22) BNNT with an intrinsic bandgap of 4.5 eV, a transverse electric field of 0.1 V Å^{-1} was found to reduce the bandgap to 2.25 eV, whereas a field of 0.19 V Å^{-1} could even eliminate the bandgap entirely.

Kral et al. [10] calculated that an electrical current should appear when a BNNT is exposed to the polarized light. Mele and Kral [129] predicted a piezoelectric behavior in BNNTs and calculated the piezoelectric constants which were found to be saturated in BNNTs. Meunier et al. [20] found that the polarization of an electric field can be used to lower the work function of BN/C NT tips and thus to enhance the field-emitting properties compared to blank CNT tips by up to two orders of magnitude. When BNNTs experience a high DC electric field, the electronic density of states is also modified. In this context, Khoo et al. [130] performed ab initio calculations within the generalized gradient approximation (GGA) of the transverse electric field effects on the BNNT energy gap. The gap was found to decrease linearly with increasing electric field. For a given DC field, the electric-field gap reduction was found to be larger for larger diameter NTs. This effect was independent of NT chirality. All these interesting theoretical predictions have stimulated the interest and paved the way to the real BNNT electrical property measurements.

The first experiment on practical evaluation of transport *I-V* and field emission (*FE*) of an individual pure multi-walled BNNT was reported by Cumings and Zettl [131] during which an electrically insulating (expected a priori) BNNT surprisingly demonstrated notable *FE* currents at relatively low voltages. In order to further evaluate the transport and *FE* prospects of BN-based NT devices, Golberg *et al.* [132] studied *I-V* and *FE* characteristics of the individual multi-walled BNNT ropes in a Fresnel projection electron microscope (also called low energy electron point source microscope, or LEEPS) [132–135]. Since the latter tubes were produced through the above-mentioned substitution reaction, a marginal amount of C (about ~5 at.%) was present in them. *FE* properties of aligned NT mats composed of multi-walled BNNT mats were also analyzed [132]. The *FE I-V* curves were independent of the constituting tube morphology, for example, homogeneously-structured BNNTs (with a C marginal doping throughout) and clearly phase-separated BN-C NTs revealed nearly analogous curves. *FE* curves recorded on multi-walled BNNT bundles (composed of several individual tubes) displayed currents up to ~2.5 μA. This is fairly close to the maximum stable *FE* currents (~2 μA) reported for individual CNTs [136]. When plotted in Fowler-Nordheim [137,138] coordinates, the *FE* curves fitted lines at low voltages proving that a conventional *FE* process was taking place. Basically, BN-rich NT ropes may be regarded as reasonable field-emitters with the emission characteristics comparable to those of standard CNT emitters. While BN emitter ability to produce a high *FE* current is similar to that of a CNT emitter, the BN-based NT ropes exhibit better environmental stability during *FE*. This is due to a higher thermal and chemical stability of BN-rich compositions when compared to C-rich ones. This opens up a new horizon for BNNTs applications in flat panel displays and/or *FE* tips for the new generation of STM and AFM microscopes.

The presence of B-B or N-N defects in BN tubes should create new energy levels within the bandgaps. Under external electric fields, a BN host may provide electrons into these newly formed levels. The field is concentrated at the tip of a tube, so that the electrons can be emitted into vacuum [9]. Such *FE* could work at very low external fields thus explaining rather good experimentally observed *FE* properties of BNNTs.

Field-effect transistors constructed from SWBNNT bundles (produced by a continuous CO_2 laser ablation) were investigated by Radosavljevic *et al.* [139]. The transistors were found to be of *p*-type. The current injection near room temperature was controlled by thermionic emission over Schottky barriers at the tube-metal contacts, unlike the usual case of CNT FETs in which carrier tunneling dominated. The Schottky barrier heights for BNNT-Ni contacts were estimated to be ca. 250–300 meV. FETs on SiO_2-covered Si substrates were also prepared from multi-walled BNNTs with or without semiconducting fillings [140]. They were also found to be of *p*-type.

A growing interest in electrical probing of BNNTs has stimulated the use of piezo-driven STM-TEM stages incorporated in TEM instruments [141–143]. The major advantage of this technique (compared to the standard transport measurements within electrical circuits preformed via optical or *e*-beam lithographies) is a unique possibility to fully characterize a given NT prior, during and after the *I-V* measurements. This is because all standard TEM operations (e.g., high-resolution microscopy at the ultimate atomic resolution, nanobeam electron diffraction and spatially-resolved chemical composition mappings using EEL or X-ray dispersion spectrometers) are selectively performed on the tube that is tested. By contrast, all previously utilized experimental setups within SEM or AFM instruments have suffered from limited magnifications and thus spatial resolutions. Series of in situ TEM informative experiments using piezo-stages was early performed on standard CNTs [141, 142], but the analogous measurements on BNNTs, including those filled with various substances, have only recently been performed by Golberg *et al.* [143]. These authors electrically tested BNNTs filled with magnesium oxides [MgO, MgO_2] and/or hydroxide [$Mg(OH)_2$] using a side-entry TEM-STM 'Nanofactory

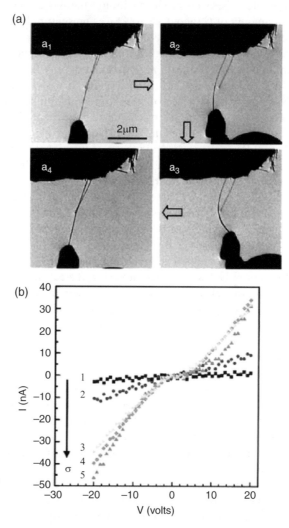

Figure 6.11 *(a) TEM images demonstrating individual multi-walled BNNT bending inside TEM using a piezo-driven TEM stage (reprinted with permission from [140] Copyright (2007) American Chemical Society); (b) consecutive I-V curves recorded under an increase in BNNT bending curvature from plots 1 to 5. The current passing through the tube is markedly increases while the tube is bent.*

Instruments AB' piezo-holder [143]. At a low bias, the BNNTs were insulating, as in the case of the measurements reported by the Berkeley group. However, at a high bias of ±30 V they showed reversible breakdown current of several dozens of nanoamperes. Under a 300 kV electron beam, BNNTs were positively charged (due to the prominent electrostatic BNNT-electrode interactions), and that allowed the authors to perform on-demand manipulation with the tubes by tuning the polarity and/or the value of a bias voltage on a gold counter-electrode (from −140 to +140 V).

Using the same holder, Bai *et al.* [140] for the first time verified experimentally a striking deformation-driven electrical transport in an individual multi-walled BNNT. Figure 6.11 shows that the insulating character of an individual BNNT has surprisingly changed under bending deformation between two gold contacts inside a TEM. A current of several dozens of nA (Figure 6.11b) may then pass through the tube when bent (Figure 6.11a). Even more surprisingly, such a transport was found to be fully reversible and could almost entirely disappear after the tube reloading. A sort of hysteresis always existed on the *I-V* curves recorded for bent BNNTs. This implies that the polarization under tube deformation plays a key role and the related piezoelectric phenomena are important for the observed BNNT electrical transport.

The insulating properties of BNNTs may also be changed by atom substitution [144], as highlighted in Figure 6.12. The incorporation of 4 at.% of fluorine into the graphene-like

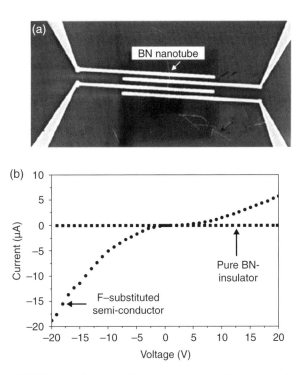

Figure 6.12 (a) A SEM image of an experimental setup used for four-probe electrical transport measurements on an individual BNNT; (b) comparative I-V curves taken at a pristine BNNT and on that substituted with 4 at.% of fluorine. The insulating behavior of BNNTs changes to the semiconducting one after substitution. Reprinted with permission from [144] Copyright (2005) American Chemical Society.

BN lattice of a pristine multi-walled BNNT made it semiconducting, as it was confirmed using four-probe electrical measurements on lithography-patterned Si/SiO$_2$ substrates. It is also possible to reduce the electronic bandgap by substituting B and N atoms by C atoms within the tubular BN structure.

6.2.4.4 Mechanical Properties

The comparative mechanical properties of single-walled C, BN, BC$_3$ and BC$_2$N nanotubes were theoretically treated by Hernandez *et al.* [145]. When using a nonorthogonal tight-binding model, the authors found that BNNTs have a 30% lower Young modulus compared to CNTs. The BNNT mechanical properties were first experimentally measured by Chopra and Zettl [146]. The Young modulus of an individual BNNT was measured using a thermal resonance technique to be ~1.1–1.3 TPa. Therefore it could be stated that BNNTs are possibly the stiffest insulating filaments ever produced. Later on, Suryavanshi *et al.* [147], using an alternating current (AC) and a mechanical resonance setup, showed values of 0.7 TPa (Figure 6.13). The authors assigned this discrepancy to the different defect status of the tubes. The former tubes in Chopra and Zettl's experiments were prepared via an arc-discharge technique, whereas those tested in the Suryavanshi *et al.*'s experiments were fabricated using a floating catalyst method.

The elastic behavior does not fully characterize the strength of a material, and calculations in a plastic region should be also performed. Dumitrica *et al.* [24] have shown that in spite of a slightly lower elastic modulus of a single-walled BNNT when compared to a CNT, BNNT resistance to thermal degradation can surpass that of CNTs. The calculations have

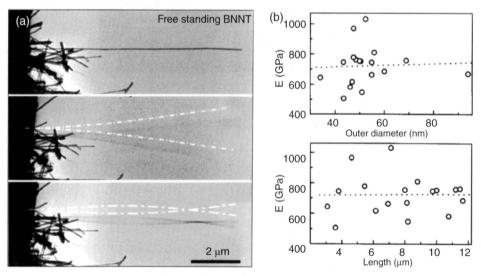

Figure 6.13 (a) TEM images showing an individual BNNT cantilever blurring under applying an AC voltage to it; (b) plots highlighting the dependence of the measured elastic modulus versus the tube length and diameter. The average elastic modulus obtained was 0.7 TPa. Reprinted with permission from [147] Copyright (2004) American Institute of Physics.

suggested that mechanically induced defects are similar to those found in CNTs, that is, at a high temperature a 5/7/7/5 dislocation dipole, the so-called Stone-Wales defect that contains two pentagons and two heptagons, must occur under the deformation through the bond rotations. Such defects were predicted to exhibit the lowest energy in a honeycomb BN lattice. It is worth noting that Stone-Wales defects could result in the formation of unfavourable B-B and/or N-N bonds. This feature makes the BN structure energetically more unstable when compared to sp^2- hybridized carbon. At room temperature and moderate deformations, CNTs appear to be stronger. When the first Stone-Wales defect is created under large strains, a CNT tube failure occurs. However, at higher temperatures (or extremely long deformation times, like during creep tests), the situation is reversed and BNNTs become more stable thermodynamically, although Stone-Wales defects still occur faster in BNNTs than in CNTs.

Qualitative and quantitative experiments related to the elastic and plastic deformations of individual BNNTs were carried out by Golberg *et al.* [105,148]. Individual multi-walled BNNTs were found to be very flexible and elastic while deformed between the two gold leads inside a HRTEM using piezo-driven 'Nanofactory Instruments AB' STM-TEM and AFM-TEM stages. The characteristics of an individual BN deformation under compression are shown in Figure 6.14. In spite of the significant corrugation and buckling of BN shells under severe bending deformation to ~70° over dozens cycles, the perfect BNNT shape was remarkably restored after the load release.

Direct measurements of the elastic modulus of multi-walled BNNTs inside TEM (using micro-Si cantilever) were also performed inside TEM on thin and thick BNNTs [105] (Figure 6.15). The values of 0.5–0.6 TPa were obtained. These numbers are in good agreement with those measured by the AC-assisted resonance technique inside TEM on multi-walled BNNTs, but are much smaller than those measured by Chopra and Zettl [146] in their pioneering work.

The stiffness and plasticity of BNNTs under compression were also investigated using generalized tight-binding molecular dynamics and ab initio total energy methods [14]. Due

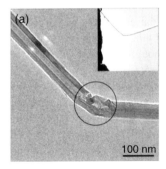 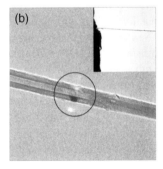

Figure 6.14 *HRTEM images displaying individual BNNT appearance after (a) severe in situ bending inside the TEM and (b) the pressure release; the insets show the corresponding low-magnification TEM images of a tube. The tube fully restores its shape independent of the notable corrugation of all constituting layers. Reprinted with permission from [148] Copyright (2007) Elsevier Ltd.*

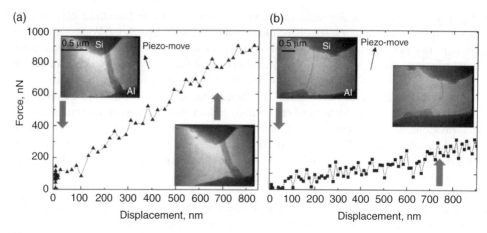

Figure 6.15 *Experimental force-displacement curves recorded inside the TEM using a piezo-driven AFM-TEM holder under bending of (a) a relatively thick and (b) a thin individual BNNT. The average elastic modulus was measured to be 0.5–0.6 TPa, thus being consistent with the data of indirect AC-resonance measurements previously shown in Figure 6.13. Reprinted with permission from [105] Copyright (2007) American Chemical Society.*

to the BN bond rotation effect, compressed zigzag BN nanotubes were found to undergo anisotropic strain release followed by anisotropic plastic buckling. The strain was preferentially released towards N atoms in the rotated BN bonds. Experimentally, kinking of BNNTs was observed by Golberg *et al.* [105] under direct compressive tests inside TEM. The kinks appeared on loading and disappeared on reloading in the exactly inverse sequences (Figure 6.16). Surprisingly, these kinks appear to be fully elastic with no traces of residual plastic deformations after the pressure was released (independent on the number of bending cycles).

Mechanical deformations of BNNTs may affect their electronic properties due to changes in the hybridization of orbitals. Kim *et al.* [120] proposed that at a high pressure BNNTs should collapse. In particular, Kim's calculations (ab initio approach with the LDA) demonstrated that for a (9,0) zigzag BNNTs, the compression causes a reduction of the bandgap from 3.5 eV to 1.0 eV, as the diameter of tube was flattened from 0.74 nm to 0.20 nm. The authors estimated that 10 GPa would be enough to fully collapse a BNNT of 3 nm in diameter. Such tubes revealed bandgaps of ~2 eV. The energy gap was found to be less pressure-sensitive for armchair BNNTs. That was reduced by only ~0.6 eV when a BNNT was flattened below the van der Waals interaction distance. From the experimental standpoint, a comparative high pressure Raman study of BNNTs up to 16 GPa and *h*-BN up to 21 GPa, was carried out at room temperature in a diamond anvil cell by Saha *et al.* [149]. BNNTs produced using a floating catalyst method underwent an irreversible phase transition at pressures of ca. 12 GPa, which is different from the values reported for CNTs (ca. 20 GPa) [150].

6.2.4.5 Magnetic Properties

Weak ferromagnetism has been observed in many graphene-based C nanosystems, such as fullerenes, tubes, foams and so on. Theoretical calculations predict that when a zigzag BNNT contains two carbon atoms replacing one B and one N atom, spin polarization effects

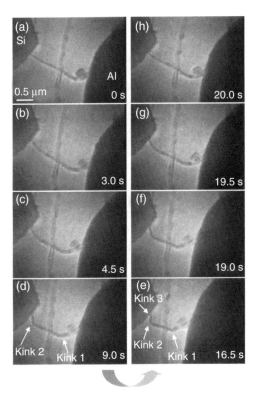

Figure 6.16 *Consecutive video frames depicting the stages on an individual multi-walled BNNT deformation inside TEM. The positions of the consecutively appearing kinks (1, 2 and 3) are marked. The kinks entirely disappear on reloading in the exact inverse sequence as they have appeared on bending and within a millisecond time frame. Reprinted with permission from [105] Copyright (2007) American Chemical Society.*

can be observed in tubes with diameters larger than those of (7,0) tubes [36]. Kang [42] has confirmed that BNNTs with holes (due to the extraction of one or two B atoms) display magnetism, if they possess a zigzag structure. This magnetism disappears for armchair BNNTs according to Kang's calculations. It was demonstrated that strong magnetic signals could be induced in BNNTs having open tips [51], a fact that has been frequently observed experimentally. The magnetic moments were found to be sensitive to the NT chirality in B-rich and N-rich tubes. In addition, B-rich ended BNNTs possess a high stability of the local spin configuration, thus implying possible utilization in future spintronic devices. The tremendous spin-splitting (>1 eV) in such pure BN *sp*-electron systems was attributed to the presence of 'conjugated' spin-polarized deep-gate states.

6.2.4.6 *Optical Properties*

The first Raman and time-resolved photoluminescence spectroscopic studies of multi-walled BNNTs were carried out by the Berkeley group [151]. Photoluminescence experiments on multi-walled BNNTs revealed the existence of a spatially indirect bandgap; result in agreement with theoretical predictions. The optical properties of BNNTs have been calculated on the level of random-phase approximation (RPA), that is, in the frame of independent particle excitations [35]. Two contradicting studies related to the optical properties of BNNTs have been reported, where the observed optical features were interpreted differently [152,153]. In an attempt to explain these discrepancies, Wirtz *et al.* [44] concluded that the optical absorption spectra of BNNTs are dominated by strongly bound excitons. The calculations by Park *et al.* [43] have also shown that excitonic effects are important in BNNTs (even more than in CNTs), and predicted an exciton with a binding energy larger than $2\,eV$ for a (8,0) zigzag BNNT. Unlike the case of CNTs, these exciton states were found to consist of coherent superposition of transitions from several various sub-band pairs, thus giving rise to unmatched optical features. In addition, BNNTs have been predicted to exhibit large second-order nonlinear optical behavior with the second-harmonic generation, and linear electro-optical coefficients being up to 30 times larger than those of a bulk layered BN. This suggests bright prospects for nonlinear BNNT optical and optoelectronic applications [35]. There has been a general agreement in the literature that BNNTs may be ideal candidates for optical devices working in the UV regime. The photoluminescence quantum yield of BNNTs was thought to surpass that of CNTs. In fact, the intense and stable UV emission has recently been observed for multi-walled BNNTs first by Wu *et al.* [151], followed by Zhi *et al.* [154].

6.2.5 Stability of BN Nanotubes to High-Energy Irradiation

BN materials may find smart applications in the fabrication of protective shields against high-energy particles, especially under appropriate B isotope enrichment. The stability of BNNTs was studied under electron beam irradiation in TEM. In addition, the effects of other high-energy sources such as Ar and He ions have also been recently elucidated [155].

BNNT shells were found to be rather sensitive to $300\,kV$ electron beams. At low irradiation doses, corresponding to normal imaging conditions inside TEM of $1–2A\ cm^{-2}$, the annealing of NT defects takes place [70]. For example, with an increase in time of electron irradiation on imperfect and defective BN layers striking changes in the plane morphology occurred. The tubular BN sheets became straight and aligned, and nearly a complete ordering of the NT shells took place. The observed movement of the dislocation-like defects under electron beam irradiation suggests the temporary presence of sp^3 bonding between the BN shells, as well as a stress gradient in the tubular walls.

In order to distinguish growth defects from those resulting from the electron beam irradiation, BNNTs were irradiated for longer times [156]. The results were similar to those observed in CNTs: longer irradiation of the entire illuminated BNNT region at a dose of $\sim30A\ cm^{-2}$ over $90\,min$ led first to a decrease in the 0002 fringe contrast sharpness and, then, to the complete destruction of NT morphology, leaving the material with the consecutive appearance of an amorphous-like BN rod and a BN rectangular 'onion'-like nanoparticle, a morphology typical in the BN systems with the octahedral symmetry.

6.2.6 Boron Nitride Meta-Nanotubes

6.2.6.1 Doped BN Nanotubes

As for CNTs (see Chapters 2 and 5), BNNTs could be doped with fluorine [144], with a maximum possible content of ca. 4 at.%. Fluorine atoms were suggested to be uniformly distributed within the outer tubular walls. This assumption was based on the tube elemental

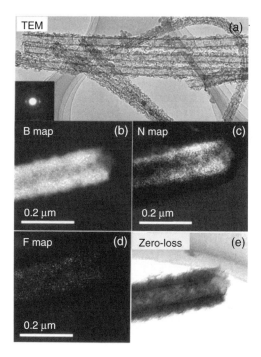

Figure 6.17 *TEM images of (a) a fluorinated BNNT rope and (e) an individual fluorinated BNNT; (b,c,d) corresponding B, N and F elemental maps displaying the uniform distribution of all constituting species throughout the whole tube. Reprinted with permission from [144] Copyright (2005) American Chemical Society.*

mapping images (Figure 6.17). Marginal F-doping was found to convert the insulating BN tube into a semiconducting one. Other doping examples (in the sense as defined in Chapter 2) of BNNTs have been scarcely reported or investigated so far.

Liu *et al.* [60] has studied the structural and electronic effects of fluorine-doped multi-walled BNNTs through ab initio calculations for an ensemble of two zigzag nanotubes, namely, (8,0) in (16,0). It was computed that F atoms could be exothermically doped into the interstitial region between two adjacent tubes or adsorbed on the exterior surface of the outmost tube. Interstitial F doping can significantly change the inter-wall interactions and lead to prominent electrical changes. Both tube walls turn into effective conducting channels, thus enhancing the overall electrical conductivity of the system. In contrast, the F adsorption on the exterior surface of the outmost tube hardly alters the inter-wall interactions and only affects the structural and electronic properties of the outer tube wall.

6.2.6.2 Functionalized BN Nanotubes

To date a vast majority of research efforts have been placed on chemical functionalization of conventional CNTs but not on BNNTs. For instance, in case of CNTs, it was established that functionalization offers an effective route to enhance the solubility of NTs in a given solvent and to diminish the aggregation into bundles. Several investigations have been performed on the sidewall fuctionalization of BNNTs. Then Xie *et al.* [157] reported on the development of soluble multi-walled BNNTs by amine-terminated oligomeric poly(ethylene

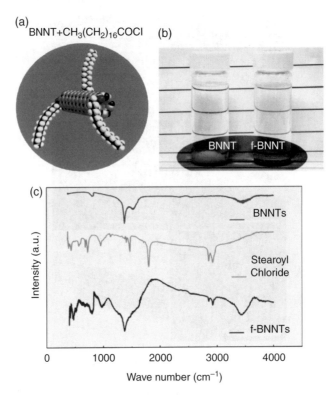

Figure 6.18 *A scheme of the covalent functionalization of a BNNT; (a) model of the functionalized BNNT obtained; (b) precipitation of pristine BNNTs in chloroform (left) and fully transparent suspension of multi-walled BNNTs in chloroform obtained after covalent functionalization of the tubes (right); (c) FTIR spectra verifying the occurrence of chemical bonds between the BNNT surface and long-chain molecules (i.e., stearoyl chloride).*

glycol) surface groups. Zhi *et al.* [158–162] synthesized stearoyl chloride-functionalized BNNTs through the interactions of COCl- groups and amino groups on multi-walled BNNT walls (Figure 6.18).

In addition to covalent functionalization, so-called non-covalent functionalization through wrapping pure multi-walled BNNTs with polymers [159] was also successfully performed by the same authors (Figure 6.19). It was found to be effective for the overall BNNT purification, as mentioned in the previous section, as well as for obtaining uniform suspensions of functionalized multi-walled BNNTs. For instance, Figure 6.19b depicts a series of transparent solutions of BNNTs obtained via noncovalent functionalization. The functionalized BNNTs could be perfectly soluble in many solvents.

The functionalization may also tune the band structure of BNNTs. Wu *et al.* [47] investigated theoretically the effects of functionalization of BNNTs with NH_3 and four other amino-functional groups (NH_2CH_3, $NH_2CH_2OCH_3$, NH_2CH_2COOH and NH_2COOH) using DFT calculations. The authors found little changes in the electronic structure of BNNTs. However, the chemical reactivity of tubes was enhanced due to the –COOH amino groups.

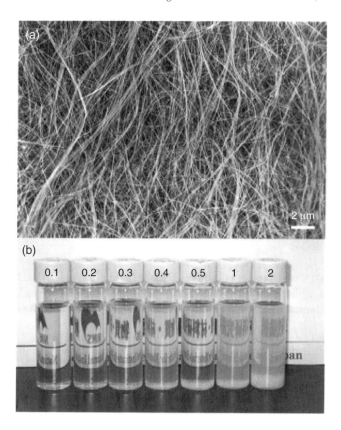

Figure 6.19 *Polymer-wrapping of multi-walled BNNTs fabricated using a floating catalyst method; (a) Representative SEM image of the as-prepared pure phase starting material; (b) a continuous row of transparent solutions of the BNNTs in 20 ml of chloroform (containing 5 mg of poly(m-phenyleneviylene)) under an increase in the BNNT loading fraction (an increasing concentration of BNNTs in the solutions in milligrams is marked on the vessel caps from left to right). As the BNNT concentration is increased some precipitation becomes visible in the suspensions.*

In contrast, functionalization with the –COCl groups was documented to have a dramatic effect on the BNNT electronic states, as confirmed by the UV-vis absorption spectroscopy. First-principle calculations have shown that BNNTs may become either *p*- or *n*-type doped, depending on the electronegativity of multi-functional molecules attached, and their energy gap could be adjusted from the UV to the visible optical range by varying concentration of functionalized species [163]. The bandgap decreased due to intertube coupling and multiple charge transfer from the B atoms to N atoms in BNNT bundles. The value of the bandgap change was dependent on the arrangement of bundles and chirality of each BNNT within a bundle [31].

Selective donor- or acceptor states in multi-walled BNNTs may also be created while functionalizing them with dyes [164]. For the effective functionalization of BNNTs it is important to control their sizes and geometry. In order to address this control,

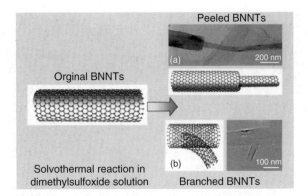

Figure 6.20 *A strategy towards (a) peeling and (b) branching multi-walled BNNTs using a solvothermal reaction in dimethylsulfoxide and the related experimental TEM images. For a better understanding of the figure, please refer to the colour plate section.*

Huang *et al.* [164] studied the possibilities of peeling and branching BNNTs during solvothermal reactions with the aid of dimethylsulfoxide (Figure 6.20). The work was successful in the sense that both desired morphologies were obtained, albeit at relatively low yields.

6.2.6.3 Decorated BN Nanotubes
The first work was published by Han and Zettl [165] in which the authors had stirred BNNTs in a $SnCl_2$ solution and obtained the full coverage of BNNTs with semiconducting SnO_2 nanoparticles of 1–5 nm in size. BNNTs may also be decorated with various nanoparticles, such as TiO_2, SnO_2, Fe, Au and so on. [166,167]. Such particles may also change the electronic properties of BNNTs [168], and/or serve as new catalyst supports for the secondary growth of various branched nanostructures on BNNT surfaces [167].

6.2.6.4 Filled BN Nanotubes
Filled BNNTs may be an exciting product for nanotechnological applications. A natural NT channel creates an attractive pathway for a variety of functional materials to be encapsulated (Figure 6.21). These nanochannels exhibit different types of physicochemical phenomena: (i) change in wetting properties of tiny encapsulates with respect to their bulk forms, (ii) variations in the melting points of the encapsulated materials, (iii) changes in the phase diagrams when compared to the bulk phases, (iv) novel high-pressure phases stabilized in the tubes cores, and so on. Contrary to multi-walled CNTs, BN tubes are electrically insulating and thus may also provide an ultimate oxidation-proof insulating shield for any placed conducting materials. The whole structure would then work as a nanocable. Initially, the energetic issues of C_{60} molecules encapsulation inside BNNTs were studied by Okada *et al.* [169] after the successful synthesis of 'peapod' structures in CNTs (see Chapter 5). The authors demonstrate that some energy gains may occur if fullerenes are encapsulated in BNNTs. Subsequently, during HRTEM imaging of straight, well-ordered multi-walled BNNTs produced through substitution reactions, an encapsulated, C-rich amorphous-like residue was accidentally observed inside the tube cores [170,171]. Under high-resolution imaging the residue was noticed to contain fullerene-like molecules. Their diameters roughly corresponded to the theoretical diameter of a C_{60} molecule ~0.7 nm. However, only recently the high-yield C_{60} and higher fullerene

Figure 6.21 *(a) to (f) TEM images showing a variety of filled multi-walled BNNTs using various filling compounds, including molten metal, solid metal alloy, and ceramics.*

filling (C_{70} etc.) of multi-walled BNNTs has been reported by the Berkeley group [172]. It was suggested that such BNNT/fullerene ensemble may create a novel nanostructural molecular device owing to an exciting possibility of shuffling/moving of C_{60} molecules inside a tube channel. Several possible switching processes were investigated taking into account various external force fields using molecular dynamic simulations [8,37].

In the case of using metallic oxides as a filling material, discrete metallic clusters were successfully encapsulated into BNNT channels. Golberg *et al.* [173] were able to embed discrete Mo clusters into BNNTs. The clusters exhibited diameters in the range of ~1–2 nm, and were deposited all along the whole length of NTs. Since the clusters did not wet the internal BN shells, they were free to move in the tubular channels thus creating natural 'pipe-lines' under applied thermal, magnetic or electric fields inside the TEM.

It is worth noting that in spite of an intriguing possibility of making insulating nanocables composed of continuous conducting metallic/semiconducting/ferromagnetic and insulating BN tubular shields, BNNTs may be hardly filled with any substance [1]. By contrast, the filling of CNTs with metal, metal oxides or metal chlorides by capillarity or wet chemistry methods has been known for more than a decade (see for instance [174–176] and Chapter 5). It is noted that well-established techniques for CNT filling do not work efficiently for BNNTs. Therefore, the research on capillarity-induced metal filling of BNNTs has not been very active. During conventional laser ablation or arc-discharge syntheses, metal-based nanoparticles (originated from electrode and/or crucible materials) were accidentally observed at the BNNT-tip-ends [4,61,62]. Corrugated BN tubular fibers with continuous ceramic cores made of SiC were produced using a thermo-chemical C template method [177]. Han *et al.* [178] embedded B$_4$C nanowires into BNNTs using a combination of CNT-confined and substitution reactions, and coated GaN nanorods with BN tubular layers [179]. Later, Han and Zettl [180] inserted potassium halide crystals into BNNT cavities and also cleaved these crystals under intense electron-beam irradiation [181]. There is also data available on the electrochemical deposition of Cu into thick BN tubular fibers [182]. The fabrication of metallic, ceramic or semiconducting nanowires inside tubular BN channels using post-synthesis high-temperature coating has been more successful when compared to wet post synthesis processes using capillarity [183–185] (see Figures 6.21 c-f).

The Bando-Golberg group at the National Institute for Materials Science in Tsukuba, Japan, invented an original way for implementing continuous BNNT filling with *3d*-transition metals by means of capillarity using a two-step high-temperature synthesis. Although the yield of the filled products was not high, the method provides a reliable pathway for inserting any metal inside BNNTs. Generally, the process involved: (i) preliminary soaking of metallic nanoparticles inside CNT channels during a hydrogen-assisted plasma CVD-process on metal substrates; (ii) a high-temperature treatment of the resultant products in B$_2$O$_3$ and N$_2$ (or NH$_3$) atmospheres during the C $\rightarrow$ BN conversion (during the substitution reactions the melting of the particles occurred) [186]. The molten material filled the tube internal channel by capillarity at high temperatures. A TEM micrograph of an Invar-filled BNNT using the latter technique is depicted in Figure 6.21a. Using either laser ablation or pyrolysis, composite structures may also be prepared: the external BN layers may cover the internal C layers, which together wrap the innermost conducting metallic nanowires, as was reported by Shelimov and Mockovits [182], Zhang *et al.* [187,188] and Ma *et al.* [189].

The properties of filled BNNTs are of special interest. From the theoretical standpoint, the electronic band structures of Cu-filled single-walled BNNTs were calculated by Zhou *et al.* [50]. The authors found that the band structure is actually the superposition of individual BN and Cu electronic bands. A charge density analysis shows that the conduction electrons are solely distributed on Cu atoms and a charge transport occurs along the encapsulated wires. The embedded Cu wires are perfectly insulated by a surrounding BN shell. Therefore, such Cu/BNNT hybrid system could work as a perfect nanocable, as explained in the beginning of this section.

As for CNTs, The BN tube core may also serve as functional nanocontainers for observing phase transformations at the nanoscale. An interesting approach for studying an inorganic compound encapsulted within a BNNT was developed by Golberg *et al.* while studying a thermo-driven behavior of MgO$_2$-filled multi-walled BNNTs [190–191]. Such tubes acted as molecular oxygen generators. The unstable magnesium dioxide phase (MgO$_2$) could be transformed into a stable Mg oxide (MgO) under moderate heating inside the TEM (up to 90°C). Since the BNNTs exhibited open tip-ends a flow of molecular oxygen was released from them.

6.2.6.5 Substituted BN Nanotubes

Due to the striking similarities between the graphene lattice for C and the graphene-like lattice for BN, the possibility of substituting BN systems with C– or C systems with BN-has attracted a lot of attention. Nanotubes from the B$_x$C$_y$N$_z$ system will however be treated later on in this chapter. On the other hand, doping with other elements, such as oxygen or silicon, has also been discussed theoretically [55,59], with a very limited amount of experimental work performed to date. In this context, Si and Xue [59] found that the Si substitution for either B or N atoms in a single-walled armchair (5,5) BNNT could induce spontaneous magnetization, which may be attributed to Si $3p$ unpaired electrons. The total magnetic moment in this doped tube was estimated to be $1.0\,\mu_B$. Xu *et al.* [192] performed Si doping in bamboo-like BNNTs. The authors used pyrolysis to synthesize these structures and then studied their phonon and photoluminescence properties. The overall amount of Si was found to be ca. 6 at.% (based on energy-dispersive X-ray spectroscopy). However, the state of Si in the structures was not understood. Some changes in the photoluminescence spectra were observed. A broad band ranging between 500–800 nm was found to appear. Such band was not present in the pristine BNNTs.

Silva *et al.* [55] used ab initio calculations of substitutional oxygen in zigzag (10,0) BNNTs, and found that the electronic band structure for O-BNTs (O substituting B) is modified to an appearance of three strongly localized states, two of whose are placed within the bandgap. In the case of O substituting N, the Fermi level shifted into the conduction band inducing a metallic character in the modified tube. The analysis of the formation energies shows that the O-B substitution is more favourable, particularly in the case of a B-rich environment. These findings are interesting and should be verified experimentally.

6.2.7 Other BN Nanomaterials

Besides NTs, other BN nanostructures have successfully been fabricated using various techniques. Huo *et al.* [193] reported the preparation of BN wires by reacting a mixture of flowing NH_3 and N_2 gases with α–FeB nanoparticles at 1100°C. HRTEM images revealed perfectly straight lattice BN fringes with interlayer spacing of ca 3.33 Å, matching the standard shell separation in BNNTs. Besides nanowires, the authors mentioned the existence of BN ribbons, but this statement needs more detailed experimental verification. Deepak *et al.* [194] demonstrated the presence of diverse nanostructures composed of B, C and/or N, including BN nanowires (BNNW). BN wires exhibiting stacked cone morphologies were obtained in high yields by the ball milling thermal method using pure B_4C powder within the Chen's group in Canberra [195]. In order to make the stacked cones fibers, the milled powders were placed on a Si wafer that has a layer of ferric nitrate $(Fe(NO_3)_3)$. Then the material was first heated to 450°C for ~1 h to activate the catalyst particles. Subsequently, the temperature was increased to 1300°C and kept over 8 h. All heat-treatments were carried out in a N_2 atmosphere. The Fe-Si alloy particles were visible on each BNNW. Recently, pure BNNWs containing turbostratic BN layers without metallic tip ends were produced in high yield by heating BI_3 powder under NH_3 atmosphere at 1100°C [196]. The same group also reported the growth of BN nanorods [197] and BNNWs by heating B and ZnO powders at 1100°C under a N_2:H_2 flow using stainless-steel foils as a substrate and catalyst [198].

Bourgeois *et al.* [100] have been working on the preparation and detailed structural characterizations of BN cones prepared through a CVD method. The cones in BN were found to exist in two morphologies: cup-stacked and those made of a continuous folding of an endless BN strip. Such conclusions were made based on the analyses of cone tip-angles

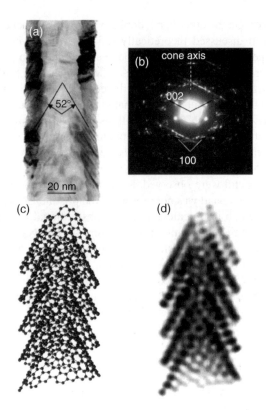

Figure 6.22 *Helical-conical BNNT; (a) a TEM image; (b) a corresponding electron diffraction pattern; (c) a structural model; (d) a computer-simulated TEM image. Reprinted with permission from [200] Copyright (2002) American Institute of Physics.*

using electron diffraction. Following the BN cone research by the Tsukuba Group, Xu *et al.* [199,200] were able to fabricate interesting helical-conical NTs. The TEM images of such tubes together with their diffraction patterns and computer models are depicted in Figure 6.22. Such tubes correspond to hollow cylindrical textures, but the walls are not parallel to the tube axis and have a sort of inclination to it (see Chapter 1). The apex angle measurements and diffraction patterns were used to conclude that the structure consisted of a BN ribbon wrapped in a helicoidal fashion. These structures were found to behave like springs as a conical angle was changed stepwise under electron beam irradiation.

BN nanocones could be synthesized by a substitution reaction process using CNTs as templates, as was reported by Bourgeois *et al.* [201]. The arc-melting methods have yielded multi-walled BN nanohorns [202] with very similar morphologies when compared to their C counterparts. The BN cone field-emission properties were treated theoretically and it was found that those could strongly depend on the tip geometry. The BN cones may be used as probes in microscopy and efficient field-emitters [203,204]. Moreover, negative Gaussian curvature introduced by the presence of octagonal rings made of alternating B and N atoms in a BN sheet was considered for BN structures [205]. A saddle shape defect could modify the electronic behavior of BN nanocones; these phenomena were also calculated by Azavedo [206]. The introduction of defects

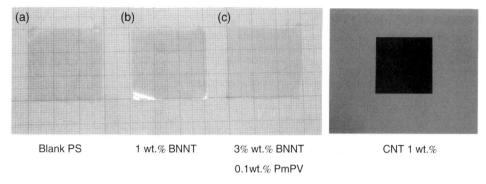

Blank PS	1 wt.% BNNT	3% wt.% BNNT 0.1wt.% PmPV	CNT 1 wt.%

Figure 6.23 (a) to (c) fully transparent polystyrene films made with the use of various loading fractions of multi-walled BNNTs; far right side: a black non-transparent film with a similar CNT loading fraction (1 wt.%) is provided for comparison. For a better understanding of the figure, please refer to the colour plate section.

and substituting atoms at the tips of the BN nanocones opens up an important pathway for modifying electronic and emission properties of BN cones [207].

6.2.8 Challenging Applications

6.2.8.1 BNNT/Polymer Composites

The physical properties of BNNTs, for example, high resistance to oxidation, superb strength, excellent thermal and chemical stabilities, good thermal conductivity and not absorbance of visible light, make them attractive for the fabrication of diverse polymeric composites. Most importantly, BNNTs may improve the polymer-matrix mechanical and thermal performance while preserving its transparency (Figure 6.23). It is noted for comparison that when 1 wt.% of CNTs is added into a given polymer, the whole material becomes blackish (Figure 6.23, right-hand side). This does not happen to BNNT-reinforced polymers even at a much higher BNNT loading fractions, up to 20–30 wt.%.

Several issues, such as dispersion of BNNTs in a given solvent, adhesion of NTs to a polymer body, BNNT-polymer interface bonding and stability, should be carefully addressed prior to the real integration of BNNT polymers into technology. The possibility of having BNNT solutions or suspensions (see above), makes BNNT mixing with polymers easier and partly solves the problem of homogeneity. The preparation of high-quality self-organized BNNT-reinforced polymer films based on polystyrene, PANI and other polymers has been achieved by Zhi *et al.* [158–162]. Tensile tests on such films indicated that both the elastic modulus and strengths were dramatically improved [208], as illustrated in Figure 6.24. BNNT composite films revealed much more pronounced reinforcement when compared to CNT composites using similar loading fractions. This effect demonstrates that it is likely that interfacial intereactions between the polymeric matrix and a tube surface are stronger for BNNTs.

However, the perfectly straight shapes of some specific BNNTs (i.e., typical needle-like appearance) as opposed to curled and wavy morphologies of entangled CVD-produced CNTs, could explain this difference. The BNNT composite films exhibited better stability to oxidation and slightly lowered glass transition temperatures compared to blank polymers.

Improving the thermal conductivity of polymers using BNNTs was suggested to be more challenging than in the case of mechanical properties. Typically, the thermal conductivities

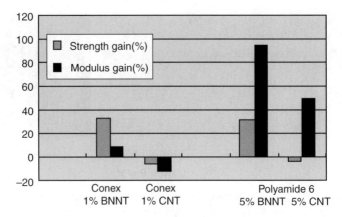

Figure 6.24 *A comparison between the effect of BNNT and CNT loading within the two sorts of polymer films (conex and polyamide 6) on the resultant film mechanical performance. The positive effect of BNNT fraction loadings (1 wt.% and 5 wt.%) notably prevails over that of CNTs in the same polymers.*

of BNNT/polymer composites have increased to only several of W mK^{-1} and failed to show dramatic percolation threshold.

Phonons within BNNTs led to high interfacial thermal resistance between individual BNNTs. It could be suggested that multi-walled BNNTs, rather than their SWNT counterparts, may exhibit the highest potential for the improvement of the overall thermal conductivity of polymer composites because of relatively low interfacial areas and less phonon scattering at the interfaces, and the existence of numerous internal shells with an originally high thermal conductance.

6.2.8.2 BNNT-Reinforced Glasses and Ceramics

Due to the outstanding high-temperature thermomechanical stability of BNNTs, they may find applications in the fabrication of engineering composite glasses and ceramics. There have been several works in this area. Historically first, a barium calcium alumosilicate glass preloaded with a ~4 wt.% of BNNT fraction was reinforced by up to 90% when compared to the pristine glass [209]. SEM examinations of the fracture surfaces revealed that BNNTs were responsible for the improvements. Huang *et al.* [210] loaded standard engineering ceramics, Al_2O_3 and Si_3N_4, with 2.5–5.0 wt.% BNNT fractions, and observed that both ceramics had become more deformable at high temperatures, revealing a superplastic behavior. For example, Al_2O_3 with a 2.5 wt.% fraction of BNNTs possessed a ~4.5 times lower yield stress and a much higher true strain to fracture than the bulk ceramic. The parameters of high temperature superplasticity were notably improved for both Al_2O_3 and Si_3N_4. Control experiments on BN micropowders added to the ceramics at the same amounts showed no analogous positive effects. This confirms that the BNNT addition was the key factor for the mechanical improvements.

6.2.8.3 Gas Adsorption

The gas adsorption capability of BNNTs has rapidly become an important issue for their common comparison with CNTs. For instance, hydrogen — the cleanest source of energy — holds promise for solving current environment pollution with zero-emission

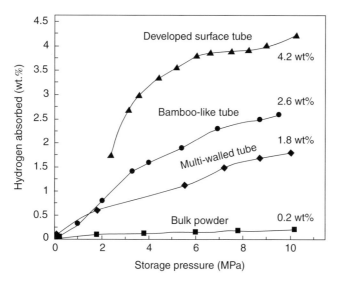

Figure 6.25 *Comparative hydrogen-uptake curves for various morphologies of a BN nanotubular product. The highest value, ca. 4.2 wt.% was obtained for the BNNTs with a highly developed surface area, so-called 'collapsed' BNNTs.*

vehicles. Safe and efficient hydrogen storage requires gravimetric hydrogen density exceeding 6 wt.% under conditions of fast and reversible kinetics of hydrogenation and dehydrogenation.

The research on hydrogen uptake in BNNTs was stimulated by the initial encouraging results for CNTs [211–213]. Single-walled CNTs were intensively studied and revealed strong interactions with hydrogen molecules. However, markedly different uptake values were reported by various groups. This may be caused by the difference in electronic interactions between hydrogen and CNTs of various textures or structures. The electrical properties of CNTs are highly sensitive to the morphology and atomic order which are yet under control. On the other hand, BNNTs are electrically uniform thus offering reliably predictable NT-hydrogen interactions.

The binding energy of chemisorbed hydrogen atoms to a zigzag (10,0) single-walled BNNT was calculated at 25, 50, 75 and 100% coverage by Han *et al.* [56] using DFT. The average binding energy (per H atom) was computed to be the highest for a 50% case when the H atoms had been adsorbed on the adjacent B and N atoms along the tube axis. This value was -53.93 kcal mol^{-1}, which is equal to half of the H_2 binding energy. Interestingly enough, the BN bandgap (which was computed to be -4.29 eV) of the tube decreased to 2.01 eV after 50% tube surface coverage.

Ma *et al.* [214] first demonstrated that BN nanostructures may absorb hydrogen at the level equal or even exceeding that of CNTs, as depicted in Figure 6.25. Multi-walled BNNTs were found to adsorb 1.8–2.6 wt.% of hydrogen under ~10 MPa at room temperature.

Novel nanostructures in BN may also be of significant importance in the field. For example, Jhi *et al.* [53] calculated that an increase in specific surface area (SSA) for BNNTs may be a solution for an increase in operating temperature and capacity for hydrogen storage. The calculated binding energy of hydrogen on activated BNNTs (those having well-developed porous surface) was found to reach as much as 22 kJ mol^{-1}, and thus to lie in the acceptable range for room-temperature hydrogen storage. The most active pores for hydrogen binding were found to be those terminated by oxygen atoms. Verifying this idea, Tang *et al.* [215] have experimentally demonstrated that significant hydrogen uptake (up to ~4.2 wt.%) took place in the so-called 'collapsed' BNNTs having a highly-developed activated external surface and overall 'cactus'-like appearance of the tube. However, in Tang's experiments, the existence of Pt nanoparticles in a BNNT product that resulted from the synthesis procedure may be a concern. In these experiments, Pt was used to actually create the developed surface of BNNT through chemical interactions with Pt at high temperature. Later on, it was indeed demonstrated that modification of BNNT surface with some metals and/or intermetallic compounds (with prominent hydrogen affinity, e.g., LaNi$_5$) may significantly improve the electrochemical hydrogen storage capacity of BNNTs [52]. Additional theoretical calculations performed by Zhou *et al.* [50] have demonstrated that a well-structured, perfect BNNT is in fact not a good candidate for hydrogen storage by either physical or chemical absorption mechanisms. This was actually demonstrated by Ma *et al.* [214] in the pioneering hydrogenation experiments. In order to find a solution to this problem, Zhou *et al.* [50] have proposed to use BNNTs as a supporting media for hydrogen-adsorbing metallic nanoclusters. The experimental possibilities of such metal cluster anchoring have indeed been documented [98]. For example, Durgun *et al.* [57] have shown that the interaction of H$_2$ molecules with the outer surface of single-walled BNNTs (normally very weak) could be dramatically enhanced in BN tubes functionalized with Ti atoms. Each Ti atom adsorbed on SWBNNT can bind up to four H$_2$ molecules.

6.2.8.4 Electrical Nanocables/Nanoelectrodes
The insulating characteristics of BNNTs open up a route towards the preparation of electrically insulated nanocables with embedded metallic or semiconducting nanowires in their internal hollow cores. Such cables may be utilized in downsized electrical devices and complex multi-cable circuits. Importantly, each cable should perform independently without a current leakage between them. The problem related to the poor wetting properties of BNNTs during filling could be effectively overcome via a two-stage processes. Another possible route towards nanocable formation is the production of BN-insulated CNTs or B-C-N NTs through sandwich-like tube formation [216–221]. Perfect insulating performance of mechanically robust BN tubular shields and the conductive properties of the internal B-C-N layers have indeed been confirmed during in situ TEM transport, and field-emission experiments inside a LEEPS microscope [134, 135].

Individual BNNT probes based on a BN-gold-polymer cable for the electrochemical and biochemical sensing could also be produced. In order to fabricate such an electrode, an individual BNNT was attached to a bigger electrode and then consecutively coated with a 10–50 nm thick Au layer, followed by a 10 nm thick insulating polymer layer. Subsequently, the cable was sliced off using a focused ion beam exposing a conducting ring of gold sandwiched between the insulating BN core and the insulating polymer ring. Such electrode could penetrate a living cell, and even pinpoint smaller cell structures, such as the nucleus or mitochondria. By functionalizing the active area of the nanoelectrode with a suitable chemical any specific chemical specie could be detected.

6.2.8.5 Field-Emitting Devices

Due to the insulating character of a BN material, the field-emission of pure BNNTs may not be of practical interest because of their intrinsically high turn-on fields, low current densities, possible arcing and resistance heating. In fact, high turn-on fields of $8.3–15.2 \text{V } \mu\text{m}^{-1}$ [222] were measured on microcrystalline BN thin films. For comparison, the best field-emitters made of aligned CNTs have typically exhibited turn-on fields in the range of $1–3 \text{V } \mu\text{m}^{-1}$. Doping of BNNTs with C seems to be prospective for lowering turn-on fields and increasing current densities. The turn-on fields of $\sim2–3 \text{V } \mu\text{m}^{-1}$ and current densities of $\sim3 \text{mA cm}^{-2}$ have been achieved on C-doped BNNT films (see Section 6.2). For comparison, the best field-emitters made of CNTs may reach a current density of 10mA cm^{-2} at a turn-on field below $4.5 \text{V } \mu\text{m}$. Therefore, the particular selection of C or BN-C NT field-emitters for a given field emission application should be determined by a trade-off between the importance of the emitter environmental stability (which is better for sandwich-like BN-C NTs than for pure CNTs) and the desired field-emission characteristics (which are better for CNTs).

6.2.8.6 Ultraviolet Lasers

The demands for compact UV lasers, substituting those made of commercial GaN materials, is the key concern in many fields including optical storage, photocatalysis, sterilization and surgery. A standard layered hexagonal BN has a direct bandgap and exhibits a stable dominant luminescence peak at $\sim215 \text{nm}$ [223]. It is worth noting that both the direct and indirect, features of hexagonal BN bandgaps have been reported. The measured band gap energies varied in a range of $3.6–7.1 \text{eV}$. This notable controversy probably stems from the fact that the crystalline quality (e.g., defect structure) of layered BN macrosamples used for the optical absorption measurements by different authors was very different. Alternatively, BNNTs, with the reported bandgaps in the range of $4–6 \text{eV}$, typically exhibit well-structured graphene-like BN layers (as opposed to many other layered BN materials which suffer from high defect concentration), and may find an important utilization in UV lasing. Such lasers, if prepared, may be then ultimately downsized.

6.2.8.7 Other Properties

Recently, diverse novel physical and chemical properties of BNNTs have been revealed. These include specific vibrational modes [224], EXAFS [225], isotopic enrichments [226,227], and immobilization of proteins on the BNNT surfaces. The latter results may be of interest for new nanomedical and nanobiology applications [228].

6.3 $B_x C_y N_z$ Nanotubes and Nanofibers

6.3.1 Tuning the Electronic Structure with C-Substituted BN Nanotubes

$B_x C_y N_z$ materials constitute another possibility for layer-by-layer systems, which are either hybrids of h-BN and graphite or intermediates between them. In the former case, they are sandwiched structures of pure BN and pure C graphene lattices combined (as in Figure 6.26). In the latter case, they are graphene lattices with $B_x C_y N_z$ composition (Figure 6.27), most likely with a nonuniform distribution of each atom species over the lattice.

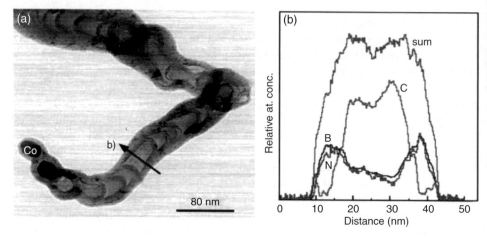

Figure 6.26 *(a) TEM image of a BC$_2$N tube and its line-scan; (b) Concentration profiles of the B, C and N across the fiber (line scan b), exhibiting two clear onsets in the total signal, indicating the position where the profile enters the internal closures (stacked cones) with a separation of ca. 25 nm. The profiles show a C double peak with the BN crest in between. An anticorrelated behavior of the C and BN signals is always maintained, thus confirming a consistent C-BN-C alternation. Reprinted with permission from [247] Copyright (1999) Elsevier Ltd.*

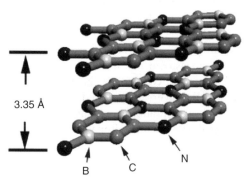

Figure 6.27 *Theoretical model of a homogenous BC$_2$N layered material. Reprinted with permission from [230] Copyright (1994) American Physical Society.*

They may exhibit semiconducting behavior and may be used as photoluminescent materials (light sources), transistors working at high temperatures, lightweight electrical conductors or high temperature lubricants [229]. These applications depend not only on the composition but also on the arrangement of the B, C and N atoms [230].

Indeed, since CNTs could behave as metals or semiconductors (depending on their helicity and diameters), and hexagonal BN is an insulator, it is therefore possible to construct hybrid structures combining BN and/or C and/or B$_x$C$_y$N$_z$ in which the electrical properties could be tuned. There have been a few theoretical predictions that layered B-C-N materials may display semiconducting properties intermediate between metallic CNTs and dielectric BNNTs. For instance, Liu *et al.* [231] calculated a 2.0 eV bandgap for a BC$_2$N stoichiometric sheet [232], whereas Zhu *et al.* [233] obtained a 0.2 eV gap for a BCN layered network.

Density of states

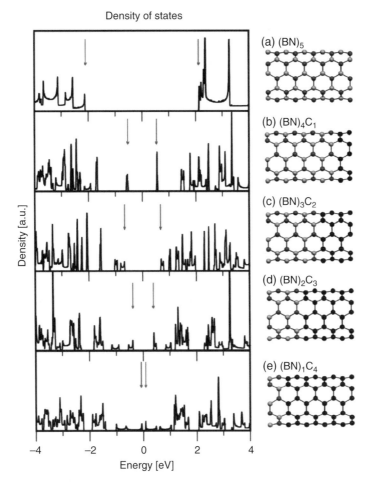

Figure 6.28 *Densities of states for C*BNNT heterojunctions with different stoichiometries, from pure BN to (BN)C$_4$. The top valence and bottom conduction states are indicated with the pairs of arrows. The energy gap in these NTs varies from 0.15–4.2 eV as the BN content is increased, with the exception of (BN)$_4$C, where two defect-like non-dispersive energy states reduce the energy gap. For a better understanding of the figure, please refer to the colour plate section.*

Hence, it is possible to reduce the electronic bandgap of BN (ca. 5.5 eV), by adding C atoms within the structure in substitution to B and/or N atoms. In this context, theoretical calculations have been performed [234,235]. It has been demonstrated that the electronic bandgap is indeed reduced as additional C atoms are introduced in the hexagonal BN network (Figure 6.28). Substituting with C was found to be an effective way to reduce the bandgap of BN tubes to ~1 eV.

There has been an experimental estimate of the bandgap in a mat of highly-defective CVD-prepared B$_{0.34}$C$_{0.42}$N$_{0.24}$ nanotubes using photoluminescence [236]. This was stated to be around 1.0 eV. The value of the bandgap experimentally determined for an individual straight perfectly-ordered BN-rich B-C-N NT rope was also close to ~1 eV [133–135].

These small values of the bandgap are thought to be due to the global reduction of the ionicity of the C-substituted BNNT network compared to pure BNNTs.

The bandgap measured for similar B-C-N bundles of aligned nanotubes was ca. 1.5 eV [134]. Recent studies have also reported the synthesis of $B_{0.45}C_{0.1}N_{0.45}$ nanotubes with homogeneous domains (C-doped BNNTs). In these samples the authors observed the presence of pyridine-like structure (hole-like structure) and bandgaps close to 2.8 eV [237]. It is therefore clear that the concentration and distribution of BN within the tubes make a clear difference in the electronic bandgaps of these B-C-N materials.

The idea of transverse BN/C heterostructured NTs (in which pure BN and C domains are placed in series along the nanotube) was developed by Blase *et al.* [234], who performed calculation using a DFT approach within the LDA. The authors estimated the formation energy of the C/BN interfaces to be around 0.4 eV per interface bond. They also studied the interdiffusion process in a C/BN interface and found that this process would have an energy cost close to 2.0 eV per atom, making it very unlikely and favoring abrupt BN/C interfaces. Similarly, heterojunctions corresponding to a (5,5) nanotube, exhibiting on one side a $(BN)C_4$ stoichiometry and a pure BN domain on the opposite side have also been calculated. The DOS were computed using the DFT-LDA approaches and nonconserving pseudopotentials implemented with the SIESTA code [238]. As the BN content increased, the energy gap increased from 0.15 eV to 4.2 eV. These results are consistent with previous reports on supercell geometries for C/BN NT heterojunctions by Choi *et al.* [239]who studied several decorations of a hexagonal lattice, and found that there was the spin polarization present on the border states of a zigzag (9,0) NT decorated with alternating zigzag rings of C and BN (doped with two holes). It is clear that the magnetic and electronic properties of BN heteroribbons and -tubes with a composition within the BCN system should exhibit unusual behavior.

The electron/hole doping may also be produced not by alloying but also through introducing defects to the tubes. In analogy with the edge states present in nanosized graphite, there are edge states present in hexagonal BN and open BNNTs. It is noteworthy that open BN edges have frequently been observed by various researchers. These edges have carefully been studied by Okada and Oshiyama [163], along with heterostructured BCN nanosized ribbons. The authors found that some BNC ribbon geometries were ferromagnetic. In a later work, Nakamura *et al.* [240] found that BN-decorated graphite ribbons had a ferromagnetic ground state caused by an unbalanced occupation of the border and edge states.

6.3.2   Production and Characterization of $B_xC_yN_z$ Nanotubes and Nanofibers

Soon after the prediction of BC_2N nanotubes, the Berkeley group reported their synthesis via the arc-evaporation of graphite electrodes filled with B, or BN powder, or, alternatively, a BCN composite anode [241]. These experiments resulted in the formation of very few $B_xC_yN_z$ nanotubes and mixtures of B_4C nanocrystals and long CNTs doped with low B concentrations [242–244]. In these experiments, few nanotubes with higher BN or B contents were observed.

Subsequently, Terrones *et al.* reported a pyrolytic route to efficiently synthesize BC_xN nanotubes [245]. This technique, when compared to arc-discharge methods, produces a uniform range of BC_2N nanoobjects, preserving the stoichiometry. In particular, the pyrolysis of $CH_3CN \cdot BCl_3$ over cobalt powder at 950–1000°C led to BC_2N nanofiberrs

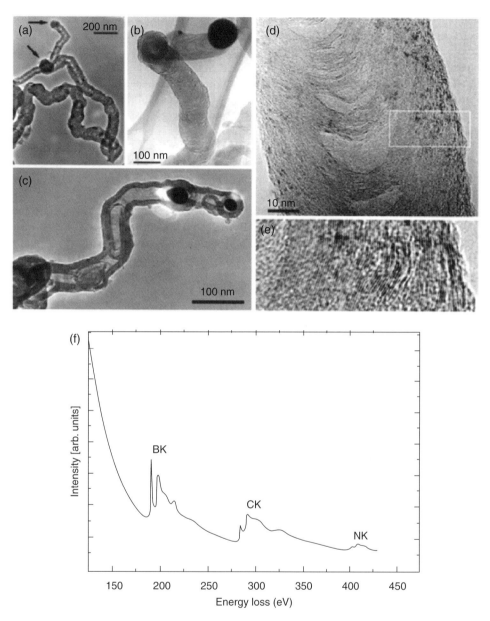

Figure 6.29 *(a-c) Low-magnification TEM images of BC₂N nanotubes and nanofibers exhibiting stacked-cone morphologies; note that metal particles are located at the tips; (d) HRTEM images of a representative nanofiber produced by pyrolysis of $CH_3CN \bullet BCl_3$ over Co at 1000°C. The filament consists of compartmentalized stacked cone structures, which adopt the metal particle morphology; (e) HRTEM image of a segment [marked in (d)] of a typical nanofiber, exhibiting layers separated by ca. 3.45Å; (f) EEL spectrum of a typical BC_xN nanofiber. The overall stoichiometry is close to BC_2N [29 ± 4 at.% B, 45 ± 6 at.% C, 26 ± 3 at.%]. Reprinted with permission from [245,247] Copyright (1996) and (1999) Elsevier Ltd.*

and nanotubes of various morphologies. These BC_2N nanofibers/nanotubes consisted of non-cylindrical layer-by-layer tubes, most of which (> 90%) exhibited stacked cone-like, somewhat hollowed, textures. The general cone-cone surface distance in the stack was close to that of graphite (Figure 6.29). Individual cones tend to adopt the shape and size of an encapsulated catalytic nanoparticle, indicating that the shapes of these particles are intimately responsible for the structural features of the tubes and fibers [245]

From the early days of $B_xC_yN_z$ nanotubes, electron energy loss spectroscopy (EELS) was used to characterize them. In 1994, Stephan *et al.* [242] managed to prepare multi-walled NTs displaying all the three types of atoms (B, C, and N). Using EELS, they have demonstrated a kind of sandwiched structure in which successive, concentric tubes can have different $B_xC_yN_z$ composition, yet the spatial resolution was not high enough at that time to provide the precise nanoscale distribution of the three elements.

Nowadays, spatially-resolved EELS and energy-filtered electron microscopy are the most useful experimental techniques for performing the chemical analysis of C-substituted BN tubes. It is interesting to note that homogeneously-structured B-C-N layers have been only observed after high-temperature syntheses. In this context, theoretical calculations have predicted the stability of homogeneous BC_2N layers and NTs, similar to the experimentally prepared homogeneous B-C-N films by Kawaguchi *et al.* [246].

Figure 6.30 compares a homogeneously-structured multi-walled $B_xC_yN_z$ NT and a heterogeneously crystallized BN-C NT prepared using different carbon precursors [132–135]. Elemental profiles recorded across the homogeneous NT display perfect correlation between the B and C intensities. Within the resolution of the energy-filtered images (~5 Å), the NT layers shown in Figure 6.30a may be interpreted as those being uniformly composed of B and C atoms. For a heterogeneous BN-C NT, the brightest intensity strips on the elemental mapping images and the corresponding concentration profiles for BN and C are clearly separated (Figure 6.30g-k). Therefore, this NT possesses a BN-rich shield and C-rich internal layers.

EELS measurements of the K-edge absorption for B, C and N were also used to estimate the stoichiometry of the nanotubes and nanofibers. Figure 6.29f shows an EELS spectrum of a typical nanofiber with ionization edges at ca. 188 eV, 284 eV and 399 eV, corresponding to the characteristic K-shells of B, C and N, respectively. Elemental quantification of the integrated EELS signal of the individual fibers and tubes revealed an overall composition commensurate with BC_2N stoichiometry (see Figure 6.29f), as expected from the following reaction: $CH_3CN + BCl_3 \rightarrow BC_2N + 3HCl$.

The careful investigation of the structure of these BC_2N nanotubes/nanofibers at the nanometer level was performed using high-spatial resolution electron energy-loss spectroscopy (HREELS). Line scans along and across the tubular structures revealed the presence of layered C-BN-C sandwich-like structures (Figure 6.26b). Therefore, B, C and N were not homogeneously distributed within the tubes, but were separated into pure C and BN domains in a manner described in Section 6.1 [247]. Pure graphene-like BN layers were always sandwiched between genuine graphene shells (Figure 6.26). In this context, theoretical calculations indicating the segregation of very stable BN/C materials have been performed [220,248] demonstrating that BC_2N systems tend to segregate into pure C and BN domains. Interestingly, Blase *et al.* pointed out that this spontaneous segregation processes may promote the formation of quantum dots or nanotube heterojunctions [248], which could then be used as nanotransistors operating at high temperatures.

Homogeneous B-C-N NT

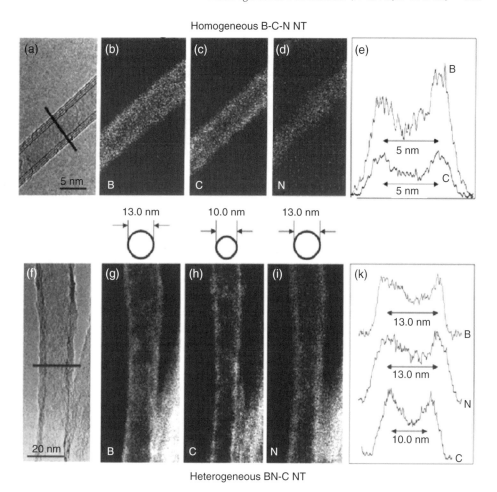

Figure 6.30   *(a) to (e): homogeneously-structured multi-walled B$_x$C$_y$N$_z$ nanotube displaying (a) a TEM image, (b) to (d) elemental maps of boron, carbon and nitrogen respectively, and (e) related elemental profiles, showing that the profile summits for B and C are aligned. (f) to (k) same series, for a heterogeneously crystallized BN-C NT, that is, with a sandwich-like display of BN-rich and C-rich layers; in the related elemental profiles in (k), the profile summits for B and N are aligned, whereas they are shifted for C. Reprinted with permission from [133] Copyright (2004) Cambridge University Press.*

The increased thickness of the filaments as a function of distance away from the catalytic particles at the tip, suggests a continuous thickening process (Figure 6.29a-c), occurring after an initial filament formation. The diameter of the inner core and the thickness of the internal stacked cones remain nearly constant, whereas the total outer wall thickness increases. This suggests a secondary growth process: extrusion and/or precipitation of a primary catalyzed filament (Co, in the case of Figure 6.29), which subsequently thickens due to further deposition from the gas phase.

Suenaga *et al.* [218] managed to produce B-C-N nanotubes with a higher BN content by arcing graphite/HfB$_2$ electrodes. These authors performed high-spatial resolution chemical analysis and found that most of the tubes possessed a sandwich structure with C layers both in the center and at the periphery, separated by BN layers (coaxial-like nanotubes). Similarly, Zhang *et al.* reported sandwich-like C-BN nanotubes produced by laser vaporization [188].

Guo *et al.* [249] were able to verify the existence of C/BN multi-walled NT heterojunctions (see Section 6.2.1) using a two-stage Hot Filament Chemical Vapor Deposition (HFCVD) synthesis by varying the flows for the C, B and N sources. If diverse BN/C heterostructures could be synthesized in a controlled way in the near future, they would definitely become valuable electronic and magnetic materials.

Johansson *et al.* have also described the generation of segregated BN-C nanoobjects termed 'nanoboxes' [250]. These structures were produced by magnetron sputtering, involving boron carbide in conjunction with a NaCl substrate in the presence of Ar and N$_2$ at lower temperatures. More recently, segregated BN-C nanotubes, sometimes filled with boron carbide, have been obtained through partial substitution using pyrolytically grown CNTs reacted with B$_2$O$_3$ and N$_2$ at elevated temperatures. These results have also confirmed the segregation of BN and C. Nevertheless, the production of homogenous BC$_2$N layers within nanotube-like objects may be possible using other approaches.

Some reports have demonstrated the successful synthesis of aligned B-C-N nanofibers and nanotubes by thermolyzing (employing hot filament chemical vapor deposition) mixtures of B$_2$H$_6$, CH$_4$, N$_2$, and H$_2$ over Ni [251]. Photoluminescence studies revealed that these materials are semiconductors exhibiting a bandgap of 1.0 eV. Additionally, BC$_2$N nanotubes and nanostructures have been proved to behave as efficient field-emitters, as well as blue and violet light-emitting materials [252–255]. In this context, it is important to note that in 1996, Watanabe *et al.* studied the bandgap of layered BC$_2$N materials, using scanning tunneling spectroscopy, and determined a semiconducting behavior with an energy gap of 2 eV, which is in agreement with the theoretical calculations for a homogenous BC$_2$N layer [255].

More recently, Yin *et al.* [256] carried out spatially resolved cathodoluminescence measurements on individual porous B-C-N nanotubular fibers and observed intense ultraviolet emission at 319 nm, suggesting the characteristics of a semiconductor with a bandgap of 3.89 eV. Therefore, the optical and electronic properties of these materials strongly depend on the way in which B/C/N domains are distributed within the nanotubes; segregation of BN-C domains will result in novel optical properties. Further elemental analyses are still needed in order to determine whether the segregation of BN and C occurs within all these structures. For example, theoretical calculations demonstrated that B$_3$C$_2$N$_3$ stoichiometries appear to be more stable than BC$_2$N structures [235,257]. These structures with large values of the electronic bandgap exhibit a B/N ratio of one, and the bandgaps could be tuned from 0 to 2.45 eV [257].

In addition to straight B-C-N nanotubes and nanofibers, several reports dealing with the production of B$_x$C$_y$N$_z$ 'Y-type' junctions and BNC-C tubular junctions have been published [249,258]. The latter structural junctions were produced via bias-assisted hot-filament chemical vapor deposition [258–260]. Interestingly, some of these junctions show typical rectifying diode behavior [258]. However, the possibility of having segregated domains of BN and C was very high. In this context, Wang and Zhang [261] observed that all CVD-produced nanotubes exhibited inner shells consisting of C and

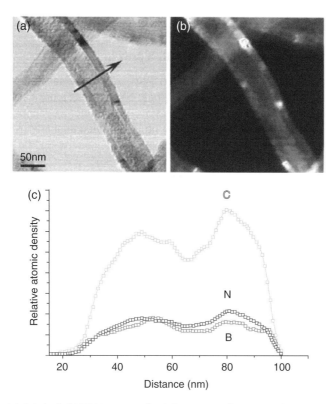

Figure 6.31 *(a) Bright-field TEM image of a BC$_2$N nanotube (outer diameter ~70 nm); (b) High-angular dark field (HADF) image of a BC$_2$N tube; (c) EELS line scans used to obtain the concentration profiles of the B, C and N across the tube. The profiles demonstrate that the C concentrations notably correlate with those of B and N.*

traces of B, whereas B and N were found mainly on the outside layers; oxygen was also present on the outside layers of studied tubes.

Terrones *et al.* reported the successful synthesis of high yields of aligned B-C-N hollow nanotubes exhibiting homogeneous concentrations of B, N and C [262]. This method involves the reaction of CN$_x$ nanotubes (instead of using pure CNTs) with B$_2$O$_3$ and CuO at ~1800 °C using an induction furnace. The B, C, and N distribution was confirmed at the nanometer level, using HREELS line-scans coupled with high-angular dark field imaging (Figure 6.31).

Therefore, the use of N*CNTs as templates for producing uniform BCN nanotubes is a clear alternative. It was actually noted that B-C-N nanotubes with homogeneous distribution of B, C and N could be obtained at 1800°C. It is important to note that if higher temperatures (> 1800°C) are used in the reaction of CN$_x$ with B$_2$O$_3$-CuO and N$_2$, the BN layers form on the outer wall of the tubes, whereas BC$_x$N domains prevail in the inner shells [132,263]. However, the reaction temperature and the way the materials are heated are crucial factors to obtain homogeneous BC$_x$N shells. Higher or lower temperatures may also result in inhomogeneous and segregated B, C and N cylinders. Therefore, these results do not contradict previous studies involving reactions between CN$_x$ NTs and B$_2$O$_3$ powders, but emphasize the importance of particular growth kinetics during substitution reactions.

6.4 B-Substituted or N-Substituted Carbon Nanotubes

6.4.1 Substituting Carbon Nanotubes with B or N

The electronic properties of N- or B-substituted SWNTs (in-plane substitution in a graphene sheet) would be different when compared to pure SWCNTs (Figure 6.32). The cylinder curvature and quantum confinement result in novel electronic, mechanical and chemical properties, which may be very different from nonsubstituted counterparts. These substitutional B and N atoms introduce localized electronic features in the valence and/or conduction bands and increase the number of electronic states at (and close to) the Fermi level (E_f) depending on the location and substitutional atom concentration [264] (Figure 6.32). Boron has one valence electron less when carbon, and when it substitutes for C atoms within a SWNT (three coordinated B) sharp localized states below the Fermi level (in the valence band) appear (Figure 6.32a). These states are caused by the presence of holes in the structure, and the tube could be considered as a *p*-type nanoconductor. From the chemical standpoint, this structure would more likely react with donor-type molecules.

For N-substituted SWNTs, two types of C-N bonding could occur (Figure 6.33a). The first is three-coordinated N atoms within the sp^2-hybridized network. These induce sharp localized states above the Fermi level due to the presence of additional electrons (Figure 6.32b). These N-substituted tubes exhibit *n*-type conduction and are more likely to strongly react with acceptor molecules. The second type of C-N substitutional bond is the pyridine-type (two

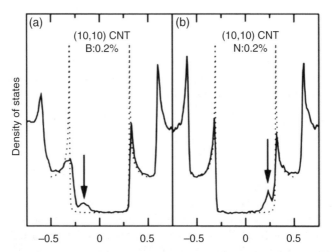

Figure 6.32 *Calculated densities of states (DOS) for: (a) an armchair (10,10) B*CNT (B = 0.2 at.%, solid line) exhibiting a clear peak in the valence band (see arrow); (b) an armchair (10,10) N*CNT (N = 0.2 at.%, solid line), in which a sharp and localized peak arises in the conduction band (see arrow). Solid lines correspond to the substituted materials whereas dotted lines are related to the pristine CNTs. Note that for all cases the presence of B introduced states in the valence band (holes), whereas N injects electrons in the conduction band (donors). The spikes shown in the DOS of the tubules are called 'van Hove' singularities and are the typical signature of one-dimensional quantum conduction, which is not present in an infinite graphite crystal (calculations performed by S. Latil).*

coordinated Ns), which can be incorporated in the SWNT lattice, provided that an additional C atom is removed from the framework (Figure 6.33). This type of defect induces the presence of localized states below and above to the Fermi level (not shown here).

Regarding the mechanical properties of individual CN_x and CB_x nanotubes, Hernández and coworkers, demonstrated that high concentrations of B and/or N within SWCNTs lower the Young's modulus to the order of 0.5–0.8 TPa [265]. This observation has been experimentally confirmed in pristine and N-doped MWCNTs [266]. The experimental Young's modulus for pure carbon and N-doped MWCNTs was measured as 0.8–1.0 TPa and ~30 GPa, respectively. The low values observed for N-doped nanotubes could be due to the relatively high N concentration (e.g., 2–5%) which introduces defects and lowers significantly the mechanical strength of the hexagonal framework. In order to observe real quantum effects in substituted CNTs with either B or N, the CNTs involved must be narrow diameter SWNTs (< 1–2 nm). However, if the N or B concentration is < 0.5%, it is expected that the mechanical properties would not be substantially altered, yet it is enough to enhance electron conduction.

In addition to MWCNTs and SWCNTs, it is of course also possible to substitute double-walled carbon nanotubes (DWCNTs) with B and/or N. In these double-layered systems one would expect that the electronic and mechanical properties change when compared to MWCNTs and SWCNTs, and novel electronic, mechanical, and vibrational behaviors could be observed.

On the other hand, besides N and B, other elements such as Si and P could also substitute C in the SWNT lattice [267–268]. In this context, a theoretical work has pointed out that substitutional Si or P induces a strong deformation of the cylindrical surface (outward) of the tubes and this bump sites lead to a more reactive surface than in the non-substituted case. However, further experimental research is required along this direction.

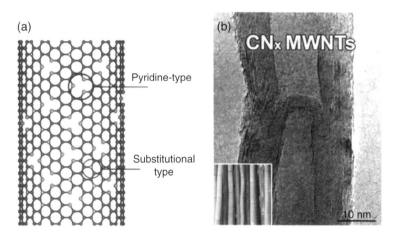

Figure 6.33 *(a) Molecular model of a N*CNT, exhibiting two types of N sites: 1) pyridine-type in which each N atom is bonded to two C atoms, responsible of creating cavities and corrugation in the nanotube structure (atoms in pink), and 2) substitutional N atoms which are bonded to three C atoms (green spheres); (b) HRTEM image of a N*MWCNT exhibiting the bamboo-type morphology, and (inset) image of natural bamboo trees whose appearance is similar to N*MWCNTs. For a better understanding of the figure, please refer to the colour plate section.*

6.4.2   Synthesis Strategies for Producing B- or N-Substituted CNTs

In the past, several authors have reported the production of B- or N-substituted CNTs using processes far from the thermal equilibrium. These techniques are listed in the following sections.

6.4.2.1   Arc-Discharge

MWCNTs substituted with B could be synthesized by arcing BN/graphite or B/graphite electrodes in an inert atmosphere (e.g., He, N_2). In this case, crystalline and relatively long ($\leq$ 100 µm) CNTs with diameters of < 20 nm could be produced [242–244]. These MWCNTs usually exhibit minute amounts of B in a C lattice. It has been demonstrated that small quantities of sulfur result in a higher incorporation rate of B in tubes [269]. Arc-experiments using pure graphite electrodes in a NH_3 atmosphere also indicate that it is extremely difficult to produce N*SWCNTs and MWCNTs, possibly because N_2 molecules are easily created and do not incorporate into the carbon lattice. However, Glerup et al. [270] reported the growth of N*SWCNTs when arcing graphite-melamine-Ni-Y anodes. These authors showed that the tubes contained low concentrations of N (< 1 at.%), and sometimes the tubes are corrugated due to the presence of N atoms in the hexagonal network. Similarly, Li et al. [271] produced B*SWCNTs by arcing graphite-Ni-Y electrodes. The authors noted that the addition of B had reduced the diameter of SWNTs. This technique should be used more frequently for better control of the level of B- and N- substituting in SWCNTs and MWCNTs.

6.4.2.2   Laser Ablation

Colliex and coworkers [272] demonstrated that B- substituted SWNTs could be generated using laser vaporization of B-graphite-Co-Ni targets. The authors found SWNTs in the products when the B content in the target material was less than 3 at.%. For higher B concentrations in a graphite target (e.g., > 3.5 at.%), graphite and metal encapsulated particles were mainly formed and only extremely low quantities of SWCNTs were obtained. More recently, Heben and coworkers have managed to produce B- substituted SWCNT by ablating NiB targets in a N_2 atmosphere [273]. These authors noted that B was incorporated in the hexagonal carbon network (1.8 wt.% of B) and a possible growth scenario was discussed. However, more energetic lasers (e.g., CO_2 lasers, femtosecond lasers, etc.) should also be tested in order to effectively generate either N- or B- substituted SWCNTs using this technique.

6.4.2.3   Chemical Vapor Deposition (with Catalyst)

The thermal decomposition of C-N organic precursors over metallic particles (e.g., Fe, Co, Ni) results in the formation of N- substituted MWCNTs ($MWCN_xNTs$) nanotubes. The first report on the formation of aligned arrays of N- substituted MWCNTs (< 1–2 at.% N) involved the pyrolysis of aminodichlorotriazine over laser-etched Co thin films at 1050 °C [274]. Soon after, melamine (triaminotriazine) was used as a precursor for CN_x and the experiments showed an increased nitrogen content (< 7 at.%) within 'corrugated' C tubular nanoobjects [275]. These observations demonstrate that it is difficult to generate crystalline and highly-ordered structures containing large concentrations of N within a hexagonal C network. Subsequently, CN_x nanotubes with low nitrogen concentrations were produced by a CVD process involving pyridine and methylpyrimidine [276].

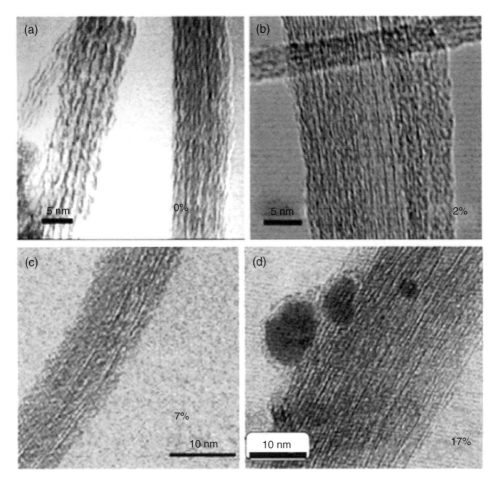

Figure 6.34 *TEM images of N*SWCNT strands synthesized with (a) 0%, (b) 2%, (c) 7 and (d) 17% N precursor (benzylamine) in the Ferrocene:Ethanol:Bencylamine (FEB) solution.*

Interestingly, these MWCN$_x$NTs are more reactive when compared to pure MWCNTs and are easily oxidized (e.g., combustion takes place at ca. 450°C in air, whereas pure CNTs do not burn in air below ca. 600 °C) [276]. The degree of perfection within graphenes is highly dependent on the nitrogen concentration (i.e., the lower the nitrogen content, the more 'graphitic' and the straighter the nanotubes become). It is also important to note that the N*MWCNTs produced this way exhibit a bamboo-type morphology (see above). In this context, Sumpter *et al.* [277] have showed from the theoretical standpoint that N-atoms are responsible for promoting compartmentalized morphologies in MWCN$_x$NTs and of narrow diameters in SWCN$_x$NTs.

Keskar *et al.* have prepared isolated N*SWCNTs through a thermal decomposition of xyelene-acetonitrile mixture over nanosized Fe catalyst particles [278]. In addition, long strands of N*SWCNT bundles (Figure 6.34) were successfully produced by pyrolyzing ferrocene ethanol solutions containing small weight ratios of benzylamine (e.g., from

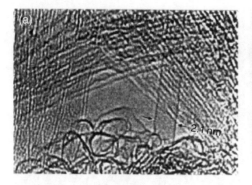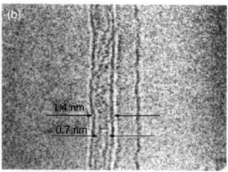

Figure 6.35   *(a) HRTEM image of an N*SWCNT bundle synthesized at 1260°C over 240 min by reacting N_2, B_2O_3 with SWCNT bundles of 1.4 nm diameter. Note the isolated SWNT exhibiting a diameter of 2.1 nm, and (b) HRTEM image of a two-tube B-substituted bundle synthesized at 1230°C over 240 min using the same reactants. The left-hand-side tube exhibits a SWNT with an inner shell (d = 0.7 nm) formed inside the outer shell (d = 1.4 nm). Reprinted with permission from [75] Copyright (2000) Elsevier Ltd.*

1–22 wt.% in ethanol) at 950°C in an Ar atmosphere [279]. These authors demonstrated that the electron conduction of the N-doped SWCNT ropes is very different when compared to pure SWCNTs, especially at temperatures lower than 20 K [279].

It is possible that, when B-containing organic precursors (e.g., boranes, boric acid, etc.) in conjunction with hydrocarbons and metal catalysts (e.g., Fe, Co, Ni) are used, MWCNTs substituted with B could be produced [280]. However, it is believed that concentrations of B and N in substituted SWCNTs have to be low. This is because the single-shell cylinders would collapse (and would not grow) if a large amount of foreign atoms were introduced into the hexagonal carbon network. However, there is still an enormous amount of experimental work to be carried out in order to enable control of the B or N contents, as well as the overall tubular structure, using the CVD approaches.

6.4.2.4   B or N Substitution Reactions

SWCNTs substituted with B or N could also be produced using partial substitution reactions of pure carbon SWCNTs in the presence of B_2O_3 vapor and N_2 at 1500–1700 K [75] (Figure 6.35a). In this case, B-substituted tubes exhibited a B/C ratio close to 0.1. However, N atoms were meanwhile incorporated into the hexagonal lattice, yet in lower amounts (N/C < 0.01). In contrast to B*MWCNTs, SWCNTs do not exhibit preferred helicities, possibly because the substitutional atoms only substitute individual C atoms within the framework, thus preserving the initial tube helicity. It is noteworthy that either B- or N-doped SWCNTs show corrugation, which could be attributed to defects created on the C surface or to electron irradiation effects (Figure 6.35b).

Borowiak-Palen and coworkers have also reported the production of SWCNTs substituted with B at high concentrations [281], in which 15% of the C atoms are replaced by B. These experiments were carried out by heating B_2O_3 in the presence of pure SWCNTs and NH_3 at 1150°C. However, additional experiments and further studies on these and similar samples should be conducted because the reported amount of B seems to be high when compared to the solubility of B found in bulk graphite (< 2 wt.% B).

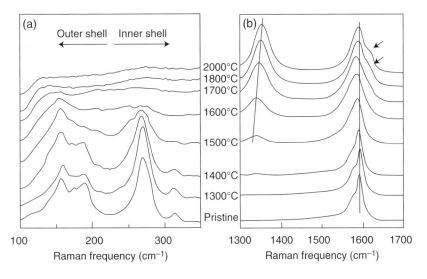

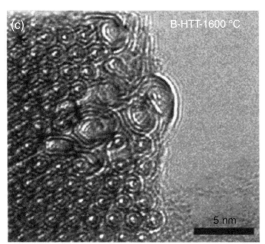

Figure 6.36 *(a) Low-frequency and (b) high-frequency Raman spectra of pristine B-doped DWNTs and those heat-treated at temperatures ranging from 1300–2000°C. The disappearance of the RBM frequency is a direct indicator of the destruction of the original DWNTs and the formation of diameter-enlarged DWNTs. It is clear that above 1500°C, B gets incorporated into the C lattice, making the D-band (at ca. 1340 cm⁻¹) very significant in all spectra; (c) High-resolution TEM image of a DWNT bundle annealed together with B powder at 1600°C.*

Endo *et al.* [282] have reported the synthesis of B-substituted DWCNTs by annealing B powders together with pure DWCNTs at different temperatures (ranging from 1200–2600°C; Figure 6.36). The authors showed that B atoms were incorporated into the carbon tube lattice using Raman spectroscopy after heat-treatment above 1600°C (Figure 6.36). In these experiments, B served as an efficient trigger for coalescing nanotubes. Theoretical simulations demonstrate that interstitial B atoms act as atomic welders and eventually get incorporated into the tube lattice. On the other hand, DWCNTs substituted with N have been produced by Ayala *et al.* using a CVD process and a carbon/nitrogen feedstock [283–284].

6.4.2.5 Plasma-Assisted CVD

Microwave plasma-enhanced chemical vapor (PECVD) deposition technique has been used to produce large areas of aligned N*MWCNTs [285–287]. These experiments involve catalytic particles of Fe and/or Ni dispersed on silica substrates. During growth at 500°C, acetylene or CH_4 and N_2 or NH_3 can be used. For producing B- (or BN-) substituted MWCNTs, other gases such as B_2H_6 in conjunction with H_2, and CH_4 could be used as a reacting gas in the PECVD process [64]. However, these methods have not yet been exploited to produce large and controlled amounts of substituted SWCNTs [288].

6.4.3 Morphology and Structure of Substituted CNTs

The techniques for studying the morphology and structure of substituted MWCNTs, DWCNTs, and SWCNTs are the same as for pure C or BN nanotubes: (a) high-resolution transmission electron microscopy (HRTEM); (b) scanning electron microscopy (SEM), and (c) scanning tunneling microscopy (STM). For determining the average crystalline structure of nanotubes, X-ray powder diffraction (XRD) becomes very powerful. Electron diffraction is also used to determine the helicity of different types of nanotubes (see Chapter 1). However, an alternative and powerful route to determine the helicity of individual nano-tubes is Raman spectroscopy, in which each single-walled nanotube has a characteristic radial breathing mode (RBM) peak associated with a particular diameter and, depending on the Raman laser line, it is possible to obtain resonance for particular tubules with specific electronic transitions. Because of the special relation between the geometric structures and the electronic structure, resonance Raman spectroscopy can be used to obtain the (n,m) indices of isolated single-walled nanotubes [289].

However, to estimate the concentration of substitutional atoms within SWCNTs, DWCNTs, and MWCNTs, analytical techniques associated with HRTEM need to be used. In this respect, EELS is a useful and powerful tool to determine the stoichiometry of elements in individual nanotubes, as well as the nature of the chemical bonds. Similarly, X-ray photoelectron spectroscopy (XPS) could be used to verify elemental stoichiometries and corresponding binding energies. Unfortunately, these techniques are generally only sensitive to elements whose content exceeds 1 at.%. Therefore, Raman spectroscopy, which is sensitive to the incorporation of foreign elements into the hexagonal carbon network, may provide another route to determine substitutional atom concentrations lower than 0.1 at.% in SWCNTs and MWCNTs (see Section 6.3.4.1). However, this technique requires additional developments regarding substituted nanotube structures before a quantitative evaluation of the heteroatom contents can be accurately performed.

6.4.3.1 Atomic Structure of N-substituted MWCNTs

CVD-grown MWCNTs substituted with N usually exhibit the following structural features: (a) unusual stacked-cone or compartmentalized morphologies (bamboo-type) (Figure 6.33b); and (b) the degree of tubular perfection decreases as a result of the increasing incor-poration of N into the hexagonal C lattice. In particular, EEL spectra of the CN_x nanotubes

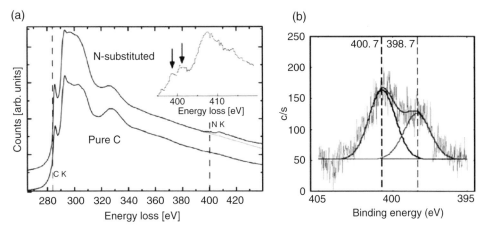

Figure 6.37 (a) EEL spectra of a typical N*MWCNT, suggesting a sp$_2$-graphite-like network due to the presence of the well defined π^* and δ^* features. Inset shows a splitting in the π^* peak of the N K-shell due to two different types of bonds at ca. 399 and 401 eV; (b) XP spectrum of a N*MWCNT sample showing two binding energies located at 398.7 and 400.7 eV.

or nanofibers indicate the presence of ionization edges at ca. 284.5 eV and 400 eV, corresponding to the C and N K-shells. In some samples researchers have observed a splitting in the π^*-*type* peak of the N-K edge. It exhibits two features at ca. 398.7 and 400.7 eV (Figure 6.37; note that these energies have also been observed in XPS). This indicates two types of bonding established between the N and C atoms within the tubular network: (i) highly coordinated N atoms replacing C atoms within the graphene sheets (ca. 401–403 eV), and (ii) pyridine-type nitrogen (ca. 399 eV; Figures 6.33a and 6.37). In addition, as the overall N content increases, the number of graphene-like walls within the nanofibers decreases and the proportion of pyridine-like N increases (remaining almost constant with respect to the number of three-coordinated N atoms). From EEL and XP spectra, it has been estimated that the N content within the tubes is usually < 10 at.%.

An astonishing result which consists in the production of highly-crystalline, thick multi-walled CNTs with a three-dimensional order (the concentric cylinders adopt the crystal structure of perfect graphite) has been reported by Koziol *et al.* [290]. These tubes were produced by CVD involving the thermolysis of toluene/1,4-diazine ($C_7H_8/C_4H_4N_2$) solutions containing 2 wt.% ferrocene (FeCp$_2$) in the presence of Ar at 760°C. Interestingly, the tubes contain 3 at.% N, but very differently to previous N-substituted nanotubes, they exhibit an extremely high crystallinity which is very similar to that of genuine 3D graphite [291]. This certainly comes with the polygonization of the tubes and facetted morphologies. N, which seems to be responsible for the dramatic structural order, was found to segregate preferentially within the nanotube cores. Therefore, researchers should try to test these novel 3D-structured substituted-MWCNTs and demonstrate that they are capable of exhibiting enhanced mechanical performance when compared with conventional corrugated N*MWCNTs.

6.4.3.2 Atomic Structure and Growth of N-Substituted SWCNTs

The structure of N*SWCNT bundles has been recently reported [272,278]. However, it was difficult to observe clear morphological changes on the N-substituted tubes when compared to pure C counterparts. As for any SWNT, whatever the composition within the BCN system, a bamboo texture is not possible since it would prevent a single tubular shell from growing continuously. Therefore, it is expected that only three-coordinated substitutional N atoms are mainly present in N*SWCNTs preferably to two-coordinated N (i.e., pyridine-like, see Figure 6.33a), so that continuous nanotube growth is achieved [279].

TEM observations of N*SWCNTs revealed tube diameters around 1.6 nm, a result that is also consistent with the Raman measurements [279]. As the N-containing precursor (benzylamine) was added to the CVD feedstock, the resulting tubes exhibited narrower diameters (e.g., 0.8 nm). Likewise, it was noted that, as the N content was increased in a laser target or a precursor for the CVD synthesis, the production of SWNTs was inhibited. Therefore, only low concentrations (below 2 wt.%) of heteroatoms could be embedded in

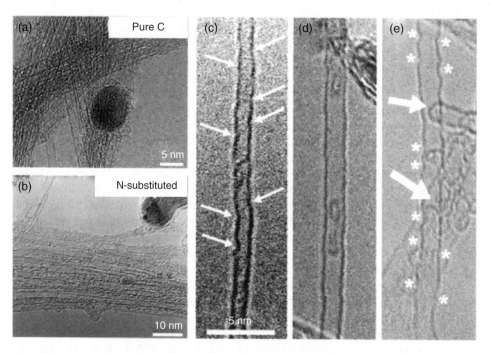

Figure 6.38 *HRTEM images of N*SWCNTs produced by the pyrolysis at 950°C of a ferrocene:benzylamine:ethanol solution with a composition ratio of 1.25:7.5:91.25 by weight: (a) and (b) SWNT bundles showing that N-substituted tubes exhibit more compacted bundles of narrow diameter tube; (c) N*SWCNT with narrow diameter (ca. 1 nm) exhibiting some degree of corrugation; (d) N*SWCNT with a 1.4 nm diameter exhibiting fullerene-like structures in its core (possibly N-substituted fullerenes that resulted from the frustrated growth of the inner tubules; see arrows); (e) highly corrugated N*SWCNT of large diameter (1.7 nm) possibly exhibiting internal bamboo-like closures (see bottom arrow) as well as a symmetric tubule cap (see top arrow).*

the graphene lattice. It is also important to mention that, in these materials, it was difficult to detect traces of N by means of EELS or XPS. This is because levels of 1–2 at.% of N are below the detection limits for these techniques. For this reason, it is very important to conduct other kinds of investigations, for example, Raman spectroscopy (see Section 6.3.4.1.), TGA analysis, and four-probe electrical conductivity measurements (see Section 6.3.4.2.). By means of each of these techniques clear changes in the structure as well as differences in the electrical response and reactivity of the N-substituted SWNT strands as a function of the N content were evidenced.

From HRTEM, Raman spectroscopy and TGA analysis [277], common structural features arising when substituting SWCNTs with N are as follows: (a) As the N content increases, the diameter of substituted SWCNTs decreases and only narrow diameter tubes are formed (Figures 6.38a and 6.38b); (b) the tubes oxidize faster when compared to the pure C counterparts; (c) Corrugations within the tube walls are sometimes observed (Figures 6.38c to 6.38e) and fullerene-like structures are formed inside the cores of the N*SWCNTs (Figure 6.38d); (d) On rare occasions, bamboo-like SWNTs could be suspected (see arrows in Figure 6.38e), yet they cannot be ascertained due to the high chances for overlapping structures; (e) The tube bundles appear to be more easily dispersed following sonication treatments when compared to pure SWCNTs (e.g., N*SWCNTs dispersed faster than pure SWCNTs using 2-propanol as a solvent, as the latter required twice as long to be dispersed). This is also true for MWNT; and (f) the entanglement of the nanotube strands is reduced as the N content increases.

Attempts to determine the helicity of substituted SWNTs (isolated or bundled) using electron diffraction have not been very successful so far, and novel techniques to reveal the helicity of individual SWNTs are being developed. In order to understand the role of N atoms during nanotube growth at various synthesis temperatures, first-principle molecular

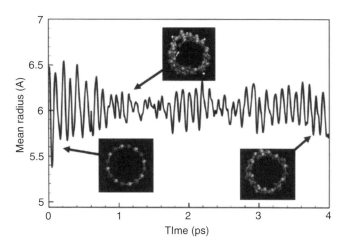

Figure 6.39 *Time dependence of the diameter for a N-substituted (8,0) SWCNT (top) (N atoms on the rim at the two-coordinated sites). The results are obtained from the first-principle MD simulation at 2500 K. Insets show snapshots of the structure for a N-doped (8,0) SWNT at T=0.0; 0.163; and 4 ps, as revealed by quantum molecular dynamics simulations. Reprinted with permission form [277] Copyright (2007) American Chemical Society.*

dynamics simulations were performed [277]. The results showed the rapid formation of pentagons (*pentagons form within 1 ps, notable in the decreased diameter) and persisted throughout the entire simulation at the growing rim. This results in the subsequent inward bending of the edge structure, leading to a defective graphene-like dome and partial closure of the tube. Figure 6.39 depicts snapshots of the dynamics and the time dependence of the average tube rim diameter when N atoms are substituting C atoms at the two-coordinated sites of the rim. After 4 ps of dynamics at 2500 K no stable pentagons have formed at the edge and those which did form during the simulation as a result of an atomic N-N bridge across the tube end (t = 0.163 ps), were relatively short-lived (<0.1 ps). Also, it clearly causes a decrease in the local diameter. This suggests that the N atoms remain included in the nanotube lattice of narrow diameter tubes.

With this experimental information at hand together with first-principles static and dynamics calculations, Sumpter *et al.* [277] demonstrated that N mediates the growth of SWCNTs by acting as a surfactant, that leads to narrow-diameter substituted tubes. The authors also showed that the tube closure, which includes N atoms embedded in the C lattice, is also preferred. The incorporation of N inside the tube lattice leads to weaker tube-tube interactions as well as providing sites where encapsulated or intercalated N can undergo chemical reactions. This may form smaller fullerene structures that become encapsulated within the hollow tube core. Metallic tubes were found to be somewhat more energetically stable for N-substitution, a behavior that has the potential to favor sample enrichment.

6.4.3.3   Atomic Structure and Growth of B-Substituted CNTs

These tubes are mainly produced using the arc-discharge method, because in CVD technique, the B precursors tend to frustrate the nanotube growth at temperatures below 900°C and only highly defective tubes can be produced. As mentioned earlier, the arc-produced B*MWCNTs are relatively long (< 100 μm), and usually exhibit open or ill-formed caps. These caps appear to contain higher B concentrations (as revealed by EELS analyses) than other parts in the tube, thus suggesting that B acts as a catalyst in the formation of long tubes. Although it has been difficult to determine the correct binding energy for B using either EELS or XPS on these systems, it is likely that substitutional B (three-coordinated) is incorporated in the hexagonal C lattice. In this context, careful EELS studies (using first and second derivative spectra) have shown that B traces (< 1 at.%) are present within the body of the tubes [292]. Electron diffraction analyses have also demonstrated that these B*MWCNTs exhibit a preferred *zigzag* or near-*zigzag* helicity (e.g., zigzag ± 3°) [293–294]. Static and dynamic ab initio and tight-binding calculations carried out by Blase *et al.* and Hernández *et al.*, respectively [293,295], revealed that B atoms also act – as N atoms - as a surfactant during the growth of long tubes, and inhibit tube closure during their formation. Therefore, in-plane substitution during nanotube growth may well control the helicity of the tubes.

Powder X-ray diffraction studies on B- substituted MWCNTs confirmed the presence of highly ordered three-dimensional graphite crystals, that is, corresponding to *ABA*... stacking. This observation has never been documented in pure MWCNTs or nanoparticles. From the *001* reflections (caused by the presence of parallel graphenes), two different average interlayer spacings were observed in these B-substituted samples (Figure 6.40). It was proposed that one family of these spacings correspond to standard CNTs/nanoparticles (ca.

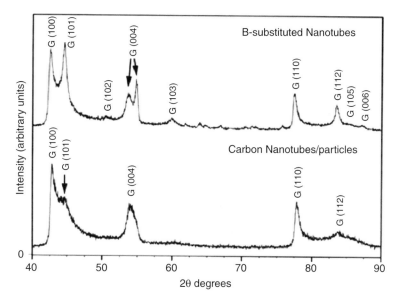

Figure 6.40 *XRD patterns from the inner core deposits obtained from the BN/graphite arc-discharge experiments compared to pure MWCNTs and nanoparticles produced using the arc-discharge technique. The pattern from 40–90 2θ degrees clearly shows the high degree of crystallinity of the B-substituted sample (note the (101), (103) and (112) planes, which denote a three-dimensional order and an ABAB stacking of the layers). Also note that the 00l reflection exhibits two peaks corresponding to 3.35Å and 3.42Å, thus confirming the presence of two different 'graphitic' structures (ABAB Bernal stacking as in graphite, possibly from facetted areas of polygonized nanotubes, and turbostratic stacking arising from concentric graphene cylinders of MWNTs).*

3.4Å) and the other type of spacing to *AB* stacked B-substituted graphenes (ca. 3.35Å). Whatever, the *AB* stacking observed in those nanotubes requires the existence of flattened tube domains (also confirmed by the interlayer spacing irregularities).

6.4.4 Properties of Substituted CNTs

6.4.4.1 Optical Properties (Raman)

CNTs display characteristic Raman spectroscopy features that are different from those observed in other forms of carbon. The main Raman features observed for SWCNTs are: (i) radial breathing mode (RBM), here all carbon atoms vibrate in phase in the radial direction of a nanotube. Its frequency ω_{RBM} is related to the nanotube diameter d_t as $\omega_{RBM} \sim 1/d_t$; (ii) tangential G-band related to in-plane Raman-active mode present in graphite at ~1580 cm^{-1}. A clear signature of SWCNTs, is that the G-band splits into G$^+$ and G$^-$, and the G$^-$ lineshape for a semiconducting SWCNT is very distinct from a metallic SWCNT; (iii) disorder-induced D-band which is related to any 'defect' that breaks translational symmetry, and (iv) its second-order harmonic, the G' -band.

Regarding N-substituted SWNTs, Keskar *et al.* [278] noted that as the N concentration in the CVD mixture increased (1 – 33 at.%), the RBM intensity decreased dramatically. In

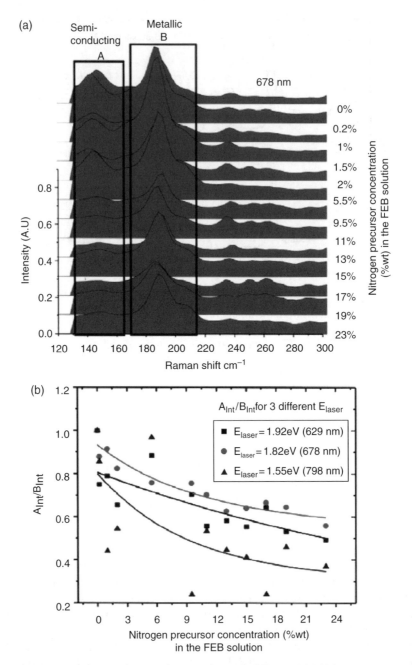

Figure 6.41   *(a) RBM spectral region for samples with different initial N precursor (benzylamine) concentrations in the Ferrocene-Ethanol-Benzylamine (FEB) solution. The laser energy is always $E_{laser} = 1.82\,eV$ (678 nm). Note how the intensity of the low frequency peak decreases as the initial N precursor concentration increases; (b) Integrated D-band to G-band intensity ratios for four different E_{laser} as a function of initial nitrogen precursor (benzylamine) concentration.*

addition, the intensity of the D-band at ~1350 cm^{-1} increased relative to the G-band intensity, indicating that the presence of N in the feedstock is resulting in the distortion of the hexagonal framework of the tubes. Other Raman studies on long strand of N*SWCNTs have been reported [277]. The corresponding RBM spectra are shown in Figure 6.41a. The authors concluded that the formation of large diameter tubes (region A) is prevented by the amount of N precursor (in accordance with the calculations shown in the previous section). The D-band to G-band ratio increases as the amount of N precursor increases, as displayed in Figure 6.41b. The I_D/I_G ratio change indicates that the nanotube wall is getting more disorder and this should be related to the incorporation of N atoms in the C lattice.

It is important to note that in agreement with Keskar *et al.* [278], Villalpando *et al.* [279] also found that the intensity of RBM, which is much lower in N*SWCNTs when compared to pure SWCNTs, and the modes associated with defects in the lattice, that is, the D- and D'-bands, become strong with increasing N concentration in the CVD feedstock.

For B*SWCNTs produced using the laser-ablation technique, McGuire *et al.* [296] noted that the overall effect of introducing B in the C lattice leads to: (i) a systematic increase in intensity of the D-band upon B substitution; (ii) a systematic downshift in the G'-band frequency due the relatively weaker C-B bond, and (iii) a non-linear variation in the RBM and G'-band intensities which is attributed to shifts in resonance conditions in the substituted tubes. It is therefore clear that resonant Raman spectroscopy reveals significant variations in the intensity of prominent features even when the concentration of substituting atoms is below the detectable limit of other analytical techniques. In addition to Raman spectroscopy, thermopower studies could provide complementary evidence for the presence of small B or N concentrations within SWCNTs.

Raman studies on substituted DWNTs have also been carried out [297]. It was observed that N and B induce shifts in the D-band and G-band of sp^2-carbon due to the incorporation of foreign atoms in the C lattice. The downshift of the G-band was correlated with the decrease of electrical resistivity of the substituted DWCNTs. Very recently, it was observed that imperfections within the graphene lattice such as substituting atoms could introduce new features in the G'-band. In this context, the presence of a defect-related feature (hence called G'$_{Def}$ band) in the immediate vicinity of the regular second-order G' band of pristine tube (so-called G'$_{Pris}$ band, located at ~2600 – 2700 cm^{-1} for E_{laser}=2.41 eV or λ=514 nm) has provided valuable information about charged defects present in SWCNTs. It was demonstrated that for substituted nanotubes, the relative intensity of the G'$_{Def}$ is proportional to the amount of substituting atoms in the tube material. However, further Raman studies on substituted SWCNTs and DWCNTs need to be carried out because the samples appear to be very different and the spectra strongly depend on the synthesis processes and the type of substituting atoms.

6.4.4.2 *Electronic and Transport Properties*

Microwave conductivity studies on bulk B*MWCNTs (produced using the arc-discharge method) have indicated that these structures are intrinsically metallic [298]. This metallicity differs from that of standard arc-produced MWCNTs, which show thermally-activated transport. Carroll *et al.* were the first to study individual B*MWCNTs and were able to determine their experimental density of states (DOS) using STM [299]. The authors found peaks in the valence band caused by the introduction of B holes (B acts as an acceptor; Figure 6.42a). Using first principles calculations, these authors suggested the presence of

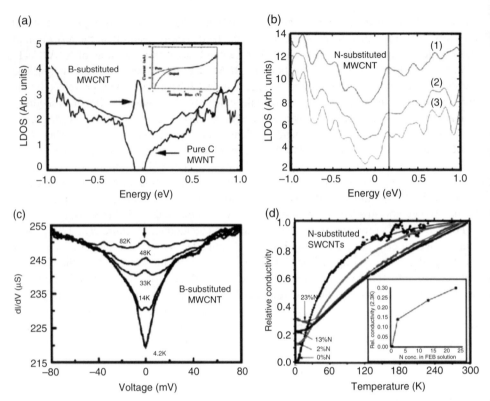

Figure 6.42 (a) LDOS of a B*MWCNT showing a peak on the valence band caused by the presence of B atoms (courtesy of P.M. Ajayan and D.L. Carroll); (b) Tunneling spectra acquired on a straight and clean section of a CN$_x$ nanotube. Note the peak at 0.18 eV (conduction band) in all spectra (see vertical line); (c) dI/dV versus V curves of an individual MWCNT substituted with B, exhibiting a small peak close to the E_f which is associated with the presence of an acceptor (B): peak due to B indicated by an arrow; (d) Experimental conductivity (squares) data points and their corresponding fitting results (continuous lines) as a function of temperature for SWNT fibers synthesized with different initial nitrogen precursor (benzylamine) concentrations (wt.%) in Ferrocene:Ethanol:Bencylamine (FEB) solution (reprinted with permission from [279] Copyright (2006) Elsevier Ltd). The parameters used to obtain the values for the donor energy levels and the carrier concentrations were derived from the fitting parameters contained in these curves (continuous lines). The inset shows the relative conductivity behavior of the SWNT strands as a function of the nitrogen precursor (benzylamine) concentration (wt.%) in the FEB solution at low temperatures (2–3 K).

BC$_3$ islands, distributed within the tubules that alter significantly the local DOS from a semimetal to an intrinsic metal.

Scanning tunneling microscopy (STM) and scanning tunneling spectroscopy (STS) studies have indicated that N*MWCNTs are metallic and exhibit a characteristic peak in the conduction band of the DOS (Figure 6.42b). For pure CNTs the valence and conduction band features appear to be symmetric about the Fermi level, whereas for the N*MWCNTs an additional electronic feature appears at ca. 0.18 eV. This result is in contrast to the B-substituted case [300]. It is noteworthy that the electron donor feature observed in the N-substituted material is always seen all along the substituted nanotubes. Therefore, both two- and three-coordinated N atoms (substitutional and pyridinic, respectively), randomly distributed within armchair and zigzag CNTs, could explain donor peaks close and/or above the Fermi level (E_f). More recently, it has been demonstrated that CN_x nanotubes exhibit a metallic-like behavior [134].

Direct electronic transport measurements on individual B*MWCNTs reveal a metallic behavior above 30 K. This has been explained due to an enhancement in conduction channels without experiencing strong back scattering [301]. The dI/dV versus V curves of these B*MWCNTs exhibited a small peak close to E_f, which is associated with the presence of acceptor states caused by BC_3 islands (Figure 6.42c). At lower temperatures, the resistance starts to increase. The results suggest that B-substitution induces a *p*-type behavior within MWCNTs.

When performing a comparative study of pure SWCNTs and N*SWCNTs using four-probe electrical measurements, clear differences were witnessed as the N content increased within the tubes [279]. Here, nonsubstituted SWCNT bundles displayed a semiconducting behavior over the whole temperature range (Figure 6.42d); the conductivity decreased exponentially with temperature. However, for strands of SWCNTs containing different N precursor contents (0%, 2%, 13% and 26%), the relative conductivity (r/r_{RT}, where RT refers to room temperature = 300 K) at very low temperatures is higher as the substitution rate is increased (Figure 6.42d, and inset). In contrast to the pure C case, the conductivity of bundles of N*SWCNTs did not continue dropping all the way to zero. The authors mentioned that it was unlikely that the mechanical stability of the electrical contacts was the cause for this behavior. They also demonstrated that the electron conduction could also be enhanced if the amount of foreign atoms is less than 0.5 wt.%. The latter suggested that below 13% of benzylamine in the ferrocene:ethanol:bencylamine (FEB) solution, N*SWCNTs could be produced with less than 0.5 wt.% of N (not detectable by TEM analytical techniques). Above 13% of benzylamine in the FEB, the generation of tubes with larger concentrations of N takes place, and this creates enhanced quantum interference effects within the tubes, so that the electrical conduction starts to decrease.

Additional electron conductivity measurements should be carried out on different types of heteronanotubes. One should also bear in mind that this method could be very efficient in detecting low substitution rates. However, low temperature measurements should be conducted in order to observe any electron conductance enhancement. One would expect that the electronic transport of substituted DWCNTs and SWCNTs will be thoroughly studied in the near future.

In addition to the previous techniques, electron paramagnetic resonance (EPR) studies have confirmed the intrinsic metallic behavior of bulk B*MWCNTs [302]. In particular, g values of 2.002 at room temperature have been observed, whereas a typical value for arc-discharge pure MWCNTs is 2.0189. Furthermore, magnetic measurements have demonstrated that the bulk B-substituted material is paramagnetic (standard arc-produced MWCNTs are strongly diamagnetic), exhibiting a weak temperature dependence of the magnetization commensurate with a metallic response.

A powerful and sensitive technique known as thermoelectric power (TEP) may reveal the carrier type sign of any material. Since TEP has a zero current transport coefficient, it is able

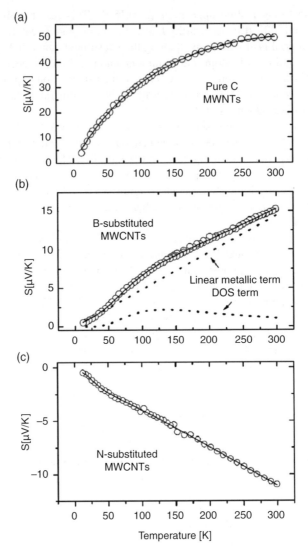

Figure 6.43 *Thermopower plots for pure C and hetero-MWNT mats. (a) TEP of pure MWCNT mat fit by heterogeneous model (solid line); (b) TEP of B*MWCNT mat and fit (solid line) indicating positive (hole) carriers. The dotted lines show the effect of the linear metallic term (straight line) and the DOS term; (c) TEP of N*MWCNT mat indicating negative (electron) carriers and fit (solid line).*

to probe the intrinsic conduction properties of nanotubes, being less influenced by randomly entangled morphologies and imperfections in the measured mats, as compared to other stand-ard conductivity measurements. TEP studies of mats of B- and N-substituted MWCNTs have been carried out [303]. The TEP measurements indicate that arc-produced B*MWCNTs have positive TEP, implying hole-like carriers. In contrast, the N*MWCNTs exhibit negative TEP over the same temperature range, suggesting electron-like conduction (Figure 6.43).

The results could also be correlated to the DOS for B- and N-substituted MWCNTs. It is important to note that as-produced pure MWCNTs exhibit positive TEP, which has been attributed to the presence of substitutional oxygen within the tube mats. Interestingly, if the tube sample is vacuumed for 94 h, the TEP signal considerably decreases. In this context, it is important to note that TEP studies for SWNTs have been able to indicate clearly alkali metal intercalation [304] and O contamination (or doping) [305].

McGuire *et al.* [296] demonstrated that B*SWCNTs (produced using the laser technique) had displayed positive TEP values, thus confirming the presence of hole carriers (as for B*MWCNTs). It is important to note that the clear changes observed in the TEP occurred for very low substitution rate (e.g., 0.05–0.1 at.% B), these are too low to detect with EELS or XPS. Therefore, TEP measurements are extremely sensitive to low substitution rates of B (or N) within SWCNTs and MWCNTs.

Thermal conductivity measurements on B- and N-substituted CNTs with various substitution rates need to be performed. By changing the substitution rate, the electronic and thermal properties would vary drastically, since the conductivity of non-substituted nanotubes is dominated by the lattice. In addition, further theoretical and experimental studies on B-substituted SWCNTs and B- (N-) substituted DWCNTs need to be carried out in order to elucidate their transport characteristics.

6.4.5 Applications of Substituted CNTs

Some possible applications of different types of substituted CNTs in diverse areas such as electronics, biology, composites and catalysis have been investigated. The key point for defining the physical-chemical properties of substituted structures is the charge transfer effects between the molecules and the nanotubes. It will be shown that substituted nanotubes (MWCNTs, SWCNTs or DWCNTs) could be more efficient when compared to their pure C counterparts when dealing with specific applications.

6.4.5.1 Field-Emission Sources

Pure C nanotubes are able to easily emit electrons from their tips when a potential is applied between a CNT and an anode. Charlier *et al.* [306] demonstrated theoretically and experimentally that B*MWCNTs could exhibit enhanced electron field-emission at a low turn-on voltage (ca. 1.4 V μm^{-1}); note that pure MWCNT turn-on voltages are ca. 3 V μm^{-1}) (Figure 6.44). This observation is believed to be due to the presence of B atoms, mainly at the nanotube tips, which results in an increased DOS close to the E_f. Theoretical tight-binding and ab initio calculations demonstrate that the work function of B*SWCNTs is much lower (1.7 eV) than that observed in pure MWCNTs (Figure 6.44, inset).

Similarly, it has been demonstrated that bundles of N*MWCNTs are able to emit electrons at relatively low turn-on voltages (2 V μm^{-1}) and high current densities (0.2–0.4 A cm^{-2}) [134]. These experimental results are in agreement with recent theoretical calculations indicating that N*SWCNTs are excellent field-emitters [307]. Additional measurements on individual N*MWCNTs have also shown excellent field-emission properties at 800 K; experimental work functions of 5 eV and emission currents of ca. 100 nA were obtained at ±10 V (Figure 6.45) [308].

Recent field-emission experiments indicated that B*MWCNTs are better suited for large-area emitter applications such as flat panel displays, whereas N*CNTs on pointed W tips would be excellent candidates for high current density single electron beam applications [309].

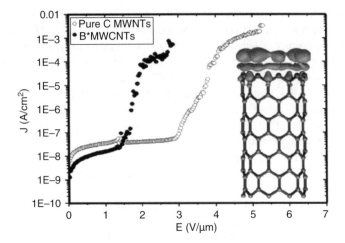

Figure 6.44 *J–E emission characteristics measured in parallel plate configuration of the B*MWCvNT films and comparable films of pure arc-produced MWCNTs. The plate area was 0.25 cm². Inset-side view of localized states at the edges of a B saturated zigzag (9,0) C nanotube computed using ab intio calculations. The work function of this tube is 1.7 eV, lower than that of the same nanotube made exclusively of C atoms.*

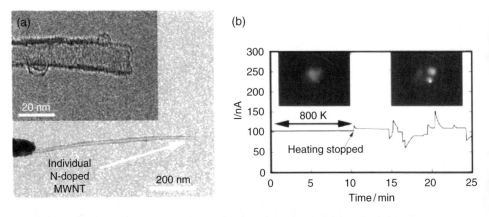

Figure 6.45 *(a) TEM image of an individual N*MWCNT taken after the field-emission measurements; (inset) image of the end cap of the nanotube shown in (a) with a radius of ca. 8 nm. The dark contour connecting the two main walls at the end of the nanotube could indicate a closed cap; (b) stable emission profile of the tube shown in (a) obtained at 800 K. Note that the current fluctuations occurred after the heating was stopped, as shown in the plot. Here, the extraction voltage was adjusted by maximal 10 V to obtain approximately 100 nA of emitted current. The corresponding emission patterns before and after heating recorded with a micro-channel plate and a phosphor screen positioned 1.5 cm in front of the emitter are also depicted.*

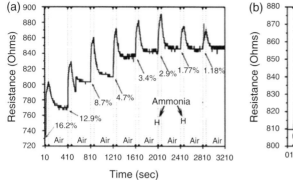 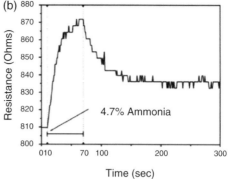

Figure 6.46 *(a) and (b) Plots of resistance versus time for N*MWCNT ammonia sensors: (a) It is clear that the sensor is sensitive to ca. 1% wt. of NH$_3$. In this case a clear chemisorption is observed, which can be attributed to the strong interactions between the pyridinic sites of the tube surface with the NH$_3$ molecules; (b) plot indicating the response time for ammonia gas (4.7% wt. concentration).*

6.4.5.2   Li-Ion Batteries

Lithium is an alkali metal, which could donate electrons from Li$^+$, and has proved to be extremely important in producing light-weight and energy-efficient batteries. When using graphite-like materials in Li-ion batteries, the ions are intercalated between the graphenes, so that Li$^+$ migrates from a graphitic-like anode to the cathode (usually LiCoO$_2$, LiNiO$_2$ and LiMn$_2$O$_4$ are in use). The Li storage capacity in graphite is ca. 372 mAh g^{-1} (LiC$_6$), and the charge and discharge phenomenon in these batteries is based upon the Li$^+$ intercalation and de-intercalation [310]. At present, electronic companies commercialize such batteries in portable computers, mobile telephones, digital cameras, and so on. Interestingly, Endo *et al.* [310] demonstrated that B-substituted vapor grown carbon fibers (VGCFs, which are CVD-grown carbon nanotubes subsequently thickened with CVD deposited pyrolytic carbon) and nanofibers are by far superior compared with any other C source present in the graphitic anode inside Li-ion batteries. This effect is because the population of Li ions has a stronger affinity in the B-substituted sites, thus resulting in higher storage efficiencies. N-substituted CNTs and nanofibers have also shown efficient reversible Li storage (480 mAh g^{-1}), that is, much higher than for C materials (330 mAh g^{-1}) [311].

6.4.5.3   Gas Sensors

It has been demonstrated by various groups [312–314] that pure SWCNTs and MWCNTs can detect toxic species, because small concentrations of these species are capable of producing significant variations in the nanotube conductance, thus shifting the Fermi level. However, N-substituted MWCNTs appear to be more efficient because they are able to display a fast response (order of milliseconds) when specific gases and organic solvents are approached [315]. In these experiments, the authors noted an increase in the electrical resistance caused by the presence of molecules strongly bound to the nitrogenated sites present within the CN$_x$ nanotubes (Figure 6.46). From the theoretical standpoint, a decrease in the density of states at E$_f$ is observed, indicative of lower conduction and chemisorption.

Other theoretical studies [316] have demonstrated that B*CNTs could be extremely efficient in detecting HCOH. Therefore, further experiments need to be performed in this direction.

6.4.5.4 Polymer-Matrix Composites with Substituted CNTs

For fabricating efficient CNT composites, the formation of stable tube-surface/polymer interfaces is crucial. Unfortunately, the surface of pure MWCNTs is often highly crystalline, tends

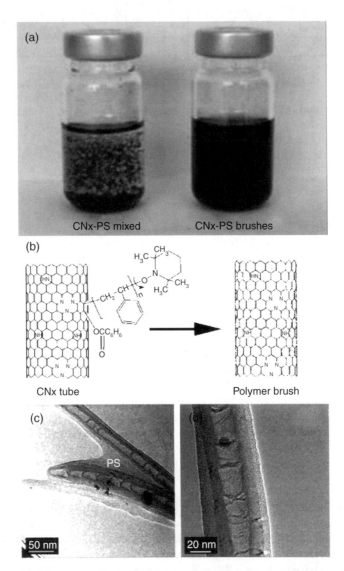

Figure 6.47 *(a) Polystyrene-CN$_x$ nanotubes suspensions: solubility differences in toluene between physically mixed PS and CN$_x$ tubes, and PS–CN$_x$ brushes obtained via the nitroxide mediated radical polymerization (NMRP) method; (b) Reaction scheme for the NMRP reaction for producing the PS-CN$_x$ tube brushes; (c) and (d) TEM micrographs showing the PS-CN$_x$ brushes obtained by the scheme showed in (b); note the uniform coating of PS on the tubes surfaces.*

to be similar to graphite, and hence is chemically 'inert'. Therefore, surface modification treatments are required, so that efficient tube-matrix interactions could be established [317–318]. In this context, the creation of nanotubes containing a few numbers of foreign atoms in the hexagonal network such as N or B, could circumvent this problem. In some cases, the mechanical properties would not be altered significantly because these 'substituted' structures would preserve their outstanding mechanical properties, since the substitution rate is low (< 1–2%). Preliminary studies on the preparation of epoxy composites using N^*MWCNTs revealed an increase of 20°C in the glass transition temperature with incorporation of 2.5 wt.% N using dynamic mechanical thermal analysis (DMTA) [319]. More recently, it has been demonstrated that it is possible to grow polystyrene (PS) on the surface of N^*MWCNTs using atomic transfer radical polymerization (ATRP) [320] and nitroxide mediated radical polymerization [321] without using any acid treatment. These polymers grown on the substituted-nanotube surfaces are also known as polymer brushes and could be dissolved in organic solvents (Figure 6.47). The latter clearly demonstrates that in-plane lattice substitution is important in the establishment of covalent bonds between the nanotube surfaces and polymer chains.

Recent mechanical and electrical tests have demonstrated that PS-grafted CN_x nanotubes exhibit enhanced properties when compared to mixtures of PS and pristine CN_x tubes (Figures 6.48a and 6.48b) [322]. These studies are now motivating further work in the fabrication of robust and conducting composites.

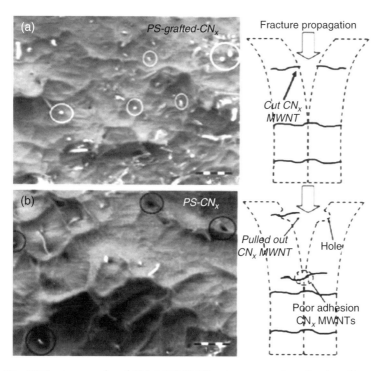

Figure 6.48 *SEM micrographs of CN_x MWNTs/PS nanocomposites after breaking at ambient temperature; (a) PS-grafted-CN_x/PS nanocomposite with white circles indicating cut tubes; (b) a-CN_x/PS nanocomposite. Black circles indicate holes or pulled out tubes. Aside to the SEM images the proposed fracture mechanisms are represented. Reprinted with permission from [322] Copyright (2008) Pergamon Press.*

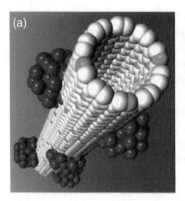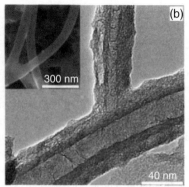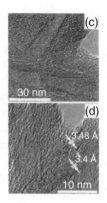

Figure 6.49 *(a) Molecular model of a CN$_x$ nanotube wearing metal clusters anchored to its surface (N atoms are gray); (b) TEM image of a 'Y'-junction consisting of a CN$_x$ MWNT inner core and an outer shell of C layers; (c-d) other examples of 'Y' junctions showing higher crystallinity of the CN$_x$ MWNT with respect to the outer shell, which has a larger interlayer spacing (by ca. 0.08Å). The material was produced by pyrolysis of toluene over Fe-coated N*MWCNTs.*

6.4.5.5 Anchoring of Molecules and Clusters

It is also possible that the substituted sites on the tube surface are used for an efficient anchoring of proteins [323], Au [324], Ag [325], Fe [326] (Figure 6.49). It has been demonstrated that the substituted tubes are much more efficient when compared to pure CNTs. In this context, Fe-coated CN$_x$ tubes have been used to grow co-axial CN$_x$ nanotube networks using a two-step CVD approach [326]. The materials could be used as damping devices or in electronics.

6.4.5.6 Toxicity and Bio-Applications

Comparative toxicological studies of N*MWCNTs and pure MWCNTs with similar residual Fe catalyst content (ca. <0.5 wt.%) on mice have been carried out [327]. Several routes of administration were tested (nasal, oral, intratracheal and intraperitoneal). In comparison with previous studies using SWCNTs, the new results demonstrated that CN$_x$ tubes appear to be far less harmful. For example, using extremely high doses of CN$_x$ nanotubes (e.g., 5 mg kg^{-1}), no lethal effects were observed on the mice, which is in contrast to previous reports using pristine MWCNTs or SWCNTs [328]. The plausible formation of cyanide groups driven by the interaction of CN$_x$ nanotubes and the mice did not occur. On the contrary, the presence of N in the graphene lattice could be the reason for the CN$_x$ nanotubes to be less injurious to a living organism such as a mouse. It is possible to have amino groups on the surface of the CN$_x$ tubes. It is therefore plausible that these amino-groups may be the reason for a better biocompatibility on the CN$_x$ tubes when compared to pristine MWCNTs. However, the two types of tubes exhibit clear structural differences. For example, MWCNTs display a high degree of crystallinity and, therefore, are more robust. This crystallinity results in stronger van der Waals interactions between tubes and the MWCNTs tend to aggregate easily. Thus MWCNTs will agglomerate faster and would create large clumps that result in the death of the mice by dyspnea (Figure 6.50).

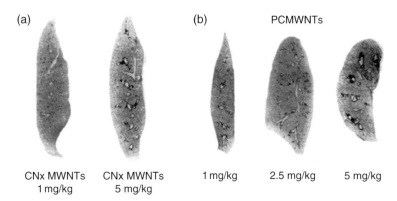

(a) (b) PCMWNTs

CNx MWNTs   CNx MWNTs   1 mg/kg   2.5 mg/kg   5 mg/kg
1 mg/kg   5 mg/kg

Figure 6.50 *Progressive pulmonary lesions caused by intratracheal instillation of: (a) CN_x MWNTs to CD1 mice. The panels show hematoxylin- and eosin-stained lung sections from mice treated with CN_x MWNTs at 1 mg kg^{-1} after 24 h and 5 mg kg^{-1} after 24 h (Original magnifications are 2X; (b) pure MWCNTs (here labelled PCMWNTs) to CD1 mice. The panels show hematoxylin- and eosin-stained lung sections from mice treated with MWNTs at 1 mg kg^{-1} after 24 h, 2.5 mg kg^{-1} after 24 h, and 5 mg kg^{-1} after 24 h. The number of bronchioles of small to large size occupied by MWNTs and the magnitude of the deposition is related to the amount of nanotubes instilled into the lungs. With 5 mg kg^{-1} dose, the majority of the bronchioles are involved.*

On the contrary, CN_x nanotubes possess a bamboo-like structure with a rougher surface, exhibiting weaker van der Waals interactions, that result in the formation of less agglomerates possibly causing much less damage to the mouse lungs (e.g., they are definitely better dispersed than pure MWCNTs). In addition, pure MWCNTs do not break as easily as CN_x nanotubes. This mechanical superiority of pristine MWCNTs could also increase the damage in the mouse tissue and therefore make them more hazardous because they damage the epithelium (Figure 6.50).

Very recently, Elias *et al.* [329] carried out a detailed cell viability study with amoebas and different types of nanotubes (pristine MWCNTs and N*MWCNTs). The authors noted that when the cells were incubated with CN_x nanotubes, they survived and there was no change in their behavior or morphology at all doses tested. In contrast, most of the amoeba population died after eight hours of incubation with pristine MWCNTs (100 μg). Samples were also characterized by optical microscopy, SEM and TEM [329].

These results imply that CN_x nanotubes could be biocompatible. Further sutides are however needed to discriminate between the role of the differences in structure and composition and the lower toxicity of N*MWCNTs with respect to pure MWCNTs. Should the lower toxicity of CN_x tubes be confirmed, they could preferably be used for especially appealing biochemical applications such as agents for drug delivery, cell transporters, supports for enhanced enzyme activities, biofilters, virus inhibitors, gene transfers, etc. However, additional and strict biosafety measures need to be developed for the production and processing of these nanotubes into new materials. In addition, the toxicological effects of other types of heteronanotubes should also be studied in the near future.

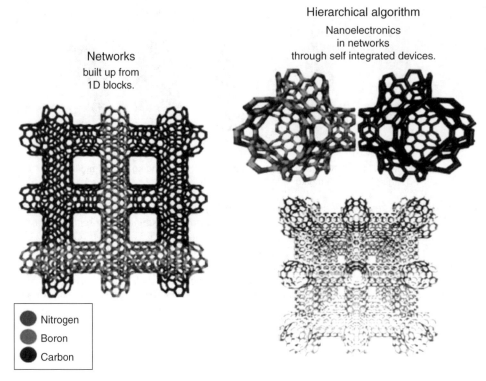

Hierarchical algorithm

Nanoelectronics
in networks
through self integrated devices.

Networks
built up from
1D blocks.

Nitrogen
Boron
Carbon

Figure 6.51 *Molecular models of possible 2D and 3D networks containing nanotubes made of C and BN fragments. These devices will certainly exhibit novel electronic and mechanical properties. For a better understanding of the figure, please refer to the colour plate section.*

6.5 Perspectives and Future Outlook

It is apparent that the BNNT, $B_xC_yN_z$, and B/N-substituted CNT related fields are rapidly developing. In spite of many difficulties involved in their preparation, structural analysis and property measurements (compared to well-established methods and techniques in the case of standard CNTs) notable breakthroughs in their synthesis, atomic structural determination and properties have recently been achieved by different research groups.

At present, the main challenge is to significantly increase the yield and purity of BN and $B_xC_yN_z$ NTs and/or to optimize their synthesis routes, as well as to demonstrate their advantageous properties over regular, pristine CNTs. In particular, BNNTs offer the following advantages: (i) uniform and homogeneous insulating properties and a wide optical direct bandgap; (ii) a higher resistance to oxidation, chemical passivity and thermal stability; (iii) good thermal conductivity; (iv) enhanced strength paired with superb elasticity and flexibility; (v) unique possibility to widely tune a bandgap energy by substituting B and/or N atoms by C, O, F, Si, and so on.

With respect to N- or B-substituted CNTs, perhaps the ternary B-C-N system is of particular interest. In the future, it may be possible to construct novel C- BN-C 1D, 2D and 3D covalent heteronetworks [330] and to explore their interesting properties (Figure 6.51). Such arrays would represent a significant step towards practical nanoelectronics where the generated networks (nanocircuits) demonstrate different electronic paths through BNNT or CNT compartments [331].

Very recently, it has been shown that mesoporous sponge-like BN nanomaterials can be fabricated using substitution reactions starting from mesoporous C templates [332]. Therefore, the adsorption of H_2 and other gases (N_2, O_2) and the catalytic properties of these novel structures should be investigated. In addition, the controlled synthesis of BN nanoribbons and/or one-atom thick graphene-like BN sheets require further experimental and theoretical efforts, since it may be possible that ribbons exhibiting zigzag edges could be metallic and possess excellent field-emission [333].

Besides BNNTs, nanowires, nanocones, ribbons and mesoporous BN materials, some alternative BN nanostructures [334–337] exhibiting novel physicochemical properties are anticipated in the following years. Especially, BNNT filled with various functional materials should be the target. We also expect that the real industrial applications of BNNTs and related nanostructures should emerge shortly, perhaps, even faster than those of CNTs. We particularly emphasize that properly *p*- or *n*-doped BNNTs could decently perform as stable low turn-on voltage long-life-time field-emitters or efficient gas accumulators and/ or sensors. Insulating, thermal conductive flexible polymeric composites containing minor fractions of multi-walled BNNTs are envisaged to appear in the market. Those are expected to possess superior properties when compared to available composites containing powder-like or particle-like BN fractions. However, kilogram quantities of BNNTs, at a reasonable cost, should be fabricated on regular basis. However, before BNNTs may be actually integrated into modern nanotechnology, the analysis of their environmental and human body-related safety issues should be thoroughly performed.

One interesting application of BNNTs could emerge, referring to the discovery of CNT nanothermometers [338]. It was found that when a liquid metal, like Ga or In, is placed inside a C nanotubular channel, the height of a metal column reflects the change of temperature of the microenvironment, for example that inside a high-vacuum TEM column, in a way similar to a standard mercury thermometer used for human body temperature measurements. However, CNT-based thermometers have temperature range limitations due to a relative ease in CNT oxidation at temperatures exceeding 400–500°C, as mentioned above, resulting in the thermometer shell degradation. BNNTs are much more stable to oxidation and may provide much better shield for a thermometer-like device. We indeed succeeded in the first preparation of BNNT microthermometers (not 'nano' as yet, due to easier filling of larger diameter BNNT) inheriting the same thermal properties as CNT nanothermometers [339]: a column of indium freely moves inside the internal channel and points out the environmental temperature inside TEM, as displayed in Figure 6.52. The utilization of such devices would require the development of temperature reading techniques other then the presently used TEM observations. Needless to say, the latter is rather expensive and time-consuming method which is unlikely to be of interest for the technologists.

From this chapter, it is also clear that B- or N-substituted CNTs are attractive for developing both basic science and nanotube-based technology. It was shown that various research groups have worked on N*MWCNTs, however, further work is needed with respect to the B-substituted nanotube systems. In order to exploit fully the novel properties, low concentrations of substitutional atoms (e.g., <0.5% at.) should be incorporated within these tubes. In this way, the electronic conductance would be significantly enhanced and

19°C 144°C 265°C 355°C 377°C 246°C 118°C 21°C

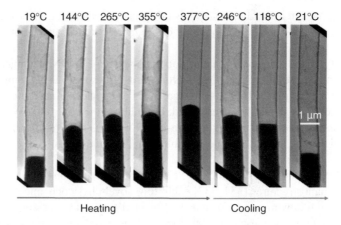

Heating   Cooling

Figure 6.52 *The ultimate performance of a BNNT microthermometer containing Indium. Indium first melts at ~140 °C and then a change in the In liquid height perfectly reflects the corresponding temperature changes in the TEM column.*

the mechanical properties would not be altered. The illustrated examples and phenomena clearly demonstrate that the science of substituting CNTs should be developed further as this is becoming a real interdisciplinary field. We should state that Raman spectroscopy, TEP studies and electron transport are very sensitive to low substitution rates. In addition, reliable methods for producing heteronanotube materials in a reproducible way are required. Finally, one should also be aware that CNT substitution may also occur unintentionally, and this will result in significant changes of the electronic, transport, chemical and vibrational properties.

In summary, this chapter reviewed significant achievements in the field of production, properties and applications of heterogeneous nanotubes within the B-C-N system. We believe that in following years, these fields will develop fast and it is likely that in some cases BN- or B- and/or N-substituted CNTs will be able to replace their pure carbon cousins on the market.

Acknowledgements

The authors are indebted to Y. Bando, T. Sato, C.Y. Zhi, C.C. Tang, T. Sasaki, L. Bourgeois, W.Q. Han, M. Eremets, O. Stephan, X.D. Bai, F.F. Xu, R.Z. Ma, J. Sloan, P. Dorozhkin, Z.C. Dong, A. Gloter, L. Chernozatonskii, T. Laude, T. Oku, Y.K. Yap, M. Terauchi, S. Saito, K. Suenaga, M. Endo, P.M. Ajayan, Z.L. Wang, Y. Chen, A. Zettl, S.G. Louie, D. Tomanek, B. Yakobson, A. Souza-Filhio, J.C. Charlier, A. Jorio, M. S. Dresselhaus, R. Saito, T. Hayashi. Y.A. Kim, H. Muramatsu, H. Terrones, F. López-Urías, E. Muñoz-Sandoval, A.R. Rao, E. Cruz-Silva, J.M. Romo-Herrera, J.A. Rodríguez-Manzo, A Zamudio, X. Blase, R. Kamalakaran, N. Grobert, Ph. Redlich, D.L. Carroll, R. Czerw, A.K. Cheetham, M. Rühle, K. McGuire, P. L. Gai, A.L. Elías, J.P. Laclette, J.C. Carrero-Sánchez, B. Fragneaud, M. De Honor, A. González-Montiel, J.Y. Cavallié, Karine Masenelli-Varlot, F. Villalpando-Páez, L. Noyola-Cherpitel and A. Loiseau for many fruitful and stimulating discussions over the years of the authors' explorations of BN and B-C-N NTs and preparation of this chapter. The financial support within the International Center for Materials Nanoarchitectonics (MANA) at the National Institute for Materials Science (NIMS) in Tsukuba is particularly acknowledged. MT thanks the Eberly College of Science at Penn State University for financial support, and JST-Japan for funding the Research Center for Exotic NanoCarbons, under the Japanese regional Innovation Strategy Program by the Excellence.

References

[1] J.J. Pouch and A. Alterovitz, Synthesis and Properties of Boron Nitride, *Mater. Sci. Forum*, Vol. **54–55**, Trans. Tech. Publ., Swizerland (1990).

[2] A. Rubio, J.L. Corkill and M.L. Cohen, Theory of graphitic boron nitride nanotubes, *Phys. Rev. B*, **49**, 5081–5084 (1994).

[3] X. Blase, A. Rubio, S.G. Louie and M.L. Cohen, Stability and band gap constancy of boron-nitride nanotubes, *Europhys. Lett.*, **28**, 335–340 (1994).

[4] N.G. Chopra, R.J. Luyken, K. Cherrey, V.H. Crespi, M. L. Cohen, S.G. Louie and A. Zettl, Boron nitride nanotubes, *Science*, **269**, 966–967 (1995).

[5] T. Ishii, T. Sato, Y. Sekikawa and M. Iwata, Growth of whiskers of hexagonal boron nitride, *J. Cryst. Growth*, **52**, 285–289 (1981).

[6] J.W.G. Wildoer, L.C. Venema, A.G. Rinzler, R.E. Smalley and C. Dekker, Electronic structure of atomically resolved carbon nanotubes, *Nature*, **391**, 59–62 (1998).

[7] L.A. Chernozatonskii, E.G. Gal'perin, I.V. Stankevich and Y.K. Shimkus, Nanotube C-BN heterostructures: electronic properties, *Carbon*, **37**, 117–121 (1999).

[8] Y.-K. Kwon, D. Tomanek and S. Iijima, "Bucky shuttle" memory device: synthetic approach and molecular dynamic simulations, *Phys. Rev. Lett.*, **82**, 1470–1473 (1999).

[9] L. Vaccarini, C. Goze, L. Henrard, E. Hernandez, P. Bernier and A. Rubio, Mechanical and electronic properties of carbon and boron nitride nanotubes, *Carbon* **38**, 1681–1690 (2000).

[10] P. Kral, E.J. Mele and D. Tomanek, Photogalvanic effects in heteropolar nanotubes, *Phys. Rev. Lett.*, **85**, 1512–1515 (2000).

[11] S. Berber, Y.-K. Kwon and D. Tomanek, Unusually high thermal conductivity of carbon nanotubes, *Phys. Rev. Lett.*, **84**, 4613–4616 (2000).

[12] P. Zhang and V.H. Crespi, Plastic deformations of boron nitride nanotubes: an unexpected weakness, *Phys. Rev. B*, **62**, 11050–11053 (2000).

[13] P.E. Lammert, V.H. Crespi and A. Rubio, Stochastic heterostructures and diodium in B/N-doped carbon nanotubes, *Phys. Rev. Lett.*, **87**, 136402-1-4 (2001).

[14] D. Srivastava, M. Menon and K. Cho, Anisotropic nanomechanics of boron nitride nanotubes: Nanostructured "skin" effect, *Phys. Rev. B*, **63**, 195413-1-5 (2001).

[15] O.E. Alon, Symmetry properties of single-walled boron nitride nanotubes, *Phys. Rev. B*, **64**, 153408-1-4 (2001).

[16] S. Erkoc, Structural and electronic properties of single-wall BN nanotubes, *J. Mol. Struct.: Theochem.*, **542**, 89–93 (2001).

[17] K.N. Kudin, G.E. Scuseria and B.I. Yakobson, C$_x$F, BN, and C nanoshell elasticity from ab initio computations, *Phys. Rev. B*, **64**, 235406-1-4 (2001).

[18] J. Kongsted, A. Osted, L. Jensen, P.O. Astrand and K.V. Mikkelsen, Frequency-dependent polarizability of boron nitride nanotubes: a theoretical study *J. Phys. Chem. B*, **105**, 10243–10248 (2001).

[19] H.F. Bettinger, T. Dumitrica, G.E. Scuseria and B.I. Yakobson, Mechanically induced defects and strength of BN nanotubes, *Phys. Rev. B*, **65**, 041406-1-4 (2002).

[20] V. Meunier, C. Roland, J. Bernholc and M.B. Nardelli, Electronic and field emission properties of boron nitride/carbon nanotube superlattices, *Appl. Phys. Lett.*, **81**, 46–48 (2002).

[21] T. Oku and I. Narita, Calculation of H-2 gas storage for boron nitride and carbon nanotubes studied from the cluster calculation, *Physica B*, **323**, 216–218 (2002).

[22] S. Okada, S. Saito and A. Oshiyama, Interwall interaction and electronic structure of double-walled BN nanotubes, *Phys. Rev. B*, **65**, 165410-1-4 (2002).

[23] N. Sai and E.J. Mele, Microscopic theory for nanotube piezoelectricity, *Phys. Rev. B*, **68**, 241405-1-4 (2003).

[24] T. Dumitrica, H.F. Bettinger, G. Scuseria and B.I. Yakobson, Thermodynamics of yield in boron nitride nanotubes, *Phys. Rev. B*, **68**, 085412-1-4 (2003).

[25] H.J. Xiang, J. Yang, J.G. Hou and Q.S. Zhu, First-principles study of small-radius single-walled BN nanotubes, *Phys. Rev. B*, **68**, 035427-1-4 (2003).

[26] A. Trave, F.J. Ribeiro, S.G. Louie and M.L. Cohen, Energetics and structural characterization of C-60 polymerization in BN and carbon nanopeapods, *Phys. Rev. B*, **70**, 205418-1-4 (2004).

[27] H. Xu, J. Ma, X. Chen, Z. Hu, K. Huo and Y. Chen, The electronic structures and formation mechanisms of the single-walled BN nanotube with small diameter, *J. Phys. Chem. B*, **108**, 4024–4034 (2004).

[28] C.-W. Chen, M.-H. Lee and S.J. Clark, Band gap modification of single-walled carbon nanotube and boron nitride nanotube under a transverse electric field, *Nanotechnol.*, **15**, 1837–1843 (2004).

[29] W.H. Moon and H.J. Hwang, Molecular-dynamics simulation of structure and thermal behaviour of boron nitride nanotubes, *Nanotechnol.*, **15**, 431–434 (2004).

[30] J. Bernholc, S.M. Nakhmanson, M.B. Nardelli and V. Meunier, Understanding and enhancing polarization in complex materials, *Comp. Sci. Eng.*, **6**, 12–21 (2004).

[31] F. Zheng, G. Zhou, S. Hao and W. Duan, Structural characteristics and electronic properties of boron nitride nanotube crystalline bundles, *J. Chem. Phys.*, **123**, 124716-1-5 (2005).

[32] T. Dumitrica and B.I. Yakobson, Rate theory of yield in boron nitride nanotubes, *Phys. Rev. B*, **72**, 035418-1-4 (2005).

[33] S.-H. Jhi, D.J. Roundy, S.G. Louie and M.L. Cohen, Formation and electronic properties of double-walled boron nitride nanotubes, *Sol. State Comm.*, **134**, 397–402 (2005).

[34] M.-F. Ng and R.Q. Zhang, Optical spectra of single-walled boron nitride nanotubes, *Phys. Rev. B*, **69**, 115417-1-4 (2004).

[35] G.Y. Guo and J.C. Lin, Second harmonic generation and linear electro-optical coefficients of BN nanotubes, *Phys. Rev. B*, **72**, 075416-1-9 (2005).

[36] G. S. Guo, W. J. Fan and R.Q. Zhang, Diameter-dependent spin polarization of injected carriers in carbon-doped boron nitride nanotubes, *Appl. Phys. Lett.*, **89**, 123103-1-3 (2006).

[37] H.J. Hwang, W.Y. Choi and J.W. Kang, Molecular dynamics simulations of nanomemory element based on boron-nitride nanotube-to-peapod transition, *Comp. Sci. Eng.*, **33**, 317–324 (2005).

[38] M. Grujicic, G. Cao and W.N. Roy, Suitability of boron-nitride single-walled nanotubes as fluid-flow conduits in nano-valve applications, *Appl. Surf. Sci.*, **246**, 149–158 (2005).

[39] S.V. Lisenkov, G.A. Vinogradov, T.Y. Astakhova and N.G. Lebedev, Nonchiral BN Haeckelite nanotubes, *J. Exp. Theor. Phys. Lett.*, **81**, 346–350 (2005).

[40] N.G. Lebedev and L.A. Chernozatonski, Quantum-chemical calculations of the piezoelectric characteristics of boron nitride and carbon nanotubes, *Phys. Sol. State*, **48**, 2028–2034 (2006).

[41] K. Harigaya, Possible charge-ordered states in boron-nitride and boron-carbon-nitride nanotubes and nanoribbons, *Jap. J. Appl. Phys. Part I*, **45**, 7237–7239 (2006).

[42] H.S. Kang, Theoretical study of boron nitride nanotubes with defects in nitrogen-rich synthesis, *J. Phys. Chem. B*, **110**, 4621–4628 (2006).

[43] C.-H. Park, C.D. Spataru and S.G. Louie, Excitons and many-electron effects in the optical response of single-walled boron nitride nanotubes, *Phys. Rev. Lett.*, **96**, 126105-1-4 (2006).

[44] L. Wirtz, A. Marini and A. Rubio, Excitons in boron nitride nanotubes: Dimensionality effects, *Phys. Rev. Lett.*, **96**, 126104-1-4 (2006).

[45] Z. Zhou, J. Zhao, Z. Chen, X. Gao, J.P. Lu, P.von R. Schleyer and C.-K. Yang, True nanocable assemblies with insulating BN nanotube sheaths and conducting Cu nanowire cores, *J. Phys. Chem. B*, **110**, 2529–2532 (2006).

[46] A. Zobelli, C.P. Ewels, A. Gloter, G. Seifert, O. Stephan, S. Csillag and C. Colliex, Defective structure of BN nanotubes: From single vacancies to dislocation lines, *Nano Lett.*, **6**, 1955–1960 (2006).

[47] X.J. Wu, W. An and X.C. Zheng, Chemical functionalization of boron-nitride nanotubes with NH_3 and amino functional groups, *J. Am. Chem. Soc.*, **128**, 12001–12006 (2006).

[48] M. Ishigami, J.D. Sau, S. Aloni, M.L. Cohen and A. Zettl, Observation of the giant Stark effect in boron-nitride nanotubes, *Phys. Rev. Lett.*, **94**, 056804-1-4 (2005).

[49] M. Ishigami, J.D. Sau, S. Aloni, M.L. Cohen and A. Zettl, Symmetry breaking in boron nitride nanotubes, *Phys. Rev. Lett.*, **97**, 176804-1-4 (2006).

[50] Z. Zhou, J. Zhao, Z. Chen, X. Gao, T. Yan, B. Wen and P.von R. Schleyer, Comparative study of hydrogen absorption on carbon and boron nitride nanotubes, *J. Phys. Chem. B*, **110**, 13363–13369 (2006).

[51] S.G. Hao, G. Zhou, W.H. Duan, J. Wu, B.L and Gu, Tremendous spin splitting effects in open boron nitride nanotubes: Application to nanoscale spintronic devices, *J. Am. Chem. Soc.*, **128**, 8453–8458 (2006).

[52] X. Chen, X.P. Gao, H. Zhang, Z. Zhou, W.K. Hsu, G.L. Pang, H.Y. Zhu, T.Y. Yan and D.Y. Song, Preparation and electrochemical hydrogen storage of boron nitride nanotubes, *J. Phys. Chem. B*, **109**, 11525–11529 (2005).

[53] S.H. Jhi, Activated boron nitride nanotubes: A potential material for room-temperature hydrogen storage, *Phys. Rev. B*, **74**, 155424-1-4 (2006).

[54] J. Zhang, K.P. Loh, J.W. Zheng, M.B. Sullivan and P. Wu, Adsorption of molecular oxygen on the walls of pristine and carbon-doped (5,5) boron nitride nanotubes: Spin-polarized density functional study, *Phys. Rev. B*, **75**, 245301-1-4 (2007).

[55] L.D.A. Silva, S.C. Guerini, V. Lemos and J.M. Filho, Electronic and structural properties of oxygen-doped BN nanotubes, *IEEE Trans. Nanotechnol.*, **5**, 517–522 (2006).

[56] S.S. Han, S.H. Lee, J.K. Kang and H.M. Lee, High coverage of hydrogen on a (10,0) single-walled boron nitride nanotube, *Phys. Rev. B*, **72**, 113402-1-4 (2005).

[57] E. Durgun, Y.R. Jang and S. Ciraci, Hydrogen storage capacity of Ti-doped boron nitride and B/Be substituted carbon nanotubes, *Phys. Rev. B*, **76**, 073413-1-4 (2007).

[58] X. Xu and H.S. Kang, First-principles study of the oxygenation of carbon nanotubes and boron nitride nanotubes, *Chem. Mater.*, **19**, 3767–3772 (2007).

[59] M.S. Si and D.S. Xue, First-principles study of silicon-doped (5,5) BN nanotubes *Europhys. Lett.*, **76**, 664–669 (2006).

[60] H.T. Liu, G. Zhou, Q.M. Yan, J. Wu, B.L. Gu, W.H. Duan and D.L. Zhao, Structural and electronic properties of fluorinated double-walled boron nitride nanotubes: effect of interwall interaction, *Phys. Rev. B*, **75**, 125410-1-6 (2007).

[61] A. Loiseau, F. Willaime, N. Demoncy, G. Hug and H. Pascard, Boron nitride nanotubes with reduced numbers of layers synthesized by arc discharge, *Phys. Rev. Lett.*, **76**, 4737–4740 (1996).

[62] M. Terrones, W.K. Hsu, H. Terrones, J.P. Zhang, S. Ramos, J.P. Hare, R. Castillo, K. Prassides, A.K. Cheetham, H.W. Kroto and D.R.M. Walton, Metal particle catalysed production of nanoscale BN structures, *Chem. Phys. Lett.*, **259**, 568–773 (1996).

[63] J. Cumings and A. Zettl, Mass-production of boron nitride double-wall nanotubes and nanococoons, *Chem. Phys. Lett.*, **316**, 211–216 (2000).

[64] D. Golberg, Y. Bando, M. Eremets, K. Takemura, K. Kurashima and H. Yusa, Nanotubes in boron nitride laser heated at high pressure, *Appl. Phys. Lett.*, **69**, 2045–2047 (1996).

[65] D.P. Yu, X.S. Sun, C.S. Lee, I. Bello, S.T. Lee, H.D. Gu, K.M. Leung, G.W. Zhou, Z.F. Dong and Z. Zhang, Synthesis of boron nitride nanotubes by means of excimer laser ablation at high temperature, *Appl. Phys. Lett.*, **72**, 1966–1968 (1998).

[66] A. Thess, P. Nikolaev, H.J. Dai, P. Petit, J. Robert, C.H. Xu, Y.H. Lee, S.G. Kim, A.G. Rinzler, D.T. Colbert, G.E. Scuseria, D. Tomanek, J.E. Fisher and R.E. Smalley, Crystalline ropes of metallic carbon nanotubes, *Science*, **273**, 483–487 (1996).

[67] R.S. Lee, J. Cavillet, M.L. de la Chapelle, A. Loiseau, J.L. Cochon, D. Pigache, J. Thibault and F.Willaime, Catalyst-free synthesis of boron nitride single-wall nanotubes with a preferred zig-zag configuration, *Phys. Rev. B*, **64**, 121405-1-4 (2001).

[68] T. Laude, Y. Matsui, A. Marraud and B. Jouffrey, Long ropes of boron nitride nanotubes grown by a continuous laser heating, *Appl. Phys. Lett.*, **76**, 3239–3241 (2000).

[69] W. Han, Y. Bando, K. Kurashima and T. Sato, Synthesis of boron nitride nanotubes from carbon nanotubes by a substitution reaction, *Appl. Phys. Lett.*, **73**, 3085–3087 (1998).

[70] D. Golberg, Y. Bando, M. Eremets, K. Takemura, K. Kurashima, T. Tamiya and H. Yusa, Boron nitride nanotube growth defects and their annealing-out under electron irradiation, *Chem. Phys. Lett.*, **279**, 191–196 (1997).

[71] D. Golberg, Y. Bando, M. Eremets, K. Takemura, K. Kurashima, T. Tamiya and H. Yusa, High-resolution analytical electron microscopy of boron nitrides laser heated at high pressure, *J. Elect. Microsc.*, **46**, 281–292 (1997).

[72] D. Golberg, Y. Bando, K. Kurashima and T. Sato, MoO₃-promoted synthesis of multi-walled BN nanotubes from C nanotube templates, *Chem. Phys. Lett.*, **323**, 181–191 (2000).

[73] W.Q. Han, W. Mickelson, J. Cumings and A. Zettl, Transformation of $B_xC_yN_z$ nanotubes to pure BN nanotubes, *Appl. Phys. Lett.*, **81**, 1110–1112 (2002).

[74] D. Golberg, Y. Bando, W. Han, K. Kurashima and T. Sato, Single-walled B-doped carbon, B/N-doped carbon and BN nanotubes synthesized from single-walled carbon nanotubes through a substitution reaction, *Chem. Phys. Lett.*, **308**, 337–342 (1999).

[75] D. Golberg, Y. Bando, L. Bourgeois, K. Kurashima and T. Sato, Large-scale synthesis and HRTEM analysis of single-walled B- and N-doped carbon nanotube bundles, *J. Carbon*, **38**, 2017–2027 (2000).

[76] O.R. Lourie, C.R. Jones, B.M. Bartlett, P.C. Gibbons, R.S. Ruoff and W.E. Buhro, CVD growth of boron nitride nanotubes, *Chem. Mater.*, **12**, 1808–1810 (2001).

[77] R.Z. Ma, Y. Bando, T. Sato and K. Kurashima, Growth, morphology, and structure of boron nitride nanotubes, *Chem. Mater.*, **13**, 2965–2971 (2001).

[78] W. L. Wang, Y. Bando, E.G. Wang and D. Golberg (2008), unpublished.

[79] W.L. Wang, X.D. Bai, K.H. Liu, Z. Xu, D. Golberg, Y. Bando and E.G. Wang, Direct synthesis of B-C-N single-walled nanotubes by bias-assisted hot filament chemical vapor deposition, *J. Amer. Chem. Soc.*, **128**, 6530–6531 (2006).

[80] J.S. Wang, V.K. Kayastha, Y.K. Yap, Z.Y. Fan, J.G. Lu, Z.W. Pan, I.N. Ivanov, A.A. Puretzky and D.B. Geohegan, Low temperature growth of boron nitride nanotubes on substrates, *Nano Lett.*, **5**, 2528–2532 (2005).

[81] Y. Chen, J. Fitz Gerald, J.S. Williams and S. Bulcock, Synthesis of boron nitride nanotubes at low temperatures using reactive ball milling, *Chem. Phys. Lett.*, **299**, 260–264 (1999).

[82] Y. Chen, L.T. Chadderton, J. Fitz Gerald, J.S. Williams and S. Bulcock, A solid-state process for formation of boron nitride nanotubes, *Appl. Phys. Lett.*, **74**, 2960–2962 (1999).

[83] Y. Chen, M. Conway, J.S. Williams and J. Zou, Large-quantity production of high-yield boron nitride nanotubes, *J.Mater. Res.*, **17**, 1896–1899 (2002).

[84] J. Fitz Gerald, Y. Chen and M.J. Conway, Nanotube growth during annealing of mechanically milled Boron, *Appl. Phys. A*, **76**, 107–110 (2003).

[85] F.Q. Ji, C.B. Cao, H. Zu and Z.Q. Yang, Mechanosynthesis of boron nitride nanotubes, *Chinese J. Chem. Eng.*, **14**, 389–393 (2006).

[86] The Australian National University, *Commercial Boron Nitride Nanotubes,* http://www.rsphysse.anu.edu.au/nanotube/commercialservices.php, (2007) [accessed 22 February 2008].

[87] L.T. Chadderton and Y. Chen, Nanotube growth by surface diffusion, *Phys. Lett. A*, **263**, 401–405 (1999).

[88] L.T. Chadderton and Y. Chen, A model for the growth of bamboo and skeletal nanotubes: catalytic capillarity, *J. Cryst. Growth*, **240**, 164–169 (2002).

[89] J. J. Velázquez-Salazar, E. Muñoz-Sandoval, J. M. Romo-Herrera, F. Lupo, M. Rühle, H. Terrones and M. Terrones, Synthesis and state of art characterization of BN bamboo-like nanotubes: Evidence of a root growth mechanism catalyzed by Fe, *Chem. Phys. Lett.*, **416**, 342–348 (2005).

[90] C.C. Tang, Y. Bando, T. Sato and K. Kurashima, A novel precursor for synthesis of pure boron nitride nanotubes, *Chem. Comm.*, **12**, 1290–1291 (2002).

[91] C.Y. Zhi, Y. Bando, C. Tang and D. Golberg, Effective precursor for high yield synthesis of pure BN nanotubes, *Sol. State Commun.*, **135**, 67–70 (2005).

[92] M. Terauchi, M. Tanaka, K. Suzuki, A. Ogino and K. Kimura, Production of zigzag-type BN nanotubes and BN cones by thermal annealing, *Chem. Phys. Lett.*, **324**, 359–364 (2000).

[93] Y. Shimizu, Y. Morioshi, H. Tanaka and S. Komatsu, Boron nitride nanotubes, webs, and coexisting amorphous phase formed by the plasma jet method, *Appl. Phys. Lett.*, **75**, 929–931 (1999).

[94] S. Komatsu, A. Okudo, D. Kasami, D. Golberg, Y.B. Li, Y. Moriyoshi, M. Shiratani and K. Okada, Electron field emission from self-organized micro-emitters of sp(3)-bonded 5H boron nitride with very high current density at low electric field, *J. Phys. Chem. B*, **108**, 5182–5184 (2004).

[95] R.Z. Ma, Y. Bando and T. Sato, Growth, morphology, and structure of boron nitride nanotubes, *Chem. Phys. Lett.*, **337**, 61–71 (2001).

[96] E. Bengu and L.D. Marks, Single-walled BN nanostructures, *Phys. Rev. Lett.*, **86**, 2385–2387 (2000).

[97] H. Chen, Y. Chen, J. Yu and J.S. Williams, Purification of boron nitride nanotubes *Chem. Phys. Lett.*,**425**, 315–319 (2006).

[98] C. Y. Zhi, Y. Bando, C. Tang and D. Golberg, Purification of boron nitride nanotubes through polymer wrapping, *J. Phys. Chem. B*, **110**, 1525–1528 (2006).

[99] Y. Saito and M. Maida, Square, pentagon, and heptagon rings at BN nanotube tips, *J. Phys. Chem. A*, **103**, 1291–1293 (1999).

[100] L. Bourgeois, Y. Bando, K. Kurashima and T. Sato, Co-produced carbon and boron nitride helical cones and the nucleation of curved BN sheets, *Phil. Mag. A*, **80**, 129–142 (2000).

[101] L. Bourgeois, Y. Bando and T. Sato, Tubes of rhombohedral boron nitride, *J. Phys. D*, **33**, 1902–1908 (2000).

[102] D. Golberg, Y. Bando, L. Bourgeois and K. Kurashima, T. Sato, Insights into the structure of BN nanotubes, *Appl. Phys. Lett.*, **77**, 1979–1981 (2000).

[103] R.Z. Ma, Y. Bando and T. Sato, Controlled synthesis of BN nanotubes, nanobamboos, and nanocables, *Adv. Mater.*, **14**, 366–368 (2002).

[104] Y.I. Kim, J.K. Jung, K.S. Ryu, S.H. Nahm and D.H. Gregory, Quantitative phase analysis of boron nitride nanotubes using Rietveld refinement, *J. Phys. D*, **38**, 1127–1131 (2005).

[105] D. Golberg, P.J.M.F. Costa, O. Lourie, M. Mitome, X.D. Bai, K. Kurashima, C.Y. Zhi, C.C. Tang and Y. Bando, Direct force measurements and kinking under elastic deformation of individual multiwalled boron nitride nanotubes, *Nano Lett.*, **7**, 2146–2151 (2007).

[106] D. Golberg, W. Han, Y. Bando, K. Kurashima and T. Sato, Fine structure of boron nitride nanotubes produced from carbon nanotubes by a substitution reaction, *J. Appl. Phys.*, **86**, 2364–2366 (1999).

[107] D. Golberg, Y. Bando, K. Kurashima and T. Sato, Ropes of BN multi-walled nanotubes, *Sol. State Comm.*, **116**, 1–6 (2000).

[108] D. Golberg, Y. Bando, K. Kurashima and T. Sato, Synthesis, HRTEM and electron diffraction studies of B/N-doped C and BN nanotubes, *Diam. Relat. Mater.*, **10**, 63–67 (2001).

[109] D. Golberg and Y. Bando, Unique morphologies of boron nitride nanotubes, *Appl. Phys. Lett.*, **79**, 415–417 (2001).

[110] A. Celik-Aktas, J.M. Zuo, J.F. Stubbins, C. Tang and Y. Bando, Structure and chirality distribution of multiwalled boron nitride nanotubes, *Appl. Phys. Lett.*, **86**, 133110-1-3 (2005).

[111] A. Celik-Aktas, J.-M. Zuo, J.F. Stubbins, C. Tang and Y. Bando, Double-helix structure in multiwall boron nitride nanotubes, *Acta Cryst. A*, **61**, 533–541 (2005).

[112] B.G. Demczyk, J. Cumings, A. Zettl and R.O. Ritchie, Structure of boron nitride nanotubules, *Appl. Phys. Lett.*,**78**, 2772–2774 (2001).

[113] R. Arenal, M. Kociak, A. Loiseau and D.-J.Miller, Determination of chiral indices of individual single- and double-walled boron nitride nanotubes by electron diffraction, *Appl. Phys. Lett.*, **89**, 073104-1-3 (2006).

[114] X. Blase, A. de Vita, J.C. Charlie and R. Car, Frustration effects and microscopic growth mechanisms for BN nanotubes, *Phys. Rev. Lett.*, **80**, 1666–1669 (1998).

[115] M. Menon and D. Srivastava, Structure of boron nitride nanotubes: tube closing versus chirality, *Chem. Phys. Lett.*, **307**, 407–412 (1999).

[116] D. Golberg, M. Mitome, Y. Bando, C.C. Tang and C.Y. Zhi, Multi-walled boron nitride nanotubes composed of diverse cross-section and helix type shells, *Appl. Phys. A*, **88**, 347–352 (2007).

[117] S. Iijima, T. Ichihashi and Y. Ando, Pentagons, heptagons and negative curvature in graphite microtubule growth, *Nature*, **356**, 776–778 (1992).

[118] X.F. Zhang, X.B. Zhang, G. Van Tendeloo, S. Amelinckx, M. Op de Beeck and J. Van Landuyt, Carbon nano-tubes – their formation process and observation by electron microscopy, *J. Cryst. Growth*, **130**, 368–382 (1993).

[119] D. Bernaerts, S. Amelinckx, Ph. Lambin and A.A. Lucas, The diffraction space of circular and polygonized multishell nanotubules, *Appl. Phys. A*, **67**, 53–64 (1998).

[120] Y.-H. Kim, K.J. Chang and S.G. Louie, Electronic structure of radially deformed BN and BC_3 nanotubes, *Phys. Rev. B.*, **63**, 205408-1-5 (2001).

[121] B.W. Smith, M. Monthioux and D. Luzzi, Encapsulated C-60 in carbon nanotubes *Nature*, **396**, 323–324 (1998).

[122] D. Golberg, Y. Bando, K. Kurashima and T. Sato, Synthesis and characterization of ropes made of BN multiwalled nanotubes, *Scripta Mater.*, **44**, 1561–1565 (2001).

[123] Y. Chen, J. Zou, S.J. Campbell and G. Le Gaer, Boron nitride nanotubes: pronounced resistance to oxidation, *Appl. Phys. Lett.*, **84**, 2430–2433 (2004).

[124] P. Kim, L. Shi, A. Majumdar and P.L. McEuen, Thermal transport measurements of individual multiwalled nanotubes, *Phys. Rev. Lett.*, **87**, 215502-1-4 (2001).

[125] C.W. Chang, A.M. Fennimore, A. Afanasiev, D. Okawa, T. Ikuno, H. Garcia, D. Li, A. Majumdar and A. Zettl, Isotope effect on the thermal conductivity of boron nitride nanotubes, *Phys. Rev. Lett.*, **97**, 085901-1-4 (2006).

[126] C.W.Chang, D. Okawa, A. Majumdar and A. Zettl, Solid-state thermal rectifier, *Science*, **314**, 1121–1124 (2006).

[127] K. Yum and M.-F. Yu, Measurement of wetting properties of individual boron nitride nanotubes with the Wilhelmy method using a nanotube-based force sensor, *Nano Lett.*, **6**, 329–333 (2006).

[128] J.-F. Jia, H.-S. Wu and H. Jiao, The structure and electronic property of BN nanotube, *Physica B*, **381**, 90–95 (2006).

[129] E.J. Mele and P. Kral, Electric polarization of heteropolar nanotubes as a geometric phase, *Phys. Rev. Lett.*, **88**, 056803-1-4 (2002).

[130] K.H. Khoo, M.S.C. Mazzoni and S.G. Louie, Tuning the electronic properties of boron nitride nanotubes with transverse electric fields: A giant dc Stark effect, *Phys. Rev. B*, **69**, 201401-1-4 (2004).

[131] J. Cumings and A. Zettl, in: *Electronic Properties of Molecular Nanostructures*, H. Kuzmany, J. Fink, M. Mehring and S. Roth (eds.), Am. Inst. Phys. Conf. Proc. Ser., New York, **591**, 577 (2001).

[132] D. Golberg, P.S. Dorozhkin, Y. Bando, Z.-C. Dong, N. Grobert, M. Reyes-Reyes, H. Terrones and M. Terrones, Cables of BN-insulated B-C-N nanotubes, *Appl. Phys. Lett.*, **82**, 1275–1277 (2003).

[133] D. Golberg, P. S. Dorozhkin, Y. Bando and Z.-C. Dong, Synthesis, analysis, and electrical property measurements of compound nantubes in the B-C-N ceramic system *Mater. Res. Soc. Bull.*, **29**, 38–42 (2004).

[134] D. Golberg, P.S. Dorozhkin, Y. Bando, Z.-C. Dong, C.C. Tang, Y. Uemura, N. Grobert, M. Reyes-Reyes, H. Terrones and M. Terrones, Structure, transport and field-emission properties of compound nanotubes: CN_x vs. BNC_x (x < 0.1), *Appl. Phys. A*, **76**, 499–507 (2003).

[135] P. Dorozhkin, D. Golberg, Y. Bando and Z.-C. Dong, Field emission from individual B-C-N nanotube rope, *Appl. Phys. Lett.*, **81**, 1083–1085 (2002).

[136] K.A. Dean and B.R. Chalamala, Current saturation mechanisms in carbon nanotube field emitters, *Appl. Phys. Lett.*, **76**, 375–377 (2000).

[137] R. Gomer, *Field Emission and Field Ionization*, Harvard University Press, Cambridge, MA (1961).

[138] J.-M. Bonard, M. Croci, C. Klinke, R. Kurt, O. Noury, N. Weiss, Carbon nanotube films as electron field emitters, *Carbon*, **40**, 1715–1728 (2002).

[139] M. Radosavljevic, J. Appenzeller, V. Derycke, R. Martel, Ph. Avouris, A. Loiseau, J.-L. Cochon and D. Pigache, Electrical properties and transport in boron nitride nanotubes, *Appl. Phys. Lett.*, **82**, 4131–4133 (2003).

[140] X.D. Bai, D. Golberg, M. Mitome, C. Tang, C.Y. Zhi and Y. Bando, Deformation-driven electrical transport of individual boron nitride nanotubes, *Nano Lett.*, **7**, 632–637 (2007).

[141] P. Poncharal, Z.L. Wang and W.A. de Heer, Electrostatic deflections and electromechanical resonances of carbon nanotubes, *Science*, **283**, 1513–1516 (1999).

[142] J. Cumings and A. Zettl, Low-friction nanoscale linear bearing realized from multiwall carbon nanotubes, *Science*, **289**, 602–604 (2000).

[143] D. Golberg, M. Mitome, K. Kurashima, C.Y. Zhi, C. Tang, Y. Bando and O. Lourie, In situ electrical probing and bias-mediated manipulation of dielectric nanotubes in a high-resolution transmission electron microscope, *Appl. Phys. Lett.*, **88**, 123101-1-3 (2006).

[144] C. Tang, Y. Bando, Y. Huang, S. Yue, C. Gu and D. Golberg, Fluorination and electrical conductivity of BN nanotubes, *J. Am. Chem. Soc.*, **127**, 6552–6553 (2005).

[145] E. Hernandez, C. Goze, P. Bernier and A. Rubio, Elastic properties of C and $B_xC_yN_z$ composite nanotubes, *Phys. Rev. Lett.*, **80**, 4502–4505 (1998).

[146] N.G. Chopra and A. Zettl, Measurement of the elastic modulus of a multi-wall boron nitride nanotube, *Sol. State Comm.*, **105**, 297–300 (1998).

[147] A.P. Suryavanshi, M.F. Yu, J. Wen, C. Tang and Y. Bando, Elastic modulus and resonance behavior of boron nitride nanotubes, *Appl. Phys. Lett.*, **84**, 2527–2529 (2004).

[148] D. Golberg, X.D. Bai, M. Mitome, C. Tang, C.Y. Zhi and Y. Bando, Structural peculiarities of in-situ deformation of a multi-walled BN nanotube inside a high-resolution analytical transmission electron microscope, *Acta Mater.*, **55**, 1293–1298 (2007).

[149] S. Saha, D.V.S. Muthu, D. Golberg, C. Tang, C. Zhi, Y. Bando and A.K. Sood, Comparative high pressure Raman study of boron nitride nanotubes and hexagonal boron nitride, *Chem. Phys. Lett.*, **421**, 86–90 (2006).

[150] S.M. Sharma, S. Karmakar, S.K. Sikka, P.V. Teredesai, A.K. Sood, A. Govindaraj and C.N.R. Rao, Pressure-induced phase transformation and structural resilience of single-wall carbon nanotube bundles, *Phys. Rev. B*, **63**, 205417-1-5 (2001).

[151] J. Wu, W.Q. Han, W. Walukievicz, J.W. Ager III, W. Shan, E.E. Haller and A. Zettl, Raman spectroscopy and time-resolved photoluminescence of BN and $B_xC_yN_z$ nanotubes, *Nano Lett.*, **4**, 647–650 (2004).

[152] J.S. Lauret, R. Arenal, F. Ducastelle, A. Loiseau, M. Cau, B. Attal-Tretout, E. Rosencher and L. Goux-Capes, Optical transitions in single-wall boron nitride nanotubes, *Phys. Rev. Lett.*, **94**, 037405-1-4 (2005).

[153] R. Arenal, O. Stephan, M. Kociak, D. Taverna, A. Loiseau and C. Colliex, Electron energy loss spectroscopy measurement of the optical gaps on individual boron nitride single-walled and multiwalled nanotubes, *Phys. Rev. Lett.*, **95**, 127601-1-4 (2005).

[154] C.Y. Zhi, Y. Bando, C.C. Tang, D. Golberg, R. Xie and T. Sekiguchi, Phonon characteristics and cathodolumininescence of boron nitride nanotubes, *Appl. Phys. Lett.*, **86**, 213110-1-3 (2005).

[155] O. Lehtinen, T. Nikitin, A.V. Krasheninnikov, L. Sun, L. Khriachtchev, F. Banhart, T. Terao, D. Golberg and J. Keinonen, Ion irradiation of multi-walled boron nitride nanotubes, *Phys. Stat. Sol. C*, **7**, 1256–1259 (2010).

[156] D. Golberg, Y. Bando, W. Han, L. Bourgeois, K. Kurashima and T. Sato, Multi- and single-walled boron nitride nanotubes produced from carbon nanotubes by a substitution reaction, *Mater. Res. Soc. Symp. Proc.*, **593**, 27–32 (2000).

[157] S.Y. Xie, W. Wang, K.A.S. Fernando, X. Wang, Y. Lin and Y.P. Sun, Solubilization of boron nitride nanotubes, *Chem. Comm.*, 3670–3672 (2005).

[158] C.Y. Zhi, Y. Bando, C.C. Tang, S. Honda, K. Sato, H. Kuwahara and D. Golberg, Covalent functionalization toward soluble boron nitride nanotubes, *Ang. Chem. Int. Ed.*, **44**, 7932–7935 (2005).

[159] C.Y. Zhi, Y. Bando, C. Tang, R. Xie, T. Sekiguchi and D. Golberg, Perfectly dissolved boron nitride nanotubes due to polymer wrapping, *J. Am. Chem. Soc.*, **127**, 15996–15997 (2005).

[160] C.Y. Zhi, Y. Bando, C. Tang and D. Golberg, Engineering of electronic structure of boron nitride nanotubes by covalent functionalization, *Phys. Rev. B*, **74**, 153413–4 (2006).

[161] C.Y. Zhi, Y. Bando, C. Tang, S. Honda, K. Sato, H. Kuwahara and D. Golberg, Mechanical properties of boron nitride nanotubes/polyaniline composites, *Angew.Chem. Int. Ed.*, **44**, 7929–7932 (2005).

[162] C.Y. Zhi, Y. Bando, C. Tang and D. Golberg, Purification of boron nitride nanotubes through polymer wrapping, *J. Phys. Chem. B*, **110**, 1525–1528 (2006).

[163] S. Okada and A. Oshiyama, Magnetic ordering in hexagonally bonded sheets with first-row elements, *Phys. Rev. Lett.*, **87**, 146803-1-4 (2001).

[164] Q. Huang, A.S. Sandanayaka, Y. Bando, C.Y. Zhi, D. Golberg, J. Zhao, Y. Araki and O. Ito, Donor-acceptor nanoensembles based on boron nitride nanotubes, *Adv. Mater.*, **19**, 934–938 (2007).

[165] W.Q. Han and A. Zettl, Functionalized boron nitride nanotubes with a stannic oxide coating: A novel chemical route to full coverage, *J. Am. Chem. Soc.*, **125**, 2062–2063 (2003).

[166] C.Y. Zhi, Y. Bando, C.C. Tang and D. Golberg, SnO_2 nanoparticle functionalized BN nanotubes, *J. Phys. Chem. B*, **110**, 8548–8550 (2006).

[167] C.Y. Zhi, Y. Bando, G.Z. Shen, C.C. Tang and D. Golberg, Boron nitride nanotubes: nanoparticles functionalization and junction fabrication, *J. Nanosci. Nanotech.*, **7**, 530–534 (2007).

[168] C.C. Tang, Y. Bando, C.Y. Zhi and D. Golberg, unpublished (2008).

[169] S. Okada, S. Saito, and A. Oshiyama, Semiconducting form of the first-row elements: C-60 chain encapsulated in BN nanotubes, *Phys. Rev. B*, **64**, 201303-1-4 (2001).

[170] D. Golberg, Y. Bando, M. Mitome, K. Kurashima, N. Grobert, M. Reyes-Reyes, H. Terrones and M. Terrones, Preparation of aligned BN and B/C/N nanotubular arrays and their characterization using HRTEM, EELS and energy-filtering TEM, *Physica B*, **323**, 60–66 (2002).

[171] Y. Bando, D. Golberg, M. Mitome, K. Kurashima and T. Sato, C to BN conversion in multi-walled nanotubes as revealed by energy-filtering transmission electron microscopy, *Chem. Phys. Lett.*, **346**, 29–34 (2001).

[172] M. Mickelson, S. Aloni, W.Q. Han, J. Cumings and A. Zettl, Packing C_{60} in boron nitride nanotubes, *Science*, **300**, 467–469 (2003).

[173] D. Golberg, Y. Bando, K. Kurashima and T. Sato, Nanotubes of boron nitride filled with molybdenum clusters, *J. Nanosci. Nanotech.*, **1**, 49–54 (2001).

[174] E. Dujardin, T.W. Ebbesen, H. Hiura and K. Tanigaki, Capillarity and wetting of carbon nanotubes, *Science*, **265**, 1850–1852 (1994).

[175] S.C. Tsang, Y.K. Chen, P.J.F. Harris and M.L.H. Green, A simple chemical method of opening and filling carbon nanotubes, *Nature*, **372**, 159–162 (1994).

[176] P.M. Ajayan, T.W. Ebbesen, T. Ichihashi, S. Iijima, K. Tanigaki and H. Hiura, Opening carbon nanotubes with oxygen and implications for filling, *Nature*, **362**, 522–525 (1993).

[177] W. Han, P. Redlich, F. Ernst and M. Ruhle, Synthesizing boron nitride nanotubes filled with SiC nanowires by using carbon nanotubes as templates, *Appl. Phys. Lett.*, **75**, 1875–1877 (1999)

[178] W.Q. Han, P. Kohler-Redlich, F. Ernst and M. Ruhle, Formation of $(BN)_xC_y$ and BN nanotubes filled with boron carbide nanowires, *Chem. Mater.*, **11**, 3620–3623 (1999).

[179] W.Q. Han and A. Zettl, GaN nanorods coated with pure BN, *Appl. Phys. Lett.*, **81**, 5051–5053 (2002).

[180] W.Q. Han and A. Zettl, Encapsulation of one-dimensional potassium halide crystals within BN Nanotubes, *Nano Lett.*, **4**, 1355–1357 (2004).

[181] W.Q. Han and A. Zettl, Nanocrystal cleaving, *Appl. Phys. Lett.*, **84**, 2644–2645 (2004).

[182] K.B. Shelimov and M. Mockovits, Composite nanostructures based on template-grown boron nitride nanotubules, *Chem. Mater.*, **12**, 250–254 (2000).

[183] Y.C. Zhu, Y. Bando, D.F. Xue, F.F. Xu and D. Golberg, Insulating tubular BN sheathing on semiconducting nanowires, *J. Am. Chem. Soc.*, **125**, 14226–14227 (2003).

[184] Y.B. Li, P. Dorozhkin, Y. Bando and D. Golberg, Controllable modification of SiC nanowires encapsulated in BN nanotubes, *Adv. Mater.*, **17**, 545–549 (2005).

[185] D. Golberg, F.F. Xu and Y. Bando, Filling boron nitride nanotubes with metals, *Appl. Phys. A*, **76**, 479–485 (2003).

[186] Y. Bando, K. Ogawa and D. Golberg, Insulating 'nanocables': Invar Fe-Ni alloy nanords inside BN nanotubes, *Chem. Phys. Lett.*, **347**, 349–534 (2001).

[187] Y. Zhang, K. Suenaga, C. Colliex and S. Iijima, Coaxial nanocable: silicon carbide and silicon oxide sheathed with boron nitride and carbon, *Science*, **281**, 973–975 (1998).

[188] Y. Zhang, H. Gu, K. Suenaga and S. Iijima, Heterogeneous growth of B-C-N nanotubes by laser ablation, *Chem. Phys. Lett.*, **279**, 264–269 (1997).

[189] R.Z. Ma, Y. Bando and T. Sato, Thin boron nitride nanotubes with unusual large inner diameters, *Chem. Phys. Lett.*, **350**, 443–440 (2001).

[190] D. Golberg, Y. Bando, K. Fushimi, C. Tang and M. Mitome, Nanoscale oxygen generators: MgO_2-based fillings of BN nanotubes, *J. Phys. Chem. B*, **107**, 8726–8729 (2003).

[191] D. Golberg, Y. Bando, M. Mitome, K. Fushimi and C. Tang, Boron nitride nanotubes as nanocrucibles for morphology and phase transformations in encapsulated nanowires of the Mg-O system, *Acta Mater.*, **52**, 3295–3303 (2004).

[192] S. Xu, Y. Fan, J. Luo, L. Zhang, W. Wang, B. Yao and L. An, Phonon characteristics and photoluminescence of bamboo structured silicon-doped boron nitride multiwall nanotubes, *Appl. Phys. Lett.*, **90**, 013115 (2007).

[193] K.F. Huo, Z. Hu, F. Chen, J.J. Fu, Y. Chen, B.H. Liu, J. Ding, Z.L. Dong and T. White, Synthesis of boron nitride nanowires, *Appl. Phys. Lett.*, **80**, 3611–3613 (2002).

[194] F.L. Deepak, C.P. Vinod, K. Mukhopadhyay, A. Govindaraj and C.N.R. Rao, Boron nitride nanotubes and nanowires, *Chem. Phys. Lett.*, **353**, 345–352 (2002).

[195] H.Z. Zhang, J Yu, Y. Chen and J. Fitz Gerald, Conical boron nitride nanorods synthesized via the ball-milling and annealing method, *J. Am. Ceram. Soc.*, **89**, 675–679 (2006).

[196] Y.J. Chen, H.Z. Zhang and Y. Chen, Pure boron nitride nanowires produced from boron triiodide, *Nanotechnology*, **17**, 786–789 (2006).

[197] H.Z. Zhang, J.D. Fitz Gerald, L.T. Chadderton, J. Yu and Y. Chen, Growth and structure of prismatic boron nitride nanorods, *Phys. Rev. B*, **74**, 045407-1/045407-9 (2006).

[198] Y.J. Chen, B. Chi, D.C. Mahon and Y. Chen, An effective approach to grow boron nitride nanowires directly on stainless-steel substrates, *Nanotechnology*, **17**, 2942–2947 (2006).

[199] F.F. Xu, Y. Bando and D. Golberg, The tubular conical helix of graphitic boron nitride, *New J. Phys.*, **5**, 118.1–118.16 (2003).

[200] F.F. Xu, Y. Bando, R.Z. Ma, D. Golberg, Y.B. Li and M. Mitome, Formation, structure, and structural properties of a new filamentary tubular form: hollow conical-helix of graphitic boron nitride, *J. Am. Chem. Soc.*, **125**, 8032–8038 (2003).

[201] L. Bourgeois, Y. Bando, W.Q. Han and T. Sato, Structure of boron nitride nanoscale cones: Ordered stacking of 240° and 300° disclinations, *Phys. Rev. B*, **61**, 7686–7691 (2000).

[202] A. Nishiwaki and T. Oku, Atomic structures of multi-walled boron nitride nanohorns, *J. Elect. Microsc.*, **54** (Supplement 1), i9-i14 (2005).

[203] M. Machado, P. Piquini and R. Mota, Charge distributions in BN nanocones: electric field and tip termination effects, *Chem. Phys. Lett.*, **392**, 428–432 (2004).

[204] S. Azevedo and F. de Brito Mota, Influence of the electric field on BN conical structures, *Int. J. Quant. Chem.*, **106**, 1907–1911 (2006).

[205] P.W. Fowler, K.M. Rogers, G. Seifert, M. Terrones and H. Terrones, Pentagonal rings and nitrogen excess in fullerene-based BN cages and nanotube caps, *Chem. Phys. Lett.*, **299**, 359–367 (1999).

[206] S. Azevedo, Stability and electronic structure of BN negative disclination, *J. Sol. State Chem.*, **178**, 3090–3094 (2005).

[207] W. Ah, X.J. Wu and X.C. Zeng, Effect of apical defects and doped atoms on field emission of boron nitride nanocones, *J. Phys. Chem B*, **110**, 16346–16352 (2006).

[208] C.Y. Zhi, Y. Bando, C. Tang, S. Honda, H. Kuwahara and D. Golberg, Boron nitride nanotubes/polystyrene composites, *J. Mater. Res.*, **21**, 2794–2800 (2006).

[209] N.P. Bansal, J.B. Hurst and S.R. Choi, Boron nitride nanotubes-reinforced glass composites, *J. Am. Ceram. Soc.*, **89**, 388–390 (2006).

[210] Q. Huang, Y. Bando, X. Xu, T. Nishimura, C.Y. Zhi, F.F. Xu and D. Golberg, in *Proc. 16ᵗʰ Int. Conf. on Compos. Mater.*, Kyoto, Japan, July 8–13, 810 (2007).

[211] A.C. Dillon, K.M. Jones, T.A. Bekkedahl, C.H. Kiang, D.S. Bethune and M.J. Heben, Storage of hydrogen in single-walled carbon nanotubes, *Nature*, **386**, 377–379 (1997).

[212] A. Chambers, C. Park, R.T. K. Baker and N.M. Rodriguez, Hydrogen storage in graphite nanofibers *J. Phys. Chem. B*, **102**, 4253–4256 (1998).

[213] C. Liu, Y.Y. Fan, M. Liu, H.T. Cong, H.M. Cheng and M.S. Dresselhaus, Hydrogen storage in single-walled carbon nanotubes at room temperature, *Science*, **286**, 1127–1129 (1999).

[214] R.Z. Ma, Y. Bando, H.W. Zhu, T. Sato, C.L. Xu and D.H. Wu, Hydrogen uptake in boron nitride nanotubes at room temperature, *J. Am. Chem. Soc.*, **124**, 7672–7673 (2002).

[215] C.C. Tang, Y. Bando, X.X. Ding, S.R. Qi and D. Golberg, Catalyzed collapse and enhanced hydrogen storage of BN nanotubes, *J. Am. Chem. Soc.*, **124**, 14550–14551 (2002).

[216] M. Terrones, W.K. Hsu, H.W. Kroto and D.R.M. Walton, in *Nanotubes: a Revolution in Materials Science and Electronics*, Springer, Berlin, 227 (1999).

[217] X. Blase, J.-C. Charlier, A. De Vita and R. Car, Theory of composite $B_xC_yN_z$ nanotube heterojunctions, *Appl. Phys. Lett.*, **70**, 197–199 (1997).

[218] K. Suenaga, C. Colliex, N. Demoncy, A. Loiseau, H. Pascard and F. Willaime, Synthesis of nanoparticles and nanotubes with well-separated layers of boron nitride and carbon, *Science*, **278**, 653–655 (1997).

[219] P. Kohler-Redlich, M. Terrones, C. Manteca-Diego, W. K. Hsu, H. Terrones, M. Ruhle, H.W. Kroto and D.R.M. Walton, Stable BC_2N nanostructures: low-temperature production of segregated C/BN layered materials, *Chem. Phys. Lett.*, **310**, 459–465 (1999).

[220] O.A. Louchev, Y. Sato, H. Kanda and Y. Bando, Coupling of kinetic and transport phenomena in self-organization of C-B-N nanotube growth into sandwich structures, *Appl. Phys. Lett.*, **77**, 1446–1448 (2000).

[221] D. Golberg, P. Dorozhkin, Y. Bando, M. Hasegawa and Z.-C. Dong, Semiconducting B-C-N nanotubes with few layers, *Chem. Phys. Lett.*, **359**, 220–228 (2002).

[222] T. Sugino, C. Kimura and T. Yamamoto, Electron field emission from boron-nitride nanofilms, *Appl. Phys. Lett.*, **80**, 3602–3604 (2002).

[223] K. Watanabe, T. Taniguchi and H. Kanda, Direct-bandgap properties and evidence for ultraviolet lasing of hexagonal boron nitride single crystal, *Nature Mater.*, **3**, 404–409 (2004).

[224] R. Arenal, A.C. Ferrari, S. Reich, L. Wirtz, J.-Y. Mevellec, S. Lefrant, A. Rubio and A. Loiseau, Raman spectroscopy of single-wall boron nitride nanotubes, *Nano Lett.*, **6**, 1812–1816 (2006).

[225] T. Hemraj-Benny, S. Banerjee, S. Sambasivan, D.A. Fisher, W. Han, J.A. Misewich and S.S. Wong, Investigating the structure of boron nitride nanotubes by near-edge X-ray absorption fine structure (NEXAFS) spectroscopy, *Phys. Chem. Chem. Phys.*, **7**, 1103–1106 (2005).

[226] J. Yu, Y. Chen, R.G. Elliman and M. Petravic, Isotopically enriched [10]BN nanotubes, *Adv. Mater.*, **18**, 2157–2160 (2006).

[227] W.Q. Han, P. Todd and M. Strongin, Formation and growth mechanism of [10]BN nanotubes via a carbon nanotube-substitution reaction, *Appl. Phys. Lett.*, **89**, 173103 (2006).

[228] C.Y. Zhi, Y. Bando, C. Tang and D. Golberg, Immobilization of proteins on boron nitride nanotubes, *J. Am. Chem. Soc.*, **127**, 17144–17145 (2005).

[229] M. Kawaguchi and N. Bartlett, In *Fluorine-Carbon and Fluoride-Carbon Materials*, Marcel Dekker, Inc., New York., 187–238 (1995).

[230] Y. Miyamoto, A. Rubio, M.L. Cohen and S.G. Louie, Chiral tubules of hexagonal BC_2N, *Phys. Rev. B*, **50**, 4976–4979 (1994).

[231] A.Y. Liu, R.M. Wetzcovitch and M.L. Cohen, Atomic arrangement and electronic-structure of BC_2N, *Phys. Rev. B*, **39**, 1760–1765 (1989).

[232] Y. Miyamoto, A. Rubio, S.G. Louie and M.L. Cohen, Electronic-properties of tubule forms of hexagonal BC_3, *Phys. Rev. B*, **50**, 18360–18366 (1994).

[233] H.-Y. Zhu, D.J. Klein, N.H. March and A. Rubio, Small band-gap graphitic CBN layers, *J. Phys. Chem. Sol.*, **59**, 1303–1308 (1998).

[234] X. Blasé, J.C. Charlier, A. De Vita and R. Car, Structural and electronic properties of composite $B_xC_yN_z$ nanotubes and heterojunctions, *Appl. Phys. A*, **68**, 293–300 (1999).

[235] M.S.C. Mazzoni, R.W. Nunes, S. Azevedo and H. Chacham, Electronic structure and energetics of $B_xC_yN_z$ layered structures, *Phys. Rev. B*, **73**, 073108, (2006).

[236] C.Y. Zhi, J.D. Guo, X.D. Bai and E.G. Wang, Adjustable boron carbonitride nanotubes, *J. Appl. Phys.*, **91**, 5325–5333 (2002).

[237] S.Y. Kim, J. Park, H.C. Choi, J.P. Ahn, J.Q. Hou and H.S. Kang, X-ray photoelectron spectroscopy and first principles calculation of BCN nanotubes, *J. Am. Chem. Soc.*, **129**, 1705–1716 (2007).

[238] J.M. Soler, E. Artacho, J.D. Gale, A. Garcia, J. Junquera, P. Ordeejon and D. Sanchez-Portal, The SIESTA method for ab initio order-N materials simulation, *J. Phys.: Cond. Matter*, **14**, 2745–2779 (2002).

[239] J. Choi, Y.H. Kim, K.J. Chang and D. Tomanek, Itinerant ferromagnetism in heterostructured C/BN nanotubes, *Phys. Rev. B*, **67**, 125421 (2003).

[240] J. Nakamura, T. Nitta and A. Natori, Electronic and magnetic properties of BNC ribbons, *Phys. Rev. B*, **72**, 205429/1-5 (2005).

[241] Z. Wengsieh, K. Cherrey, N.G. Chopra, X. Blase, Y. Miyamoto, A. Rubio, M.L. Cohen, S.G. Louie, A. Zettl and A.R. Gronsky, Synthesis of $B_xC_yN_z$ nanotubules, *Phys. Rev. B*, **51**, 11229–11232 (1995).

[242] O. Stephan, P.M. Ajayan, C. Colliex, Ph. Redlich, J.M. Lambert, P. Bernier and P. Lefin, Doping graphitic and carbon nanotube structures with boron and nitrogen, *Science*, **266**, 1683–1685 (1994).

[243] M. Terrones, W.K. Hsu, S. Ramos, R. Castillo and H. Terrones, The role of boron nitride in graphite plasma arcs, *Full. Sci. Technol.*, **6**, 787–800 (1998).

[244] P. Redlich, J. Loeffler, P.M. Ajayan, J. Bill, F. Aldinger and M. Rühle, B-C-N nanotubes and boron doping of carbon nanotubes, *Chem. Phys. Lett.*, **260**, 465–470 (1996).

[245] M. Terrones, A.M. Benito, C. Manteca-Diego, W.K. Hsu, O.I. Osman, J.P. Hare, D.G. Reid, H. Terrones, A.K. Cheetham, K. Prassides, H.W. Kroto and D.R.M. Walton, Pyrolytically grown $B_xC_yN_z$ nanomaterials: nanofibres and nanotubes, *Chem. Phys. Lett.*, **257**, 576–582 (1996).

[246] M. Kawaguchi, T. Kawashima and T. Nakajima, Syntheses and structures of new graphite-like materials of composition BCN(H) and BC_3N(H), *Chem. Mater.*, **8**, 1197–1201 (1996).

[247] P. Kohler-Redlich, M. Terrones, C. Manteca-Diego, W.K. Hsu, H. Terrones, M., Rühle, H.W. Kroto and D.R.M. Walton, Stable BC_2N nanostructures: low-temperature production of segregated C/BN layered materials, *Chem. Phys. Lett.*, **310**, 459–465 (1999).

[248] X. Blase, Properties of composite $B_xC_yN_z$ nanotubes and related heterojunctions, *Comp. Mater. Sci.*, **17**, 107–114 (2000)

[249] J.D. Guo, C.Y. Zhi, X.D. Bai and E.G. Wang, Boron carbonitride nanojunctions, *Appl. Phys. Lett.*, **80**, 124–126 (2002).

[250] M.P. Johansson, K. Suenaga, N. Hellgren, C. Colliex, J.E. Sundgren and L. Hultman, Template-synthesized BN:C nanoboxes, *Appl. Phys. Lett.*, **76**, 825–827 (2000).

[251] J. Yu, J. Ahn, S.F. Yoon, Q. Zhang, Rusli, B. Gan, K. Chew, M.B. Yu, X.D. Bai and E.G. Wang, Semiconducting boron carbonitride nanostructures: Nanotubes and nanofibers, *Appl. Phys. Lett.*, **77**, 1949–1951 (2000).

[252] X.D. Bai, J.D. Guo, J. Yu, E.G. Wang, J. Yuan and W.Z. Zhou, Synthesis and field-emission behavior of highly oriented boron carbonitride nanofibers, *Appl. Phys. Lett.*, **76**, 2624–2626 (2000).

[253] X.D. Bai, E.G. Wang, J. Yu and H. Yang, Blue–violet photoluminescence from large-scale highly aligned boron carbonitride nanofibers, *Appl. Phys. Lett.*, **77**, 67–69 (2000).

[254] Y. Chen, J.C. Barnard, R.E. Palmer, M.O. Watanabe and T. Sasaki, Indirect band gap of light-emitting BC_2N, *Phys. Rev. Lett.*, **83**, 2406–2408 (1999).

[255] M.O. Watanabe, S. Itoh, T. Sasaki and K. Mizushima, Visible-light-emitting layered BC_2N semiconductor, *Phys. Rev. Lett.*, **77**, 187–189 (1996).

[256] L.W. Yin, Y. Bando, D. Golberg, A. Gloter, M.S. Li, X. Yuan and T. Sekiguchi, Porous BCN nanotubular fibers: growth and spatially resolved cathodoluminescence, *J. Am. Chem. Soc.*, **127**, 16354–16355 (2005).

[257] S. Azevedo and R. de Paiva, Structural stability and electronic properties of carbon-boron nitride compounds, *Europhys. Lett.*, **75**, 126–132 (2006).

[258] L. Liao, K. Liu, W. Wang, X. Bai, E.G. Wang, Y. Liu, J. Li and C. Liu, Multiwall boron carbonitride/carbon nanotube junction and its rectification behavior, *J. Am. Chem. Soc.*, **129**, 9562–9563 (2007).

[259] W. L. Wang, X. D. Bai, K. H. Liu, Z. Xu, D. Golberg, Y. Bando, and E. G. Wang, Direct synthesis of B-C-N single-walled nanotubes by bias-assisted hot filament chemical vapor deposition, *J. Am. Chem. Soc.*, **128**, 6530–6531 (2006).

[260] C.Y. Zhi, X.D. Bai and E.G. Wang, Boron carbonitride nanotubes, *J. Nanosci. Nanotechol.*, **4**, 35–51 (2004).

[261] R. M. Wang and H. Z. Zhang, Analytical TEM investigations on boron carbonitride nanotubes grown via chemical vapour deposition, *New J. Phys.*, **6**, 78-1-78-11 (2004).

[262] M. Terrones, D. Golberg, N. Grobert, T. Seeger, M. Reyes-Reyes, M. Mayne, R. Kamalakaran, P. Dorozhkin, Z.C. Dong, H. Terrones, M. Rühle and Y. Bando, Production and state-of-the-art characterization of aligned nanotubes with homogeneous BC_xN ($1 \leq x \leq 5$) compositions, *Adv. Mater.*, **15**, 1899–1903 (2003).

[263] D. Golberg, Y. Bando, M. Mitome, T. Kurashima, N. Grobert, M. Reyes-Reyes, H. Terrones and M. Terrones, Nanocomposites: synthesis and elemental mapping of aligned B–C–N nanotubes, *Chem. Phys. Lett.*, **360**, 1–7 (2002).

[264] C.P. Ewels and M. Glerup, Nitrogen doping in carbon nanotubes, *J. Nanosci. Nanotechnol.*, **5**, 1345–1363 (2005).

[265] E. Hernández, C. Goze, P. Bernier and A. Rubio, Elastic properties of single-wall nanotubes, *Appl. Phys. A.*, **68**, 287–292 (1999).

[266] R. P. Gao, Z. L. Wang, Z. G. Bai, W. A. de Heer, L. M. Dai and M. Gao, Nanomechanics of individual carbon nanotubes from pyrolytically grown arrays, *Phys. Rev. Lett.*, **85**, 622–625 (2000).

[267] R. J. Baierle, S. B. Fagan, R. Mota, A. J. R. da Silva and A. Fazzio, Electronic and structural properties of silicon-doped carbon nanotubes, *Phys. Rev. B*, **64**, 085413-1-085413-4 (2001).

[268] E. Cruz-Silva, D. Cullen, L., Gu, J.M. Romo-Herrera, E. Muñoz-Sandoval, F. López-Urías, B.G. Sumpter, V. Meunier, J.C. Charlier, D.J. Smith, H. Terrones and M. Terrones, Heterodoped nanotubes: theory, synthesis, and characterization of phosphorus?nitrogen doped multiwalled carbon nanotubes, *ACS Nano*, **2**, 441–448 (2008).

[269] S.M.C. Vieira, O. Sptephan and D.L. Carroll, Effect of growth conditions on B-doped carbon nanotubes, *J. Mater. Res.*, **21**, 3058–3064 (2006).

[270] M. Glerup, J. Steinmetz, D. Samaille, O. Stephan, S. Enouz, A. Loiseau, S. Roth and P. Bernier, Synthesis of N-doped SWNT using the arc-discharge procedure, *Chem. Phys. Lett.*, **387**, 193–197 (2004).

[271] L.J. Li, M. Glerup, A.N. Khlobystov, J.G. Wiltshire, J.-L. Sauvajol, R.A. Taylor and R.J. Nicholas, The effects of nitrogen and boron doping on the optical emission and diameters of single-walled carbon nanotubes, *Carbon*, **44**, 2752–2757 (2006).

[272] P.L. Gai, O. Stephan, K. McGuire, A. M. Rao, M. S. Dresselhaus, G. Dresselhaus and C. Colliex, Structural systematics in boron-doped single wall carbon nanotubes, *J. Mater. Chem.*, **14**, 669–675 (2004).

[273] J. L. Blackburn, Y. Yan, C. Engtrakul, P. A. Parilla, K. Jones, Th. Gennett, A. C. Dillon and M. J. Heben, Synthesis and characterization of boron-doped single-wall carbon nanotubes produced by the laser vaporization technique, *Chem. Mater.*, **18**, 2558–2566 (2006).

[274] M. Terrones, N. Grobert, J. Olivares, J.P. Zhang, H. Terrones, K. Kordatos, W.K. Hsu, J.P. Hare, P.D. Townsend, K. Prassides, A.K. Cheetham, H.W. Kroto and D.R.M. Walton, Controlled production of aligned-nanotube bundles, *Nature*, **388**, 52–55 (1997).

[275] M. Terrones, P. Redlich, N. Grobert, S. Trasobares, W.K. Hsu, H. Terrones, Y.Q. Zhu, J.P. Hare, C.L. Reeves, A.K. Cheetham, M. Ruhle, H.W. Kroto and D.R.M. Walton, Carbon nitride nanocomposites: Formation of aligned C_xN_y nanofibers, *Adv. Mater.*, **11**, 655–658 (1999).

[276] R. Sen, B.C. Satishkumar, S. Govindaraj, K.R. Harikumar, M.K. Renganathan and C.N.R. Rao, Nitrogen-containing carbon nanotubes, *J. Mater. Chem.*, **7**, 2335–2337 (1997).

[277] B.G. Sumpter, V. Meunier, J.M. Romo-Herrera, E. Cruz-Silva, D.A. Cullen, H. Terrones, D.J. Smith and M. Terrones, Nitrogen-mediated carbon nanotube growth: Diameter reduction, metallicity, bundle dispersability, and bamboo-like structure formation, *ACS Nano*, **1**, 369–375 (2007).

[278] G. Keskar, R. Rao, J. Luo, J. Hudson, J. Chen and A.M. Rao, Growth, nitrogen doping and characterization of isolated single-wall carbon nanotubes using liquid precursors, *Chem. Phys. Lett.*, **412**, 269–273 (2005).

[279] F. Villalpando-Paez, A. Zamudio, A.L. Elias, H. Son, E.B. Barros, S.G. Chou, Y.A. Kim, H. Muramatsu, T. Hayashi, J. Kong, H. Terrones, G. Dresselhaus, M. Endo, M. Terrones and

M.S. Dresselhaus, Synthesis and characterization of long strands of nitrogen-doped single-walled carbon nanotubes, *Chem. Phys. Lett.*, **424**, 345–352 (2006).

[280] Z. Wang, C.H. Yu, D.C. Ba and J. Liang, Influence of the gas composition on the synthesis of boron-doped carbon nanotubes by ECR-CVD, *Vacuum*, **81**, 579–582 (2007).

[281] E. Borowiak-Palen, T. Pichler, G.G. Fuentes, A. Graff, R.J. Kalenczuk, M. Knupfer and J. Fink, Efficient production of B-substituted single-wall carbon nanotubes, *Chem. Phys. Lett.*, **378**, 516–520 (2003).

[282] M. Endo, H. Muramatsu, T. Hayashi, Y.A. Kim, G. Van Lier, J.C. Charlier, H. Terrones, M. Terrones and M.S. Dresselhaus, Atomic nanotube welders: Boron interstitials triggering connections in double-walled carbon nanotubes, *Nano Lett.*, **5**, 1099–1105 (2005).

[283] P. Ayala, A. Grueneis, T. Gemming, D. Grimm, C. Kramberger, M.H. Ruemmeli, F.L. Freire, H. Kuzmany, R. Pfeiffer, A. Barreiro, B. Buechner and T. Pichler, Tailoring N-doped single and double wall carbon nanotubes from a nondiluted carbon/nitrogen feedstock, *J. Phys. Chem. C*, **111**, 2879–2884 (2007).

[284] S.Y. Kim, J. Lee, C.W. Na, J. Park, K. Seo and B. Kim, N-doped double-walled carbon nanotubes synthesized by chemical vapor deposition, *Chem. Phys. Lett.*, **413**, 4–6, 300–305 (2005).

[285] E.G. Wang, Z.G. Guo, J. Ma, M.M. Zhou, Y.K. Pu, S. Liu, G.Y. Zhang and D.Y. Zhong, Optical emission spectroscopy study of the influence of nitrogen on carbon nanotube growth, *Carbon*, **41**, 1827–1831 (2003).

[286] K.B.K. Teo, M. Chhowalla, G.A.J. Amaratunga, W. I Milne, D.G. Hasko, G. Pirio, P. Legagneux, F. Wyczisk and D. Pribat, Uniform patterned growth of carbon nanotubes without surface carbon, *Appl. Phys. Lett.*, **79**, 1534–1536 (2001).

[287] K.B.K. Teo, D.B. Hash, R.G. Lacerda, N.L. Rupesinghe, M.S. Bell, S.H. Dalal, D. Bose, T.R. Govindan, B.A. Cruden, M. Chhowalla, G. A. J. Amaratunga, J.M. Meyyappan and W.I. Milne, The significance of plasma heating in carbon nanotube and nanofiber growth, *Nano Lett.*, **4**, 921–926 (2004).

[288] J. Yu, X.D. Bai, J. Ahn, S.F. Yoon and E.G. Wang, Highly oriented rich boron B-C-N nanotubes by bias-assisted hot filament chemical vapor deposition, *Chem. Phys. Lett.*, **323**, 529–533 (2000).

[289] A. Jorio, M.A. Pimenta, A.G. Souza Filho, R. Saito, M.S. Dresselhaus and G. Dresselhaus, Raman spectroscopy for probing chemically/physically induced phenomena in carbon nanotubes, *New J. Phys.*, **5**, 157.1–157.15 (2003).

[290] K. Koziol, M.S. Shaffer and A.H. Windle, Three-dimensional internal order in multiwalled carbon nanotubes grown by chemical vapor deposition, *Adv. Mater.*, **17**, 760–763 (2005).

[291] C. Ducati, K. Koziol, S. Friedrichs, T.J.V. Yates, M.S. Shaffer, P.A. Midgkey and A.H. Windle, Crystallographic order in multi-walled carbon nanotubes synthesized in the presence of nitrogen, *Small*, **2**, 774–784 (2006).

[292] P. Kohler-Redlich and M. Terrones, unpublished results.

[293] X. Blase, J.-C. Charlier, A. De Vita, R. Car, P. Redlich, M. Terrones, W.K. Hsu, H. Terrones, D.L. Carroll and P.M. Ajayan, Boron-mediated growth of long helicity-selected carbon nanotubes, *Phys. Rev. Lett.*, **83**, 5078–5081 (1999).

[294] W.K. Hsu, S. Firth, P. Redlich, M. Terrones, H. Terrones, Y.Q, Zhu, N. Grobert, A. Schilder, R.J.H. Clark, H.W. Kroto and D.R.M. Walton, Boron-doping effects in carbon nanotubes, *J. Mater. Chem.*, **10**, 1425–1429 (2000).

[295] E. Hernández, P. Ordejón, I. Boustani, A. Rubio and J.A. Alonso, Tight binding molecular dynamics studies of boron assisted nanotube growth, *J. Chem. Phys.*, **113**, 3814–3821 (2000).

[296] K. McGuire, N. Gothard, P.L. Gai, M.S. Dresselhaus, G. Sumanasekera and A.M. Rao, Synthesis and Raman characterization of boron-doped single-walled carbon nanotubes, *Carbon*, **43**, 219–227 (2005).

[297] Q.H. Yang, P.X. Hou, M. Unno, S. Yamauchi, R. Saito and T. Kyotani, Dual Raman features of double coaxial carbon nanotubes with N-doped and B-doped multiwalls, *Nano Lett.*, **5**, 2465–2469 (2005).

[298] M. Terrones, W.K. Hsu, A. Schilder, H. Terrones, N. Grobert, J.P. Hare, Y.Q. Zhu, M. Schwoerer, K. Prassides, H.W. Kroto and D.R.M. Walton, Novel nanotubes and encapsulated nanowires, *Appl. Phys. A*, **66**, 307–317 (1998).

[299] D.L. Carroll, P. Redlich, X. Blase, J. C. Charlier, S. Curran, P.M. Ajayan, S. Roth and M. Ruhle, Effects of nanodomain formation on the electronic structure of doped carbon nanotubes, *Phys. Rev. Lett.*, **81**, 2332–2335 (1998).

[300] R. Czerw, M. Terrones, J.-C. Charlier, X. Blase, B. Foley, R. Kamalakaran, N. Grobert, H. Terrones, D. Tekleab, P.M. Ajayan, W. Blau, M. Rühle and D.L. Carroll, Identification of electron donor states in N-doped carbon nanotubes, *Nano Lett.*, **1**, 457–460 (2001).

[301] K. Liu, P. Avouris, R. Martel, W.K. Hsu, Electrical transport in doped multiwalled carbon nanotubes, *Phys. Rev. B*, **63**, 161404-1-161404-4 (2001).

[302] W.K. Hsu, S.Y. Chu, E. Munoz-Picone, J.L. Boldu, S. Firth, P. Franchi, B.P. Roberts, A. Schilder, H. Terrones, N. Grobert, Y.Q. Zhu, M. Terrones, M.E. McHenry, H.W. Kroto and D.R.M. Walton, Metallic behaviour of boron-containing carbon nanotubes, *Chem. Phys. Lett.*, **323**, 572–579 (2000).

[303] Y. M. Choi, D.S. Lee, R. Czerw, P.W. Chiu, N. Grobert, M. Terrones, M. Reyes-Reyes, H. Terrones, J.-C. Charlier, P.M. Ajayan, S. Roth, D.L. Carroll and Y.W. Park, Nonlinear behavior in the thermopower of doped carbon nanotubes due to strong, localized states, *Nano Lett.*, **3**, 839–842 (2003).

[304] L. Grigorian, G.U. Sumanasekera, A.L. Loper, S. Fang, J.L. Allen and P.C. Eklund, Transport properties of alkali-metal-doped single-wall carbon nanotubes, *Phys. Rev. B*, **58**, 4195–4198 (1998).

[305] K. Bradley, S.H. Jhi, P.G. Collins, J. Hone, M.L. Cohen, S.G. Louie and A. Zettl, Is the intrinsic thermoelectric power of carbon nanotubes positive?, *Phys Rev Lett.*, **85**, 4361–4364 (2000).

[306] J.-C. Charlier, M. Terrones, M. Baxendale, V. Meunier, T. Zacharia, N.L. Rupesinghe, W. K. Hsu, N. Grobert, H. Terrones and G.A. J. Amaratunga, Enhanced electron field emission in B-doped carbon nanotubes, *Nano Lett.*, **2**, 1191–1195 (2002).

[307] L. Qiao, W.T. Zheng, H. Xu, L. Zhang and Q. Jiang, Field emission properties of N-doped capped single-walled carbon nanotubes: A first-principles density-functional study, *J. Chem. Phys.*, **126**, 164702-1-7 (2007).

[308] M. Doytcheva, M. Kaiser, M. Reyes-Reyes, M. Terrones and N. de Jonge, Electron emission from individual nitrogen-doped multi-walled carbon nanotubes, *Chem. Phys. Lett.*, **396**, 126–130 (2004).

[309] R.B. Sharma, D.J. Late, D.S. Joag, A. Govindaraj and C.N.R. Rao, Field emission properties of boron and nitrogen doped carbon nanotubes, *Chem. Phys. Lett.*, **428**, 102–108 (2006).

[310] M. Endo, Y. A. Kim, T. Hayashi, K. Nishimura, T. Matusita, K. Miyashita and M.S. Dresselhaus, Vapor-grown carbon fibers (VGCFs) – Basic properties and their battery applications, *Carbon*, **39**, 1287–1297 (2001).

[311] D.Y. Zhang, G.Y. Zhang, S. Liu, E.G. Wang, Q. Wang, H. Li and X.J. Huang, Lithium storage in polymerized carbon nitride nanobells, *Appl. Phys. Lett.*, **79**, 3500–3502 (2001).

[312] J. Kong, N.R. Franklin, C.W. Zhou, M.G. Chapline, S. Peng, K.J. Cho and H.J. Dai, Nanotube molecular wires as chemical sensors, *Science*, **287**, 622–625 (2000).

[313] S.S. Wong, E. Joselevich, A.T. Woolley, C.L. Cheung and C.M. Lieber, Covalently functionalized nanotubes as nanometre-sized probes in chemistry and biology, *Nature*, **394**, 52–55 (1998).

[314] P.G. Collins, K. Bradley, M. Ishigami and A. Zettl, Extreme oxygen sensitivity of electronic properties of carbon nanotubes, *Science*, **287**, 1801–1804 (2000).

[315] F. Villalpando-Páez, A.H. Romero, E. Munoz-Sandoval, L.M. Martinez, H. Terrones and M. Terrones, Fabrication of vapor and gas sensors using films of aligned CN_x nanotubes, *Chem. Phys. Lett.*, **386**, 137–143 (2004).

[316] R. Wang, D. Zhang, Y. Zhang and C. Liu, Boron-doped carbon nanotubes serving as a novel chemical sensor for formaldehyde, *J. Phys. Chem. B*, **110**, 18267–18271 (2006).

[317] P. Calvert, Nanotube composites – A recipe for strength, *Nature*, **399**, 210–211 (1999).

[318] P.M. Ajayan and J.M. Tour, Materials science – nanotube composites, *Nature*, **447**, 1066–1068 (2007).

[319] A. Eitan, L.S. Schadler, J. Hansen, P.M. Ajayan, R.W. Siegel, M. Terrones, N. Grobert, M. Reyes-Reyes, M. Mayne and H. Terrones, Processing and thermal characterization of

nitrogen doped MWNT/Epoxy composites, *Proc. Tenth US-Japan Conf. Compos. Mater.*, 634–640 (2002).

[320] B. Fragneaud, K. Masenelli-Varlot, A. González-Montiel, M. Terrones and J.Y. Cavaillé, Efficient coating of N-doped carbon nanotubes with polystyrene using atomic transfer radical polymerization, *Chem. Phys. Lett.*, **419**, 567–573 (2006).

[321] M. Dehonor, K. Masenelli-Varlot, A. González-Montiel, C. Gauthier, J.Y. Cavaillé, H. Terrones and M. Terrones, Nanotube brushes: Polystyrene grafted covalently on CN$_x$ nanotubes by nitroxide-mediated radical polymerization, *Chem. Comm.*, 5349–5351 (2005).

[322] B. Fragneaud, K. Masenelli-Varlot, A. González-Montiel, M. Terrones, J.-Y. Cavaillé, Mechanical behavior of polystyrene grafted carbon nanotubes/polystyrene nanocomposites, *Compos. Sci. Technol.*, **68**, 3265–3271 (2008).

[323] K.Y. Jiang, L.S. Schadler, R.W. Siegel, X.J. Zhang, H.F. Zhang and M. Terrones, Protein immobilization on carbon nanotubes via a two-step process of diimide-activated amidation, *J. Mat. Chem.*, **14**, 37–39 (2004).

[324] K.Y. Jiang, A. Eitan, L.S. Schadler, P.M. Ajayan, R.W. Siegel, N. Grobert, M. Mayne, M. Reyes-Reyes, H. Terrones and M. Terrones, Selective attachment of gold nanoparticles to nitrogen-doped carbon nanotubes, *Nano Lett.*, **3**, 275–277 (2003).

[325] A. Zamudio, A.L. Elías, J.A. Rodríguez-Manzo, F. López-Urías, G. Rodríguez-Gattorno, F. Lupo, M. Rühle, D.J. Smith, H. Terrones, D. Díaz and M. Terrones, Efficient anchoring of silver nanoparticles on N-doped carbon nanotubes, *Small*, **2**, 346–350 (2005).

[326] X. Lepró, Y. Vega-Cantú, F.J. Rodríguez-Macías, Y. Bando, D. Golberg and M. Terrones, Production and characterization of coaxial nanotube junctions and networks of CN$_x$/CNT, *Nano Lett.*, **7**, 2020–2026 (2007).

[327] J.L. Carrero-Sánchez, A.L. Elías, R. Mancilla, G. Arellín, H. Terrones, J.P. Laclette and M. Terrones, Biocompatibility and toxicological studies of carbon nanotubes doped with nitrogen, *Nano Lett.*, **6**, 1609–1616 (2006).

[328] D.B. Warheit, B.R. Laurence, K.L. Reed, D.H. Roach, G.A.M. Reynolds and T.R. Webb, Comparative pulmonary toxicity assessment of single-wall carbon nanotubes in rats, *Toxicol. Sci.*, **77**, 117–125 (2004).

[329] A.L. Elías, J.C. Carrero-Sánchez, H. Terrones, M. Endo, J.P. Laclette and M. Terrones, Viability studies of pure carbon- and nitrogen-doped nanotubes with Entamoeba histolytica: From amoebicidal to biocompatible structures, *Small*, **3**, 1723 (2007).

[330] P. Dibandjo, L. Bois, F. Chassagneux and P. Miele, Thermal stability of mesoporous boron nitride templated with a cationic surfactant, *J. Eur. Ceram. Soc.*, **27**, 313–317 (2007).

[331] J.M. Romo-Herrera, M. Terrones, H. Terrones, S. Dag and V. Meunier, Covalent 2D and 3D networks from 1D nanostructures: Designing new materials, *Nano Lett.*, **7**, 570–576 (2007).

[332] W.Q. Han, J. Cumings and A. Zettl, Pyrolytically grown arrays of highly aligned B$_x$C$_y$N$_z$ nanotubes, *Appl. Phys. Lett.*, **78**, 2769–2771 (2001).

[333] M. Terrones, J.-C. Charlier, A. Gloter, E. Cruz-Silva, E. Terres, Y.B. Li, A. Vinu, Z. Zanolli, J.M. Dominguez, H. Terrones, Y. Bando and D. Golberg, Experimental and theoretical studies suggesting the possibility of metallic boron nitride edges in porous nanourchins, *Nano Lett.*, **8**, 1026–1032 (2008).

[334] A. Vinu, M. Terrones, D. Golberg, S. Hishita, K. Ariga and T. Mori, Synthesis of mesoporous BN and BCN exhibiting large surface areas via templating methods, *Chem. Mater.*, **17**, 5887–5890 (2005).

[335] J. Yu, Y. Chen, R. Wuhrer, Z.W. Liu and S.P. Ringer, In situ formation of BN nanotubes during nitriding reactions, *Chem. Mater.*, **17**, 5172–5176 (2005).

[336] D. Golberg, Y. Bando, O. Stephan and K. Kurashima, Octahedral boron nitride fullerenes formed by electron beam irradiation, *Appl. Phys. Lett.*, **73**, 2441–2443 (1998).

[337] L.W. Yin, Y. Bando, Y.C. Zhu, D. Golberg and M.S. Li, A two-stage route to coaxial cubic-aluminum-nitride-boron-nitride composite nanotubes, *Adv. Mater.*, **16**, 929–933 (2004).

[338] Y.H. Gao and Y. Bando, Carbon nanothermometer containing gallium – Gallium's macroscopic properties are retained on a miniature scale in this nanodevice, *Nature*, **415**, 599–599 (2002).

[339] J.H. Zhan, Y.H. Gao, Y. Bando, J.Q. Hu and D. Golberg, unpublished (2008).

Index

Carbon Meta-Nanotubes: Synthesis, Properties and Applications, First Edition. Edited by Marc Monthioux.
© 2012 John Wiley & Sons, Ltd. Published 2012 by John Wiley & Sons, Ltd.